Quantum Dynamics of Molecules

The New Experimental Challenge to Theorists

NATO ADVANCED STUDY INSTITUTES SERIES

A series of edited volumes comprising multifaceted studies of contemporary scientific issues by some of the best scientific minds in the world, assembled in cooperation with NATO Scientific Affairs Division.

Series B: Physics

RECENT VOLUMES IN THIS SERIES

Volume 46 – Nondestructive Evaluation of Semiconductor Materials and Devices
edited by Jay N. Zemel

Volume 47 – Site Characterization and Aggregation of Implanted Atoms in Materials
edited by A. Perez and R. Coussement

Volume 48 – Electron and Magnetization Densities in Molecules and Crystals
edited by P. Becker

Volume 49 – New Phenomena in Lepton-Hadron Physics
edited by Dietrich E. C. Fries and Julius Wess

Volume 50 – Ordering in Strongly Fluctuating Condensed Matter Systems
edited by Tormod Riste

Volume 51 – Phase Transitions in Surface Films
edited by J. G. Dash and J. Ruvalds

Volume 52 – Physics of Nonlinear Transport in Semiconductors
edited by David K. Ferry, J. R. Barker, and C. Jacoboni

Volume 53 – Atomic and Molecular Processes in Controlled Thermonuclear Fusion
edited by M. R. C. McDowell and A. M. Ferendeci

Volume 54 – Quantum Flavordynamics, Quantum Chromodynamics, and Unified Theories
edited by K. T. Mahanthappa and James Randa

Volume 55 – Field Theoretical Methods in Particle Physics
edited by Werner Rühl

Volume 56 – Vibrational Spectroscopy of Molecular Liquids and Solids
edited by S. Bratos and R. M. Pick

Volume 57 – Quantum Dynamics of Molecules: The New Experimental Challenge to Theorists
edited by R. G. Woolley

This series is published by an international board of publishers in conjunction with NATO Scientific Affairs Division

A Life Sciences B Physics	Plenum Publishing Corporation London and New York
C Mathematical and Physical Sciences	D. Reidel Publishing Company Dordrecht, Boston and London
D Behavioral and Social Sciences E Applied Sciences	Sijthoff & Noordhoff International Publishers Alphen aan den Rijn, The Netherlands, and Germantown U.S.A.

Quantum Dynamics of Molecules

The New Experimental Challenge to Theorists

Edited by
R. G. Woolley
University of Cambridge
Cambridge, United Kingdom

SPRINGER SCIENCE+BUSINESS MEDIA, LLC

Library of Congress Cataloging in Publication Data

Nato Advanced Study Institute on Quantum Dynamics of Molecules.
Quantum dynamics of molecules.

(Nato advanced study institutes series: Series B, Physics; v. 57)
"Proceedings of the institute held at Trinity Hall, Cambridge, United Kingdom, September 15–29, 1979."
Includes index.
1. Quantum chemistry–Congresses. 2. Molecular dynamics–Congresses. 3. Chemistry, Physical and theoretical–Congresses. I. Woolley, R. G. II. North Atlantic Treaty Organization. Division of Scientific Affairs. III. Title. IV. Series.
QD462.A1N17 1979 541.2'8 80-16321

ISBN 978-1-4684-3739-3 ISBN 978-1-4684-3737-9 (eBook)
DOI 10.1007/978-1-4684-3737-9

Proceedings of the NATO Advanced Study Institute on Quantum Dynamics of Molecules: The New Experimental Challenge to Theorists, held at Trinity Hall, Cambridge, United Kingdom, September 15–29, 1979.

Originally published by Plenum Press, New York in 1980
Softcover reprint of the hardcover 1st edition 1980

-Did you ever study the Mollycule Theory when you were a lad? he asked. Mick said no, not in any detail.
-That is a very serious defalcation and an abstruse exacerbation, he said severely, but I'll tell you the size of it. Everything is composed of small mollycules of itself and they are flying round in concentric circles and arcs and segments and innumerable various other routes too numerous to mention collectively, never standing still or resting but spinning away and darting hither and thither and back again, all the time on the go. Do you follow me intelligently? Mollycules?
-I think I do.

From "The Dalkey Archive", Flann O'Brien

1976, Pan Books Ltd., London

(Courtesy Granada Publishing Ltd.)

PREFACE

The Advanced Study Institute on "Quantum Dynamics of Molecules: The New Experimental Challenge to Theorists," which was sponsored by the Scientific Affairs Division of NATO, was held at Trinity Hall, Cambridge, England from September 15th till September 29th, 1979. In all, a total of 79 lecturers and students attended the meeting: they had diverse backgrounds in chemistry, physics and mathematics.

In my proposal to NATO requesting financial support for an Advanced Study Institute, I suggested that molecular physics was facing a qualitatively new experimental situation in which the exploration of previously inaccessible dynamical phenomena would become of increasing importance. At the same time I was aware that in recent years powerful theoretical techniques, that might prove crucial tools for the interpretation of the new experiments, have been developed in mathematics and theoretical physics. The aim of the ASI was to review at an advanced level these recent developments, juxtaposing new theory with new experimental possibilities in the hope that the participants in the Institute would through their subsequent work increase the awareness of the whole molecular theory community of the changing nature of chemical physics.

The recent developments in laser spectroscopy, particle scattering experiments and molecular beam technology imply that an entirely new class of phenomena involving molecules in gasses and liquids can now be investigated. While it is arguable that there are a few theoretical surprises left in the domain of diatomic molecules and perhaps the ground electronic states of simple molecules polyatomics (to use conventional parlance), the quantum states of polyatomic molecules commonly provide evidence for both molecular non-rigidity and/or electron-nuclear coupling (nonadiabaticity). These new experiments lead to results that really demand a new formulation of molecular theory going beyond the traditional approximations such as the Born-Oppenheimer and rigid-rotor, small-amplitude vibrator description of polyatomic molecules. We can at last begin to transcend the traditional but artificial distinction between structural and dynamical aspects of molecular theory.

Typical examples of the new experimental techniques are: the laser spectroscopy of hypersonic nozzle beams which permit the observation of individual quantum stationary states of atomic clusters and Van der Waals molecules; ultra-high resolution spectroscopy (Lamb dip, laser magnetic resonance, Doppler free) that permits measurement of slow intramolecular tunneling rates in molecules such as PF_5; angular distribution, polarization and angular correlation measurements in scattering and ionization that expose dynamics in compound states that typically cannot be characterized by conventional rigid-rotor or Born-Oppenheimer models. These techniques are also exceptionally appropriate for studying coherence properties in compound and excited states. Very precise experiments, some of them successful, have been conducted recently in a search for evidence for parity violations in atomic systems due to a novel interaction between electrons and nuclei: this interaction which is mediated by the weak neutral current may, for example, lead to novel dynamical consequences in optically active molecules. Finally, recent achievements in picosecond spectroscopy have led to much new information about vibrational relaxation and the transfer of energy among bonds in polyatomic molecules: in turn this has stimulated considerable theoretical activity in the description of radiationless transitions, and focussed renewed interest on the relationship between the classical and quantum dynamics of non-linear coupled oscillator systems.

It is cogent to remark that there has been a renewed interest in many of the longstanding problems in atomic and molecular physics among mathematical physicists, particularly with respect to the formulation of the quantum theory itself, and the analysis of the N-body Schrödinger equation. The recent formal developments derived from modern mathematics, especially functional analysis, are now being applied fruitfully to significant atomic and molecular problems. Among these developments one should mention the generalized formulations of quantum theory that go beyond the traditional methods of the pioneers; the progress in the abstract N-body electronic problem in atoms and its extension to molecules; the characterization of the spectral properties of the electronic Hamiltonian for atoms and molecules, and the rigorous analysis of approximate theories of electronic structure in atoms, molecules and solids such as the Thomas-Fermi and Hartree-Fock theories. There is a new understanding of the nature of the assumptions that are implied by the introduction of the concept of molecular structure into molecular quantum theory, for example through the use of a fixed Eckart framework for the equilibrium configuration of a polyatomic molecule, by the introduction of artificial high symmetries and their associated unitary groups to permit phenomenological analysis of non-rigidity, and by establishing useful distinctions between the stationary states of isolated molecules and the quantum states of molecules coupled to their environment. It is also appropriate in this context to draw attention to the

Generator Coordinate Method, originally developed in nuclear physics, but now being applied to molecular quantum theory with considerable success; it is noteworthy that the GCM formalism is a non-adiabatic theory based on inertial (space-fixed) coordinate axes, and in these respects could hardly be more different from conventional quantum chemistry.

The lectures given in the Institute were addressed to many of the topics surveyed above, and stimulated much discussion among the participants. This book contains the texts of the 12 lecture courses on which the Institute was based and I hope that it will prove to be of as much interest to the reader as the sessions were to the participants in the Institute.

October, 1979

R.G. Woolley
Cavendish Laboratory
Cambridge, England

Au terme dernier, vous m'apprenez que cet univers prestigieux et bariolé se réduit a l'atome et que l'atome lui-même se réduit a l'électron. Tout ceci est bon et j'attends que vous continuiez. Mais vous me parlez d'un invisible système planétaire où des électrons gravitent autour d'un noyau. Vous m'expliquez ce monde avec une image. Je reconnais alors que vous en êtes venus à la poésie: je ne connaîtrai jamais. Ai-je le temps de m'en indigner? Vous avez déjà changé de théorie. Ainsi cette science qui devait tout m'apprendre finit dans l'hypothèse, cette lucidité sombre dans la métaphore, cette incertitude se résout en oeuvre d'art.

From *Un raisonnement absurde* in "Le mythe de Sisyphe"

Albert Camus, 1942, Éditions Gallimard

CONTENTS

The Eckart Hamiltonian for Molecules - A Critical Exposition 1
B.T. Sutcliffe

Foundations of Theoretical Chemistry 39
H. Primas

Supersonic Molecular Beams and van der Waals Molecules 115
D.H. Levy

A General Phenomenology for Small Clusters, However Floppy 143
R.S. Berry

The Generator Coordinate Method in Molecular Physics 197
P. Van Leuven and L. Lathouwers

Generator Coordinate Theory of Diatomic Systems 221
L. Lathouwers and P. Van Leuven

Irreversible and Non-Linear Dynamics of Open Systems 239
E.B. Davies

Quasiperiodic and Stochastic Intramolecular Dynamics: The Nature of Intramolecular Energy Transfer 257
S.A. Rice

Manifestations of Parity Violations in Atomic and Molecular Systems 357
R.A. Harris

High Resolution Laser Spectroscopy of Molecules 377
W. Demtröder

Spectral Properties of Atomic and Molecular Systems 435
J.M. Combes and R. Seiler

Theory of Electron Molecule Collisions 483
P.G. Burke

List of Participants 549

Subject Index 553

THE ECKART HAMILTONIAN FOR MOLECULES - A CRITICAL EXPOSITION

B. T. Sutcliffe

Department of Chemistry
University of York
York, YO1 5DD, England

1. INTRODUCTION

For molecules, other than diatomics, the whole notion of molecular structure rests, at present, upon the Eckart Hamiltonian, in the sense that the domain of validity of the Eckart Hamiltonian consists of that collection of molecular wave functions to which the notions of structure can be applied in a widely agreed and unambiguous way. The notion of structure in the Eckart Hamiltonian arises from the peculiar position that the framework co-ordinates occupy in the formalism. It is traditional to interpret these framework co-ordinates as being the equilibrium nuclear positions as expressed in a carefully defined reference frame, which can be called the Eckart frame.

The material presented here is designed to provide a coherent and collected exposition of the genesis of the Eckart Hamiltonian and to indicate the theoretical difficulties that arise in its construction and use. The present state of the art in the resolution of these difficulties will be the subject matter (at least in part) of some of the other discussions. In particular it will form part of the discussions on "floppy clusters" and of the generator co-ordinate method. It may also be discussed in passing when the theory of collisions (particularly atom-atom collisions) is discussed.

The aim therefore is principally a straightforward expository one. As such the material presented is not especially novel and every effort will be made to include references to typical or review papers in the areas discussed, to provide a way into the literature for those who wish to pursue these matters further, or who wish to place what is presented here in a broader context.

The process of obtaining the Eckart Hamiltonian is essentially a reference frame transformation and it is of course possible to present such transformations using the methods of tensor algebra. It is also possible to present the transformation using the rather more elementary methods of matrix and vector algebra. In either case the manipulations involved are intricate and tedious so that it is neither possible nor appropriate to give all the details. In these circumstances it seems preferable to use vector and matrix methods, but to provide references to tensor approaches for comparison purposes.

2. THE ORIGIN OF THE ECKART HAMILTONIAN

To understand the genesis of the Eckart Hamiltonian it must be remembered that in the very early days of quantum mechanics it was found possible, as a semi-empirical procedure, to understand the rotational levels of a molecule in a non-degenerate vibrational state, in terms of the eigenvalues of a rotational Hamiltonian,

$$H_{ROT} = C_x J_x^2 + C_y J_y^2 + C_z J_z^2 \qquad (2.1)$$

where the constants C_α, were to a good approximation given by

$$C_\alpha = (2I^o_{\alpha\alpha})^{-1} \qquad (2.2)$$

Here the $I^o_{\alpha\alpha}$ are the principal moments of inertia of the equilibrium structure of the molecule. By equilibrium structure is meant just that geometrical structure that corresponds to a conventional chemical picture of a molecule as composed of bonds with characteristic lengths and angles between them, and in terms of which structure, X-ray and electron diffraction experiments and the like can be interpreted. This geometrical picture is supplemented here by the assumption that bonds are terminated by point masses, proportional to the atomic weight of the atoms supposed to be bonded.

The first thorough discussion of this Hamiltonian was given by Casimir (1931), in the context of some very early discussions of normal co-ordinates by Brester (1923) and Wigner (1930). However, Casimir did not attempt to derive his Hamiltonian from the exact Hamiltonian for the problem, but shortly afterwards two attempts were made to do just this, the first in classical mechanics by Eckart (1934) and the second in quantum mechanics by Hirschfelder and Wigner (1935). Both these attempts failed, in the sense that in neither approach could a Hamiltonian even remotely like (2.1) be derived for the rotational part of the full

Hamiltonian. The consequences of this failure were called by Van Vleck (1935) "Eckart's paradox" since it seemed to be the case that molecular structures assigned from spectroscopic information on the basis of (2.1) could not possibly be correct if Eckart's derived Hamiltonian was correct.

However, Eckart had another try and in Eckart (1935), came up with a classical Hamiltonian that resolved the paradox. Wilson and Howard (1936) were then able to transform this classical Hamiltonian into quantum mechanical form using the methods of Podolsky (1928) and it is this approach that can be found, in all essentials, in chapter 11 of Wilson, Decius and Cross (1955).

However, it was not all plain sailing after this because, for reasons that will become apparent, argument soon developed about whether or not the Wilson-Howard quantum mechanical Hamiltonian was the correct one.

Darling and Dennison (1940) came up with a somewhat different form for the quantum mechanical Hamiltonian by use of essentially the same techniques as Wilson and Howard, and asserted that their form was superior to that of Wilson and Howard, because it was an Hermitian form and Wilson and Howard's was not. Though one can find this remark repeated in a recent review (Makushkin and Ulenikov (1977)), it is undoubtedly mistaken. There is no doubt that the forms of the Hamiltonian as offered in both approaches are the same when properly reconciled. This point was first taken by Nielsen (1951) and subsequently by Watson (1968) and by Meyer and Günthard (1968). Very clear discussions of what is involved in the reconciliation can be found in Jørgensen (1978) and (using tensor methods) in Essén (1977, 1978 and 1979).

In order to avoid becoming involved in this area of discussion, the course taken here will be to derive the rotation-vibration Hamiltonian using classical methods, then to put it into Eckart form so that the nature of the Eckart framework can be clearly exhibited. This derivation can be done very simply and quickly. Instead of then becoming involved in the process of converting this Hamiltonian to a quantum mechanical form, a completely fresh account of the genesis of the Hamiltonian will be given as derived by transformation from the Schrödinger equation in the laboratory fixed frame. This account will be in the spirit of the work of Louck (1976) but somewhat more general.

This method of exposition has the advantage of exhibiting the "target form" for the Hamiltonian in a rather straightforward way, and then of exhibiting what restrictions the achievement of this target form lays on the co-ordinate transformation in quantum mechanics. This in turn allows a clear discussion to be provided of the domain of validity of the quantum mechanical Hamiltonian.

3. THE CLASSICAL HAMILTONIAN

Consider a system of N point-mass particles, interacting only with one another, described in a right-handed laboratory fixed co-ordinate frame. Let the cartesian unit vectors that define this frame be denoted by $\hat{e}_\alpha$ (α = x,y,z) so that any point in this frame can be written as a vector

$$\underset{\sim}{x}_i = \hat{e}\, x_i \equiv \sum_\alpha \hat{e}_\alpha x_{\alpha i} \tag{3.1}$$

where $\hat{e}$ is a row matrix of unit vectors and x_i a column matrix of co-ordinates. In this system the kinetic energy of the particles is just

$$K = \tfrac{1}{2} \sum_{i=1}^{N} m_i \left|\frac{d\underset{\sim}{x}_i}{dt}\right|^2 \equiv \tfrac{1}{2} \sum_{i=1}^{N} m_i |\dot{\underset{\sim}{x}}_i|^2 \tag{3.2}$$

and the Lagrangian is

$$L = K - V \tag{3.3}$$

where the potential function V is a function of the interparticle co-ordinates only. The Hamiltonian for the system can be obtained, in the usual way, from the Lagrangian, in terms of momenta expressed as derivatives of the Lagrangian with respect to the $\dot{x}_{\alpha i}$. Since V depends only on the interparticle co-ordinates it will not contribute terms to the momenta, so attention can be concentrated <u>pro. tem.</u> on the kinetic energy alone.

Suppose now that it is desired to represent the system in terms of a centre-of-mass co-ordinate.

$$X = M^{-1} \sum_{i=1}^{N} m_i x_i, \qquad M = \sum_{i=1}^{N} m_i \tag{3.4}$$

and in a co-ordinate system specified by a cartesian system $\hat{\varepsilon}_\alpha$, related to the space fixed system by the transformation

$$\hat{\varepsilon} = \hat{e}\, C \tag{3.5}$$

where C is an orthogonal matrix. It is convenient to introduce a set of "intermediate" cartesian co-ordinates, y_i, in this representation so that

$$x_i = X + C\, y_i \tag{3.6}$$

where the y_i are imagined to be functions of 3N-6 internal co-ordinates q_1 alone. The precise definition of these co-ordinates will be a matter for further consideration, but it is clear from (3.6) that they must be such that

$$\sum_{i=1}^{N} m_i y_i = 0 \qquad (3.7)$$

is an identity when the $y_{\alpha i}$ are expressed in terms of the q_1. There must also be another three relations among the $y_{\alpha i}$ which depend on the precise definition of C. It is convenient to think of C as the general orthogonal matrix defined by three Euler angles ϕ_i, i = 1,2,3 and the three, as yet undefined, relations between the $y_{\alpha i}$, as arising from the definition of the Euler angles.

From (3.6) it follows that

$$\frac{dx_i}{dt} = \frac{dX}{dt} + \frac{dC}{dt} y_i + C \frac{dy_i}{dt} \qquad (3.8)$$

Because C is orthogonal

$$\frac{dC^T}{dt} C + C^T \frac{dC}{dt} = 0 \qquad (3.9)$$

so that

$$\frac{dC^T}{dt} C = \Omega \qquad (3.10)$$

where Ω is a skew-symmetric matrix that can be written as

$$\Omega = \begin{bmatrix} 0 & -\omega_z & \omega_y \\ \omega_z & 0 & -\omega_x \\ -\omega_y & \omega_x & 0 \end{bmatrix} \qquad (3.11)$$

Thus (3.8) becomes

$$\frac{dx_i}{dt} = \frac{dX}{dt} + C\,\Omega^T y_i + C\,y_i \ : \qquad (3.12)$$

multiplying from the left by the row matrix $\hat{e}$, using (3.1) and (3.5) and recognising that the components of $\hat{e}$ are not functions of time, (3.12) can be written in vector notation as

$$\frac{d\underset{\sim}{x}_i}{dt} = \frac{d\underset{\sim}{X}}{dt} + \underset{\sim}{y}_i \times \underset{\sim}{\omega} + \underset{\sim}{\dot{y}}_i \qquad (3.13)$$

where $\underset{\sim}{\omega}$ is the vector whose components are the elements of Ω, so that

$$\underset{\sim}{\omega} = \hat{\varepsilon}\,\omega, \quad \underset{\sim}{y}_i = \hat{\varepsilon}\, y_i \text{ etc.} \qquad (3.14)$$

It is now an easy matter to show the standard result

$$2K = M\left|\frac{d\underset{\sim}{X}}{dt}\right|^2 + \sum_{i=1}^{N} m_i |\dot{\underset{\sim}{y}}_i|^2 + \sum_{i=1}^{N} m_i (\underset{\sim}{y}_i \times \underset{\sim}{\omega}) \cdot (\underset{\sim}{y}_i \times \underset{\sim}{\omega})$$

$$+ 2\sum_i m_i \dot{\underset{\sim}{y}}_i \cdot (\underset{\sim}{y}_i \times \underset{\sim}{\omega}) \qquad (3.15)$$

where vector products are denoted by "×" and scalar products by ".". This can be re-written in matrix form using the ordinary results of vector algebra and (3.14) as

$$2K' = 2K - M\left|\frac{d\underset{\sim}{X}}{dt}\right|^2$$

$$= \sum_{i=1}^{N} m_i \dot{y}_i^T \dot{y}_i + \omega^T I\, \omega + 2\omega^T \Lambda \qquad (3.16)$$

where I is the usual inertia tensor with elements

$$I_{\alpha\beta} = \sum_{i=1}^{N} m_i (|\underset{\sim}{y}_i|^2 \delta_{\alpha\beta} - y_{\alpha i}\, y_{\beta i}) \qquad (3.17)$$

and where Λ is a column matrix of three rows with

$$\Lambda_\alpha = \sum_{i=1}^{N} m_i (\dot{\underset{\sim}{y}}_i \times \underset{\sim}{y}_i)_\alpha \qquad (3.18)$$

The angular momentum of the system, relative to the centre-of-mass may be written as

$$\underset{\sim}{L} - \underset{\sim}{L}^{CM} = \sum_{i=1}^{N} m_i (\underset{\sim}{y}_i \times (\underset{\sim}{y}_i \times \underset{\sim}{\omega})) + \sum_{i=1}^{N} m_i \underset{\sim}{y}_i \times \dot{\underset{\sim}{y}}_i \qquad (3.19)$$

with

$$\underset{\sim}{L}^{CM} = M\,\underset{\sim}{X} \times \frac{d\underset{\sim}{X}}{dt}$$

by using (3.13), (3.5) and (3.6) in the usual expression for angular momentum in the laboratory fixed frame.

It is convenient to re-write this in matrix form by defining

$$\underset{\sim}{L} = \hat{e}\,L \tag{3.20}$$

and after a little algebra (3.19) may be re-written as

$$L - L^{CM} = -\,C\,L^{F} \tag{3.21}$$

where

$$L^{F} = I\,\omega + \Lambda \tag{3.22}$$

Clearly L^F can be regarded as the negative of the angular momentum relative to the transformed frame.

It is convenient too at this stage to introduce the 3N-6 internal co-ordinates q_k and, assuming that the y_i are functions of these co-ordinates alone, to write

$$\dot{y}_{\alpha i} = \sum_{k=1}^{3N-6} \frac{\partial y_{\alpha i}}{\partial q_k}\frac{dq_k}{dt} \equiv \sum_{k=1}^{3N-6} l_{\alpha ik}\dot{q}_k \tag{3.23}$$

Introducing the matrices Z and F with elements*

$$Z_{\alpha k} = \sum_i m_i\,(\underset{\sim}{l}_{ik} \times \underset{\sim}{y}_i)_\alpha \tag{3.24}$$

and

$$F_{kl} = \sum_i \sum_\alpha m_i\, l_{\alpha ik}\, l_{\alpha il} \tag{3.25}$$

---

*In order to avoid excessive use of superscripts it will be understood from now on that sums over greek subscripts (α, β etc) go over x, y, z, sums over i, j go from 1 to N and sums over k, l go from 1 to 3N-6.

the kinetic energy (relative to the centre-of-mass kinetic energy) becomes, from (3.16)

$$2K' = \dot{q}^T F \dot{q} + \omega^T I \omega + 2\omega^T Z \dot{q} \tag{3.26}$$

and the angular momentum becomes

$$L^F = I \omega + Z \dot{q} \tag{3.27}$$

where q denotes a column matrix of the 3N-6 co-ordinates q_k. It should be noticed that, by definition, F and I are symmetric matrices.

The components of momentum required are just

$$p_k = \frac{\partial K'}{\partial \dot{q}_k} = (F \dot{q})_k + (\omega^T Z)_k$$

or

$$p = F \dot{q} + Z^T \omega \tag{3.28}$$

It is convenient now, using (3.27) and (3.28), to write (3.26) in semi-transformed form, as

$$2K' = \dot{q}^T p + \omega^T L^F \tag{3.29}$$

To get (3.29) into Hamiltonian form it is sufficient to notice that if F^{-1} exists then (3.28) can be re-written as

$$Z F^{-1} Z^T \omega = Z F^{-1} p - Z \dot{q} \tag{3.30}$$

and eliminating for ω using (3.27) it follows that

$$\omega = W^T p + M^{-1} L^F \tag{3.31}$$

Here the symmetric matrix M is given by

$$M = (I - Z F^{-1} Z^T) \tag{3.32}$$

and

$$W = - F^{-1} Z^T M^{-1} \tag{3.33}$$

Similarly substituting (3.31) into (3.28)

$$\dot{q} = G p + W L^F \tag{3.34}$$

where the symmetric matrix G is given by

$$G = F^{-1} + F^{-1} Z^{T} M^{-1} Z F^{-1} \qquad (3.35)$$

Substituting (3.31) and (3.34) into (3.29) yields for the kinetic energy in Hamiltonian form

$$2K' = L^{FT} M^{-1} L^{F} + p^{T} G p + L^{FT} W^{T} p + p^{T} W L^{F} \qquad (3.36)$$

Introducing the column matrix π defined as

$$\pi = - Z F^{-1} p \qquad (3.37)$$

equation (3.36) may be re-written as

$$2K' = (L^{F} + \pi)^{T} M^{-1} (L^{F} + \pi) + p^{T} F^{-1} p \qquad (3.38)$$

The π_α are often called the components of vibrational angular momentum but this is something of a misnomer (see Eckart (1935), Watson (1968)). Since the last two terms in (3.36) represent the Coriolis force and π is introduced to absorb these terms the use of the rather more neutral phrase Coriolis term for π is perhaps preferable.

The derivation presented above is a simple extension of Eckart's (1935) derivation and has many similarities with the exposition of Ferigle and Weber (1953). Equation (3.36) is the parallel of Eckart's equation (12). Those wishing to make correspondence in the derivation might care to note that the matrices denoted α, β and γ by Eckart correspond to M^{-1}, $-W$ and G in the present derivation.

Of course, since the Euler angles have not been fully defined (3.36) and (3.38) are only formal expressions. Nevertheless, it will be at once recognised that the difficulties that will be encountered in turning either of them into quantum mechanical form, whatever the Euler angle definition, will be truly formidable. This will perhaps make the reluctance to take this path understandable.

Before proceeding to discuss precisely how Eckart assigned the Euler angles, it is perhaps worth pointing out that the discussion given above is quite general in that the x_i could represent electrons and nuclei. However, it could not of course be applied to a two particle system.

4. THE ECKART CHOICE OF EULER ANGLES

In this section it is convenient for clarity of exposition to imagine that the system under consideration consists only of N_n nuclei, moving in a potential which is invariant under translations and rigid rotations. It is technically rather easier to face up to the implications of including electrons at a later stage, and this will be done in the context of the quantum mechanical Hamiltonian. It will therefore be assumed that there are just N_n particles and $3N_n-6$ internal co-ordinates so that sums over these variables will be taken accordingly. To emphasise the fact that only nuclei are being considered and that all distances are relative now to the centre-of-nuclear-mass the symbol r_i will be used in this section to denote position co-ordinates. All the results of the previous section of course carry through, with y_i replaced by r_i, M by M_n (the total nuclear mass) and X by $\bar{X}$ (the centre of nuclear mass co-ordinate).

In his 1935 paper Eckart points out that the Hamiltonian kinetic energy (3.36) will yield a rotational Hamiltonian like (2.1) if the internal co-ordinates are chosen so that Z vanishes. In this case W vanishes and M^{-1} becomes I^{-1}, and if I is substantially constant then all is well. But he also points out that it is extremely unlikely that the requirement that Z vanishes, defines just $3N_n-6$ internal co-ordinates uniquely and unambiguously. The requirement that Z vanishes comes down to solving the $3N_n-6$ simultaneous differential equation

$$\sum_i m_i \left(\frac{\partial \underset{\sim}{r}_i}{\partial q_k} \times \underset{\sim}{r}_i\right)_\alpha = 0 \qquad (4.1)$$

for the r_i as functions of the q_k. It is overwhelmingly likely that this set of equations possess more than six integrals independent of the q_k. (There may in fact be just six in the case of the three particle system, but this is still an open question.) If there were just six integrals independent of the q_k then it would be possible to secure the requirement that $\Lambda_\alpha = 0$ (where Λ is a function of q_k and $\dot{q}_k$) on a definition of the q_k which depends solely on particle positions and this, as Eckart says "is not reasonable".

However, it is possible to require that Z vanishes for one definite set of values of the q_k, without restricting the definition of the r_i to an impossible extent. It is then reasonable to suppose, in classical mechanics at least, that the terms linear in L^F in (3.36) are small for values of q_k near the chosen set. If this is the region (classically) in which the motion of the system takes place then the objective has been accomplished as well as it can be.

How Eckart achieves this objective and makes his choice of q_k, is the subject of part 3 of his paper. It is a masterpiece of subtle physical reasoning and well repays careful study, as does the review article by Louck and Galbraith (1976) which is devoted almost entirely to an explication of the symmetry implications of Eckart's part 3. For the present, however, a rather bald account of what the choice actually is, must suffice.

Eckart supposes that there is a minimum in the potential function for the nuclei which can be represented in the chosen frame by

$$r_i = a_i \qquad i = 1, 2 \ldots N_n \tag{4.2}$$

so that the a_i can be thought of as the equilibrium nuclear (cartesian) co-ordinates in the chosen frame. The full co-ordinates are then written as

$$r_i = a_i + \sum_k l_{ik} q_k \tag{4.3}$$

where the l_{ik} are column matrices of constant elements $l_{\alpha ik}$. In order to satisfy the equivalent of (3.7) the a_i and l_{ik} must be such that

$$\sum_i m_i a_i = 0 \quad (4.4a), \qquad \sum_i m_i l_{ik} = 0, \text{ all } k \tag{4.4b}$$

and l_{ik} are then determined (up to a linear transformation) by the requirement that Z shall vanish when the q_k are zero. From (3.24) this leads to the condition

$$\sum_i m_i(\underset{\sim}{l}_{ik} \times \underset{\sim}{a}_i) = \underset{\sim}{0} \tag{4.5}$$

This condition is often called the second Eckart condition, or the rotational Eckart condition, for it in fact determines the Euler angle definition, in a way that is discussed both in Eckart (1935) and in Louck and Galbraith (1976) and will be dealt with in more detail later when the quantum mechanical Hamiltonian is considered. The conditions (4.4) are often jointly referred to as the first or translational Eckart condition, and the unit vectors $\hat{\varepsilon}_\alpha$ defined by (3.5) are often said to define the Eckart frame of reference.

Using (4.3), (4.4) and (4.5) then F is exactly as defined in (3.25) with the $l_{\alpha ik}$ just the constants from (4.3). The matrix Z becomes

$$Z_{\alpha k} = \sum_l \sum_i m_i(\underset{\sim}{l}_{ik} \times \underset{\sim}{l}_{il})_\alpha q_l \tag{4.6}$$

and it is customary to re-define this in terms of the elements of the Coriolis coupling coefficients

$$\xi^{\alpha}_{lk} = \sum_i m_i (\underset{\sim}{l}_{il} \times \underset{\sim}{l}_{ik})_{\alpha} \tag{4.7}$$

and so re-write Z as

$$Z_{\alpha k} = - \sum_l \xi^{\alpha}_{lk} q_l \; ; \tag{4.8}$$

with this definition

$$M_{\alpha\beta} = I_{\alpha\beta} - \sum_{klmn} \xi^{\alpha}_{km} (F^{-1})_{mn} \xi^{\beta}_{ln} q_k q_l \tag{4.9}$$

and

$$\pi_{\alpha} = \sum_{lmk} \xi^{\alpha}_{lm} F^{-1}_{mk} q_l p_k \tag{4.10}$$

It is also an easy matter to invert (4.3) to obtain q in terms of cartesian co-ordinate which satisfy the Eckart conditions, by noting that

$$\sum_k \sum_i \sum_{\alpha} m_i l_{\alpha il} l_{\alpha ik} q_k = \sum_i \sum_{\alpha} m_i l_{\alpha il}(r_{\alpha i} - a_{\alpha i}) \tag{4.11}$$

so that

$$q_l = \sum_k F^{-1}_{lk} \sum_i \sum_{\alpha} m_i l_{\alpha ik}(r_{\alpha i} - a_{\alpha i}) \tag{4.12}$$

The expressions (4.9), (4.10) and (4.12) may be simplified somewhat by noticing that if the l_{ik} satisfy (4.4b) and (4.5) then any linear combination of them (over the index k) also satisfies (4.4b) and (4.5). Thus it is possible to choose a linear combination of these such that $F_{kl} = \delta_{kl}$.

It should be emphasised that $r_{\alpha i}$ in (4.12) are not simple cartesian co-ordinates, they are constrained by the requirement that they are given by (4.3) with a_i and l_{ik} satisfying (4.4) and (4.5). To emphasise this (4.12) is often expressed as being subject to the conditions

$$\sum_i m_i \underset{\sim}{r}_i = \underset{\sim}{0} \quad (4.13), \qquad \sum_i m_i \underset{\sim}{r}_i \times \underset{\sim}{a}_i = \underset{\sim}{0} \tag{4.14}$$

which are clearly equivalent to the first and second Eckart conditions.

In practice the internal co-ordinates q_l are chosen in a way described in detail in chapter 4 of Wilson, Decius and Cross (1955). The a_i for a problem are chosen initially to satisfy (4.4a) and the l_{ik} are then determined according to special formulae, one formula for each type of intuitively conceived internal motion, in terms of cartesian displacements from the a_i, so as to satisfy (4.4b) and (4.5).

The initial choice of a_i usually corresponds to an intuitive view of the equilibrium nuclear positions, as explained previously. In consequence it is often thought plausible to regard the potential function as expandable in a Taylor series in terms of the q_k about these positions. In this case π should be negligibly small so that rotational motion is decoupled from the internal motion and, to a good approximation, the internal motion Hamiltonian will be

$$H = p^T F^{-1} p + \tfrac{1}{2} q^T V q \qquad (4.15)$$

where V contains the quadratic terms in the Taylor series expansion of the potential. In this case, as is well known it is always possible to find a co-ordinate transformation $q \rightarrow Q$ that diagonalises V and reduces F^{-1} to a unit matrix. (Unfortunately this technique is often known in the trade as the "F-G matrix" technique where, alas, what has here been called F^{-1} is there called G and what here is called V is there called F. Apologies for the possible confusion so created.) The Q in this case are always called normal co-ordinates and the eigenvalues of V corresponding to these conditions are often called the normal mode vibrational energies of the system.

The interpretation of vibrational spectra then, consists essentially of assuming (4.15) correct as a first shot, with a_i and l_{ik} and V assigned on an intuitive basis. The normal modes that correspond to this intuition are then found and the spectrum assigned as well as can be. Iterative fitting of V then follows, if necessary, to determine the best possible parameters.

These parameters then form the basis for the construction of π and of the Hamiltonian including rotational motion so that the whole rotation-vibration structure of a spectrum may be interpreted. Iterative fitting of the a_i is then performed if necessary.

Of course this description of the process of interpreting spectra is a crude oversimplification of what spectroscopists actually do. It under-represents to a ludicrous extent the subtlety and refinement of their techniques. The description offered here is offered, not because it is felt to be not well known, nor is it offered as a slight on spectroscopists. It is offered in this crude and bald form to emphasise the point that the facts of molecular structure do not emerge from the results of spectroscopic observation. Rather the assumption of molecular structure is built into the theoretical scheme in terms of which spectroscopic observations are interpreted. It is possible with this scheme, in many cases, to develop a view of the structure of a molecule which is consistent with chemical intuition and with the results of the experimental procedures.

It is clear from the above discussion that the molecule must, in some sense, be rigid for the Eckart approach to work at all. Thus it could not be expected to work very well for molecules that have, in the conventional picture, single bonds about which free rotation can occur. Neither would it work well for molecules that are conventionally believed to have a low inversion barrier. In short the Eckart approach will fail in the case of large amplitude internal motions. The problem of such motions was first dealt with by Sayvetz (1939), a pupil of Eckart, and the scheme that he developed has formed the basis of almost all the work in this area since then. However, once non-rigid molecules are considered in detail, the problems of nuclear permutational symmetry must also be considered, and such considerations are best deferred until a quantum mechanical Hamiltonian has been obtained.

It is also clear that the Eckart approach will not work for molecules whose intuitive equilibrium geometry is linear, since the inertia tensor at this geometry is singular. This problem seems first to have been considered by Welker (1936) and also by Sayvetz (1939).

It turns out that the Eckart approach can be "rigged" to deal with this case and the method of rigging it, now most widely adopted is due to Watson (1970). Since consideration of this in detail would not lead to anything new, the linear molecule is simply ignored here. However, it is of course the case that if a non-linear molecule exhibits an internal motion in which it becomes linear then the Eckart approach will break down anyway.

To consider these (and other) problems properly it is necessary to derive a form for the quantum mechanical Hamiltonian and this will now be done. The following two sections (5) and (6) parallel closely sections (3) and (4). There will be some repetition of material here, for which no apology is made, since it is hoped in this way to emphasise the similarities in the two approaches.

5. THE GENERAL CO-ORDINATE TRANSFORMATION

Consider a laboratory fixed co-ordinate frame with cartesian unit vectors $\hat{e}_\alpha$ (α = x,y,z) such that any point in this frame may be written as a vector

$$x_i = \hat{e}x_i \tag{5.1}$$

with $\hat{e}$ a row matrix of the unit vectors and x_i a column matrix of co-ordinates. In this frame the Hamiltonian operator for a system of N point masses may be set up in an unambiguous manner. It is often useful to pass from this frame however to one which moves with centre of mass of the system. This may be done by defining a centre of mass co-ordinate

$$X = M^{-1} \sum_{i=1}^{N} m_i x_i, \qquad M = \sum_{i=1}^{N} m_i \tag{5.2}$$

and 3N-3 internal co-ordinates, s_1, s_2 s_l s_{3N-3}. There is of course no unique prescription for choosing the s_l, but a proper co-ordinate transformation $(x) \rightarrow (X,s)$ exists, whatever the s_l if the jacobian matrix $\partial(x)/\partial(X,s)$ and its inverse exists. For any choice of s, the jacobian and its inverse may only exist locally, and in that case the co-ordinate transformation exists only locally.

If it is assumed that the jacobian and its inverse exist at least locally then it is possible to write

$$x_{\alpha i} = X_\alpha + \bar{x}_{\alpha i}(s_1 \ldots s_{3N-3}) \tag{5.3}$$

where $\bar{x}_{\alpha i}$ is a function of 3N-3 variables above and is such that

$$\sum_{i=1}^{N} m_i x_{\alpha i} \equiv 0 \tag{5.4}$$

is an identity when expressed in the s_l.

The fact that the product of the jacobian and its inverse must yield a unit matrix gives the following pair of identities

$$\sum_{i=1}^{N} \frac{\partial s_l}{\partial x_{\alpha i}} \equiv 0, \qquad \sum_{i=1}^{N} m_i \frac{\partial x_{\alpha i}}{\partial s_l} \equiv 0 \qquad (5.5a) \qquad (5.5b)$$

by noting that

$$\frac{\partial X_\delta}{\partial x_{\alpha i}} = \delta_{\alpha\delta}\, m_i M^{-1} \quad \text{and} \quad \frac{\partial x_{\alpha i}}{\partial X_\delta} = \delta_{\alpha\delta} \qquad (5.6a)\ (5.6b)$$

The condition (5.5b) may of course be regarded as a consequence of (5.4) and the fact that the s_1 and X_α are independent co-ordinates, while the condition (5.5a) can be seen as a consequence of the requirement that the s_1 must be at most functions of $(x_i - X)$. They may also be looked upon as tests appropriate for determining whether a postulated set of internal co-ordinates is a proper set. By writing*

$$\frac{\partial}{\partial x_{\alpha i}} = m_i M^{-1} \frac{\partial}{\partial X_\alpha} + \sum_1 \frac{\partial s_1}{\partial x_{\alpha i}} \frac{\partial}{\partial s_1} \qquad (5.7)$$

it is easy to see that (5.5a) is the identity required for the internal motion to separate from the centre-of-mass motion in the quantum mechanical kinetic energy operator.

If the inverse jacobian $\partial(X,s)/\partial(x)$ is denoted by J with row indices labelling the transformed co-ordinates and column indices labelling the original co-ordinates then the matrix representing the inverse metric tensor is

$$J\, m^{-1} J^{T} = g^{-1} \qquad (5.8)$$

(where m^{-1} is a diagonal matrix of particle masses). This matrix is blocked and it follows at once that the formal condition for the existence of the jacobian and its inverse is that G^{-1} exists, where G is the matrix with elements

$$G_{1m} = \sum_\alpha \sum_i m_i^{-1} \frac{\partial s_1}{\partial x_{\alpha i}} \frac{\partial s_m}{\partial x_{\alpha i}} \qquad (5.9)$$

In systems containing three or more point masses, it is often convenient to recognise that three of the internal co-ordinates may be chosen as Euler angles, ϕ_1, ϕ_2 and ϕ_3.

It is usual of course to re-define the internal co-ordinate s_1 in terms of the ϕ_i and 3N-6 new independent internal co-ordinates $q_1 \ldots q_{3N-6}$ which are independent of the ϕ_i. Again for such a

---

* As in sections 3 and 4, to avoid use of superscripts it will be understood from now on that sums over greek subscripts (α, β etc.) go over x, y, z, sums over i, j go from 1 to N and sums over k, 1, go from 1 to 3N-3 or to 3N-6.

transformation to exist the jacobian matrix and its inverse $(s) \rightarrow (\phi,q)$ must exist. However, it is convenient for the moment to avoid explicit discussion of this and to regard the transformation as one going directly from $(x) \rightarrow (X,\phi,q)$ so that the ϕ_i and the q_1 can be regarded as special cases of the s_1.

The Euler angles may be thought of as defining an orthogonal transformation of the space fixed frame to a new frame $\hat{\varepsilon}_\alpha$, $\alpha = x,y,z$ by

$$\hat{\varepsilon} = \hat{e}\, C \tag{5.10}$$

and for definiteness it may be supposed that C is defined as

$$C = R_{e_z}(\phi_1)\, R_{e'_y}(\phi_2)\, R_{e''_z}(\phi_3) \tag{5.11}$$

where $R_e(\phi)$ denotes a positive rotation about e by ϕ. Thus $\hat{e}$ is brought into $\hat{\varepsilon}$ by a positive rotation about e_z by ϕ_1, followed by a positive rotation about the new y-axis (e'_y) by ϕ_2 and finally by a positive rotation about the new z axis $(e''_z \equiv \hat{\varepsilon}_3)$ by ϕ_3 (Brink and Satchler (1968)).

Whatever the precise choice of C it is an easy, if somewhat tedious matter, to show that

$$\frac{\partial\phi_m}{\partial x_{\alpha i}} = \Sigma_\gamma\, \omega_\gamma^{\alpha i} D_{\gamma m} \tag{5.12a}$$

where $D_{\gamma m}$ is an element of a square matrix D whose elements are all simple trigonometrical functions of the Euler angles. For the choice of C given above

$$D \quad \begin{bmatrix} \dfrac{\cos\phi_3}{\sin\phi_2} & -\sin\phi_3 & -\cot\phi_2\cos\phi_3 \\ \dfrac{-\sin\phi_3}{\sin\phi_2} & -\cos\phi_3 & +\cot\phi_2\sin\phi_3 \\ 0 & 0 & -1 \end{bmatrix} \tag{5.13}$$

The $\omega_\gamma^{\alpha i}$ are elements of the skew-symmetric matrix

$$\Omega^{\alpha i} = \begin{bmatrix} 0 & -\omega_z^{\alpha i} & \omega_y^{\alpha i} \\ \omega_z^{\alpha i} & 0 & -\omega_x^{\alpha i} \\ -\omega_y^{\alpha i} & \omega_x^{\alpha i} & 0 \end{bmatrix} \tag{5.14}$$

that is defined by the relationship

$$\frac{\partial C^T}{\partial x_{\alpha i}} C = \Omega^{\alpha i} \tag{5.15}$$

That $\Omega^{\alpha i}$ must be skew symmetric follows from the orthogonality requirement on C, but the precise form of the $\omega_\gamma^{\alpha i}$ depends naturally on the precise definition of the ϕ_i.

In order to obtain an expression extending (5.6b) in the same way as (5.12a) extends (5.6a) it is convenient to introduce an intermediate set of cartesian co-ordinates y_i such that

$$\bar{x}_{\alpha i} = \sum_\gamma C_{\alpha\gamma} y_{\gamma i} \tag{5.16}$$

where y_i are assumed functions of the 3N-6 co-ordinates q_l quoted above. It follows after a little involved algebra that

$$\frac{\partial x_{\alpha i}}{\partial \phi_m} = (CY^i D^{-T})_{\alpha m} \tag{5.12b}$$

where Y^i is the matrix defined like Ω in (5.14) but with y_{zi} in place of ω_z and so on.

The set of identities extending (5.5) by forming the product of the jacobian and its inverse are now

$$\sum_i \frac{\partial q_l}{\partial x_{\alpha i}} = 0 \qquad \sum_i m_i \frac{\partial x_{\alpha i}}{\partial q_l} = 0 \qquad (5.17a)\ (5.17b)$$

$$\sum_i \omega_\gamma^{\alpha i} = 0 \qquad \sum_{i\alpha} \omega_\gamma^{\alpha i} (CY^i)_{\alpha\eta} = \delta_{\gamma\eta} \qquad (5.18a)\ (5.18b)$$

$$\sum_{i\alpha} \frac{\partial q_1}{\partial x_{\alpha i}} (CY^i)_{\alpha\gamma} = 0 \qquad \sum_{i\alpha} \omega_\gamma^{\alpha i} \frac{\partial x_{\alpha i}}{\partial q_1} = 0 \tag{5.19a, 5.19b}$$

The identities (5.18a) and (5.18b) were apparently first noticed by Hirschfelder and Wigner (1935) but these authors used the equivalent form for (5.18b)

$$\sum_i (\bar{x}_{\gamma i}\omega_\delta^{\alpha i} - \bar{x}_{\alpha i}\omega_\delta^{\alpha i}) = -C_{\beta\delta} \tag{5.18c}$$

$$\gamma\alpha\beta = xyz$$

and cyclic

Extending (5.7) to include the angular co-ordinates explicitly one obtains

$$\frac{\partial}{\partial x_{\alpha i}} = m_i M^{-1} \frac{\partial}{\partial X_\alpha} + \sum_m \sum_\gamma \omega_\gamma^{\alpha i} D_{\gamma m} \frac{\partial}{\partial \phi_m} + \sum_1 \frac{\partial q_1}{\partial x_{\alpha i}} \frac{\partial}{\partial q_1} \tag{5.20}$$

The second term in (5.20) can be written as

$$\frac{i}{\hbar} \sum_\gamma \omega_\gamma^{\alpha i} L_\gamma^F \tag{5.21}$$

where the L^F are angular momentum-like terms which with the present choice of Euler angles, are given by

$$\frac{i}{\hbar} L_x^F = \frac{\cos\phi_3}{\sin\phi_2} \frac{\partial}{\partial\phi_1} - \sin\phi_3 \frac{\partial}{\partial\phi_2} - \cot\phi_2 \cos\phi_3 \frac{\partial}{\partial\phi_3} \tag{5.22a}$$

$$\frac{i}{\hbar} L_y^F = -\frac{\sin\phi_3}{\sin\phi_2} \frac{\partial}{\partial\phi_1} - \cos\phi_3 \frac{\partial}{\partial\phi_2} + \cot\phi_2 \sin\phi_3 \frac{\partial}{\partial\phi_3} \tag{5.22b}$$

$$\frac{i}{\hbar} L_z^F = -\frac{\partial}{\partial\phi_3} \tag{5.22c}$$

A little algebra shows that the quantities $-\sum_\gamma C_{\alpha\gamma} L_\gamma^F$ are the usual expressions for the angular momentum operators in the space-fixed frame expressed in Euler angles (see e.,., Brink and Satchler (1968)).

In fact the result is quite general. If the total angular-momentum operator L, is constructed from (5.20) then after some algebra making use of (5.18) and (5.19) it is seen that

$$(L_\alpha - L_\alpha^{CM}) = -\sum_\gamma C_{\alpha\gamma} L_\gamma^F \qquad (5.23)$$

where L^{CM} is the centre-of-mass angular momentum.

The inverse metric matrix can be developed just as in (5.8). It is again blocked and its determinant is, after a little manipulation

$$|Jm^{-1}J^T| = M^{-3}|D|^2|M^{-1}||G - WMW^T| \qquad (5.24)$$

where

$$M_{\alpha\beta}^{-1} = \sum_{i\gamma} m_i^{-1} \omega_\alpha^{\gamma i} \omega_\beta^{\gamma i} \qquad (5.25a)$$

$$W_{l\alpha} = \sum_{i\gamma} m_i^{-1} \frac{\partial q_l}{\partial x_{\gamma i}} \omega_\alpha^{\gamma i} \qquad (5.25b)$$

$$G_{kl} = \sum_{i\gamma} m_i^{-1} \frac{\partial q_k}{\partial x_{\gamma i}} \frac{\partial q_l}{\partial x_{\gamma i}} \qquad (5.25c)$$

M^{-1} has the dimensions of an inverse inertia tensor and is, as can be seen from (5.18b) a function of the q_l alone. Its precise significance depends of course on the choice of the $\omega_\gamma^{\alpha i}$. Similarly, it can be seen from (5.19a) that both W and G must be functions of the q_l alone.

It is now a straightforward matter to construct the kinetic energy operator for the problem renormalised to incorporate the jacobian. It can be done either directly from (5.20) or alternatively by a simple extension of the analysis of Kemble (1937) in either case making use of the identities (5.17a) and (5.18a). To use Kemble's analysis it is just necessary to note that the elements of $Jm^{-1}J^T$ are the quantities Kemble denotes as g^{lk} and the determinant (5.24) is the inverse of Kemble's determinant g. Since it is not the custom to renormalise for the angular variables it is necessary to extend Kemble's analysis slightly to allow for the fact that $|D|^{-1} d\phi_1 d\phi_2 d\phi_3$ remains the volume element for angular integrations. For the choice (5.13) for D, $|D|^{-1}$ is just $\sin\phi_2$.

The results are

$$K = K_1 + K_2 + U \tag{5.26}$$

with

$$K_1 = \frac{-\hbar^2}{2M}\sum_{\varepsilon}\frac{\partial^2}{\partial X_{\varepsilon}^2} + \tfrac{1}{2}\sum_{\gamma\varepsilon} M^{-1}_{\gamma\varepsilon}L^F_{\gamma}L^F_{\varepsilon} \tag{5.27a}$$

$$\frac{-\hbar^2}{2}\sum_{kl} G_{kl}\frac{\partial^2}{\partial q_k \partial q_l} + \frac{\hbar}{i}\sum_{\varepsilon}\sum_{l} W_{l\varepsilon}L^F_{\varepsilon}\,\partial/\partial q_l$$

$$K_2 = \tfrac{1}{2}\sum_{\varepsilon} \nu_{\varepsilon}\, L^F_{\varepsilon} + \frac{\hbar}{2i}\sum_{k}\tau_k\,\partial/\partial q_k \tag{5.27b}$$

Here

$$\nu_{\varepsilon} = \sum_{i=1}^{3} \nu_{\varepsilon}^{(i)} \tag{5.28}$$

with

$$\nu_{\varepsilon}^{(1)} = \sum_{i}\sum_{\alpha\gamma} m_i^{-1}\,\omega_{\gamma}^{\alpha i}L^F_{\gamma}\omega_{\varepsilon}^{\alpha i} \tag{5.29a}$$

$$\nu_{\varepsilon}^{(2)} = \frac{-g^{-1}}{2}\frac{\hbar}{i}\sum_{k} W_{k\varepsilon}\,\partial g/\partial q_k \tag{5.29b}$$

$$\nu_{\varepsilon}^{(3)} = \frac{\hbar}{i}\sum_{\alpha}\sum_{k}\sum_{j} m_j^{-1}\frac{\partial q_k}{\partial x_{\alpha j}}\frac{\partial \omega_{\varepsilon}^{\alpha j}}{\partial q_k} \tag{5.29c}$$

Also

$$\tau_k = \sum_{i=1}^{3}\tau_k^{(i)} \tag{5.30}$$

with

$$\tau_k^{(1)} = \sum_{\alpha\gamma}\sum_{i} m_i^{-1}\,\omega_{\gamma}^{\alpha i}L^F_{\gamma}\frac{\partial q_k}{\partial x_{\alpha i}} \tag{5.31a}$$

$$\tau_k^{(2)} = \frac{-g^{-1}}{2}\frac{\hbar}{i}\sum_{l} G_{lk}\frac{\partial g}{\partial q_l} \tag{5.31b}$$

$$\tau_k^{(3)} = \frac{\hbar}{i} \sum_\alpha \sum_l \sum_i m_i^{-1} \frac{\partial q_l}{\partial x_{\alpha i}} \frac{\partial}{\partial q_l} \frac{\partial q_k}{\partial x_{\alpha i}} \tag{5.31c}$$

And

$$U = \frac{\hbar^2}{8} \sum_{i=1}^{3} U_i \tag{5.32}$$

with

$$U_1 = g^{-1} \frac{i}{\hbar} \sum_k (\tau_k^{(1)} + \tau_k^{(3)}) \frac{\partial g}{\partial q_k} \tag{5.33a}$$

$$U_2 = g^{-1} \sum_{kl} G_{lk}\, \partial^2 g/\partial q_k \partial q_l \tag{5.33b}$$

$$U_3 = \frac{-5}{4} g^{-2} \sum_{kl} G_{lk} \frac{\partial g}{\partial q_l} \frac{\partial g}{\partial q_k} \tag{5.33c}$$

and

$$g = |M|\,|G - W M W^T|^{-1} \tag{5.34}$$

It should be noted that quite generally

$$\frac{i}{h} C^T L_\gamma^F C = H_\gamma \tag{5.35}$$

where L^F operates on every element of C and where

$$(H_\gamma)_{\alpha\beta} = \begin{cases} +1 & \gamma\alpha\beta = \text{xyz and cyclic} \\ -1 & \gamma\alpha\beta = \text{xzy and cyclic} \\ 0 & \text{otherwise} \end{cases} \tag{5.36}$$

Thus, because of (5.18) and (5.19) and (5.36) both the ν_ε and τ_k of (5.29) and (5.31) can be written to show that they are functions of the q_l only and hence U is also a function of the q_l only.

The analogy between (5.27a) and (3.36) is readily apparent; but to move to the form (3.38) some further manipulation is necessary. To do this it is necessary to introduce the metric $\mathcal{J}^{-T} m \mathcal{J}^{-1}$ which is inverse to the metric of (5.8) so $\mathcal{J}^{-1}$ is $\partial(x)/\partial(X,\phi,q)$. The analogues of G and W in this metric are F and Z with elements

$$F_{kl} = \sum_{\alpha} \sum_{i} m_i \frac{\partial x_{\alpha i}}{\partial q_k} \frac{\partial x_{\alpha i}}{\partial q_l} \tag{5.37}$$

$$Z_{\alpha l} = \sum_{\beta} \sum_{i} m_i (CY^i)_{\beta\alpha} \partial x_{\beta i}/\partial q_l \tag{5.38}$$

By constructing the product of the metric and its inverse and requiring it to equal the unit matrix, the following identities are obtained:

$$M = (I - ZF^{-1}Z^T) \tag{5.39}$$

$$G = F^{-1} + F^{-1}Z^T M^{-1} Z F^{-1} \tag{5.40}$$

$$F^{-1} = G - WMW^T \tag{5.41}$$

$$W = -F^{-1}Z^T M^{-1} \tag{5.42}$$

with I the conventional inertia tensor given by

$$I = \sum_{i} m_i Y^{iT} Y^i \tag{5.43}$$

It should also be noticed that (5.41) enables (5.34) to be rewritten as

$$g = |M|\,|F| \tag{5.44}$$

Equation (5.27a) can then be rewritten in the form (3.36) by introducing

$$\pi_\alpha = \frac{-\hbar}{i} \sum_{kl} Z_{\alpha l} F^{-1}_{lk}\, \partial/\partial q_k = \frac{\hbar}{i} \sum_{\gamma} \sum_{k} M_{\alpha\gamma} W_{k\gamma}\, \partial/\partial q_k \tag{5.45}$$

and extending τ_k to include an extra term

$$\tau_k^{(4)} = -\frac{\hbar}{i} \sum_{\delta\mu} \sum_{l} W_{l\delta} \frac{\partial}{\partial q_l} (M_{\delta\mu} W_{k\mu}) \tag{5.46}$$

Then K_1 becomes

$$K_1 = \frac{-\hbar^2}{2M}\sum_{\varepsilon}\frac{\partial^2}{\partial X_\varepsilon^2} + \frac{1}{2}\sum_{\gamma\varepsilon} M_{\gamma\varepsilon}^{-1}(L_\gamma^F + \pi_\gamma)(L_\varepsilon^F + \pi_\varepsilon) - \frac{\hbar^2}{2}\sum_{kl} F_{kl}^{-1}\,\partial^2/\partial q_k\partial q_l \qquad (5.47)$$

and τ_k becomes

$$\tau_k = \sum_{i=1}^{4} \tau_k^{(i)} \qquad (5.48)$$

and so K_1 is in the required form.

A rather elegant outline of what is involved in passing from the classical to the quantum mechanical form, using tensor methods, can be found in Essen (1978), and also in Marques and Bordé (1971a, 1971b). Many of the problems that arise in this passage are due to the different possible ways of treating the jacobian. In the working given above the appropriate jacobian is $|J^{-1}|$ and is such that if the laboratory-fixed Hamiltonian operates on square integrable functions $\psi(x_i)$, then the initially transformed (but un-renormalised) operator operates on functions $\psi'(X,\phi,q)$ such that

$$\int |\psi(x_i)|^2 dx = \int |\psi'(X,\phi,q)|^2 |J^{-1}| dX d\phi dq \qquad (5.49)$$

From (5.24) and (5.34) $|J^{-1}|$ is just $M^{3/2}\prod_i m_i^{-3/2} g^{\frac{1}{2}}|D|^{-1}$ so that a wave function $\psi'' = g^{\frac{1}{4}}\psi'$ can be introduced such that

$$\int |\psi(x_i)|^2 dx = M^{3/2}\prod_i m_i^{-3/2} \int |\psi''(X,\phi,q)|^2 |D|^{-1} d\phi dX dq \qquad (5.50)$$

The matrix element of the Hamiltonian is then such that

$$\int \psi^*(x_i) H \psi(x_i) dx = M^{3/2}\prod_i m_i^{-3/2} \int \psi''(X,\phi,q) H'' \psi''(X,\phi,q) d\phi dX dq \qquad (5.51)$$

where H" is the renormalised Hamiltonian ($g^{1/4}H'g^{-1/4}$), with H' as the initially transformed Hamiltonian. Clearly the constant mass term will have no effect in expectation values and can be ignored.

6. THE ECKART CO-ORDINATES

In this section, as in section 4, a distinction of notation will be used to indicate co-ordinates relative to the centre of nuclear mass. However, in this section electrons will be considered explicitly and will be assumed counted as $i = N_n + 1, N_n + 2, \ldots N$ from the upper limit of the nuclear index and this will be explicitly indicated in sums involving electrons. The internal co-ordinates q_l will be assumed defined entirely in terms of the nuclear variables and thus l will run only up to $3N_n-6$.

In this case any particle co-ordinate in the laboratory fixed frame can be written as

$$x_{\alpha i} = \bar{X}_\alpha + (Cr_i)_\alpha, \quad i = 1, 2, \ldots N_n, \ldots N \tag{6.1}$$

where $\bar{X}_\alpha$ is the centre of nuclear mass.

$$\bar{X}_\alpha = M_n^{-1} \sum_i m_i x_{\alpha i}, \quad M_n = \sum_i m_i \tag{6.2}$$

Here and in future, sums over i, etc., without an upper limit indicated will run to N_n (assumed > 3).

As in (4.3)† the nuclear co-ordinates may be written as

$$r_{\alpha i} = a_{\alpha i} + \sum_k l_{\alpha ik} q_k \tag{6.3}$$

and, as in (4.13) and (4.14) the r_i are constrained by the requirements

$$\sum_i m_i r_{\alpha i} = 0 \tag{6.5}$$

$$\sum_i m_i A^i r_i = 0 \tag{6.6}$$

as an identity in the internal co-ordinates, where A^i is a matrix like Y^i of (5.12b) but with the element $a_{\alpha i}$ instead of $y_{\alpha i}$.

From (6.5) it follows that

$$\sum_i m_i a_{\alpha i} = 0 \quad (6.7a) \qquad \sum_i m_i l_{\alpha ik} = 0, \text{ all } k \quad (6.7b)$$

just as in (4.4a) and (4.4b) and from (2.7) it follows that

† Here as in (4.3) the $l_{\alpha ik}$ are constants, specializing (3.23).

$$\sum_{i=1} m_i A^i l_{ik} = 0, \text{ all k} \tag{6.8}$$

as an identity.

Of course for any given choice of the a_i (6.7b) and (6.8) do not define a unique choice of the l_{ik}, and for every different choice of the a_i different possible sets of l_{ik} are defined. However, rather simple arguments can be used to show that just so long as the inverse of

$$I^o = \sum_i m_i A^{iT} A^i \tag{6.9}$$

exists then any internal co-ordinates defined with the aid of (6.7b) and (6.8), generate $3N_n$- dimensional vectors of the form $(r_{xi}\ r_{yi}\ \ldots\ r_{zN_n})$ which lie in a $3N_n$-6 dimensional subspace orthogonal (in mass weighted norm) to the 6 - dimensional space defined by the 3N-vectors.

$$t^x = (100\ 100\ \ldots\ 100) \text{ and similarly for } t^y \text{ and } t^z \tag{6.10}$$

and

$$r^x = (0\ a_{z_1}\ -a_{y_1}\ 0\ a_{z_2}\ -a_{y_2}\ \ldots\ a_{y_N}) \text{ and similarly for } r^y \text{ and } r^z \tag{6.11}$$

so that all possible sets of l_{ik} are related by simple linear transformations.

The relationship (6.6) in fact defines the matrix C, as can be seen by substituting (6.1) into (6.6) to yield the relationship

$$C^T F - F^T C = 0 \tag{6.12}$$

with

$$F_{\alpha\beta} = \sum_{i=1} m_i (x_{\alpha i} - X_\alpha) a_{\beta i} \tag{6.13}$$

giving the polar decomposition

$$C = F(F^T F)^{-\frac{1}{2}} \tag{6.14}$$

where (6.14) is constructed on the assumption that the positive symmetric square root is to be taken.

C will not exist unless F^TF is positive definite, or equivalently, F is non-singular. However, if C exists and hence the $r_{\alpha i}$ can be defined, it is easy to show by substituting (6.1) into (6.13), that F may be re-written as

$$F = C\bar{B}, \quad \bar{B}_{\alpha\beta} = \sum_{i=1} m_i r_{\alpha i} a_{\alpha i} \tag{6.15}$$

If the $r_{\alpha i}$ are expressed in terms of the q_k given by (6.3) with (6.7) and (6.8) satisfied then $\bar{B}$ is a symmetric matrix and (6.14) is then simply an identity. This means that the q_k are independent of the rotational co-ordinates, as required.

Looked at from the point of view of the local co-ordinates $\bar{B}$ may be regarded as a diagnostic for the existence of C. If for any chosen set of internal co-ordinates $\bar{B}$ is not symmetric then not all the constraints on the r_i required by (6.8) have been imposed. If $\bar{B}$ is symmetric, but singular for some values of the q_k, then C does not exist at the implied configuration. As has been mentioned before (section 4) this implies that the a_i may not be chosen so that the complete set specifies a straight line. It also seems to imply that the a_i cannot define a plane, but this is not a real restriction in practice, because a conventional choice (see e.g., Louck and Galbraith (1976)) can be made for the initial vector $\hat{e}_z$ of (5.10). Perhaps this is most easily seen by noting that if the a_i are taken to define the x-y plane, then I^o_{zz} is non-zero and so I^o can be non-singular.

Again assuming the existence of C, by differentiating (6.12) with respect to $x_{\alpha i}$ and after a little re-arrangement, an expression for $\Omega^{\alpha i}$ is obtained, which yields:

$$\omega^{\alpha i}_{\gamma} = m_i \sum_{\delta} C_{\alpha\delta} (A^i B^{-1})_{\delta\gamma} \tag{6.16}$$

where

$$B = \sum_i m_i A^{iT} R^i \tag{6.17}$$

where R^i is like A^i but with elements $r_{\alpha i}$. On substituting from (6.3), (6.17) becomes

$$B = I^o + \sum_k b^k q_k \tag{6.18}$$

where

$$b^{k} = \sum_{i} m_{i} A^{iT} L_{k}^{i} \tag{6.19}$$

and where L_k^i is like A^i but with elements of $l_{\alpha ik}$. From (6.8) it follows that b^k must be a symmetric matrix. It also follows that

$$B = bE - \bar{B}, \quad b = \mathrm{tr}\bar{B} \tag{6.20}$$

so that (providing a suitable conventional choice is made in the "planar" case) then, from the point of view of the internal co-ordinates, C and hence $\omega^{\alpha i}$ exist, if in the chosen set of internal co-ordinates B is symmetric and non-singular.

If it is supposed that a set of a_i and also a set of l_{ik} satisfy (6.7) and (6.8), then it is easily seen (cf. section 4) that given any such set, the $l_{\alpha ik}$ may be chosen so that

$$\sum_{i} \sum_{\alpha} m_{i} l_{\alpha ik} l_{\alpha il} = \delta_{kl} \tag{6.21}$$

so that (6.3) can be inverted to yield

$$q_{l} = \sum_{\alpha} \sum_{i} m_{i} (r_{\alpha i} - a_{\alpha i}) l_{\alpha il} \tag{6.22}$$

If there were only nuclear positions to be considered, it would be possible to pass at this stage to the expression for the Hamiltonian, by equating the $r_{\alpha i}$ of (6.3) with the $y_{\alpha i}$ of (5.16). However, there are electrons to consider and although these may be included in any mumber of ways, here the traditional (see, e.g., Howard and Moss (1970)) approach is used in which electron variables are expressed relative to the centre of nuclear mass, but are not included in the definition of F. Thus (6.1) defines the electron variables completely since C depends only on the nuclear variables, so that the $r_{\alpha i}$ ($i = N_n + 1, \ldots N$) can be treated directly as the internal co-ordinates for the electrons.

If the electrons are to be included in this way then it is sufficient to note that

$$\bar{X} = X - M^{-1} m\, C \sum_{j=N_n+1}^{N} r_j \tag{6.23}$$

counting the electrons from N_n+1 onwards. Thus the intermediate cartesian co-ordinates introduced in (5.6) are related to the r_i by

$$y_{\alpha i} = r_{\alpha i} - M_n^{-1} \sum_{j=N_n+1}^{N} r_{\alpha j} \tag{6.24}$$

and using (6.22), (6.1) and (6.7)

$$\frac{\partial q_k}{\partial x_{\varepsilon j}} = \Delta_j (m_j \sum_\alpha C_{\varepsilon\alpha} l_{\alpha jk} + \sum_l \sum_\alpha \omega_\alpha^{\varepsilon j} \xi_{lk}^\alpha q_l) \tag{6.25}$$

where $\Delta_j = 0$ unless j is a nucleus and where

$$\xi_{lk}^\alpha = \sum_i \sum_\gamma m_i (L_l^i)_{\alpha\gamma} l_{\gamma ik} \tag{6.26}$$

Using (6.1) and (6.6)

$$\frac{\partial r_{\alpha i}}{\partial x_{\varepsilon j}} = (\delta_{ij} - \Delta_j m_j M_n^{-1}) C_{\varepsilon\alpha} + \Delta_j (\Omega^{\varepsilon j} r_i) \tag{6.27}$$

It also follows that

$$\frac{\partial x_{\varepsilon j}}{\partial q_l} = \Delta_j \sum_\gamma C_{\varepsilon\gamma} l_{\gamma jl} \tag{6.28}$$

$$\frac{\partial x_{\varepsilon j}}{\partial r_{\alpha i}} = (\delta_{ij} - M^{-1} m) C_{\varepsilon\alpha} \tag{6.29}$$

and that the definition of $\omega_\gamma^{\alpha i}$ given in (6.16) must be extended sc that it is zero if i is an electron.

It is an easy matter to check now that all the identities required by (6.17), (6.18) and (6.19) are satisfied.

The algebra required to produce the Hamiltonian from the above results is tedious and rather intricate, but leads to the following result for the kinetic energy, relative to the centre of mass kinetic energy

$$K' = \frac{1}{2}\sum_{\alpha\beta}\mu_{\alpha\beta}(L^F_\alpha + \pi_\alpha + L^e_\alpha)(L^F_\beta + \pi_\beta + L^e_\beta)$$

$$-\frac{\hbar^2}{2}\sum_k \frac{\partial^2}{\partial q_k^2} - \frac{\hbar^2}{2m}\sum_{i=N_n+1}^{N} \nabla^2(i)$$

$$-\frac{\hbar^2}{2M_n}\left|\sum_{i=N_n+1}^{N} \nabla(i)\right|^2 - \frac{\hbar^2}{8}\sum_\alpha \mu_{\alpha\alpha} \qquad (6.30)$$

In this equation L^e_α represents the α'th component of the electronic angular momentum in the Eckart frame (i.e., expressed in the usual way, in terms of the $r_{\alpha i}$) and where the final term arises entirely from the term U, (5.32). The traditional notation for the generalised inverse inertia tensor has been used, too, so that

$$\mu_{\alpha\beta} \equiv M^{-1}_{\alpha\beta} = (B^{-1}I^0B^{-1})_{\alpha\beta} \qquad (6.31)$$

Thus the transformation to an Eckart frame exists if B is a symmetric positive definite matrix, given that I^0 is also a symmetric positive definite matrix and that the l_{ik} can be chosen to satisfy (6.21). As seen above, this is equivalent to saying that an Eckart frame exists if C exists.

7. EXISTENCE OF AN ECKART FRAMEWORK CHOICE

So far the existence of C has been implicitly regarded as a global existence problem. However, it is clear from (6.14) and (6.13) that, treating the variables classically, there must be variable ranges in which C certainly does not exist. The characterisation of the regions from this point of view is however not a useful one in a quantum mechanical approach since the variables cannot be treated classically. From a quantum mechanical point of view the interest is in the valid domain for the transformed operator in a Hilbert space of square-integrable functions, constructed on the transformed variables assuming that these transformed variables exist for their complete ranges. Essentially this comes down to determining the subspace of functions that can be chosen so that the expectation value of the Hamiltonian in internal co-ordinates, exists. This in turn, in the Eckart approach, must be intimately related to the possible choices for the a_i, which determine the singularity properties of M. In particular the question arises as to whether or not there is, in any given problem, a unique framework choice of the a_i such that the valid domain of the Hamiltonian is over wavefunctions compatible with the traditional picture of a molecule with definite structure, rotating and vibrating and so on.

Combes (1977) and Aventini and Seiler (1975) have made a detailed and delicate study of this problem in the case of the diatomic molecule. In his discussion Combes adopted the adiabatic approach (see, e.g., Born and Huang (1954)) to the Hamiltonian in internal co-ordinates and he was able to show that if any particular electronic state yielded, for the nuclear motion equation, a potential with certain properties, chief among them being that it had a global minimum at a point (the equilibrium internuclear distance) then the internal nuclear motion functions could be chosen to be asymptotically localised about this point. In particular he was able to show that to first order they could be chosen to be harmonic-oscillator-type functions and that even then these functions are sufficiently localised to enable the singularities that arise from treatment of the rotation operator to be avoided. Thus it can be shown that there could be a subspace of eigenstates of the full molecular Hamiltonian for a diatomic molecule, that has traditional molecular structure properties, and that whether or not such states actually exist depends essentially on whether the potential energy curve possesses a minimum. The result of this work provides a firm mathematical foundation for what has long been the practical intuition of molecular physicists.

However, a discussion for the polyatomic molecule like that given for the diatomic molecule would be extremely difficult basically for two reasons. First because of the many-dimensional nature of the potential energy surface which will in general possess multiple minima and second because of increased problems of separating off rotational motion with the consequent permutational symmetry problems. It would be particularly difficult to give such a discussion in terms of the Eckart Hamiltonian since here the assumption of a framework is built in rather than emerging, as the internuclear distance does in the approach for diatomic molecules.

To see the kind of problems that might arise it is simply necessary to note that for any given choice of the framework parameters a_i it is possible to choose a set of nuclear vibration functions which are sufficiently well localised about the framework so as to avoid singularities in the potential energy integrals in the problem. Thus, for example, the internuclear distance operator x_{nm} is

$$x_{nm} = |(a_m - a_n) + \sum_k (l_{mk} - l_{nk}) q_k| \qquad (7.1)$$

and this clearly becomes zero for some set of values of q_k and hence will cause singular potential energy integrals if the wave function remains non-zero at these points. But a set of trial functions can be chosen to avoid this possibility and the expansion attempted in terms of such trial functions. The only problem that would then arise would be whether for that set of trial functions divergences arise in the full Eckart Hamiltonian expectation value, from the matrix elements of M^{-1}. The elements of this inverse go like $(q - b)^{-2}$ as q tends to some fixed value b and thus diverge more strongly than the potential energy integrals. It might be that the divergences lie within the range in which the functions are appreciable, but it is very difficult to give a detailed analysis of this point. However, if it is supposed that a particular set of framework parameters and an associated manifold of functions have been chosen, so that convergence has been obtained (it might be imagined here by use of traditional Born-Oppenheimer approach to determine equilibrium framework values) then for this choice the series for M^{-1}.

$$M^{-1} = \sum_n (-1)^n (I^o b)^n I^{o-1}, \quad b = \sum_k b^k q_k \tag{7.2}$$

will converge in expectation value on the chosen manifold of functions. For a small change in the framework parameters and a correspondingly small change in the associated manifold of functions the series will also converge. This small change can be quite arbitrary in nature and so at least to first order the framework choice is not unique. Indeed this result is neither counter-intuitive nor contrary to practical perceptions since the eigen-values of the nuclear motion Eckart Hamiltonian when treated by standard approximate methods are not sensitive to small changes in framework choice. Since even small changes can break the symmetry of the framework this observation renders the notion of framework symmetry as advanced by Louck and Galbraith (1976) somewhat problematical.

Whether or not there is a greater or less degree of arbitrariness in the framework choice, it might be argued that the problem could be resolved by treating the framework as a set of generator co-ordinates, in a way analogous to that suggested recently for diatomic molecules by Lathouwers (1978). However, there are difficulties with that approach too, in the polyatomic molecule. These arise because (as will be remembered) it is necessary to integrate over the generator co-ordinates and in the process of integration values will inevitably be encountered that make I^o singular; with all the attendant problems arising therefrom. Whether or not it would be possible to design the weight function, to avoid these singularities and still remain a useful function, is a matter that would require detailed further investigation.

In the Eckart approach there remain also, the problems raised by permutational invariance, a subject that has been widely discussed in the context of the non-rigid molecule (for recent papers with references to earlier work see Ezra (1979), Günthard et al. (1978), see also the books by Altmann (1978) and Bunker (1978)). In the context of the present discussion the problems raised by this may easily be stated. Suppose that P is a permutation operator for identical nuclei, which can be realised in the laboratory fixed frame by the correspondence

$$P\, x_i = x_{i'} \tag{7.3}$$

where the i' are a re-ordering of the i. If P is now considered as operating on the elements of F as given by (6.13) then (where the sum over i implies the sum over i')

$$PF_{\alpha\beta} \equiv F'_{\alpha\beta} = \sum_i m_i (x_{\alpha i'} - \bar{X})a_{\beta i} \tag{7.4}$$

since $\bar{X}$ is invariant under a permutation of like particles, the $m_i = m_{i'}$ by hypothesis and the $a_{\alpha i}$ are regarded as arbitrary constants. In terms of the matrix F' it is possible to define a new C matrix C' by

$$C' = F'(F'^{T}F')^{-\frac{1}{2}} \equiv PC \tag{7.5}$$

Thus the basic effect of a permutation of like particles is to re-define the Eckart frame. However the original and "permuted" Eckart frames $\hat{e}$ and $\hat{e}'$ are clearly related, using (5.10), by

$$\hat{e} = \hat{e}'D, \qquad D = C'^{T}C \tag{7.6}$$

where D is an orthogonal matrix (see also Louck and Galbraith (1976)). Now considering the effect of P upon x_i as given by (6.1) it follow that

$$\begin{aligned} Px_i = x_{i'} &= P\bar{X} + PCr_i \\ &= \bar{X} + C'Pr_i \end{aligned} \tag{7.7}$$

A consistent interpretation of (7.7) may be obtained by noting that, since $x_{i'}$ is one of the x_i then

$$x_{i'} = \bar{X} + Cr_{i'} \tag{7.8}$$

so that

$$Pr_i \equiv r'_i = Dr_{i'} \tag{7.9}$$

Thus P induces in the Eckart frame cartesians, the same permutation as in the laboratory fixed frame, but followed by a rotation and of course, in general, r_i' is not among the original variables.

The result (7.9) can be used to determine the effect of a permutation on the internal co-ordinates q_1, using (6.22) and (7.9) it follows that

$$Pq_1 = q_1' = \sum_\alpha \sum_i m_i(r_{\alpha i'} - a'_{\alpha i})l'_{\alpha i1} \tag{7.10}$$

where

$$l'_{i1} = D^T l_{i1} \quad \text{and} \quad a'_i = D^T a_i \tag{7.11}$$

Thus the effect of the permutation operator in terms of the internal co-ordinates is to produce, in general, a completely new-set of internal co-ordinates. Of course if l'_{i1} is $l_{i'k}$ and a'_i is $a_{i'}$ so that q_1' is just q_k then the new internal co-ordinates are simply permutations of the old, and in consequence the Eckart Hamiltonian is invariant in a conventional way under particle label permutations. To discover what happens in general, however, it is necessary to use (7.10) to establish a relationship between the q_1' and the old internal co-ordinates. Substituting for $r_{i'}$ from (6.3) causes no particular trouble here but the difficulty arises in re-expressing D. It follows from (7.4) that

$$F' = C \sum_i m_i r_{i'} a_i^T = C \bar{B}' \tag{7.12}$$

and that therefore

$$D = (\bar{B}'^T \bar{B}')^{-\frac{1}{2}} \bar{B}'^T \tag{7.13}$$

which yields D as a function of the q_k on substituting (6.3) into $\bar{B}'$. However, $\bar{B}'$ is not in general a symmetric matrix when expressed in the old internal co-ordinates so (unsurprisingly, as explained below (6.15)) the old internal co-ordinates are not independent of the new rotation co-ordinates and vice-versa. From a technical point of view since D is only defined in terms of the q_k via a matrix square root, the derivatives of D with respect to the q_k are not necessarily well defined so that the jacobian from the old to the new (permuted) co-ordinates does not necessarily exist.

Thus in general each permutation requires the setting up of a completely new Eckart Hamiltonian with a new definition of the Euler angles and the consequent very severe difficulties involved in constructing wave functions that satisfy the appropriate permutation symmetry requirements.

It is not, of course, suggested that this unpleasant behaviour of the Eckart Hamiltonian is a consequence of the physics, but simply that it is a consequence of incorporating a framework into a formal account of the molecule. In this context it may perhaps be thought that some of the difficulties could be avoided by relaxing the rigidity requirement on the framework, in the manner suggested by Sayvetz (1939). Unfortunately this is not so, for the condition Sayvetz proposes to replace (6.8) and hence to define the "non-rigid" rotating frame in practice run into similar problems. Indeed <u>any</u> method of defining a rotating co-ordinate system will run into similar difficulties unless the Euler angles can be defined in such a way that they are invariant under permutations of like particles.

It should be noticed that in this discussion the a_i are treated as formal parameters of the theory as are the l_{i1} and as such, not subject to permutations. However, there is a problem closely related to the present one in which it is necessary to consider permutations of the a_i. Suppose that two distinct calculations on the same system are attempted independently and the framework choices are different from each other, but related (as they must be) by a permutation of indices on the a_i. To see the connection, it is noticed that (7.4) can be re-written as

$$F'_{\alpha\beta} = \sum_i m_i(x_{\alpha i} - X_\alpha)a_{\beta i''} \qquad (7.14)$$

where i" depicts a permutation inverse to P. Since the two frameworks are related by permutation, P^{-1} can be chosen to permute the original framework into the new one, so that it can be seen that the effect of permuting the framework is "equal and opposite to" the effect of permuting the x_i. Thus the relationship between the internal co-ordinates, referred to the old and the new frameworks, is precisely analogous to the relationship given by (7.10) with P interpreted as the permutation inverse to the framework permutation.

8. CONCLUSIONS

The aim of this chapter has been to show how, without making any special assumptions, it is possible to develop a formal Hamiltonian for an assembly of electrons and nuclei that is strongly reminiscent of the Eckart Hamiltonian for a molecule. The formal Hamiltonian (equation (3.38) or (5.27)) is, of course, of no practical use until some specific assumptions have been made: thus we require a clear definition of the unit vectors $\hat{e}_\alpha$ (equation (3.5) or (5.10)) together with subsidiary definitions of the remaining internal co-ordinates, the q_k. The difficulties associated with such definitions have come to be known as the problem of embedding the rotating frame. In this discussion it has been shown that a particular embedding, namely the Eckart choice, can, at best, be only a local embedding and leads to some nasty problems. I conjecture that any other embedding would also only be local and lead to similar difficulties.

The point however is not the difficulty of embedding, but the process of embedding itself, which in no way arises naturally from within the quantum mechanics of the problem. The embedding rules are imported into the quantum mechanics from an outside theory: in the case of Eckart's embedding that theory is the "classical" theory of molecular structure. This observation is usually objected to on the grounds that the potential energy surface of the system determines the embedding. It is perhaps sufficient to point out that there must be some difficulty inherent in this assertion, because the embedding must be made before the problem whose solution is the potential energy surface can be posed. It is not enough to assert that a fixed nucleus calculation can provide the potential energy surface, at least not until (and if) a suitable connection can be provided between the fixed nucleus description and the full problem for polyatomic molecules, such has been provided for diatomics by Combes and Seiler (Ch. 11, this volume). It is not impossible that the quantum mechanics could determine the embedding; however it is still not clear that it does.

REFERENCES

Altmann, S.L., 1978, "Induced Representations in Crystals and Molecules", Academic Press.
Aventini, P. and Seiler, R., 1975, Commun. Math. Phys. 41, 119.
Brester, C.J., 1923, "Kristallsymmetrie und Restrahlen", Dissertation, Utrecht.
Brink, D.M. and Satchler, G.R., 1968, "Angular Momentum" 2nd Ed., Clarendon Press.
Born, M. and Huang, K., 1954, "Dynamical Theory of Crystal Lattices", Clarendon Press.
Bunker, P.R., 1978, "Molecular Symmetry and Spectroscopy", Academic Press.
Casimir, H.B.G., 1931, "The Rotation of a Rigid Body in Quantum Mechanics", J.B. Wolters, Den Haag.
Combes, J-M., 1977, Acta Phys. Austriaca. Suppl.XVII, 139.
Darling, B.T. and Dennison, D.M., 1940, Phys. Rev. 57, 128.
Eckart, C., 1934, Phys. Rev. 46, 384.
Eckart, C., 1935, Phys. Rev. 47, 552.
Essén, H., 1977, "Quantization and Independent Co-ordinates", USIP Report 77-34, Stockholm.

Essén, H., 1978, Am. J. Phys., 46, 983.
Essén, H., 1979, "Topics in Molecular Mechanics", USIP Report 79-08, Stockholm.
Ezra, G.S., 1979, Molec. Phys. 38, 863.
Ferigle, S.M. and Weber, A., 1953, Am. J. Phys. 21, 102.
Günthard, H. H., Frei, H., Groner, P. and Bauder, A., 1978, Molec. Phys. 36, 1469.
Hirschfelder, J.O. and Wigner, E., 1935, Proc. Nat. Acad. Sci. 21, 113.
Howard, B.J. and Moss, R.E., 1970, Molec. Phys. 19, 433.
Jørgensen, F., 1978, "On the Molecular Vibration-Rotation Problem", Dissertation, Copenhagen.
Kemble, E.C., 1937, "The Fundamental Principles of Quantum Mechanics", McGraw-Hill (Rep. Dover 1955).
Lathouwers, L., 1978, Phys. Rev. A18, 2150.
Louck, J.D., 1976, J. Molec. Spec. 61, 107.
Louck, J.D. and Galbraith, H.N., 1976, Rev. Mod. Phys. 48, 69.
Makushkin, Yu. S. and Ulenikov, O.N., 1977, J. Mol. Spec. 18, 1.
Marques, N.M. and Bordé, J., 1971a, Molec. Phys., 21, 635.
Marques, N.M. and Bordé, J., 1971b, Molec. Phys., 22, 809.
Meyer, R. and Günthard, H.H., 1968, J. Chem. Phys. 49, 1510.
Nielsen, H.H., 1951, Rev. Mod. Phys. 23, 90.
Podolsky, B., 1928, Phys. Rev. 32, 812.
Sayvetz, A., 1939, J. Chem. Phys., 7, 383.
Van Vleck, J.H., 1935, Phys. Rev. 47, 487.
Watson, J.K.G., 1968, Molec. Phys. 15, 479.
Watson, J.K.G., 1970, Molec. Phys. 19, 465.
Welker, H., 1936, Zeits. F. Physik 101, 95.
Wigner, E., 1930, Göttingen Nachricten, 133.
Wilson, E.B. and Howard, J.B., 1936, J. Chem. Phys. 4, 262.
Wilson, E.B., Decius, J.C. and Cross, P.C., 1955, "Molecular Vibrations", McGraw-Hill.

FOUNDATIONS OF THEORETICAL CHEMISTRY

H. Primas

Laboratory of Physical Chemistry
Swiss Federal Institute of Technology
Zürich, Switzerland

PREFACE

The objective of these lectures is to present a unified approach to the modern theory of molecular matter. Careful attention is paid to the fundamentals and questions of interpretation, the emphasis is upon ideas, concepts and a realistic link between theory and experiment.

As quantum mechanics has developed, the basic ideas have changed, sometimes changed in a dramatic way. Quantum theory started in 1900 with Planck's discovery of the quantum of action, yet modern quantum logic is independent of it. Quantizations of classical theories and Bohr's correspondence principle have played a crucial role in the development of quantum mechanics but they do not any longer represent the modern way of thinking; they have been replaced by the representation theory of the Galilei group. The close relationship between classical point mechanics and elementary quantum mechanics is due to the fact that both theories have a common kinematical group structure. Bohr's deep but difficult concept of complementarity has found a precise formulation in terms of a non-Boolean logic of properties. In Born's interpretation, quantum mechanics has been conceived as a statistical theory, so that it is notable that from a modern point of view probabilities as such do not necessarily enter at the most fundamental level but only as a derived concept.

In exact sciences, every theory has a philosophical, mathematical and empirical content. All these three aspects are equally important; if we neglect one of them, we sooner or later get into difficulties. The experience of half a century with quantum mechanics and the enormous literature on its interpretation has shown that good mathematics makes not yet good physics, it is a necessary but not sufficient condition. The rejection of metaphysics by the logical positivists of the Vienna Circle influenced the development of quantum mechanics very much, even though the operationalists' thesis that every scientifically meaningful concept must be capable of full definition in terms of performable physical operations led to a philosophical plight. Positivism and operationalism have neither solved the great philosophical problems as formulated by Parmenides and Plato, nor are they a tenable basis for modern science. Operationalism is a pseudo-philosophy which fails as a foundation for a theory of matter. The lack of well-founded philosophical basis has damaging consequences in research, teaching and culture. Even our best texts still stick to operationalism and define a state as "the result of a series of physical manipulations on the system" (Jauch, 1968, p.92), and not by the properties of the system. Nevertheless, everybody (including the theoreticians) believes in some kind of realism. It is disturbing and inconsistent to expect that a scientist accepts the existence of an external reality in the world of everyday experience and in the laboratory but not if he is working on his theories. *It is the duty of the theoreticians to make such a choice of metaphysical regulative principles that the reasonable requirements of realism are fulfilled.*

A conceptual recasting of quantum mechanics that depends neither on the particle concept nor on classical mechanics is necessary if we like to start new things rather than to improve or elaborate. In many respects, the conceptual foundations of pioneer quantum mechanics are superseded by the applications of the results of this theory in experimental and engineering science. The modern development of axiomatic quantum mechanics has recognized the quantum mechanics of the pioneers as too narrow and inadequate for our theoretical understanding of nature. The wider framework of quantum logic and W*-systems seem particularly well-suited to accommodate the needs of molecular sciences.

Moreover, a scientific theory needs an ontology and a referent. If we consider realism as useful and legitimate for our everyday experience, then it is sensible to adopt a mode of thinking that accepts realism under all those circumstances where it is not contradicted by empirical facts. It is difficult to imagine that we get a satisfactory theory of chemical and biological phenomena if we do not accept the idea that our theory is about real objects, existing quite independently of our mind and our experiments. Hence I reject operationalism that confuses being with methods of testing. In quantum logic it is possible to describe individual systems of any kind by their properties. Therefore I will adopt an individual and ontic interpretation of quantum mechanics which posits that the referents of quantum mechanics are real objects, existing independently of our mind, our knowledge and our experiments.

In these lectures I will concentrate on the technical aspects of the theory of molecular matter. A detailed discussion of the important conceptual and philosophical problems of the molecular theory and a review of the historical development of pioneer quantum mechanics and the modern versions of axiomatic quantum mechanics is presently in preparation (Primas, 1980).

1. BOOLEAN FRAMES OF REFERENCE

1.1 Einstein-Podolsky-Rosen Correlations

The discovery of Einstein, Podolsky and Rosen (1935) that quantum mechanics allows situations where noninteracting and spatially separated systems are nontrivially correlated is one of the greatest contributions to the theory of matter. Nowadays, it is experimentally established beyond any reasonable doubt (compare the review by Clauser and Shimony, 1978) that an essentially noninteracting and spatially well-separated several-particle system may be in a definite pure state without the individual particles being in pure states. The existence of such Einstein-Podolsky-Rosen correlations implies that nature is holistic. The whole-part relationship of quantum mechanics casts severe doubts on the existence of isolated systems. In contrast to classical theories, quantum mechanics predicts an entanglement of every system with its surroundings under the influence of even extremely weak interactions. An interacting system is never represented by a product state but by a nontrivial linear combination of product states.

The existence of Einstein-Podolsky-Rosen correlations implies that the classical way of looking at phenomena is no longer defensible. We can no longer adhere to the old dream of the existence of a single frame of reference that eliminates the pluralism of physical, chemical and biological theories. Every description of nature depends on the division of the world into a part whose effects are to be considered, and into a part whose effects can be ignored. Such a division is delicate and is impeded by the omnipresence of Einstein-Podolsky-Rosen correlations. Only if we abstract deliberately from Einstein-Podolsky-Rosen correlations between subsystems can we investigate the world by compartmentalization. What counts as a phenomenon depends on the breaking of the holistic symmetry of the world by division and abstraction. Adopting different partitions, we will in general observe different phenomena. Each division creates its own reality.

1.2 Contexts as Boolean Frames of Reference

We accept the following two tenets whose importance for the description of nature has been stressed by Niels Bohr:

(i) the notion phenomenon should be limited "to refer exclusively to observations obtained under specified circumstances, including an account of the whole experiment" (Bohr, 1948);

(ii) "however far the phenomena transcend the scope of classical physical explanation, the account of all evidence must be expressed in classical terms" (Bohr, 1949, p.209).

We interpret Bohr's phrase "in classical terms" to mean in terms of propositions fulfilling the laws of two-valued classical logic.

A part of the world which is created by a well-defined set of prior conceptions, and whose ontological structure is amenable to the application of classical two-valued logic will be called a context. Contexts are reference frames relative to which observations and measurements can be made and a description of the world can be given. A context constitutes a set of necessary and sufficient conditions for the existence of facts.

Classical logic is based on the law of the excluded middle, asserting that every proposition is either true or false, and can be represented by a Boolean algebra $\mathcal{B}$ and a truth function $\tau:\mathcal{B} \to \{0,1\}$. Accordingly, we characterize a context description by a Boolean frame of reference:

Boolean frame of reference

A context is represented by a Boolean frame of reference $(\mathcal{B},\tau)$, consisting of a Boolean algebra $\mathcal{B}$ and a truth function $\tau:\mathcal{B} \to \{0,1\}$. The elements of $\mathcal{B}$ are called propositions and are interpreted as a two-valued Boole-Whitehead propositional calculus with the truth values

$\tau(F) = 1$ iff F is true,

$\tau(F) = 0$ iff F is false

for every $F\varepsilon\mathcal{B}$

Before discussing the mathematical structure of Boolean algebras, we should have some idea of what propositions can refer to.

1.3 Epistemic and Ontic Propositions

If the propositions of a Boolean frame of reference relate to our knowledge of the system or to measurements on the system, we speak of epistemic propositions. If the propositions relate to the properties of the system, we speak of ontic propositions.

The calculus of propositions of classical theories (like Newtonian mechanics, Maxwell's electrodynamics or Einstein's relativity theories) and of all phenomenological theories (like the thermodynamics of Clausius, or network and system theory) is represented by a Boolean frame of reference. If in a classical theory the properties are taken as a primitive concept, and if every proposition of the associated Boolean algebra has a definite truth value and a definite ontic meaning, then we speak of an ontic interpretation of the theory. An ontic interpretation posits that the theory is about real objects, existing independently of our mind, our knowledge, or our measurements. As a paradigmatic example of a classical theory in an ontic interpretation we can take celestial mechanics.

Note that the objectivity implicit in any ontic interpretation is compatible with the subjective element introduced by the choice of a particular Boolean frame of reference. The problem of reality is the question of the status of objects while they are not perceived. If we speak of reality of an entity, it is without exception a concept relative to a given context. Since we perceive nature always filtered through a Boolean frame, every phenomenon is conditional. This fact is uninteresting in classical physics which assumed the existence of an universal Boolean frame of reference. The existence of Einstein-Podolsky-Rosen correlations implies that such a universal frame of reference does not exist in quantum mechanics. There are no phenomena without breaking the holistic unity of nature. In a holistic world, reality is created by adopting a particular point of view which deliberately neglects some Einstein-Podolsky-Rosen correlations. Nevertheless, what makes the propositions of a chosen context true or false is something external, it is independent of the structure of our minds and independent of our knowledge. There are always factors occurring outside the context in question. Engineers refer to such factors as "noise" and include some of their effects by discussing stochastic systems. Again, the notions "noise" and "signal" are context-dependent. *What is noise and what is signal depends on the questions we ask.*

Example: The Boolean frame of reference of celestial mechanics. Newtonian mechanics gives a reasonable ontological description of the movements of the planets. However, do the planets really exist as individually existing objects? Planets are composed of elementary particles with an inseparably associated electromagnetic radiation field. There exists no non-arbitrary decomposition of the electromagnetic field into a part associated with the Sun, a part associated with Mercury, and so on. Furthermore, all planets are entangled via the quantized radiation field. *Moral:* Reality is created by omitting all that what is considered as irrelevant. The Boolean frame of reference of celestial mechanics does not include the radiation field of matter; the Einstein-Podolsky-Rosen correlations between the planets are deliberately neglected. Adopting this viewpoint, we can say planets exist objectively, independently of our mind and our experiments.

1.4 The Propositional Calculus of Classical Logic

In order to discuss the connection between Boolean algebras and classical logic, we examine how sentences are combined by means of sentential connectives. This so-called propositional calculus deals with propositions irrespective of their truth. The set $\mathcal{B}$ of propositions is assumed to be closed under the operations of negation, conjunction and disjunction. If F is a proposition, then the negation "not F", denoted by $F^{\perp}$ is also a proposition. If F,G are two propositions, then the conjunction "F and G", is also a proposition which is denoted by $F \wedge G$. The proposition "F or G, or both" is called the disjunction ("the inclusive or") and is denoted by $F \vee G$. If a proposition F is logically equivalent to a proposition G, we write F=G.

The following three "laws of thought" are traditionally treated as basic to classical logic: (i) the law of contradiction saying that nothing can be both F and not-F; (ii) the law of excluded middle, saying that anything must either be F or not-F; (iii) the law of identity, saying that if anything is F, then it is F. According to Boole's (1847, 1854) third law, the propositions of classical logic form a Boolean algebra.

The law of contradiction suggests the introduction of the absurd proposition 0 defined by $0 = F \wedge F^{\perp}$ for any $F \in \mathcal{B}$. The negation of the absurd proposition is called the trivial proposition $E = 0^{\perp}$, which fulfills $E = F \vee F^{\perp}$ for every $F \in \mathcal{B}$. After identification of logically equivalent propositions, the mathematical structure $\{\mathcal{B}, E, 0, \wedge, \vee, \perp\}$ defined by the propositional calculus of classical logic is called a Boolean algebra, and will be discussed in more detail in the next section.

In a Boolean logic, one can define another three important logical operations: the "exclusive or", the implication, and the binary rejection. The proposition, "F or G, but not both" is called the "exclusive or", and is denoted by $F \triangle G$. This proposition is logically equivalent to the proposition $(F \wedge G^{\perp}) \vee (F^{\perp} \wedge G)$ so that we can write

$$F \Delta G \overset{\text{def}}{=} (F \wedge G^{\perp}) \vee (F^{\perp} \wedge G)$$

The proposition $F^{\perp} \vee G$, "either not-F or G", is called the logical implication "if F, then G", and is denoted by $F \Rightarrow G$,

$$F \Rightarrow G \overset{\text{def}}{=} F^{\perp} \vee G.$$

Note that $F \Rightarrow G$ is an element of the Boolean algebra, and not a relation between the elements F and G. The binary rejection $F^{\perp} \wedge G^{\perp}$, "neither F nor G", is denoted by the Sheffer stroke operation $|$, defined by

$$F | G \overset{\text{def}}{=} F^{\perp} \wedge G^{\perp}.$$

Remarkably, all the operations of a Boolean algebra can be expressed in terms of the Sheffer stroke operation:

$$F \wedge G = (F|F)|(G|G) \quad ,$$
$$F \vee G = (F|G)|(G|F) \quad ,$$
$$F \Rightarrow G = ((F|F)|G)|(G|(F|F)) \quad ,$$
$$F^{\perp} = F|F \quad ,$$
$$0 = F|\ (F|F) \quad .$$

This reduction of the Boolean operations to a single one is of importance in the design of digital computers.

To summarize, in Boolean logic we have the following correspondences:

$F^{\perp}$	means	"not F"	,
$F \wedge G$	means	"F and G"	,
$F \vee G$	means	"F or G, or both"	,
$F \Delta G$	means	"F or G, but not both"	,
$F \Rightarrow G$	means	"if F, then G"	,
$F \| G$	means	"neither F, nor G"	.

In classical logic, every proposition is either true or false. The proposition $F^{\perp}$ is true if and only if the proposition F is false. The proposition $F \wedge G$ is true if and only if both F and G are true. The proposition $F \vee G$ is true if and only if either F or G, or F and G are true. Such a truth evaluation is, mathematically speaking, a homomorphism τ of the Boolean algebra $\mathcal{B}$ of propositions onto the Boolean algebra $\mathcal{B}_2$ consisting of the two elements 1 and 0, together with the convention that a proposition is true if and only if it is mapped onto the element 1, i.e.,

$$\tau(F) = 1 \quad \text{if F is true} \quad ,$$

$$\tau(F) = 0 \quad \text{if F is false} \quad .$$

The value $\tau(F)$ of a proposition F is called the truth value of F. The algebraic rules in the Boolean algebra $\mathcal{B}_2 = (1,0)$ of the truth values 1,0 are

$$0^{\perp} = 1, \qquad 1^{\perp} = 0 \; ,$$

$$0 \wedge 0 = 0 \; , \qquad 0 \wedge 1 = 0 \; , \qquad 1 \wedge 1 = 1 \; ,$$

$$0 \vee 0 = 0 \; , \qquad 0 \vee 1 = 1 \; , \qquad 1 \vee 1 = 1 \; ,$$

$$0 \Delta 0 = 0 \; , \qquad 0 \Delta 1 = 1 \; , \qquad 1 \Delta 1 = 0 \; .$$

A homomorphism $\tau : \mathcal{B} \to \mathcal{B}_2$ of a Boolean algebra $\mathcal{B}$ onto the Boolean algebra $\mathcal{B}_2$ of truth values is called a truth functional. Since a homomorphism is a structure-preserving mapping, a truth functional respects Boolean operations, e.g.

$$\tau(F^{\perp}) = \tau(F)^{\perp} \; ,$$

$$\tau(F \wedge G) = \tau(F) \wedge \tau(G) \; ,$$

$$\tau(F \vee G) = \tau(F) \vee \tau(G) \; ,$$

$$\tau(F \Delta G) = \tau(F) \Delta \tau(G) \; ,$$

$$\tau(F \Rightarrow G) = \tau(F) \Rightarrow \tau(G) \; ,$$

$$\tau(F | G) = \tau(F) | \tau(G) \; .$$

A table showing the truth values of a compound proposition for every possible combination of truth values of its constituent is called a truth table. The following truth table is easy to verify:

$\tau(F)$	$\tau(G)$	$\tau(F^{\perp})$	$\tau(G^{\perp})$	$\tau(F \wedge G)$	$\tau(F \vee G)$	$\tau(F \Delta G)$	$\tau(F \Rightarrow G)$	$\tau(G \Rightarrow F)$	$\tau(F \mid G)$
1	1	0	0	1	1	0	1	1	0
1	0	0	1	0	1	1	0	1	0
0	1	1	0	0	1	1	1	0	0
0	0	1	1	0	0	0	1	1	1

1.5 Boolean Algebras as Rings

Boolean algebras have an embarrassingly rich structure and there are many different axiom systems characterizing them. The usual text-book definitions of Boolean algebras as systems $(\mathcal{B}, E, O, \wedge, \vee, ^{\perp})$ are somewhat boring and highly redundant. However, a discussion of this redundancy is remarkably involved (Huntington, 1933). The shortest definition is in terms of the theory of rings: *a Boolean algebra is a ring with unit in which every element is multiplicatively indempotent (Stone, 1936)*. Thus we consider a set $\mathcal{B}$ together with two binary operations written Δ (the addition) and $\wedge$ (the multiplication), a zero element 0 (such that for all $F \in \mathcal{B}$ we have $F \Delta 0 = F$), and a unit element E (such that for all $F \in \mathcal{B}$ we have $F\ E = F$). A ring $(\mathcal{B}, E, 0, \Delta, \wedge)$ is a Boolean algebra if $F \wedge F = F$ for every $F \in \mathcal{B}$.

The introduction of a binary operation $\vee$ (called join) and a unary operation $^{\perp}$ (called complementation) through the equations

$$F \vee G \overset{\text{def}}{=} F \Delta G \Delta (F \wedge G)$$

$$F^{\perp} \overset{\text{def}}{=} F \Delta E$$

converts the Boolean algebra $(\mathcal{B}, E, 0, \Delta, \wedge)$ into an algebraic system $(\mathcal{B}, E, 0, \wedge, \vee, ^{\perp})$ in which

$$F \vee G = G \vee F$$

$$F \vee (G \vee H) = (F \vee G) \vee H$$

$$(F^{\perp} \vee G^{\perp})^{\perp} \vee (F^{\perp} \vee G)^{\perp} = F$$

holds. On the other hand, if $(\mathcal{B}, E, 0, \wedge, \vee, ^{\perp})$ is an algebraic system obeying these three relations, then it is a Boolean algebra; and the introduction of a new operation Δ through the equation

$$F \Delta G \overset{\text{def}}{=} (F \wedge G^{\perp}) \vee (F^{\perp} \wedge G)$$

converts it into the ring $(\mathcal{B}, E, 0, \Delta, \wedge)$ (Stone, 1936).

1.6 Boolean Algebra as Lattices

In every Boolean algebra $\mathcal{B}$ there is a natural order relation $\leq$, defined by

$$F \leq G \quad \text{iff} \quad F \wedge G = F.$$

This order relation is fully compatible with the algebraic structure of $\mathcal{B}$, in particular we have

$$0 \leq F \leq E \quad \text{for every} \quad F \in \mathcal{B}$$

$$F \leq G \quad \text{iff} \quad G^{\perp} \leq F^{\perp}$$

The fact that Boolean algebras also can be defined in terms of this partial order is important for the generalization of Boolean logic to quantum logic.

Recall that a partially ordered set is a system $(L,\leq)$ consisting of a set L and a relation $\leq$ satisfying the following postulates:

(i) $F\leq G$ and $G\leq F$ hold if and only if $F=G$
(ii) if $F\leq G$ and $G\leq H$, then $F\leq H$.

An element U of a partially ordered set L is said to be an upper bound for a subset $S\subset L$ if $F\leq U$ for every $F\in S$. If S is an upper bound and $S\leq U$ for every upper bound U of S, then S is called a least upper bound, or a supremum. Analogous definitions hold for lower bounds. A greatest lower bound is called an infimum. If the maximal and minimal elements of a partially ordered set L exist, then they are unique and will be denoted by E and O, respectively,

$$O\leq F\leq E \quad \text{for every} \quad F\in L$$

A lattice is a partially ordered set L in which every pair of elements $F,G\in L$ has a supremum (denoted by $F\vee G$) and an infimum (denoted by $F\wedge G$). A lattice L is called complete if every subset of L has a supremum in L, and σ-complete if every countable set has a supremum. (Note that the prefix σ always refers to the possibility of defining a property by countably many operations).

An element $A\neq O$ in a lattice L such that $F\leq A$ implies either $F=O$ or $F=A$ is then called an atom. A lattice is called atomic if every nonzero element has an atom under it. A lattice containing no atoms is called atom-free (synonyma we will not use: continuous, nonatomic). Note that there are lattices which are neither atomic nor atom-free.

An involution is a map $F\to F'$ satisfying $(F')' = F$, $(F\wedge G)' = F'\vee G'$ and $(F\vee G)' = F'\wedge G'$.

A lattice L with O and E is called complemented if for every $F\in L$ there exists an element $G\in L$ such that $F\wedge G=O$ and $F\vee G=E$; G is called a complement of F. An orthocomplementation is an involution $F\to F^{\perp}$ such that $F^{\perp}$ is a complement of F.

A lattice L is called distributive if for every triplet of elements F,G,H the relation $F\wedge(G\vee H) = (F\wedge G)\vee(F\wedge H)$ and the dual relation $F\vee(G\wedge H) = (F\vee G)\wedge(F\vee H)$ hold.

A complemented and distributive lattice is called a <u>Boolean lattice</u>. The distributivity property implies that every element F of a Boolean lattice has a unique complement $F^{\perp}$, and that $F\to F^{\perp}$ is an orthocomplementation.

Let $(L,\leq,^{\perp})$ be a Boolean lattice with the maximal element E and the minimal element O, and define

$$F \vee G \overset{\text{def}}{=} \sup \{F,G\}, \qquad F \wedge G \overset{\text{def}}{=} \inf \{F,G\},$$

then $(L,E,0,\wedge,\vee,^{\perp})$ is a Boolean algebra. On the other hand, every Boolean algebra $(\mathcal{B},E,0,\wedge,\vee,^{\perp})$ is via its natural order relation

$$F \leq G \quad \text{iff} \quad F \wedge G = F,$$

a Boolean lattice $(\mathcal{B},\leq,^{\perp})$ with $0 \leq F \leq E$ for every $F \in \mathcal{B}$. Accordingly, we will use the notions Boolean algebra and Boolean lattice as synonyma.

A σ-complete Boolean lattice is called a Boolean σ-algebra. A Boolean σ-algebra $\mathcal{B}$ is complete if and only if every pairwise disjoint subset of $\mathcal{B}$ is at most countable.

A Boolean homomorphism is a structure-preserving mapping $\phi:\mathcal{B}\to\mathcal{B}'$ from a Boolean algebra $\mathcal{B}$ to a Boolean algebra $\mathcal{B}'$, such that

$$\phi(F \wedge G) = \phi(F) \wedge \phi(G) \, ,$$

$$\phi(F \vee G) = \phi(F) \vee \phi(G) \, ,$$

$$\phi(F^{\perp}) = \phi(F)^{\perp}.$$

A homomorphism which is one-to-one and onto is called an isomorphism. An isomorphism of $\mathcal{B}$ onto $\mathcal{B}$ is called an automorphism of $\mathcal{B}$.

Examples for Boolean algebras

(i) The set of all subsets of a given set Ω is a complete Boolean algebra if we define $F \wedge G$ to be the intersection of the subsets F and G, $F \vee G$ to be the union of F and G, and $F^{\perp}$ to be the complement of F in Ω. The natural order relation $\leq$ is given by the set-theoretical inclusion. The unit E in this Boolean algebra is given by the set Ω, the zero 0 is the empty set.

(ii) Let $\mathcal{B}$ be a family of commuting, idempotent linear operators acting on some Banach space, such that (i) $1 \in \mathcal{B}$, $0 \in \mathcal{B}$, (ii) $F \in \mathcal{B}$ implies $1-F \in \mathcal{B}$, (iii) $F,G \in \mathcal{B}$ implies $FG \in \mathcal{B}$. If we define $F \wedge G = FG$, $F \vee G = F+G-FG$ and $F^{\perp} = 1-F$, then $(\mathcal{B},1,0,\wedge,\vee,^{\perp})$ is a Boolean algebra. The partial ordering in $\mathcal{B}$ can be expressed in the form $F \leq G$ iff $FG=G$.

Reference works on Boolean algebras

Very elementary but useful introductions into the theory of lattices and Boolean algebras can be found in the texts by MacLane and Birkhoff (1967), and by Birkhoff and Bartee (1970). The standard reference work on lattice theory is Birkhoff (1940), the basic reference on Boolean algebras is Sikorski (1960).

1.7 Observations and Countable Finiteness

Every directly perceptible phenomenon refers to some Boolean frame of reference. Clearly, no system can be observed if it is not interacting with the outside world. In contradistinction to the measurements of pioneer quantum mechanics (which change the state and bring into existence the value of the observable to be measured), we define an observation as a procedure that only determines what is already the case, and that can be performed without significant disturbance of the state of the system. That is, an observation is understood as the perception of something that objectively exists. This may be a difficult engineering problem but it does not pose any deep philosophical problem like the measuring problem of pioneer quantum mechanics.

The result of every observation can be expressed by a finite number of binary decisions, or by an integer. This elementary but fundamental fact is much in evidence in modern technology which favours digital data processing and instruments with digital readings. Furthermore, we can carry out a finite number of observations only. Accordingly, the set of all empirical data is finite and every theory is corroborated or rejected on the basis of finite empirical data.

From this empirical point of view, a description of a context should be possible in terms of a finite Boolean algebra. Since in principle every experiment can be refined, any prescription of the number of elements of the Boolean algebra characterizing a context would be highly arbitrary and unsatisfactory.

It is reasonable to adopt a generous position and admit as feasible a countable number of experiments, but not more. The notion "countable" is used here in the sense of "finite or denumerably infinite". That is, we admit that a Boolean algebra describing a context is in general infinite, but we require that it be generated by a countable set of elements and that it be closed under countable Boolean operations. A Boolean algebra is called σ-decomposable (synonym: countably decomposable) if every family of mutually disjoint nonzero elements is countable. Every σ-complete and σ-decomposable Boolean algebra is complete.

To summarize: *Directly observable phenomena refer always to some context and can be understood as the perception of something that objectively exists. A context is characterized by a Boolean frame of reference* $(\mathcal{B},\tau)$ *consisting of a σ-decomposable and σ-complete Boolean algebra and a truth function* $\tau:\mathcal{B}\to\mathcal{B}_2$, *which is interpreted in terms of two-valued classical logic.*

2. BOOLEAN ALGEBRAS FOR EXPERIMENTS

2.1 Boolean σ-algebras of Point Sets

The theoretical description of phenomena arising in a Boolean frame of reference is the task of system theory. Every variant of system theory is crucially based upon the law of contradiction and the law of excluded middle. In a system-theoretical setting, the law of contradiction states that if F is some system specification, and $F^{\perp}$ is a contradiction of F, then F and $F^{\perp}$ cannot exist at the same time. The law of excluded middle states that either F or $F^{\perp}$ is true, no other possibility is permissible. Every modern system analysis is based on a set-theoretical representation, using some set Ω and a distinguished Borel algebra Σ of subsets which is closed under countable Boolean operations.

Let Ω be an arbitrary non-empty set. The elements of Ω (which is also called a space) are called points, and the subsets of Ω point sets. The subsets of Ω will be denoted by capital letters $A,B,C, \ldots$; the subset having no element is called the empty set and is denoted by ϕ. The union of the subsets A and B is written $A \cup B$, it is the set of all elements which belong either to A or to B or to both. The intersection of the subsets A and B is written $A \cap B$, it is the set of all elements which below to both A and B. Two sets A and B are called disjoint if $A \cap B = \phi$. The set of elements of A which are not members of B is called the difference of A and B and is denoted by $A \backslash B$. The difference $\Omega \backslash A$ is called the complement of A and is denoted by A^c. The symmetric difference of two sets A and B is denoted by $A \Delta B$, it is the set of all elements in A, or in B, but not in both. That is, $A \Delta B = (A \cap B^c) \cup (A^c \cap B)$.

A σ-algebra is a collection Σ of subsets of a given set Ω that is closed under complementation and formation of countable unions. That is,

(i) $\phi \in \Sigma$

(ii) if $A \in \Sigma$ then $A^c \in \Sigma$

(iii) if $\{A_n\}$ is a countable collection of sets $A_n \in \Sigma$, then also their union $\cup A_n$ and their intersection $\cap A_n$ belong to Σ.

From this definition it is plain that $(\Sigma, \Omega, \phi, \cap, \cup, {}^c)$ is a σ-complete Boolean algebra.

A set Ω together with a σ-algebra Σ of subsets of Ω such that Ω belongs to Σ is called a measurable space (Ω, Σ). A subset A of Ω is called measurable (or more precisely: Σ-measurable) if $A \in \Sigma$. If Σ_1 and Σ_2 are two σ-algebras of subsets of Ω, then we say that Σ_1 is refined by Σ_2 (or equivalently, that Σ_1 is coarser than Σ_2, or that Σ_2 is finer than Σ_1) if every element of Σ_1 is also element of Σ_2, i.e. if $\Sigma_1 \subset \Sigma_2$.

Examples of σ-algebras

(i) The coarsest σ-algebra of subsets of Ω turning Ω into a measurable space is the pair $\{\phi,\Omega\}$.

(ii) The finest σ-algebra of subsets of Ω that turns Ω into a measurable space is the collection of all subsets of Ω.

(iii) Every other σ-algebra that turns Ω into a measurable space lies between the two extremes (i) and (ii). If Ω is a topological space, then we often use the Borel algebra, defined as the σ-algebra generated by the collection of all open subsets of Ω.

2.2 Representation of σ-complete Boolean Algebras

Every Boolean algebra having a finite number **m** of elements is isomorphic to the Boolean algebra of all subsets of a set Ω. It is necessary that **m** is of the form 2^n, where **n** is the number of elements in Ω. Accordingly, two Boolean algebras with a finite number of elements are isomorphic if and only if they have the same number of elements.

The infinite-dimensional case is more complicated. Stone's (1936) representation theorem says that an infinite Boolean algebra $\mathcal{B}$ is isomorphic to a Boolean algebra of all subsets of some set Ω if and only if $\mathcal{B}$ is complete and atomic.

On the other hand, not every σ-complete Boolean algebra has a representation by a σ-complete Boolean algebra of point sets. A morphism that preserves countable suprema and infima is called a σ-morphism. For σ-complete Boolean algebras we have the following fundamental representation theorem due to Loomis (1947) and Sikorski (1960, p.117): *Every σ-complete Boolean algebra is σ-isomorphic to a σ-complete Boolean algebra of point sets modulo a σ-ideal in that algebra.*

A subset Δ of a σ-algebra Σ is said to be an ideal in Σ if

(i) $A \varepsilon \Delta$, $B \varepsilon \Delta$ imply $A \cap B \varepsilon \Delta$

(ii) $A \varepsilon \Delta$, $B \varepsilon \Delta$, $B \subseteq A$ imply $B \varepsilon \Delta$

An ideal Δ is called a σ-ideal if for every countable collection $\{A_n\}$ of sets $A_n \varepsilon \Delta$ we have $A_n \varepsilon \Delta$. One can easily prove that for a σ-algebra Σ and a σ-ideal Δ, the quotient Boolean algebra is also a σ-algebra (cf. Sikorski, 1960, p.74). The representation theorem by Loomis and Sikorski says that for every Boolean α-algebra $\mathcal{B}$ there exists a σ-algebra Σ of point sets and a σ-ideal Δ of Σ such that $\mathcal{B}$ is isomorphic to Σ/Δ.

2.3 The State Space of Experiments

Every experimental investigation of natural phenomena inevitably depends on some context. We call observations connected with a Boolean frame of reference an experiment. To every experiment there corresponds a Boolean algebra $\mathcal{B}$ of possible outcomes of the experiment. We call the unit $E\epsilon\mathcal{B}$ the sure event and the zero $0\epsilon\mathcal{B}$ the impossible event. Every $F\epsilon\mathcal{B}$ is called a possible event of the algebra, it corresponds to an experimentally possible outcome of the experiment.

Usually, to an experiment there is associated in a natural way a state space Ω such that some subsets of Ω correspond to the possible outcomes of the experiment. The points of the state space are called elementary events, they may or may not correspond to experimentally distinguishable possible outcomes of the experiment.

Example: yes-no experiments
The simplest experiment is a yes-no experiment with only two outcomes, say 0 and 1. The natural state space associated to a yes-no experiment is the set $\Omega=\{0,1\}$, containing the two points 0 and 1. In this case, the elementary events are also experimentally possible events.

Example: temperature observation
Consider a context in which the concept of temperature is well-defined. Using the Kelvin scale, the natural state space of temperature observations are the positive real numbers $\mathbb{R}^+=\{T|T\geq 0\}$. The elementary events associated to $\Omega=\mathbb{R}^+$ are the positive real numbers. Since no experiment whatsoever can distinguish between rational and irrational numbers, the elementary events of the state space $\Omega=\mathbb{R}^+$ do not correspond to possible outcomes on an experiment. The outcome of every practical observation is of the type: "the temperature measured lies in the interval $[a,b]\subset\mathbb{R}^+$.

Consider an experiment associated with a state space Ω. Every experiment within this context can be characterized by a Boolean algebra Σ of subsets of Ω. An experiment characterized by the pair (Ω,Σ_1) is called finer than an experiment (Ω,Σ_2) if the Boolean algebra Σ_2 is contained in the Boolean algebra $\Sigma_1, \Sigma_2\subset\Sigma_1$. Is there a finest feasible experiment within a context? The first idea is perhaps that the finest feasible experiment corresponds to the Boolean algebra of all subsets of the state space Ω. However, in many examples (like in the case of temperature observations, or in Newtonian mechanics) the phase space Ω contains uncountably many points (i.e., Ω is an uncountable space) so that it is unreasonable to consider experiments whose outcomes are in a one-to-one correspondence with the points of Ω. Only if the state space Ω is discrete (i.e. if Ω contains a countable number of points) is such a viewpoint tenable.

2.4 The Selection of Boolean Algebras of Events

The most important Boolean frames of reference have an associated natural phase space containing uncountably many points. The paradigmatic examples of uncountable spaces are the unit interval, $\Omega=[0,1]$, the real axis, $\Omega=\mathbb{R}$, or the n-dimensional Euclidean space, $\Omega=\mathbb{R}^n$.

Example: Classical mechanics
Consider a Newtonian n-particle system. Assuming the forces to be known completely, the instantaneous state of this system can be described by giving the values of the n position vectors together with the corresponding n velocity vectors, so that the state can be described by a point in a 6n-dimensional Euclidean space. Equivalently, we can describe the state by giving the values of $F=3n$ position coordinates and F momentum coordinates, again a point in a 2F-dimensional space, called the canonical phase space of the system, More generally, the phase space of a classical Hamiltonian system is a smooth even-dimensional manifold.

In the case of an uncountable phase space the family of all subsets of the phase space is much too large as to qualify for a Boolean algebra of experimentally distinguishable events. It is an atomic algebra, and the points of the phase space would correspond to experimentally decidable questions. According to the Loomis-Sikorski representation theorem, the Boolean σ-algebra $\mathcal{B}$ of events is isomorphic to Σ/Δ, where Σ is a σ-algebra of point sets and Δ is some σ-ideal. So we have the following problem: *For a given phase space Ω, we have to select a σ-algebra Σ of subsets of Ω and an ideal Δ of negligible subsets such that Σ/Δ is isomorphic to the Boolean σ-algebra $\mathcal{B}$ of the experimentally decidable questions.* This is a nontrivial task.

First, consider the paradigmatic example where the phase space Ω is the real axis, $\Omega=\mathbb{R}$, or the unit interval, $\Omega=[0,1]$. An ideal experiment can determine at most whether at some instant the state of the system lies in some small interval of Ω, or not. In this case, the idealization that all intervals of Ω are experimentally accessible events is reasonable. The σ-algebra generated by all intervals is exactly the Borel algebra of Ω. It is well known that there are many subsets of $\mathbb{R}$ which are not in the Borel algebra of $\mathbb{R}$, but there are good reasons to believe that the only reasonable subsets are indeed the Borel sets (compare the remarks by Birkhoff, 1940, in the 3rd edition, 1967, on p.258). In the more general case where Ω is an arbitrary topological space, the Borel algebra Σ is the smallest σ-algebra of subsets of Ω which contains all open subsets of Ω. The elements of the Borel algebra Σ are called Borel sets.

Remark: Baire σ-algebra
Another natural σ-algebra is the Baire σ-algebra, defined as the smallest σ-algebra with respect to which all continuous functions on the phase space are measurable. If the phase space is locally compact, the Baire σ-algebra is contained in the Borel σ-algebra. In most practical cases, the topology of the phase space Ω can be induced by a metric, such spaces are called metrizable. If Ω is a compact metrizable space, then the Baire and Borel σ-algebras coincide. If Ω is only a metrizable locally compact Hausdorff space, then a one-point compactification removes the necessity to distinguish between Baire and Borel sets.

Of course, the points of Ω are Borel sets which correspond to experimentally inaccessible events. They have to be eliminated by a proper choice of the ideal Δ.

The only reasonable choice for the finest Boolean algebra of events was proposed by von Neumann (1932c, cf. in particular pp.595-598): for the case $\Omega=\mathbb{R}$ one should take the Boolean algebra $\mathcal{B}$ of events to be σ-isomorphic to the σ-algebra Σ/Δ where Σ is the σ-algebra of Borel sets of Ω and Δ is the σ-ideal of Borel sets of Lebesgue measure zero. This proposal is closely related to Kolmogorov's (1933) foundation of mathematical probability theory. This choice of the algebra of events for a Boolean frame of reference with a phase space Ω is well motivated by some isomorphism theorems, to be discussed in the next section.

2.5 Measure Spaces and Measure Algebras

A real-valued set-function μ defined on a σ-algebra Σ of point sets is called σ-additive if it is countably additive in the sense that for every sequence $\{B_n\}$ of pairwise disjoint subsets of Σ we have $\mu(\cup B_n)=\Sigma\mu(B_n)$. A measure on a measurable space (Ω,Σ) is a nonnegative, extended real-valued, σ-additive set function μ defined on the σ-algebra Σ and satisfying the requirement $\mu(\phi)=0$. A measurable space (Ω,Σ) together with a measure μ on (Ω,Σ) is called a measure space and will be denoted by (Ω,Σ,μ). If $\mu(\Omega)<\infty$, then the measure μ and the measure space (Ω,Σ,μ) are said to be finite. If $\mu(\Omega)=1$, then μ is called a probability measure, and (Ω,Σ,μ) a probability space. A measure space (Ω,Σ,μ) is called σ-finite, if Ω is the union of a countable collection of measurable sets of finite measure. If $\mu(\{\omega\})>0$ for some $\omega\in\Omega$, we call the point ω an atom. We say that μ is free of atoms if there are no points $\omega\in\Omega$ such that $\mu(\{\omega\})>0$.

Let (Ω,Σ,μ) be a measure space. If the phase space Ω is uncountable, it may happen that there exist nonempty measurable sets of measure zero, the so-called null-sets. The class of all μ-null sets

$$\Delta_\mu \overset{\text{def}}{=} \{N | N \in \Sigma \ , \ \ \mu(N) = 0\}$$

forms a σ-ideal in the σ-algebra Σ. The Boolean quotient algebra Σ/Δ_μ is the relevant Boolean σ-algebra associated with the measure space (Ω,Σ,μ).

The use of the quotient algebra Σ/Δ_μ is equivalent to the use of the measure space (Ω,Σ,μ) if null sets are ignored. Given some property $\mathbf{p}(.)$ defined on the points of a measure space (Ω,Σ,μ), one says that this property holds almost everywhere with respect to μ (or: μ-almost everywhere) if there exists a null set $N \in \Sigma$ such that $\mathbf{p}(\omega)$ holds for every ω in $\Omega \backslash N$. For example, two functions $\mathbf{f}$ and $\mathbf{g}$ on Ω are equal μ-almost everywhere if there exist a null set $N \in \Sigma$ such that $\{\omega | \omega \in \Omega, \mathbf{f}(\omega) \neq \mathbf{g}(\omega)\} \subseteq N$. Similarly, a function $\mathbf{f}$ is said to be μ-measurable on Ω if there exists a Σ-measurable function $\mathbf{g}$ that is defined everywhere on Ω such that $\mathbf{f}$ equals $\mathbf{g}$ μ-almost everywhere. In spite of this somewhat clumsy way of speaking, it is convenient to work with the measure space (Ω,Σ,μ) instead of the conceptually relevant Boolean σ-algebra Σ/Δ_μ.

A measure algebra $(\mathcal{B},\mu)$ is a Boolean σ-algebra $\mathcal{B}$ together with a postive, countably additive measure μ on $\mathcal{B}$. By a measure on a Boolean algebra we understand a function defined over all elements of the algebra, assuming nonnegative numbers as values (but not identically vanishing), and satisfying the condition of countable additivity. A measure μ on $\mathcal{B}$ is called strictly positive if it vanishes for the zero element only. To every measure space (Ω,Σ,μ) one can associate a unique measure algebra $(\mathcal{B},\mu)$ with a strictly positive measure μ. Let $\mathcal{B}$ be the Boolean quotient σ-algebra of the measurable subsets of Ω modulo the μ-null sets,

$$\mathcal{B} \overset{\text{def}}{=} \Sigma/\Delta_\mu \quad , \quad \Delta_\mu \overset{\text{def}}{=} \{N | N \in \Sigma \ \mu(N) = 0\}.$$

The relation

$$A \sim B \ \text{ iff } \ \mu(A \Delta B) = 0 \quad , \quad A,B \in \Sigma \ ,$$

defines an equivalence relation on Σ, and we denote by $[A]$ the equivalence class of the measurable set $A \in \Sigma$,

$$[A] \overset{\text{def}}{=} \{B | B \in \Sigma \ , \ B \sim A\} \ .$$

The collection of all equivalence classes equals $\mathcal{B}$, and with the definitions of the binary operations Δ, $\wedge$ and $\vee$ on $\mathcal{B}$ by

$$[A] \Delta [B] \overset{\text{def}}{=} [A \Delta B] \quad ,$$

$$[A] \wedge [B] \overset{\text{def}}{=} [A \cap B] \quad ,$$

$$\bigvee_{n=1}^{\infty} [A_n] \overset{\text{def}}{=} [\bigcup_{n=1}^{\infty} A_n] \quad ,$$

the set $\mathcal{B}$ becomes a Boolean σ-algebra. Since $A \sim B$ implies $\mu(A)=\mu(B)$, the relation

$$\mu([A]) \overset{\text{def}}{=} \mu(A)$$

defines unambiguously a strictly positive measure μ on the Boolean algebra $\mathcal{B}$. The measure algebra $(\mathcal{B},\mu)$ is called the measure algebra of the measure space (Ω,Σ,μ).

An atom in a measure algebra $(\mathcal{B},\mu)$ is an element $A \neq 0$ such that $B \leq A$ implies either $B=0$ or $B=A$. If (Ω,Σ,μ) is a measure space whose associated measure algebra $(\mathcal{B},\mu)$ contains no atoms, then the measure algebra $(\mathcal{B},\mu)$, the measure space (Ω,Σ,μ) and the measure μ are said to be atom-free (synonym: purely nonatomic). In an atom-free measure space, every measurable set of strictly positive measure contains measurable subsets of smaller strictly positive measure.

Every measure algebra $(\mathcal{B},\mu)$ has a natural metric ρ, defined by

$$\rho([A],[B]) \overset{\text{def}}{=} \mu([A \Delta B]) \text{ for all } [A]\,,\, [B] \in \mathcal{B}.$$

It is easy to see that $\mathcal{B}$ is a complete space under this metric. A measure space (Ω,Σ,μ) is said to be separable if its associated metric space $\mathcal{B}$ is separable. Every σ-finite measure space (Ω,Σ,μ) with a σ-decomposable σ-algebra Σ is separable (cf. Halmos, 1950, p.168).

Example:
Let $\Omega=\mathbb{R}^n$, Σ the σ-field of Borel sets of $\mathbb{R}^n$, and μ the Lebesgue measure. Then (Ω,Σ,μ) is separable.

Additional references:
The basic reference on this subject is the book by Halmos (1950). Furthermore, the texts by Royden (1963) and by Brown and Pearcy (1977) are highly recommended.

2.2 Transformations between Measure Spaces

Conceptually, all morphisms between measure spaces are defined via the associated measure algebras. Consequently, the corresponding mappings between measure spaces are defined only modulo null sets. For example, we speak of a measure-theoretic isomorphism if after removal of suitable sets of measure zero we obtain an isomorphism in the strict sense.

Let (Ω,Σ,μ) and (Ω',Σ',μ') be two σ-finite measure spaces, and consider a mapping $\phi:\Omega\to\Omega'$. To say that ϕ is measurable means that $\phi^{-1}(B') \overset{\text{def}}{=} \{\omega|\phi(\omega)\varepsilon B'\}\varepsilon\Sigma$ for every $B'\varepsilon\Sigma'$, so that ϕ^{-1} maps Σ' into Σ. A measurable ϕ of a measure space (Ω,Σ,μ) into a measure space (Ω',Σ',μ') induces a measure μ'_ϕ in (Ω',Σ'), defined by

$$\mu'_\phi(B') \overset{\text{def}}{=} \mu(\phi^{-1}B') \quad , \quad B'\varepsilon\Sigma' .$$

A measurable transformation $\phi:(\Omega,\Sigma,\mu)\ (\Omega',\Sigma',\mu')$ is called absolutely continuous (synonym: null-preserving) if μ'_ϕ is absolutely continuous with respect to μ', $\mu'_\phi<<\mu'$, that is if $\mu'_\phi(B')=0$ whenever $\mu'(B')=0$. A transformation ϕ is said to be measure preserving if it is measurable with respect to Σ and Σ' and if $\mu'_\phi=\mu'$.

A transformation ϕ is called invertible if except for sets of measure zero, maps Ω one-to-one onto Ω', and ϕ^{-1} has the same properties (measurable and absolutely continuous) as ϕ. An invertible, absolutely continuous measurable mapping is called a measure-theoretic isomorphism. An isomorphism is called measure preserving if $\mu'_\phi=\mu'$. Measure spaces which can be isomorphically transformed one on another are called isomorphic. An isomorphism of a measure space onto itself is called an automorphism; the group of all automorphisms of a measure space (Ω,Σ,μ) is denoted by $\mathrm{Aut}(\Omega,\Sigma,\mu)$.

Example:
Let Σ_B be the σ-algebra of the Borel sets of $\mathbb{R}^n$, let Σ_L be the σ-algebra of Lebesgue-measurable sets in $\mathbb{R}^n$, let μ_L the Lebesgue measure and μ_B its restriction to Σ_B. Then the measure space $(\mathbb{R}^n,\Sigma_B,\mu_B)$ is isomorphic to $(\mathbb{R}^n,\Sigma_L,\mu_L)$.

A remarkable result in measure theory states that two separable measure spaces are isomorphic if and only if they have the same number of atoms and if both are atomic or atom-free. More precisely, a famous theorem by Halmos and von Neumann (1942) says that *every separable and atom-free measure space* (Ω,Σ,μ) *with* $0<\mu(\Omega)\leq\infty$ *is measure-theoretically isomorphic to the measure space consisting of the interval* $[0,\mu(\Omega)]$ *equipped with the Lebesgue-Borel measure.*

Example:
Let $[0,1]^n$ denote the Cartesian product of n copies of the unit interval and let μ_n denote the product Lebesgue measure on $[0,1]^n$. Then the Borel measure space $([0,1]^n,\mu_n)$ is measure-theoretically isomorphic to the measure space consisting of the unit interval $[0,1]$ and the Lebesgue-Borel measure on $[0,1]$.

A measure space (Ω,Σ,μ) is said to be a Lebesgue space if it is a measure-theoretically isomorphic to a measure space consisting of the unit interval $[0,1]\subset\mathbb{R}$ and the Lebesgue measure on the Lebesgue-measurable sets in $[0,1]$. Lebesgue spaces are a rather wide class of spaces, containing all spaces of interest in the theory of time-continuous classical dynamical systems.

The Halmos-von Neumann theorem implies that every separable atom-free probability space (Ω,Σ,μ) is isomorphic to the Lebesgue space consisting of the unit interval and the Lebesgue measure. One can ask whether this measure-theoretic isomorphism is induced by an invertible point map between Ω and the unit interval $[0,1]$. In general the answer is no. It was shown by von Neumann (1932b) that the answer is yes if Ω is a polish space and μ is a σ-finite measure on the Borel subsets of Ω.

Remark: Polish spaces
A polish space is a separable space whose topology can be defined by a metric which makes it complete. Every separable and metrizable locally compact space is polish. Presumably, it most often happens in cases of interest that a phase space is polish.

References:
For all omitted details, see Halmos (1950), Royden (1963), Brown and Pearcy (1977).

2.7 Classical Dynamical Systems

A one-parameter group $\{\theta_t|t\in\mathbb{R}\}$ of automorphisms $\theta_t\in\mathrm{Aut}(\Omega,\Sigma,\mu)$ such that θ_o is the identity mapping and $\theta_{t+s}=\theta_t\theta_s$ holds for all $t,s\in\mathbb{R}$, is called a flow on the measure space (Ω,Σ,μ). Note that the measure μ is not assumed to be invariant. A semiflow is a semigroup $\{\theta_t|t\geq 0\}$ of null preserving (i.e. absolutely continuous) transformations $\theta_t:\Omega\to\Omega$ such that θ_o is the identity mapping and $\theta_{t+s}=\theta_t\theta_s$ holds for all $t,s\geq 0$. A σ-finite measure space together with a semiflow is called a (measure-theoretical) classical dynamical system. If the time evolution is given by a flow we speak of an invertible classical dynamical system. An isomorphism of two dynamical systems $(\Omega,\Sigma,\mu,t\to\theta_t)$ and $(\Omega',\Sigma',\mu',t\to\theta_t')$ is an isomorphism $\phi:(\Omega,\Sigma,\mu)\to(\Omega',\Sigma',\mu')$ of the measure spaces such that $\phi\circ\theta_t=\theta_t\circ\phi$.

Measure-theoretical classical dynamical systems generalize the problem of studying ordinary differential equations. The measure-theoretical properties of dynamical systems exhibit best the stochastic properties that may be associated with deterministic systems of ordinary differential equations.

Example: Classical point mechanics
Historically, classical point mechanics has been conceived in the framework of differentiable dynamical systems, dealing with a smooth canonical manifold Ω (e.g. $\Omega=\mathbb{R}^{2F}$) and a smooth Hamiltonian flow $\{\theta_t|t\in\mathbb{R}\}$ on the phase space Ω.

Hamiltonian's equations of motion, $dq/dt=\partial h/\partial p$ and $dp/dt=-\partial h/\partial q$ where h is the Hamiltonian function and $\omega=(p,q)\in\Omega$, generate a one-parameter group $\{\theta_t | t\in\mathbb{R}\}$ of transformations of the phase space. The transformation θ_t shifts each point of the phase space t units of time along its trajectory, $\theta_t(\omega_o)=\omega_t$. The group properties $\theta_t\theta_s=\theta_{t+s}$, $\theta_o=1$, $\theta_{-t}=\theta_t^{-1}$, follow directly from the uniqueness of the solution for the Hamiltonian Cauchy problem.

Liouville's theorem asserts that the Hamiltonian flow preserves the Lebesgue measure μ_L on the phase space Ω. Choosing Σ as the σ-algebra of Borel sets of Ω, the quadruple $(\Omega,\Sigma,\mu_L,t\to\theta_t)$ becomes an invertible, measure preserving, classical dynamical system. The measure space (Ω,Σ,μ_L) of classical mechanics is separable and contains no atoms.

According to the usual interpretation (say of celestial mechanics), an ontic state at time t is represented by the point $\omega_t=(p_t,q_t)$ of the phase space Ω. Since no experimental procedure can distinguish between rational and irrational numbers, the ontic states of classical mechanics are neither directly observable nor accessible to any experimental method. The result of every experimental attempt to determine the state of the system at time t can be specified by some Borel set B_t of Ω, having nonvanishing Lebesgue measure, $\mu_L(B_t)\neq 0$, and allowing the interpretation: the phase point ω_t characterizing the ontic state of the system at time t lies with certainty in B_t, $\omega_t\varepsilon B_t$.

Let (Ω,Σ,μ) be a finite, separable measure space and $\theta\varepsilon\mathrm{Aut}(\Omega,\Sigma,\mu)$ an automorphism. Since Ω splits into an atomic and an atom-free part, both of which are necessarily invariant under θ, we can consider separately the cases of atomic and atom-free measure spaces. For an atomic measure space, an automorphism is just a permutation of the atoms, so that the automorphism group is a subgroup of the symmetric group S_n, where n is the number of atoms in Ω. A flow acting continuously on an atomic measure space must necessarily leave all atoms invariant. That is, the atomic part of a classical dynamical system is uninteresting. We may therefore assume without loss of generality that the measure space of a classical dynamical system is atom-free. In view of the isomorphism theorem by Halmos and von Neumann, we may even assume that (Ω,Σ,μ) is a Lebesgue space.

2.8 Stochasticity within Determinism

Classical point mechanics is deterministic in the sense that given an initial point ω in the phase space, the ontic state ω_t at all other times is uniquely determined by the equations of motion. One of the most important results of contemporary classical dynamics is the proof that the deterministic differential equations of some smooth classical Hamiltonian systems have solutions exhibiting stochastic behaviour. Classical Hamiltonian systems can exhibit every kind of behaviour ranging from complete integrability to complete stochasticity.

The first clue in this direction came from celestial mechanics. A theorem by Bruns (1884) states that in the usual perturbation expansion of the canonical equations of celestial mechanics the points of absolute convergence and the points of divergence are equally dense. More generally, Henri Poincaré proved in 1892 the following fundamental theorem: leaving out some exceptional cases, the canonical equations of celestial mechanics do not admit any analytical and uniform integrals besides the energy integral (Poincaré, 1892, Vol.1, chapt.5). That is, on this basis the stability of the solar system cannot be proven. This fact led a whole generation to the mistaken belief that most Hamiltonian systems are ergodic. A Hamiltonian system is said to be ergodic if it explores the whole of the available energy surface in the phase space. The stability of the usual model of the solar system could be settled by Arnold (1962). This important results stems from a major achievement in classical mechanics that has become known as the KAM theorem, due to Kolmogorov (1957), Arnold (1963a) and Moser (1962). Roughly formulated, the KAM theorem says that a small perturbation of an integrable system only slightly modifies most orbits (yet some stochastic regions may appear). Arnold (1963b) showed that in the n-body problem of celestial mechanics (i.e. a system with one large mass and n-1 sufficiently small masses) there exist quasi-periodic motions near a periodic motion. Moreover, these solutions are not exceptional in the sense of Lebesgue measure: the orbits which avoid collisions and infinities for all time form a set of non-zero Lebesgue measure.

While the solar system provides an example of a classical dynamical system with a stable behaviour of the solutions of its differential equations (except for a set of initial conditions of small measure), there also exist classical Hamiltonian systems exhibiting chaotic motions in spite of the perfectly deterministic nature of the equations of motion. Heuristically, the extreme sensitivity of some Hamiltonian systems to initial conditions is well illustrated by an old example due to Borel (1914). Suppose we construct a purely classical mechanical model of an ideal gas and assume that we have a relative change of the gravitational potential of 10^{-100} not under our control. Such a change corresponds to an external perturbation due to a translation by 1cm of a particle of mass 1gram located on Sirius (distance from Earth: $8.3\ 10^{16}$m). Then the prediction of classical mechanics for the initial positions of the molecules in a macroscopic sample becomes completely wrong after 10^{-6} seconds. A major landmark is Sinai's (1963) rigorous proof of the complete instability of the Boltzmann model of a hard-sphere gas (consisting of hard spheres contained in a finite rectangular box and colliding elastically with each other and the walls). This model is ergodic and even a K-system. The distance between two trajectories having an arbitrarily small difference in the initial conditions increases exponentially as a function of time.

If we consider the time evolution $t\to\theta_t$ as a continuous action of the topological group $\mathbb{R}$ on a topological space Ω, we speak of a topological dynamical system. The theory of topological dynamical systems studies the time evolution $\omega\to\omega_t=\theta_t(\omega)$ of a point in the phase space Ω. If Ω is a metric space, one can study the average time dependence of the distance between two trajectories originating from two nearby points. For ordered motions, this distance increases linearly with the time. We call a topological dynamical system with only ordered motions a robust topological system. The robustness characterizes the continuity of the motion in its dependence on initial parameters. A more rapid growth of the distance between neighbouring trajectories indicates an instability of the motion under small changes of the initial conditions. In particular, an exponential growth of this distance is associated with a stochastic motion of the system; such systems are referred to as stochastic deterministic topological dynamical systems.

In stochastic systems the time evolution of the points in phase space still is deterministic but neighbouring initial points exhibit a <u>qualitatively</u> different behaviour. In order to discuss this instability, the trajectories of the theory of topological dynamical systems have to be replaced by bundles of trajectories, corresponding to arbitrarily small open subsets of the phase space as initial state. Mathematically this means that we replace the topological dynamical system by a measure-theoretical dynamical system. If a deterministic topological dynamical system is not robust with respect to the initial conditions, then the corresponding measure-theoretical dynamical system is no longer deterministic. Although the bundle of trajectories can be chosen to be initially as narrow as we like, it spreads in course of time exponentially. That is, given an epistemic initial state (i.e. an open subset of Ω), we cannot make precise predictions for arbitrarily long times.

To summarise: In a stochastic invertible classical dynamical system, the time evolution of every ontic state is deterministic and reversible, but not robust. In the same system, the time evolution of every epistemic state is robust but irreversible.

References:
That classical mechanics is everything but a dead subject may be seen by browsing through the books "Foundations of Mechanics" by Ralph Abraham (1967) and "Ergodic Problems of Classical Mechanics" by V.I. Arnold and A. Avez (1968). For a review of the modern theory of stable and random motions in dynamical systems, we recommend Moser (1973).

2.9 Irreversibility: Mixing and K-systems

Consider a measure-theoretical, invertible classical dynamical system $(\Omega,\Sigma,\mu,t\to\theta_t)$, where (Ω,Σ,μ) is a finite measure space. This system is called ergodic, if every measurable set invariant under the flow $t \to \theta_t$ either has measure zero or is the complement of a set of measure zero. The system is called mixing (more precisely: strongly mixing), if for every two measurable sets **A** and **B** we have

$$\lim_{t\to\infty} \mu\{\theta_t(A) \cap B\} = \frac{\mu(A)\mu(B)}{\mu(\Omega)} .$$

It turns out that mixing implies ergodicity, but there exist ergodic flows which are not mixing. In fact, the property of mixing is appreciably stronger than that of ergodicity. Mixing transformations were introduced by Eberhard Hopf (1934) as a mathematical abstraction of the processes occurring when solutes are homogeneously dispersed in solvents by prolonged stirring. As Hopf pointed out, ergodicity is not sufficient for the intuitive notion of irreversibility. On the other hand, mixing transformations have dispersive effects characteristic of irreversible processes. On the average, a mixing transformation hardly ever maps a set of positive measure into a set of smaller diameter.

Amplification:
If Ω is a metric space with a distance function $d:\Omega\times\Omega\to\mathbb{R}^+$, one defines the diameter of a subset **A** of Ω by

$$\mathrm{diam}(A) = \sup\{d(\omega,\omega') \mid \omega,\omega' \in A\}.$$

If **A** is a measurable subset of a metric measure space (Ω,Σ,μ), then the essential diameter of **A** is defined by

$$\mathrm{ess\ diam}\ (A) = \{\inf\ \mathrm{diam}\ (A') \mid A' \in \Sigma, A' \subseteq A, \mu(A')=\mu(A)\}.$$

Let $(\Omega,\Sigma,\mu,t\to\theta_t)$ be a mixing time reversible dynamical system with Ω being a metric space, Σ containing all Borel sets and μ having the property that $\mu(Q)>0$ for all open balls **Q** in Ω. Then Erber et al. (1973) have shown that for every positive number **d** less than **ess diam** (Ω) and every $A\in\Sigma$ with $\mu(A)>0$ there exists a positive time **T(A)** such that

$$\mathrm{ess\ diam}\ \{\theta_t(A)\} > d \quad \text{for} \quad t>T(A).$$

Note that with θ_t the inverse $\theta_t^{-1}=\theta_{-t}$ is also mixing, so that this theorem applies also to θ_{-t}. That is, we encounter dispersion if we follow a Borel set of small diameter and positive measure into the distant past as well as into the distant future.

A class of systems showing even stronger stochastic properties than mere mixing systems are the so-called K-systems, introduced by Kolmogorov (1958). A Kolmogorov system is an invertible dynamical system $(\Omega,\Sigma,\mu,t\to\theta_t)$ for which there exists a sub-σ-algebra $\Sigma_o \subseteq \Sigma$ such that with $\Sigma_t \overset{\mathrm{def}}{=} \theta_t(\Sigma_o)$:

(i) $\Sigma_s \subset \Sigma_t$ if $s<t$,

(ii) $\bigvee_{t\in\mathbb{R}} \Sigma_t = \Sigma$,

(iii) $\bigwedge_{t\in\mathbb{R}} \Sigma_t = \{\Omega,\phi\}$ (the trivial algebra) .

It follows easily that K-systems are ergodic and mixing. In the theory of stochastic processes K-systems are called completely nondeterministic, they are of crucial importance for the theory of statistical prediction and for the study of irreversible classical dynamical systems. On the ontic level, K-systems are perfectly deterministic, epistemically they show perfect randomness on every experimentally attainable level. Most probably, K-systems possess the kind of irreversibility which is required for thermodynamical systems.

K-systems have a simple characterization in terms of an information-theoretical entropy. Consider a finite partition $\mathcal{A}=(A_1, \ldots, A_m)$ of the phase space Ω into disjoint measurable sets A_i

$$\Omega = \bigcup_{i=1}^{m} A_i \quad , \quad A_i \cap A_j = \phi \quad \text{for } i\neq j \ ,$$

and associate to this partition an information-theoretical entropy $H(\mathcal{A})$

$$H(\mathcal{A}) \overset{\text{def}}{=} - \sum_{i=1}^{m} \mu(A_i) \log \mu(A_i) \ .$$

For a system of finite partitions $\mathcal{A}_1, \mathcal{A}_2, \ldots, \mathcal{A}$, we denote by $\bigvee_{k=1}^{n} \mathcal{A}_k$ the finite partition of Ω generated by them. The members of this partition have the form $A_1 \cap A_2 \cap \ldots \cap A_n$ with $A_k \in \mathcal{A}_k$. The average information entropy per step over the whole flow relative to the partition $\mathcal{A}$ is given by

$$H(\mathcal{A},\theta_t) \overset{\text{def}}{=} \lim_{n\to\infty} \frac{1}{n} H(\mathcal{A} \vee \theta_t \mathcal{A} \vee \theta_{2t} \mathcal{A} \vee \ldots \vee \theta_{nt} \mathcal{A}).$$

The supremum of $H(\mathcal{A},\theta_t)$ over all finite measurable partitions of the space Ω is called the Kolmogorov-Sinai entropy,

$$H(\theta_t) \overset{\text{def}}{=} \sup_{\mathcal{A}} H(\mathcal{A},\theta_t) \ .$$

For a K-flow it turns out that $H(\theta_t)$ is strictly positive and proportional to t,

$$H(\theta_t) = t/T \quad , \quad 0 \leq T < \infty$$

The positive constant T is called the relaxation time of the K-system.

References:
There are many outstanding texts on ergodic theory. From these we recommend in particular: Halmos (1956), Arnold and Avez (1968), Ornstein (1974), Brown (1976), Sinai (1976).

2.10 Conclusions

Separable σ-finite measure spaces (Ω,Σ,μ) give a unified and operative scheme for theoretical and experimental investigations of classical systems of any kind. Their Boolean frames of reference are characterized by the measure algebras $(\mathcal{B},\mu)$ associated with the measure spaces (Ω,Σ,μ).

The points $\omega\varepsilon\Omega$ of the phase space Ω are in one-to-one correspondence to the ontic states of the system. In a nontrivial dynamical system $(\Omega,\Sigma,\mu,t\to\theta_t)$, the phase space Ω must contain uncountably many points. For every initial state $\omega\varepsilon\Omega$, we call $\{\theta_t\omega\}$ the orbit of ω under the action of the time evolution $t\to\theta_t$. The point $\omega_t=\theta_t\omega$ corresponds to the ontic state at time t.

The Boolean algebra $\mathcal{B}$ contains the propositions which are in principle experimentally decidable. The finest σ-algebra Σ whose elements are of experimental interest is the σ-algebra of the Borel sets of Ω. An actual experiment may be cruder; in this case the feasible experimental outcomes are theoretically characterized by some sub-σ-algebra Σ_{exp} of the Borel σ-algebra Σ. The Borel sets of Ω are in one-to-one correspondence with the epistemic states of the system.

Purely atomic measure spaces are of no interest for the discussion of dynamical systems having a continuous flow $t\to\theta_t$. Since every separable σ-finite measure space can be expressed uniquely as the direct sum of a purely atomic and an atom-free measure space, the discussion of classical dynamical systems can be specialized without loss of generality to separable, atom-free measure spaces, hence to Lebesgue spaces.

If (Ω,Σ,μ) is a Lebesgue space, then for every experiment there exists a properly finer experiment. Every experiment determines an epistemic state such that the ontic state is with certainty within this epistemic state. Nevertheless, there exists no sequence of epistemic states that converges sequentially to an ontic state. That is, for Lebesgue systems ontic states are not accessible by experiments. This fact is responsible for the objectively cryptodeterministic character of classical mechanics.

The term crypto-deterministic has been coined by Whittaker (1943) to describe phenomena that are in reality deterministic, although we cannot fortell their outcome on account of our lack of information. In Lebesgue systems ontic states are not only inaccessible to us in practice but they are inaccessible in principle. Hence the crypto-deterministic phenomena in Lebesgue systems are not due to our ignorance but they are objective features.

The concept of probability is of central importance for the theory of matter, but there are many discordant interpretations of probability theory. The great controversial questions are the following ones: Are probabilities subjective or objective quantities? Are probabilities given *a priori* or *a posteriori*? Are probabilities related to ensembles or to single phenomena? Are probabilities reducible or irreducible quantities?

The concept of probability is to an extraordinary extent charged with predilections which can only be understood from the historical development. We cannot enter here into this discussion but we stress that probability is a sophisticated notion that has no single unique meaning. The subjective interpretations consider probability as a measure for the personal belief that the event in question occurs. A probability interpretation is called objective if the probabilities are assumed to be independent or dissected from any human consideration.

As pointed out by Smoluchowski (1918), the concept of probability can be defined and the fundamental laws of probability can be derived from the theory of strictly deterministic but non-robust classical dynamical systems. The modern theory of mixing and K-systems elucidate clearly the true origin of the laws of probability. The probability distribution arising in these systems can be traced back to a causal origin. The probability concept used to discuss the stochastic behaviour of an invertible dynamical Lebesgue system refers to single isolated systems (as opposed to a virtual ensemble); it does not measure our ignorance but it has a strictly objective meaning.

3. CLASSICAL W^*-SYSTEMS

3.1 Random Variables, Observables and Stochastic Processes

In probability theory one assumes that the events of a statistical experiment constitute a Boolean algebra and defines probability as an additive positive function on the events. More precisely, the mathematical theory of probability consists of the study of normed, strictly positive σ-additive measures on σ-complete Boolean algebras. As remarked by Kolmogorov (1948), the assumption of the denumerable additivity of the measure and of the algebra is not vital since for every Boolean algebra $\mathcal{B}_0$ with a strictly positive measure μ_0 there exists a σ-algebra $\mathcal{B}$, unique up to isomorphisms, with a strictly positive σ-measure μ such that $\mathcal{B}$ is an extension of $\mathcal{B}_0$, μ is an extension of μ_0, and $\mathcal{B}$ itself is the least σ-subalgebra of $\mathcal{B}$ which contains $\mathcal{B}_0$.

According to the fundamental representation theorem for Boolean σ-algebras by Loomis and Sikorski (cf. sect.2.2), there exists to every Boolean σ-measure μ a probability space (Ω,Σ,μ) with $\mu(\Omega)=1$, such that $(\mathcal{B},\mu)$ is isomorphic with the σ-algebra Σ of subsets of Ω, reduced by identification according to sets of measure zero, and the value of μ for an event $F\varepsilon\mathcal{B}$ is identical with the value of the measure for the corresponding subset $F\varepsilon\Sigma$. This triple (Ω,Σ,μ) is the basis of Kolmogorov's (1933) exiomatic foundation of probability theory, where an experiment is characterized by a measurable space (Ω,Σ). The basic set Ω is called the space of elementary events, its points $\omega\varepsilon\Omega$ are called elementary events. The σ-algebra Σ of distinguished subsets of Ω corresponds to the σ-algebra of observable events: a measurable set $A\varepsilon\Sigma$ is called an observable event. The full set $\Omega\varepsilon\Sigma$ is the certain event, the empty set $\phi\varepsilon\Sigma$ the impossible event. The probability measure μ is interpreted as the probability $\mu(\mathbf{F})$ for the occurrence of the event $F\varepsilon\Sigma$. Note that the points in Ω play no fundamental role: the basic objects of mathematical probability theory are the elements of the quotient algebra Σ/Δ_μ, where Δ_μ is the ideal of all measurable sets of measure zero. A set of probability zero is called a null set. A property which holds on Ω except a null set, is said to hold with probability one.

Intuitively, a random variable is a function whose values are determined by chance. There has been a long way from the early intuitive concepts to the sophisticated notions used in modern mathematical probability theory. According to Kolmogorov's (1933) axiomatization, a random variable is a (real- or complex-valued) measurable function defined on the basic set Ω of a probability space (Ω,Σ,μ). The terminology random variable is usual but quite misleading: a "random variable" is a fixed function on a fixed space, hence neither "random" nor a "variable".

A real-valued random variable $\mathbf{x}:\Omega\to\mathbb{R}$ can be thought to be associated to an observable in an experiment whose outcome is governed by the probability measure μ. Recall that a function $\mathbf{x}:\Omega\to\mathbb{R}$ is said to be measurable if for every Borel set B of the Borel algebra $\Sigma_\mathbb{R}$ of the real axis $\mathbb{R}$ the inverse image $\mathbf{x}^{-1}(B)$ belongs to Σ,

$$\mathbf{x}^{-1}(B) \overset{\text{def}}{=} \{\omega|\omega\varepsilon\Omega,\mathbf{x}(\omega)\varepsilon B\}\ \varepsilon\Sigma \quad , \quad B\varepsilon\Sigma_\mathbb{R}$$

The correspondence set up by the random variable $\mathbf{x}:\Omega\to\mathbb{R}$ through

$$\xi(B) \overset{\text{def}}{=} \mathbf{x}^{-1}(B) \quad , \quad B\in\Sigma_{\mathbb{R}}$$

induces a homomorphism $\xi:\Sigma_{\mathbb{R}}\to\Sigma$ of the algebra of Borel sets in $\mathbb{R}$ in the σ-algebra Σ. Sikorski (1949) has proved that every homomorphism $\xi:\Sigma_{\mathbb{R}}\to\Sigma$ is induced by some measurable function $\mathbf{x}:\Omega\to\mathbb{R}$. Accordingly a random variable $\mathbf{x}:\Omega\to\mathbb{R}$ and the homomorphism $\xi:\Sigma_{\mathbb{R}}\to\Sigma$ induced by $\mathbf{x}$ are mathematically equivalent notions. Both, $\mathbf{x}$ and ξ are referred to as random variables. In quantum mechanics, the term observable is used instead of random variable. In traditional quantum mechanics, the concept of an observable is associated with the function $\mathbf{x}:\Omega\to\mathbb{R}$, while modern quantum logic prefers the induced random variable $\xi:\Sigma_{\mathbb{R}}\to\Sigma$ as the basic quantity. In quantum logic, an observable is a σ-homomorphism from the Boolean logic Σ of the Borel sets of the real axis into the lattice L of the propositions of the system. In classical systems, L is again a Boolean logic which can be represented by a Boolean σ-algebra $\mathcal{B}$, hence by a measure space (Ω,Σ,μ).

Let $\mathbf{x}:\Omega\to\mathbb{R}$ be a random variable and $\xi:\Sigma_{\mathbb{R}}\to\Sigma$ its associated observable. Then we can transfer the probability measure μ of (Ω,Σ) to a probability measure $F_{\mathbf{x}}$ on the real line by

$$F_{\mathbf{x}}(B) \overset{\text{def}}{=} \mu\{\xi(B)\} \quad , \quad B\in\Sigma_{\mathbb{R}} \; .$$

The probability measure $F_{\mathbf{x}}:\Sigma_{\mathbb{R}}\to[0,1]$ is called the distribution function of the random variable x. One thinks of $F_{\mathbf{x}}(B)$ as the probability that an observation of $\mathbf{x}$ will lead to a value in the Borel set B. The integral

$$E\{\mathbf{x}\} \overset{\text{def}}{=} E\{\xi\} \overset{\text{def}}{=} \int_{\mathbb{R}} \lambda F_{\mathbf{x}}(d\lambda) \quad ,$$

(if it exists) is called the expectation value of the random variable $\mathbf{x}$, or the expectation value of the observable ξ. The variance is defined by

$$E\{(\mathbf{x}-m)^2\} = E\{(\xi-m)^2\} = \int_{\mathbb{R}} (\lambda-m)^2 F_{\mathbf{x}}(d\lambda) \quad ,$$

where m is the expectation value of $\mathbf{x}$.

An indexed family $\{\mathbf{x}_t : t\in\mathbb{R}\}$ of random variables $\mathbf{x}_t:\Omega\to\mathbb{R}$, all defined on the same probability space (Ω,Σ,μ), is called a stochastic process. In our applications, the index t represents the physical time. For each fixed $\omega\in\Omega$, $t\to\mathbf{x}_t(\omega)$ defines a nonrandom function of t, it is called a realization of the stochastic process $\{\mathbf{x}_t\}$ corresponding to the event ω.

Let $\{x_t : t\in\mathbb{R}\}$ be a stochastic process on a probability space (Ω,Σ,μ). Let $T=(t_1,t_2,..,t_n)$ be a finite sequence of distinct times, $t_i\in\mathbb{R}$. Consider the map $x_T \overset{\text{def}}{=} (x_{t_1},x_{t_2},..,x_{t_n})$ of Ω into $\mathbb{R}^n$, and let F_{x_T} be the n-dimensional probability distribution determined by x_T,

$$F_{x_T}(B) = \mu\{x_T^{-1}(B)\} \quad \text{for } B\in\Sigma_{\mathbb{R}^n}$$

where $\Sigma_{\mathbb{R}^n}$ is the σ-algebra of Borel sets of $\mathbb{R}^n$. Let $\mathcal{T}$ be the collection of all T's. Then $\{F_{x_T} | T\in\mathcal{T}\}$ is called the system of finite-dimensional probability distributions determined by the stochastic process $\{x_t | t\in\mathbb{R}\}$. Two stochastic processes having the same finite-dimensional distribution functions for all $T\in\mathcal{T}$ are called equivalent. A stochastic process is said to be stationary (more precisely: stationary in the strict sense) if $F_{x_T}=F_{x_{T+\tau}}$ for each $T\in\mathbb{R}$, where $T+\tau=(t_1+\tau,t_2+\tau,...,t_n+\tau)$.

References:
For all omitted details, see for example Parthasarathy (1977) and Doob (1953).

3.2 The Lebesgue Spaces L_p

Let (Ω,Σ,μ) be a measure space, and $p\geq 1$ a fixed real number. We write $\mathcal{L}_p(\Omega,\Sigma,\mu)$ for the set of all measurable functions $f:\Omega\to$ such that $|f|^p$ is integrable with respect to μ. For $f\in\mathcal{L}_p(\Omega,\Sigma,\mu)$

$$\|f\|_p \overset{\text{def}}{=} \{\int_\Omega |f(\omega)|^p \mu(d\omega)\}^{1/p} \quad ,$$

so that $\mathcal{L}_p(\Omega,\Sigma,\mu)$ is a complex vector space with respect to the pointwise linear operations, and the function $f\to\|f\|_p$ is a seminorm on $\mathcal{L}_p(\Omega,\Sigma,\mu)$. Let η denote the linear space of all those measurable complex-valued functions on Ω that vanish outside some set of μ-measure zero, then η is the zero space of the seminorm $\|\ \|_p$ for all $p\geq 1$,

$$\eta(\Omega,\Sigma,\mu) = \{f | f\in\mathcal{L}_p(\Omega,\Sigma,\mu) \quad , \quad \|f\|_p = 0\}$$

Set $\eta_p \overset{\text{def}}{=} \eta\cap\mathcal{L}_p$, then the quotient space

$$L_p(\Omega,\Sigma,\mu) \overset{\text{def}}{=} \mathcal{L}_p(\Omega,\Sigma,\mu)/\eta_p(\Omega,\Sigma,\mu)$$

is the space of the equivalence classes of measurable functions defined almost everywhere on Ω such that $|f|^p$ is integrable with respect to μ; it is a Banach space in the quotient topology. The Banach spaces $L_p(\Omega,\Sigma,\mu)$ are known as the Lebesgue spaces on the measure space (Ω,Σ,μ). In particular, $L_2(\Omega,\Sigma,\mu)$ is a Hilbert space with inner product

$$(f|g) \overset{\text{def}}{=} \int f^*(\omega)g(\omega)\mu(d\omega) \quad , \quad f,g\in L_2(\Omega,\Sigma,\mu)$$

When there is no ambiguity, we shall write L_p instead of $L_p(\Omega,\Sigma,\mu)$. Although the elements of an L_p-space are equivalence classes of functions, one normally treats them as functions and says that the functions constituting L_p are "only defined up to sets of measure zero".

The Banach space of the equivalence classes of μ-essentially bounded functions on Ω, with the μ-essential supremum norm, is denoted by $L_\infty(\Omega,\Sigma,\mu)$. Recall that a μ-measurable complex-valued function f on Ω is said to be μ-essentially bounded, if there exists a real constant $M<\infty$ such that $|f|<M$ μ-almost everywhere. The smallest number M with the property $|f|\leq M$ μ-almost everywhere is called the μ-essential supremum of f, and is denoted by μ-ess sup(f). Hence the norm on $L_\infty(\Omega,\Sigma,\mu)$ is given by

$$||f||_\infty \overset{\text{def}}{=} \mu\text{-ess sup}|f| \ .$$

The use of the subscript "∞" is well motivated by the fact that for $f\in L_p$ for some $p\geq 1$ implies that $\lim_{q\to\infty}||f||_q$ exists and equals $||f||_\infty$, provided we allow ∞ as a possible value of the limit.

A Lebesgue space $L_p(\Omega,\Sigma,\mu)$ with $1\leq p<\infty$ is separable if and only if the measure space (Ω,Σ,μ) is separable. Note that $L_\infty(\Omega,\Sigma,\mu)$ is as a rule nonseparable. In fact, $L_\infty(\Omega,\Sigma,\mu)$ is separable if and only if (Ω,Σ,μ) is purely atomic and consists of a finite number of atoms.

References:
For further details, we recommend the texts by Halmos (1950), Bartle (1966), Royden (1968) and Brown and Pearcy (1977).

3.3 The Observables of Classical Systems

Let $(\mathcal{B},\mu)$ be the measure algebra of a Boolean frame of reference and let (Ω,Σ,μ) be the associated σ-finite measure space. A real-valued measurable function on this measure space will be called a classical observable associated with the Boolean frame of reference. A real-valued element of $L_\infty(\Omega,\Sigma,\mu)$ will be called a bounded classical observable. Since by an appropriate relabeling of the scale every observable can be compressed to a bounded one, the restriction of our discussion to bounded observables is not a severe one. Since the Banach space L_∞ is in fact a Banach algebra, we call L_∞ the algebra of the observables associated with the Boolean frame of reference.

If f and g are elements of L_∞, then the pointwise product is well-defined, it is in L_∞, and $||fg||_\infty\leq||f||_\infty||g||_\infty$. Thus L_∞ is a commutative Banach algebra. With the involution $f \to f^*$, defined by complex conjugation $f^*(\omega)=\{f(\omega)\}^*$, L_∞ is a Banach *-algebra. In L_∞, there is a strong link between the involution and the norm, namely $\|f^*f\|=\|f\|^2$. A Banach *-algebra whose norm satisfies this condition is called a C*-algebra.

The fact that the Banach space L_∞ is the dual space of the Banach space L_1, $(L_1)^*=L_\infty$, is of crucial importance for physical applications. A C*-algebra A is called a W*-algebra if it is a dual space of a Banach space. That is, the C*-algebra A is a W*-algebra if there exists a Banach space A_* such that $(A_*)^*=A$, where $(A_*)^*$ is the dual Banach space of A_*. Such a Banach space A_* is called the predual of A. As discussed in chapter 2, the measure space (Ω,Σ,μ) of a Boolean frame of reference is separable, so that the predual $L_1(\Omega,\Sigma,\mu)$ of the W*-algebra $L_\infty(\Omega,\Sigma,\mu)$ is separable. Accordingly, the algebra of observables of a Boolean frame of reference is a commutative W*-algebra L_∞ having a separate predual L_1.

For f in L_∞ define a multiplication operator F on L_2 such that

$$\{F\Psi\}(\omega) \overset{\text{def}}{=} f(\omega)\Psi(\omega) \quad \text{for every} \quad \Psi \in L_2 \quad .$$

Obviously F is a bounded linear operator on L_2 with $\|F\|=\|f\|_\infty$. Furthermore, the complex conjugate function f^* generates the adjoint multiplication operator F^*, so that F is self-adjoint if and only if f is real. Let $A \overset{\text{def}}{=} \{F|f\in L_\infty\}$, then the mapping $f \to F$ is an isometrical *-isomorphism of L_∞ onto A. Since A is a commutative W*-algebra, every F commutes with F^*, so that all operators of A are normal. Recall that a self-adjoint operator P is called a projector if it is idempotent ($P=P^2$). An operator $F\in A$ is idempotent if and only if $f=f^2$. Hence, the projections are those F for which f is an indicator function.

References:
Details and proofs can be found in Douglas (1972) or in Sakai (1971).

Digression: L_∞ *as algebra of continuous functions*
The algebra $L_\infty(\Omega,\Sigma,\mu)$ is isomorphic and isometric with the algebra $C(X)$ of all continuous functions on some compact Hausdorff space X. It is illuminating to discuss this isomorphism by reducing measurable functions modulo the null sets to continuous functions. Let (Ω,Σ,μ) be a measure space, and let Δ_μ be the σ-ideal of μ-null sets of Σ. It can be shown (Sikorski, 1960, p.206/207) that there exists a canonical map $L_\infty(\Omega,\Sigma,\mu) \to C(X)$ where $C(X)$ is the linear space of all continuous complex-valued functions defined on the Stone space X of the Boolean σ-algebra Σ/Δ_μ. [A Stone space is a compact totally disconnected Hausdorff space. Every Boolean algebra is isomorphic to the algebra of all open and closed subsets of a Stone space.

This Stone space is uniquely determined up to homeomorphisms; cf. Sikorski, 1960, p.24.] Let $f \to \tilde{f}$, $f \in L_\infty(\Omega,\Sigma,\mu)$, $\tilde{f} \in C(X)$, be this canonical map, then $\sup\{\tilde{f}(x) | x \in X\} = \|f\|_\infty$. The canonical map $f \to \tilde{f}$ preserves the natural partial ordering, the algebraic operations, and the uniform convergence.

3.4 The Propositions of Classical Systems

Let (Ω,Σ,μ) be the σ-finite measure space of a classical system. Any attempt to determine the ontic state $\omega \in \Omega$ of this system will result at best in the proposition that ω lies with certainty in some Borel subset B of Ω. Instead of this subset B we can equally well use the indicator function χ_B of B. The indicator function of a set B is the function χ_B which assumes the value 1 at all points of $B \subset \Omega$, and the value 0 at all points of the complement of B in Ω. Thus we define $\chi_B : \Omega \to \mathbb{R}$ by

$$\chi_B(\omega) = 1 \quad \text{if } \omega \in B \; ,$$

$$\chi_B(\omega) = 0 \quad \text{otherwise.}$$

Clearly, every subset of Ω has an indicator function, and every function $\chi : \Omega \to \{0,1\}$ is the indicator function of some set, namely the set $\{\omega | \omega \in \Omega, \chi(\omega) = 1\}$. The set-theoretical operations can be expressed by the characteristic functions as follows

$$\chi_{A \cap B} = \inf\,(\chi_A,\chi_B) = \chi_A\chi_B \overset{\text{def}}{=} \chi_A \wedge \chi_B$$

$$\chi_{A \cup B} = \sup\,(\chi_A,\chi_B) = \chi_A + \chi_B - \chi_A\chi_B \overset{\text{def}}{=} \chi_A \vee \chi_B$$

$$\chi_{A \Delta B} = |\chi_A - \chi_B| \overset{\text{def}}{=} \chi_A \Delta \chi_B$$

$$\chi_{\Omega \setminus A} = 1 - \chi_A \overset{\text{def}}{=} \chi_A^{\perp}$$

Every indicator function χ_B of a measurable set $B \in \Sigma$ defines a projection operator $E(B)$, acting on the Hilbert space $L_2(\Omega,\Sigma,\mu)$:

$$\{E(B)\Psi\}(\omega) \overset{\text{def}}{=} \chi_B(\omega)\Psi(\omega) \quad , \quad \Psi \in L_2 \; .$$

Clearly this assignment has the following properties

(i) $E(B) = \{E(B)\}^* = \{E(B)\}^2$,

(ii) $E(B) \geq 0$ for each $B \in \Sigma$,

(iii) $E(A \cap B) = E(A)E(B)$,

(iv) $E(A \cup B) = E(A) + E(B) - E(A)E(B)$,

(v) $\sup E(B_n) = E(\sup B_n)$ whenever $\{B_n\}$ is an increasing sequence of sets in Σ whose union is also in Σ.

(vi) $E(\Omega) = 1$.

An operator-valued set function having these six properties is called a normalized spectral measure. The spectral measure generated by the measurable sets of a measure space (Ω,Σ,μ) is called the standard spectral measure associated with the measure space (Ω,Σ,μ). The following uniqueness theorem holds (Brown, 1974, p.134): *Let* μ_1 *and* μ_2 *be two finite measures on the measurable space* (Ω,Σ). *Then the standard spectral measures on* $L_2(\Omega,\Sigma,\mu_1)$ *and* $L_2(\Omega,\Sigma,\mu_2)$ *are unitarily equivalent if and only if the measures are equivalent,* $\mu_1 \sim \mu_2$. *Moreover, such unitary mapping is necessarily implemented by a unitary multiplication operator.*

A standard spectral measure $\{E(B)|B\in\Sigma\}$ constitutes a Boolean algebra $\mathcal{P}$ if we define negation and conjunction respectively by

$$E(B)^{\perp} = 1-E(B) ,$$

$$E(A)\wedge E(B) = E(A)E(B) ,$$

$$E(A)\vee E(B) = E(A)+E(B)-E(A)E(B) .$$

A partial ordering of $\mathcal{P}$ is defined by setting

$$E(A) \leq E(B) \quad \text{whenever} \quad E(A)E(B) = E(A) ,$$

so that $E(A)\leq E(B)$ if and only if $A\subseteq B$. With this ordering, the Boolean lattice $\mathcal{P}$ is complete, provided $\{E(A)|A\in\Sigma\}$ is the standard spectral measure of a separable measure space (Ω,Σ,μ). Since the mapping $B\to E(B)$ is faithful, the Boolean projection lattice $\mathcal{P}=\{E(B)|B\in\Sigma\}$ is isomorphic to the Boolean algebra $\mathcal{B}=\Sigma/\Delta_\mu$ of the measure space (Ω,Σ,μ).

Let (Ω,Σ,μ) be a separable σ-finite measure space, and let $a:\Omega\to C$ be a μ-essentially bounded measurable function. Then every operator A of the form

$$A \overset{\text{def}}{=} \int_\Omega a(\omega)E(d\omega)$$

is a multiplication operator acting on $L_2(\Omega,\Sigma,\mu)$, $A\in\mathcal{A}$. Conversely, every operator in $\mathcal{A}$ admits a spectral resolution in terms of the standard spectral measure $\{E(B)|B\in\Sigma\}$, so that

$$\mathcal{A} = \{A|A = \int_\Omega a(\omega)E(d\omega) \quad , \quad a\in L_\infty(\Omega,\Sigma,\mu)\} .$$

We say, that $\mathcal{A}$ is the W*-algebra generated by the standard spectral measure. The commutative algebra is maximal commutative in $L_2(\Omega,\Sigma,\mu)$, that is, every bounded operator acting on L_2 and commuting with every element of $\mathcal{A}$ belongs itself to $\mathcal{A}$.

With this, we have shown that the logic of propositions of every Boolean frame of reference can be represented by the complete Boolean projection lattice $\mathcal{P}(\mathcal{A})$ of a commutative W*-algebra $\mathcal{A}$ having a separable predual $\mathcal{A}_*$. The converse is also true: every commutative W*-algebra has a Boolean lattice structure. Let $\mathcal{A}$ be an abstract commutative W*-algebra having a separable predual. According to a classical result by von Neumann (1929), every commutative W*-algebra is generated by a single self-adjoint operator. Let Ω be the spectrum and $\{\mathbf{E}(\mathbf{B})|\mathbf{B}\in\Sigma\}$ the spectral measure of a generating element of a commutative W*-algebra $\mathcal{A}$ with separable predual $\mathcal{A}_*$, where Σ is the σ-algebra of the Borel sets of Ω. Then there exists a regular measure μ having support Ω such that $\mathcal{A}$ is *-isomorphic to the W*-algebra $\mathbf{L}_\infty(\Omega,\Sigma,\mu)$. Accordingly, the Boolean projection lattice $\mathcal{P}(\mathcal{A})$,

$$\mathcal{P}(\mathcal{A}) \overset{\text{def}}{=} \{\mathbf{P}|\mathbf{P}\in\mathcal{A} \quad , \quad \mathbf{P} = \mathbf{P}^2 = \mathbf{P}^*\}$$

is isomorphic to the Boolean algebra $\mathcal{B}=\Sigma/\Delta_\mu$ of the measure space (Ω,Σ,μ).

References:
A good discussion of the spectral theorems and the extended functional calculus can be found in chapt.4 of Douglas (1972). For the theory of abstract commutative W*-algebras, compare chapt.1 of Sakai (1971).

3.5 The States of Classical Systems

Let $\mathbf{B}$ be a measurable subset of the phase space Ω of a Boolean frame of reference, and let $\mathbf{E}(\mathbf{B})\in\mathcal{P}$ the corresponding projection operator on $L_2(\Omega,\Sigma,\mu)$. The projection $\mathbf{E}(\mathbf{B})$ represents a particular kind of observable which is related to the question: does the ontic state ω lie in the subset $\mathbf{B}$? The truth functional $\tau:\mathcal{P}\to\{0,1\}$ of the system associates to every projection $\mathbf{E}(\mathbf{B})$ of $\mathcal{P}$ the values 0 or 1 according as the proposition associated to $\mathbf{B}$ is true or false. In the ontic interpretation of a Boolean frame of reference, all propositions of $\mathcal{P}$ are truth-definite so that at every instant every proposition of $\mathcal{P}$ is either true or false, *tertium non datur*. That is, at every instant of time t there exists a truth-function $\tau_t:\mathcal{P}\to\{0,1\}$ which for every proposition $\mathbf{P}\in\mathcal{P}$ says whether it is true or not,

$$\tau_t(\mathbf{P}) = 1 \quad \text{iff} \quad \mathbf{P} \text{ is true at time t} \quad ,$$

$$\tau_t(\mathbf{P}) = 0 \quad \text{iff} \quad \mathbf{P} \text{ is false at time t} \quad .$$

This functional $\tau_t:\mathcal{P}\to\{0,1\}$ is called the ontic state of the system at time t.

It is convenient to extent the concept of an ontic state from the Boolean projection lattice $P(A)$ to the whole commutative W*-algebra A. Let A be generated by the spectral measure $\{E(B)|B\varepsilon\Sigma\}$, so that every element A of A can be written in the form

$$A = \int_{\Omega} a(\omega)E(d\omega)$$

where the integral is an abstract Radon-Stieltjes integral with respect to the strong topology of A (cf. Sakai, 1971, sect.1.1). With this, the ontic state on $P(A)$ carries over to the operators A in the algebra A,

$$\tau_t(A) \overset{\text{def}}{=} \int_{\Omega} a(\omega)\tau_t\{E(d\omega)\}$$

where the integral is Radon's integral with respect to the finitely additive measure $B\mapsto\tau_t\{E(B)\}$, $B\varepsilon\Sigma$ (cf. Yosida, 1965, sect.IV.9, example 5). The functional $\tau_t:A\to C$ is linear, nonnegative (i.e. $\tau_t(A^*A)\geq 0$ for every $A\varepsilon A$) and normalized (i.e. $\tau_t(1)=1$). The mathematicians call such a functional a state on the algebra A.

Remark: Mathematical states vs. physical states
Unfortunately, the mathematicians have borrowed the term state from the statistical interpretation of quantum mechanics as synonym for a "normalized positive linear functional". This terminology has become so common that we will accept it in spite of the fact that it is confusing. In our context, the concept of a state is a physical rather than a mathematical notion. To avoid conceptual confusions, we use the term ontic state if we are referring to the totality of properties actualized in a system at a certain instant. An epistemic state refers to an experiment designed to determine the ontic state. Both, ontic and epistemic states refer to physical situations; in the framework of *-systems they are represented by states in the sense of positive linear functionals.

A linear functional on an algebra A is a mapping $\rho:A\to C$ such that $\rho(c_1A_1+c_2A_2)=c_1\rho(A_1)+c_2\rho(A_2)$ for every $A_1,A_2\varepsilon A$ and $c_1,c_2\varepsilon C$. A linear functional $\rho:A\to C$ on a Banach algebra A is called bounded, if there exists a constant c such that $|\rho(A)|\leq c||A||$ for all $A\varepsilon A$. The space of all bounded linear functionals on a Banach algebra A is called the dual space, and is denoted by A^*. A linear functional $\rho\varepsilon A^*$ is said to be positive if $\rho(A^*A)\geq 0$ for all $A\varepsilon A$. A positive linear functional $\rho\varepsilon A^*$ with $\rho(1)=1$ is called a state on A. A state is called pure if it cannot be expressed as a nontrivial convex combination of other states. That is, $\rho\varepsilon A^*$ is pure if and only if $\rho=c\rho_1+(1-c)\rho_2$ with $0<c<1$ and $\rho_1,\rho_2\varepsilon A^*$ implies $\rho_1=\rho_2$. A state $\rho\varepsilon A^*$ is called dispersion-free on A if $\rho(A^2)=\rho(A)^2$ for all

$A\varepsilon\mathcal{A}$. In a dispersion-free state ρ, every observable A has the definite value $\rho(A)$. If ρ is a dispersion-free state on an arbitrary C*-algebra $\mathcal{A}$, then $\rho(AB)=\rho(BA)=\rho(A)\rho(B)$ for every $A,B\varepsilon\mathcal{A}$ (Misra, 1967). Since a dispersion-free state on $\mathcal{A}$ is a one-dimensional representation of $\mathcal{A}$, every dispersion-free state is pure (Segal, 1947a). On the other hand, if every pure state on a W*-algebra $\mathcal{A}$ is dispersion-free, then $\mathcal{A}$ is commutative. To every point a in the spectrum of an arbitrary observable $A\varepsilon\mathcal{A}$ there is a dispersion-free pure state $\rho_a\varepsilon\mathcal{A}^*$ such that $\rho_a(A^n)=a^n$, $n=0,1,2,\ldots$. Clearly, the ontic states of every classical system are dispersion-free states, and every pure state on the W*-algebra of observables is a feasible ontic state of the system.

A state $\rho\varepsilon\mathcal{A}^*$ on a commutative W*-algebra $\mathcal{A}$ is said to be normal if it is completely additive on the Boolean projection lattice $\mathcal{P}(\mathcal{A})$. That is, a state ρ is normal if and only if $\rho(\Sigma_n P_n)=\Sigma_n \rho(P_n)$ for every family $\{P_n\}$ of mutually orthogonal projectors $P_n\varepsilon\mathcal{A}$. Since every family of mutually orthogonal nonzero projections in a W*-algebra $\mathcal{A}$ with separable predual $\mathcal{A}_*$ is countable (cf. Sakai, 1971, p.80), complete additivity is the same as σ-additivity in our case. It is known that a state $\rho\varepsilon\mathcal{A}^*$ is normal if and only if ρ is σ-continuous on $\mathcal{A}$, that is, if and only if $\rho\varepsilon\mathcal{A}_*$, where $\mathcal{A}_*$ denotes the predual of the W*-algebra $\mathcal{A}$ (cf. e.g. Kadison, 1976).

Since every commutative W*-algebra $\mathcal{A}$ with a separable predual $\mathcal{A}_*$ is *-isomorphic to the algebra $L_\infty(\Omega,\Sigma,\mu)$ of essentially bounded functions of some separable σ-finite measure space (Ω,Σ,μ), we may assume without loss of generality that $\mathcal{A}=L_\infty(\Omega,\Sigma,\mu)$. The dual space $\mathcal{A}^*$ of $\mathcal{A}=L_\infty(\Omega,\Sigma,\mu)$ is the space of all finitely additive set functions $\nu:\Sigma\to\mathbb{C}$ of bounded total variation, (that is, $\sup\{|\nu(B)| :B\varepsilon\Sigma\}<\infty$, compare Yoshida, 1965, sect.IV.9, example 5). The predual of $L_\infty(\Omega,\Sigma,\mu)$ is the Banach space $L_1(\Omega,\Sigma,\mu)$. Let $a:\Omega\to\mathbb{R}$ be the representer of an observable $A\varepsilon\mathcal{A}$. Then the expectation $\rho(A)$ of the observable A with respect to a normal state $\rho\varepsilon\mathcal{A}_*$ can be written as

$$\rho(A) = \int_\Omega f(\omega)a(\omega)\mu(d\omega) \quad ,$$

where the positive and normalized function $f\varepsilon L_1$,

$$f(\omega) \geq 0 \quad , \quad \int_\Omega f(\omega)\mu(d\omega) = 1 \quad ,$$

is called the probability density associated to the normal state ρ.

The normal states on a commutative W*-algebra correspond exactly to the probability measures of Kolmogorov's axiomatization of probability theory. Let $\mathcal{A}$ be a commutative W*-algebra generated by the spectral measure $E:\Sigma \to P(A)$. A state $\rho \varepsilon A^*$ induces a σ-additive measure $\mu_\rho : \Sigma \to [0,1]$

$$\mu_\rho(A) \overset{\text{def}}{=} \rho\{E(A)\} \quad , \quad A \varepsilon \Sigma \quad ,$$

if and only if ρ is normal, $\rho \varepsilon A_*$. As we discussed in chapter 2, the epistemic states of a classical system are in one-to-one correspondence to the Borel sets of the phase space Ω, hence to the indicator functions in $L_1(\Omega,\Sigma,\mu)$. That is, in the W*-model, every epistemic state is represented by a normal state.

In every commutative W*-system, the normal states admit a statistical interpretation. If the outcome of an ideal observation with the same initial conditions exhibit variations from experiment to experiment which no amount of control can remove, we speak of stochastic experiments. Statistical probability theory discusses such stochastic experiments that can be repeated many times under well-controlled conditions and whose outcomes stabilize as the number of experiments is increased. These nonrandom regularities are called statistical properties. The limit theorems of mathematical probability theory show that the relative frequency of occurrence of events under large-scale repetitions are given by the mathematical concept of probability. Statistical probability theory is an ensemble theory, all statistical properties are to be obtained by averaging over an ensemble of a large number of identically prepared systems. We call a state referring to such an ensemble a statistical epistemic state. It is true but by no means plain that the statistical epistemic states of a commutative W*-system are represented by the normal states.

In the weak-* topology the set of all states on a W*-algebra A is a compact convex subset of A^*, called the state space K of A. The state space K is a simplex if and only if A is commutative (cf. e.g. Alfsen and Shultz, 1976, proposition, 11.8). That is, each state on A is the resultant of a unique measure supported by the pure states on A if and only if A is commutative. That is, for classical systems the non-pure states can be interpreted as mixtures of ontic states. In the corresponding statistical ensemble interpretation, to every statistical epistemic state there is a unique ensemble, which can be obtained by mixing individual systems in appropriate ontic states. If ρ is a normal statistical state, then $\rho\{E(B)\}$ can be interpreted as the statistical probability that the ontic state $\omega \varepsilon \Omega$ lies in the Borel subset $B \subset \Omega$.

Remark:
Since the state space of the algebra of all bounded operators acting on some Hilbert space is not a simplex, the statistical ensemble interpretation of pioneer quantum mechanics is not well-founded. In pioneer quantum mechanics (and more generally, in

any noncommutative W*-system), the knowledge of a non-pure statistical state does not imply the knowledge of the kind of ensemble to which it refers. There exist infinitely many different ensembles having the same state, so that a consistent statistical interpretation is possible only if we go back to the interpretation by individual states. The common use of the name "mixed states" in noncommutative systems is unfortunate, because one does not know what the components of the mixture are.

Most classical W*-systems of interest are continuous in the sense that their Boolean lattices are atom-free. Since an atom-free W*-algebra has no dispersion-free normal states (Plymen, 1968), the ontic states of a continuous classical system are never normal. That is, while the experimentally accessible states of a classical dynamical system are given by the elements of the predual $A_* \subset A^*$, the ontic states are not in A_*, but in the dual A^*. Since non-normal states cannot be sequentially approximated by normal states (Dell'Antonio, 1967), the ontic states of classical dynamical systems are not accessible by experiments. Note that we have defined the ontic states as the states in which the system can be, but that we have not assumed that there exists an experimental arrangement that can prepare or measure them. Remarkably, the paradigmatic example that this is not always possible comes from the modern development of classical mechanics!

3.6 Dynamical Classical W*-Systems

Measure-theoretical dynamical systems and dynamical commutative W*-systems are in one-to-one correspondence. A measure-theoretical dynamical system is a quadruple $(\Omega,\Sigma,\mu,t\to\theta_t)$, consisting of a σ-finite measure space (Ω,Σ,μ), and a semiflow $\{\theta_t | t\in\mathbb{R}\}$ of null-preserving transformations $\theta_t:\Omega\to\Omega$ such that θ_0 is the identity mapping, and the semigroup property $\theta_{t+s}=\theta_t\theta_s$ holds for all $t,s\geq 0$.

Remark: Operators on L_p*-spaces*
For the convenience of the reader we recall some notions from the theory of operators acting on L_p-spaces. Let $L_p=L_p(\Omega,\Sigma,\mu)$, $1\leq p\leq\infty$, the Lebesgue spaces on the σ-finite measure space (Ω,Σ,μ), and let $\|\cdot\|_p$ be the norm of L_p. For every continuous linear operator $T:L_p\to L_p$, $1\leq p<\infty$, there is a well-defined continuous linear operator $T^*:L_q\to L_q$, where $1/p+1/q=1$ and $q=\infty$ for $p=1$. The operator T^* is called the adjoint of T, and is determined by the equation

$$\int_\Omega \{Tf\}(\omega)g(\omega)\mu(d\omega) = \int_\Omega f(\omega)\{T^*g\}(\omega)\mu(d\omega) \quad ,$$

for all $f\in L_p$ and $g\in L_q$. The case $p=\infty$ is exceptional. If $T:L_\infty\to L_\infty$, then the preadjoint $T_*:L_1\to L_1$ is defined by

$$\int_{\Omega} \{T_* f\}(\omega) a(\omega) \mu(d\omega) = \int_{\Omega} f(\omega) \{Ta\}(\omega) \mu(d\omega) \quad ,$$

for all $f \in L_1$ and $a \in L_\infty$. A linear operator $T: L_p \to L_p$, $1 \leq p \leq \infty$, is said to be positive if $f \in L_p$, $f \geq 0$ implies $Tf \geq 0$; it is called a contraction if $\|T\|_p \leq 1$. If T is a positive contraction on L_1, then its adjoint T^* is a positive contraction on L_∞. A positive linear operator T on a L_1-space is called a Markov operator if $T(1)=1$, where 1 represents the function on Ω which is identically equal to one. Dually, a positive linear operator T on a L_1-space is called stochastic if $\|Tf\|_1 = \|f\|_1$ for all $f \geq 0$. An operator $T: L_1 \to L_1$ is stochastic if and only if its adjoint T^* is Markovian. A positive linear operator $T: L_1 \to L_1$ is called doubly stochastic if $T(1)=1$ and $T^*(1)=1$. An operator T on a L_1-space is doubly stochastic if and only if $T(1)=1$ and $\|T\|_1 \leq 1$ (Chong, 1976). A linear operator $T: L_p \to L_p$ is called isometric if $\|Tf\|_p = \|f\|_p$ for every $f \in L_p$. Every isometry T is invertible, since $Tf=0$ implies $f=0$, but the domain of T^{-1} need not be the whole space. An isometric operator on L_2 is called unitary if the range of T is the whole of L_2.

Remark: Semigroups and groups on L_p*-spaces*
A family $\{T_t | t \geq 0\}$ of positive contractions T_t is called a positive contraction semigroup if $T_t T_s = T_{t+s}$, for all $t,s \geq 0$. A semigroup $t \to T_t$ on L_p is called strongly continuous if for all $f \in L_p$ we have $\| T_t f - f\|_p \to 0$ for $t \to +0$. A semigroup $t \to T_t$ on L_∞ is called σ-weakly continuous if $t \to \int f(\omega)\{T_t a\}(\omega)\mu(d\omega)$ is continuous for all $a \in L_\infty$ and all $f \in L_1$. If $t \to T_t$ on L_1 is strongly continuous, then the adjoint semigroup $t \to T_t^*$ on L_∞ is σ-weakly continuous. Conversely, if $t \to T_t$ on L_∞ is σ-weakly continuous, the preadjoint semigroup $t \to T_{*t}$ on L_1 is strongly continuous. A family $\{T_t | t \in \mathbb{R}\}$ of positive isometries T_t is called a positive isometric one-parameter group if $T_t T_s = T_{t+s}$ for all $t,s \in \mathbb{R}$. Note that $(T_t)^{-1} = T_{-t}$ implies that every T_t has a bounded inverse so that $\{T_t | t \in \mathbb{R}\}$ is a one-parameter group of automorphisms of the corresponding Lebesgue space. For $p=2$, T_t are unitary operators acting on L_2, and $\{T_t | t \in \mathbb{R}\}$ is called a unitary one-parameter group.

Every null-preserving transformation $\theta: \Omega \to \Omega$ induces a positive contraction $T: L_1(\Omega,\Sigma,\mu) \to L_1(\Omega,\Sigma,\mu)$, specified uniquely by the condition that

$$\int_{\theta^{-1}B} f(\omega)\mu(d\omega) = \int_{B} \{Tf\}(\omega)\mu(d\omega)$$

for each $f \in L_1(\Omega,\Sigma,\mu)$ and each $B \in \Sigma$. Let $t \to \theta_t$ be a semiflow on (Ω,Σ,μ) that depends measurably on t, then the contraction operators $T_t: L_1 \to L_1$ induced by θ_t form a strongly continuous semigroup $\{T_t | t \geq 0\}$ of positive contractions satisfying $T_t T_s = T_{t+s}$ for $t,s \geq 0$ (the continuity follows from the results in chapt.8 of Bourbaki, 1963). The adjoint $T_t^*: L_\infty \to L_\infty$ generates a σ-weakly continuous semigroup $\{T_t^*:t\geq 0\}$ of positive contractions with the property $\{T_t^*\,1\}(\omega)=1(\omega)$, where $\omega\to 1(\omega)$ is the function on Ω which is identically one. Moreover, T_t^* is normal in the sense that if $a_n\uparrow a$, then $T_t^*(a_n)\uparrow T_t^*(a)$. Let A be a multiplication operator generated by $a\in L_\infty$

$$\{A\Psi\}(\omega) = a(\omega)\Psi(\omega) \quad , \quad \Psi\in L_2 \quad ,$$

$a\to A$ the isometrical *-isomorphism of L_∞ onto the multiplication algebra $\mathcal{A}$, and define a map $\alpha_t:\mathcal{A}\to\mathcal{A}$ by

$$\{\alpha_t(A)\Psi\}(\omega) = \{T_t^*a\}(\omega)\Psi(\omega) \;,\quad \Psi\in L_2 \quad ,$$

then $\{\alpha_t|t\geq 0\}$ is a σ-weakly continuous semigroup of positive normal maps α_t of $\mathcal{A}$ into itself such that α_0 is the identity map, and $\alpha_t(1)=1$ for all $t\geq 0$. To say that $t\to\alpha_t$ is σ-weakly continuous means that $\rho\{\alpha_t(A)\}\to\rho(A)$ for $t\to +0$ for all $A\in\mathcal{A}$ and all $\rho\in\mathcal{A}_*$. A positive linear map α_t is said to be normal if $A_n\uparrow A$ implies $\alpha_t(A_n)\uparrow\alpha_t(A)$.

For an invertible dynamical system, the flow $\{\theta_t|t\in\mathbb{R}\}$ is a one-parameter group of automorphisms, $\theta_t\in\mathrm{Aut}(\Omega,\Sigma,\mu)$. In this case θ_t induces a positive isometry T_t from L_p onto L_p, $1\leq p\leq\infty$,

$$\{T_tf\}(\omega) \overset{\mathrm{def}}{=} \left(\frac{d\mu_t}{d\mu}\right)^{1/p}(\omega)f(\omega_t) \quad , \quad f\in L_p \;,$$

where $\omega_t \overset{\mathrm{def}}{=} \tau_t^{-1}(\omega)$ and $\mu_t(B) \overset{\mathrm{def}}{=} \mu(\tau_t{}^{-1}B)$ for all $B\in\Sigma$. Note that the Radon-Nikodym derivatives $d\mu_t/d\mu$ and $d\mu/d\mu_t$ exist as positive elements of $L_1(\Omega,\Sigma,\mu)$.

Remark: Does every isometry arise in this way?
It is not true that every isometry arises from a point map. But if Ω is a Polish space, and if μ is a σ-finite measure on the Borel subsets of Ω, then every isometry on the space $L_p(\Omega,\Sigma,\mu)$, $p\neq 2$, is induced by an invertible bimeasurable point map of Ω (Banach, 1932, von Neumann, 1932b, Halmos and von Neumann, 1942, Lamperti, 1958). However, not every unitary operator on $L_2(\Omega,\Sigma,\mu)$ is induced by an automorphism of the underlying measure space. A unitary operator T acting on the Hilbert space $L_2(\Omega,\Sigma,\mu)$, where (Ω,Σ,μ) is a σ-finite measure space, is induced by a nonsingular transformation θ on (Ω,Σ,μ) if and only if $TL_\infty(\Omega,\Sigma,\mu)T^{-1}\subseteq L_\infty(\Omega,\Sigma,\mu)$ (Sato and Oka, 1974).

If $t\to\theta_t$ depends measurably on t, then $t\to T_t$ is strongly continuous on L_p for $1\leq p<\infty$. On L_∞, the isometry T_t is given by

$$\{T_ta\}(\omega) \overset{\mathrm{def}}{=} a(\omega_t) \quad , \quad a\in L_\infty \quad ,$$

and $t\to T_t$ is only σ-weakly continuous. That is, a measurable automorphism group $\{\theta_t|t\in\mathbb{R}\}$ on (Ω,Σ,μ) yields a σ-weakly continuous one-parameter group $\{\alpha_t | t\varepsilon\mathbb{R}\}$ of automorphisms α_t of the commutative W*-algebra A by

$$\{\alpha_t(A)\Psi\}(\omega) = a(\tau_t^{-1}\omega)\Psi(\omega) \quad , \quad \Psi\varepsilon L_2 \quad ,$$

where $a\varepsilon L_\infty$ is the representer of $A\varepsilon A$. On the other hand, if $t\to\alpha_t$ is a σ-weakly continuous one-parameter group of automorphisms of a commutative W*-algebra $A=L_\infty(\Omega,\Sigma,\mu)$, then there exists a measurable flow $t\to\theta_t$ of automorphisms of a σ-finite measure space (Ω,Σ,μ) such that θ_t induces α_t.

Since positive contractions on L_p-spaces have positive dilations to positive bounded-invertible isometries, every non-invertible classical dynamical system can be regarded as arising through projection from a larger invertible classical dynamical system.

Amplification: Dilation of non-invertible to invertible systems
Let $\{T_t | t\geq 0\}$ be a strongly continuous semigroup of positive contractions on $L_1(\Omega,\Sigma,\mu)$. Then there exists another measure space $(\tilde{\Omega},\tilde{\Sigma},\tilde{\mu})$ and a projection P of norm one of $L_1(\tilde{\Omega},\tilde{\Sigma},\tilde{\mu})$ onto $L_1(\Omega,\Sigma,\mu)$, and a strongly continuous group $\{\tilde{T}_t | t\varepsilon\mathbb{R}\}$ of positive isometries with bounded inverses such that $P\tilde{T}_t f = T_t f$ for all $f\varepsilon L_1(\Omega,\Sigma,\mu)$ (Stroescu, 1973; Akcoglu, 1975). By duality, this dilation theorem can be transferred to the corresponding commutative W*-algebras L_∞ with σ-weakly continuous dynamics.

Combining the above remarks, we arrive at the following definition of a classical dynamical W*-system:

Classical Dynamical W-Systems*

A closed dynamical classical W*-system is a pair $(A, t\to\alpha_t)$ consisting of a commutative W*-algebra A having a separable predual, and a σ-weakly continuous semigroup $\{\alpha_t | t\geq 0\}$ of positive, identity-preserving normal maps α_t of A into itself.

If every α_t is an automorphism of A, then the semigroup $\{\alpha_t | t\geq 0\}$ can be extended to a σ-weakly continuous one-parameter group $\{\alpha_t | t\varepsilon\mathbb{R}\}$ with $\alpha_{-t} \overset{\text{def}}{=} \alpha_t^{-1}$. In this case $(A, t\to\alpha_t)$ is called an invertible classical W*-system. Every non-invertible classical W*-system $\{A, t\to\alpha_t\}$ is the restriction of an invertible classical W*-system $\{\tilde{A}, t\to\tilde{\alpha}_t\}$ with $A\subset\tilde{A}$ and $\alpha_t(A)=\pi\tilde{\alpha}_t(A)$ for all $A\varepsilon A$, where π is a projection of norm one from $\tilde{A}$ onto A.

Example 1: Classical Hamiltonian mechanics
Consider a classical Hamiltonian system with F degrees of freedom, and the phase space $\Omega=\mathbb{R}^{2F}$. An ontic initial state is described by a point $\omega=(p_1,..,p_F,q_1,..,q_F)$ of the phase space Ω; its time evolution is governed by Hamilton's equation of motion, $\dot{p}_j=-\partial h/\partial q_j$, $\dot{q}_j=\partial h/\partial p_j$, $j=1,..,F$, where $h:\Omega\to\mathbb{R}$ is the twice differentiable Hamiltonian function of the system. If for every initial value $\omega\in\Omega$ there exists a unique global solution $\omega_t=\{p_1(t),..,p_F(t),$ $q_1(t),..,q_F(t)\}$, $t\in\mathbb{R}$, then $\omega\to\omega_t \overset{\text{def}}{=} \theta_t(\omega)$ defines an invertible flow $t\to\theta_t\in\text{Aut}(\Omega,\Sigma,\mu)$ where Σ is the σ-algebra of Borel sets of $\Omega=\mathbb{R}^{2F}$, and μ is the Lebesgue measure. Since the Hamiltonian flow preserves the volume element in the phase space, the flow $t\to\theta_t$ is measure preserving, $\mu_t=\mu$, so that the induced unitary one-parameter group $t\to T_t$ on $L_2(\Omega,\Sigma,\mu)$ is given by $\{T_t\Psi\}(\omega)=\Psi(\omega_t)$ for all $\Psi\in L_2$ (Koopmann, 1931). By Stone's theorem, the infinitesimal generator H , defined by $T_t=\exp(-itH)$ is an unbounded self-adjoint operator in the Hilbert space $L_2(\Omega,\Sigma,\mu)$. It is given by

$$H\Psi = \sum_{j=1}^{F} \frac{\partial\Psi}{\partial q_j}\frac{\partial h}{\partial p_j} - \frac{\partial\Psi}{\partial p_j}\frac{\partial h}{\partial q_j}$$

for all $\Psi\in S(\mathbb{R}^{2F})\subset L_2$, where $S(\mathbb{R}^{2F})$ is the Schwartz space of rapidly decreasing smooth functions on $\mathbb{R}^{2F}$. To every bounded observable $a\in L_\infty(\Omega,\Sigma,\mu)$ there corresponds a self-adjoint operator A in the multiplication algebra $\mathcal{A}$ of $L_2(\Omega,\Sigma,\mu)$. The Heisenberg-type time evolution $A\to A_t$ of a classical observable is given by a one-parameter group of automorphisms α_t of $\mathcal{A}$, defined by $\alpha_t(A)=T_t^*AT_t$ for all $A\in\mathcal{A}$. With this, to every classical global Hamiltonian system there corresponds a unique invertible classical W*-system $(\mathcal{A},t\to\alpha_t)$.

Example 2: Stationary stochastic processes
Let $\{x_t|t\in\mathbb{R}\}$ be a real-valued stationary stochastic process on a separable probability space (Ω,Σ,μ). Assume that all moments $E\{x_t^n\},n=1,2,..$ are finite (this is no essential restriction of the generality since one can always apply an instantaneous nonlinear transformation to obtain a derived process with all moments finite). A bounded stochastic process $t\to x_t\in L_\infty(\Omega,\Sigma,\mu)\subset L_2(\Omega,\Sigma,\mu)$ induces a unitary shift operator T_t on $L_2(\Omega,\Sigma,\mu)$ by $T_t(x_s)=x_{t+s}$ for all $s,t\in\mathbb{R}$, and therewith generates a strongly continuous one-parameter group $\{T_t|t\in\mathbb{R}\}$ of unitary operators acting on L_2. Let $\mathcal{A}$ be the multiplication algebra of $L_2(\Omega,\Sigma,\mu)$ and let $X_t\in\mathcal{A}$ be the classical observable corresponding to $x_t\in L_\infty(\Omega,\Sigma,\mu)$. Define an automorphism $\alpha_t(A)=T_t^*AT_t$ for all $A\in\mathcal{A}$. Then the invertible classical W*-system $(\mathcal{A},t\to\alpha_t)$ generates a stochastic process $t\to X_t=\alpha_t(X)$ which is equivalent to the process $t\to x_t$.

Let Σ_t be the smallest σ-algebra with respect to which every member of $\{x_s|s\leq t\}$ is measurable. Clearly we have $\Sigma_s\subseteq\Sigma_t$ for every $s<t$. The σ-algebra Σ_t represents the state of our knowledge of the

stochastic process $\{x_t\}$ at time t, and Σ_t is referred to as "the past at time t of the process $\{x_t\}$". The limiting case $\Sigma_{-\infty} \overset{def}{=} \cap_{t\in\mathbb{R}} \Sigma_t$ exists and is called the remote past. If Σ_t is independent of t (i.e. if $\Sigma_{-\infty}=\Sigma_{\infty}$), the process is called singular. Since stationary singular processes can (in principle!) be predicted with vanishing least-square prediction error by using information of its past only, singular processes are also called deterministic processes. If $\Sigma_{-\infty}$ is trivial, $\Sigma_{-\infty}=\{\Omega,\phi\}$, the process is called regular or purely nondeterministic. Let A_t be the smallest W*-subalgebra of A containing all observables X_s with $s\leq t$. Then the stochastic process $T\to X_t$ is deterministic if and only if $A_t=A$ for all $t\in\mathbb{R}$; it is completely nondeterministic if and only if $A_{-\infty}=\cap_{t\in\mathbb{R}}A_t$ is trivial, $A_{-\infty}=1\cdot\mathbb{C}$. A stationary purely nondeterministic stochastic process is a K-system.

Example 3: K-systems
The measure-theoretical definition of a Kolmogorov system $(\Omega,\Sigma,\mu,t\to\theta_t)$ (compare sect.2.9) can be formulated algebraically by choosing A as the multiplication algebra on $L_2(\Omega,\Sigma,\mu)$, and $t\to\theta_t$ as the one-parameter group of automorphisms $\alpha_t\in Aut(A)$ induced by the reversible flow $t\to\theta_t$ of automorphisms $\theta_t\in Aut(\Omega,\Sigma,\mu)$. Every W*-system which is *-isomorphic to this system is called a Kolmogorov W*-system. With this we have the following characterization: A Kolmogorov W*-system $(A,t\to\alpha_t)$ is an invertible classical W*-system for which there exists a W*-subalgebra $A_o\subset A$ such that with $A_t \overset{def}{=} \alpha_t(A_o)$: (i) $A_s\subset A_t$ if $s<t$, (ii) $\vee_{t\in\mathbb{R}}A_t=A$, (iii) $\wedge_{t\in\mathbb{R}}A_t=1\cdot\mathbb{C}$ (the trivial algebra). Here $\vee A_t$ denotes the smallest W*-subalgebra of A containing all W*-algebras A_t, and $\wedge A_t$ denotes the largest W*-subalgebra contained in all W*-algebras A_t.

3.7 Conclusions

There are a number of important conclusions to be drawn:

(i) Every Boolean frame of reference and every classical system can be represented by a commutative W*-algebra having a separable predual A_*.

(ii) The lattice $P(A)$ of all projections in A is a complete Boolean algebra. The elements of $P(A)$ represent the propositions about the properties of the system.

(iii) The W*-algebra A is generated by a standard spectral measure $E:\Sigma\to P(A)$ of the associated measure space (Ω,Σ,μ). The elements of A represent the bounded observables, the unbounded observables can be obtained by integration of measurable functions on Ω with respect to the spectral measure E.

(iv) The pure states on A are dispersion-free for every observable. they represent the ontic states of the system. Every observable $A\in A$ has at every instant t the definite value $\tau_t(A)$ in an objective and realistic sense, where $\tau_t \varepsilon A^*$ is the ontic state of the system at time t.

(v) In atom-free classical systems the ontic states are not normal. Since all epistemic states are normal, the ontic states describing the ultimate truth are not accessible by experiments. This fact is the deeper reason for the stochastic behaviour of non-robust continuous classical systems.

(vi) Two classical W*-systems are equivalent if and only if their Boolean projection lattices are isomorphic. This is the case if and only if the corresponding commutative W*-algebras are *-isomorphic.

(vii) Every classical dynamical system (deterministic or stochastic) can be represented by a pair $(A, t \to \alpha_t)$, where A is a commutative, atom-free W*-algebra, and $t \to \alpha_t$ is a σ-weakly continuous one-parameter group of automorphisms $\alpha_t \varepsilon \mathrm{Aut}(A)$.

4. GENERAL W*-SYSTEMS

4.1 The Logic of the World is Non-Boolean

The syntax of colloquial speech is not Boolean (compare Blau, 1973, 1977). Moreover, it has long been recognized that many forms of human inference do not fit into Boolean logic (see e.g. Watanabe, 1969). It is an empirical fact that the propositions of human experience cannot be fully comprehended by the rules of classical logic. It is the introduction of compartmentalization as <u>the</u> method of science that leads to the recognition of domains of experience in which two-valued logic is valid. Clearly, any compartmentalization separates things from their natural embedding. If we want a theoretical description of how, fundamentally, the world is, the compartmentalization has to be balanced by an attempt to grasp things in their interrelations, their conflicts and contradictions, that is by some kind of dialectical thinking. Of course, there are no contradictions in nature but logical conflicts may arise by using different Boolean frames of reference.

In spite of the importance of dialectical thinking in the informal discussion of human experience and scientific experimentation, it contributes little to a formal unification of the compartmentalized domains of scientific knowledge. At least within the domain of molecular phenomena we have nowadays a more powerful method: the embedding of the local descriptions given by the Boolean frames of reference into a comprehensive global description, called <u>quantum logic</u>. Since there exists incompatible frames of reference, quantum logic is in general non-Boolean. Boolean frames of reference only givea partial description of the world, nevertheless they play a distinguished role in quantum logic. The different frames of reference necessary for a full description of nature can be pasted together to structured families of Boolean algebras, called Boolean atlases, such that quantum logic can be represented in terms of Boolean atlases (Domotor, 1974).

At the risk of belabouring the obvious, we point out that the metalogic of quantum theory is still the usual Boolean logic (together with the usual dialectical handling of different contexts). The metalogic is the syntax of the metalanguage, that is the language which we use if we speak about molecular theories. Thus the rules according to which the formalism of "quantum logic" is handled are the rules of two-valued classical logic.

4.2 From Boolean Atlases to Orthomodular Logic

Science has always be searching for separability. Classical physics believed in a universal frame of reference, a universal context that permits independent variations of the elements. Quantum mechanics taught us that the hunt for a universal Boolean frame of reference is in vain. The crucial difference between classical and quantum theories concerns the existence of incompatible properties in nonclassical systems. Note that the existence of incompatible qualities has *a priori* nothing to do with Planck's constant of action.

Nature is holistic and separability is a matter of context. Whatever partition we make of nature, it gives an answer which is not the whole answer. There exist side by side antithetical contexts which give rise to complementary descriptions. In 1928, Niels Bohr introduced the term complementarity to account for situations where different conditions of observation yield conclusions that are classically incompatible. Complementary descriptions are not contradictory since they refer to different contexts. Since different contexts do not necessarily refer to different objects but to different ways of perceiving the same object, incompatible properties reflect different points of view.

In order to grasp the many aspects of nature, we must try to get everything back into connection with everything. Occasionally, several Boolean frames of reference can be unified into a single but larger Boolean frame of reference. As a rule, however, different Boolean frames of reference are not embeddable into a larger theory having a Boolean propositional calculus. But it is possible to paste together the Boolean algebras of classical theories having referents of the same category into a single partial Boolean algebra. The notion of a partial Boolean algebra L (introduced by Specker, 1960; cf. also Kochen and Specker, 1965a, 1965b, 1967) is a generalization of the concept of a Boolean algebra $(\mathcal{B},E,0,\wedge,\vee,^{\perp})$, in the sense that the Boolean operations $\wedge$ and $\vee$ are only defined for elements having the same domain. If two elements F,G have the same domain, they are called compatible, written $F \leftrightarrow G$. Elements which belong to a common Boolean subalgebra are by definition compatible. The compatibility is a binary relation of L, and is required to have the following properties:

(i) every proposition is compatible with itself, i.e. $F \leftrightarrow F$ for all $F \in L$,

(ii) the trivial propositions 0 and E are compatible with every proposition, i.e. $0 \leftrightarrow F$ and $E \leftrightarrow F$ for every $F \in L$,

(iii) if F is compatible with G, then G is compatible with F, i.e. $F \leftrightarrow G$ iff $G \leftrightarrow F$,

(iv) if F,G,H are mutually compatible, then $F \wedge G \leftrightarrow H$ $F \vee G \leftrightarrow H$ $F^{\perp} \leftrightarrow H$,

(v) if $F \leftrightarrow G$, then the smallest lattice containing $F, F^{\perp}, G, G^{\perp}$ is Boolean.

That is, the compatibility $\leftrightarrow$ is a reflexive and symmetric binary relation on L, closed under the Boolean operations $\wedge$, $\vee$ and $^{\perp}$.

Clearly, every Boolean algebra is a partial Boolean algebra, but not the other way round. In a nontrivial partially Boolean algebra the compatibility relation is not transitive, i.e. $F \leftrightarrow G$ and $G \leftrightarrow H$ do not imply $F \leftrightarrow H$. Nevertheless, it is possible to introduce into a partially Boolean algebra a transitive structure: a partial Boolean algebra is called transitive if $F \leq G$ and $G \leq H$ imply $F \leftrightarrow H$, and therefore $F \leq H$ (Kochen and Specker, 1965a). Since a universe of discourse is more than a collection of all its Boolean frames of reference, it is natural to base the logic of any universe of discourse on the class of transitive partial Boolean algebras. This assumption introduces a partial ordering into the set of L of all propositions, guaranteeing an intrinsic coherence of the Boolean atlases.

Let $\{B_\alpha | \alpha \in \P\}$ with $\P$ an index set, be an indexed family of σ-complete Boolean algebras $(B_\alpha, E_\alpha, 0_\alpha, \leq_\alpha, {}^{\perp\alpha})$ fulfilling the following consistency relations:

(i) $E_\alpha = E_\beta$ and $0_\alpha = 0_\beta$ for all $\alpha, \beta \in \P$,

(ii) if $F \in B_\alpha \cap B_\beta$, then $F^{\perp\alpha} = F^{\perp\beta}$,

(iii) if $F, G \in B_\alpha \cap B_\beta$, then $F \leq_\alpha G$ iff $F \leq_\beta G$.

On such a consistent set $L = \cup_{\alpha \in \P} B_\alpha$ we can define:

(i) a map $F \to F^{\perp}$ by asserting that $F^{\perp} = F^{\perp\alpha}$ for every B_α which contains F,

(ii) a partial order $\leq$ by writing $F \leq G$ if and only if there is a $\alpha \in ¶$ such that $F \leq_\alpha G$.

With $E \stackrel{def}{=} E_\alpha$ and $O \stackrel{def}{=} O_\alpha$ for all $\alpha \in ¶$, $(\mathcal{B}, E, O, \leq, ^\perp)$ becomes an orthomodular poset (Finch, 1969; Czelakowski, 1974). Since every component Boolean logic $\mathcal{B}_\alpha$ of L is a σ-decomposable Boolean σ-algebra, it is sensible to postulate that every subset of mutually orthogonal element in L is countable and has a supremum in L. With this, L becomes a separable σ-orthocomplete orthomodular poset, hence a separable σ-complete orthomodular lattice (Holland, 1970a). Since every separable σ-complete orthomodular lattice is complete (Varadarajan, 1968, p.183), the logic of L of any universe of discourse is given by a separable complete orthomodular lattice.

Two orthomodular proposition systems L_1 and L_2 are considered to be logically equivalent if there exists a bijective map from L_1 onto L_2 that preserves the logical structure of these theories. Accordingly, an isomorphism between two orthomodular lattices L_1 and L_2 is defined as a one-to-one map α from L_1 onto L_2 having the properties (i) $\alpha(O)=O$, $\alpha(E)=E$, (ii) $\alpha(F) \leq \alpha(G)$ iff $F \leq G$, (iii) $\alpha(F^\perp)=\alpha(F)^\perp$.

Memento: orthomodular lattices
A poset (i.e. a partially ordered set) is a set L with a binary relation $\leq$ on L that is reflexive (i.e. $F \leq F$), antisymmetric (if $F \leq G$ and $G \leq F$, then $F=G$), and transitive (if $F \leq G$ and $G \leq H$, then $F \leq H$). A lattice is a poset with the infimum and supremum defined for every pair of elements. The infimum of F and G is denoted by $F \wedge G$, the supremum by $F \vee G$. An orthomodular lattice is a lattice L with universal bounds E and O, and an orthocomplementation $F \to F^\perp$ such that for every $F, G \in L$: (i) $(F^\perp)^\perp = F$, (ii) $F \wedge F^\perp = O$, $F \vee F^\perp = E$, (iii) $F \leq G$ implies $G^\perp \leq F^\perp$, (iv) $F \leq G$ implies $G = F \vee (G \wedge F^\perp)$. A subset of an orthomodular lattice L is called an orthomodular sublattice of L if it is a sublattice of L containing E and O, and is closed under the orthocomplementation of L. Two elements F,G of an orthomodular lattice are said to be compatible if $F=(F \wedge G) \vee (F \wedge G^\perp)$. Every Boolean lattice is orthomodular, and an orthomodular lattice is Boolean if and only if all its elements are mutually compatible. Two elements F,G of an orthomodular lattice are compatible if and only if the sublattice generated by $F, F^\perp, G, G^\perp$ is Boolean. Two elements F,G of an orthomodular lattice are said to be orthogonal (synonym: disjoint), written $F \perp G$, if $F \leq G^\perp$. Evidently $F \perp G$ if and only if $G \perp F$. An orthomodular lattice L is called separable, if every subset of mutually orthogonal elements in L is countable. *Standard reference work on orthomodular lattices:* Maeda and Maeda (1970).

4.3 The Interpretation of Orthomodular Logics

The interpretation of an orthomodular lattice as a generalized logic of properties is inherited from its interpreted partial

Boolean structure. Every Boolean sublattice $\mathcal{B}$ of the orthomodular lattice L associated to a universe of discourse is interpreted in terms of classical two-valued logic as a feasible frame of reference.

The existence of incompatible properties implies that at some fixed instant not all propositions of L are either true or false. That is, the elements of L represent the potential properties of the system under discussion. If L is not Boolean, then not all potential properties can be actualized at the same instant, so that in a nonclassical system the set of actualized properties is properly contained in the set L of potential properties.

A proposition $F \in L$ is called truth-definite at time t if F is either true or false at time t. Propositions that are true at time t correspond to properties the system has at time t. Propositions which are not truth-definite at time t correspond to potential properties not actualized at time t.

In the following we are going to define the ontic state at the instant t via the set of all propositions of L that are true at time t. We start with the remark that for simultaneously truth-definite propositions the lattice operations have a straightforward logical interpretation. Assume that F and G are two propositions which are truth-definite at time t. Then we interpret

(i) $F^{\perp}$ as the proposition which is true if F is false, and false if F is true;

(ii) $F \vee G$ as the proposition which is true if F, or G, or F and G are true, and false if F and G are false;

(iii) $F \wedge G$ as the proposition which is true if F and G are true, and false if F, or G, or F and G are false.

Note that the logical interpretation of the lattice operations $\vee$ and $\wedge$ are strictly restricted to simultaneously truth-definite propositions. Since in an orthomodular lattice comparable propositions are compatible (i.e $F \leq G$ implies $F \leftrightarrow G$, the order relation $\leq$ in L can be interpreted as implication:

$F \leq G$ means F logically entails G.

The set of all propositions of L which are truth-definite at time t is called the truth-definite set T_t of the system at time t. It easily follows that T_t is an orthomodular sublattice of L (having the same universal bounds as L). By construction of T_t, there exists an ortholattice homomorphism $\rho_t : T_t \to \mathcal{B}_2$ from the truth-definite set T_t onto the Boolean algebra $\mathcal{B}_2 = (0,1)$ consisting of the two elements 0 and 1, and characterized by

$$\rho_t(F) = 1 \quad \text{iff} \quad F \in T_t \text{ is true at time t ,}$$

$$\rho_t(F) = 0 \quad \text{iff} \quad F\varepsilon T_t \quad \text{is false at time t .}$$

We call an orthomodular sublattice of L assessable if it is the pre-image of some ortholattice homomorphism onto the Boolean algebra B_2 consisting of the two truth values 0 and 1. An assessable orthomodular sublattice of L is called maximal if it is not properly contained in any other assessable orthomodular sublattice of L. Note that a maximal assessable orthomodular sublattice of L is in general neither distributive nor σ-complete. The ontic interpretation of orthomodular logic posits that at every instant t the truth-definite set T_t is a maximal assessable orthomodular sublattice of L. Roughly, this means that at every instant as many potential properties as possible are actualized. The truth function $\rho_t:T_t\to B_2=(0,1)$ will be called the ontic state of the system at time t.

The set of all those propositions of L that are compatible with every proposition of L is called the centre $Z(L)$ of L,

$$Z(L) \overset{\text{def}}{=} \{Z|Z\varepsilon L, Z \leftrightarrow F \text{ for every } F\varepsilon L\}$$

The centre of a separable σ-complete orthomodular lattice L is a separable σ-complete Boolean lattice; it is contained in every maximal assessable orthomodular sublattice of L. As we have discussed in the preceding chapters, all theories of classical physics are Boolean, so that for these $Z(L)=L$. Accordingly we call a system <u>classical</u> if $Z(L)=L$. For a nonclassical system, the centre $Z(L)$ represents the classical part of the system; the propositions of the centre and their corresponding properties are called classical. A classical property can be characterized as being (in principle) observable to an arbitrary accuracy without thereby inducing perturbations which would invalidate predictions based on this observation. If a system has no classical properties, then the centre $Z(L)$ of its logic L is trivial, $Z(L)=\{0,E\}$; such systems will be called <u>factor systems</u>. For example, pioneer quantum mechanics knows no classical properties, hence it represents factor systems.

Since every classical proposition is compatible with every proposition of L, the centre $Z(L)$ is contained in every truth-definite set T_t,

$$Z \subseteq T_t \quad \text{for every } t\varepsilon\mathbb{R} \text{ .}$$

Accordingly, every classical proposition is truth-definite at every instant so that every classical property is actualized at every instant.

References:

There exists an extensive literature on orthomodular logics from miscellaneous philosophical points of view. We cannot review these approaches here but refer to Primas (1980). A part of the relevant literature is also discussed in Jammer (1974). A reprint collection of the pioneering papers in the area of quantum logic has been edited by Hooker (1975).

4.4 W*-Quantum Logic

There exist orthomodular lattices having quite pathological properties. For example, there exist infinitely many nonisomorphic orthomodular lattices which admit no states at all (Greechie, 1971). Thus the class of all orthomodular lattices overgeneralize the logical systems useful in natural science. In the following we single out a special class of nice orthomodular logics, the so-called W*-quantum logics whose lattice L of propositions are orthoisomorphic to the projection lattice of W*-algebras.

Memento: W-algebras*
Recall that a Banach algebra is a Banach space A over the complex numbers, with a multiplication making it an associative algebra and satisfying $\|AB\| \leq \|A\| \cdot \|B\|$ for all $A, B \in A$. A C*-algebra is a Banach algebra A having an involution (i.e. a conjugate-linear map $A \to A$ satisfying $A^{**}=A$ and $(AB)^* = B^* A^*$ for all $A, B \in A$) satisfying $\|A^*A\| = \|A\|^2$. Every C*-algebra is isometrically *-isomorphic to a norm-closed *-subalgebra of the algebra $B(H)$ of all bounded operators acting on some Hilbert space H.

A W*-algebra is a C*-algebra which is the dual space of a Banach space. This Banach space is called the predual of the W*-algebra A and is denoted by A_*, so that $(A_*)^* = A$. The predual A_* is a norm-closed subspace of the dual A^* of the W*-algebra A. The predual A_* induces in A the weak-*-topology $\sigma(A, A_*)$ (i.e. the weakest topology on A for which all the functionals in A_* are continuous), which is called the σ-topology on A (synonyma: σ-weak topology, ultraweak topology). Every W*-algebra is σ-weakly continuously *-isomorphic to a weakly closed *-subalgebra of $B(H)$ containing the identity. Such concrete W*-algebras of operators acting on some Hilbert space H are also called von Neumann algebras. Every von Neumann algebra A on a separable Hilbert space has a separable predual A_*. *Standard reference works on C*- and W*-algebras:* Dixmier (1957, 1964), Sakai (1971).

The W*-model of quantum logic assumes that the lattice L of temporal propositions of a system is given by the lattice $P(A)$ of projections of some W*-algebra A,

$$P(A) \overset{\text{def}}{=} \{P \mid P \in A \,,\; P = P^2 = P^*\}$$

The partial ordering in $P(A)$ is given by $F \leq G$ iff $FG = F$. The orthocomplementation $F \to F^{\perp}$ in $P(A)$ is given by $F^{\perp} \overset{\text{def}}{=} 1 - F$. With this

$(P,1,0,\leq,^{\perp})$ becomes a complete orthomodular and semimodular lattice (Kaplansky, 1951; Topping, 1967; Holland, 1970b). The orthomodular lattice $P(A)$ is separable if and only if the predual A_* is separable. Note that not every complete, orthomodular and semimodular lattice is a projection lattice of a W*-algebra; a purely lattice-theoretical characterization of the projection lattices of W*-algebras is not yet known (compare Holland, 1964). The main motivation for using the W*-projection lattices as possible logics of molecular systems is the fact that - to the best of our knowledge - every system whatsoever having relevance for molecular science can be represented as a W*-system. Famous and well-understood examples are the classical theories (where A is commutative W*-algebra), the pioneer quantum mechanics (where A is chosen to be the algebra of all bounded operators on some Hilbert space; von Neumann, 1932a), quantum mechanics with superselection rules (where A is chosen to be a W*-algebra of type I), thermodynamics (where A is a W*-algebra of type III_1).

If M is an arbitrary nonempty subset of a W*-algebra A, then there is a unique smallest W*-subalgebra containing M, called the W*-algebra generated by M. It is a most remarkable fact that every W*-algebra is generated by the projections it contains. For merely historical reason, the W*-algebra generated by the lattice of temporal propositions is called the algebra of observables. The self-adjoint elements of the algebra of observables are called bounded observables.

W*-quantum logics are characterized by a close connection between lattice-theoretical and algebraic morphisms. Roughly formulated: two W*-systems are logically isomorphic if and only if the projection lattices are isomorphic, and this is the case (besides some technicalities) if and only if the algebras are isomorphic.

Memento: Morphisms of W-algebra*
A mapping $\phi:A_1\to A_2$ from a W*-algebra A_1 into a W*-algebra A_2 is called a *-homomorphism if it preserves the algebraic structure, that is if for all $A,B\in A_1$ and $\lambda\in C$ we have: (i) $\phi(\lambda A)=\lambda\phi(A)$, (ii) $\phi(A+B)=\phi(A)+\phi(B)$, (iii) $\phi(AB)=\phi(A)\phi(B)$, (iv) $\phi(A^*)=\phi(A)^*$. If (iii) is replaced by (iii') $\phi(AB)=\phi(B)\phi(A)$, we call ϕ an anti-homomorphism. A faithful *-homomorphism from A_1 onto A_2 is called a *-isomorphism. A *-isomorphism from A onto A is called a *-automorphism. A morphism $\phi:A_1\to A_2$ is called isometric if $\|\phi(A)\|=\|A\|$ for all $A\in A_1$. A morphism $\phi:A_1\to A_2$ is called normal if it is completely additive, i.e. if $\phi(\sup_\alpha\{E_\alpha\})=\Sigma_\alpha\phi(E_\alpha)$ for each orthogonal family $\{E_\alpha\}$ of projections $E_\alpha\in A_1$. Algebraic *-morphisms of W*-algebra exhibit rather strong continuity properties, in particular we have: every *-homomorphism ϕ is a contraction, i.e. $\|\phi(A)\|\leq\|A\|$. Every *-isomorphism ϕ is isometric, i.e. $\|\phi(A)\|=\|A\|$. Every normal *-homomorphism is σ-weakly continuous. Every *-isomorphism is normal. Warning: not every *-homomorphism

of a W*-algebra into a W*-algebra is normal. However, the image of a normal homomorphism is a W*-algebra.

A fundamental theorem by Dye (1955) allows us to characterize logically equivalent W*-logics in purely algebraic terms. Recall that two orthomodular propositional systems L_1 and L_2 are called logically equivalent if there exists a bijective map ϕ from L_1 onto L_2 that preserves the partial ordering and the orthocomplementation. Let A_1 be a W*-algebra without direct summands of type I_2, then every lattice-orthoisomorphism from the projection lattice $P(A_1)$ to the projection lattice $P(A_2)$ of a W*-algebra A_2 is implemented by a Jordan-isomorphism $A_1 \to A_2$ which for its part can be represented by a direct sum of a *-isomorphism and a *-anti-isomorphism (Dye, 1955). That is, the logical symmetries of W*-quantum logics are given by *-isomorphisms and *-anti-isomorphisms of the W*-algebras of observables.

4.5 Classifications of W*-Systems

Since every W*-algebra is generated by its projection lattice and by virtue of Dye's theorem, W*-quantum logics can be rephrased in purely algebraic terms. As a rule, the algebraic characterizations are more convenient (and more popular!) than the lattice-theoretical descriptions. All classifications of physical systems are classifications of W*-algebras, and every characterization of a particular W*-algebra is at the same time a characterization of a particular physical system. The classifications of W*-algebras are invariant under *-isomorphisms. Theories having *-isomorphic algebras of observables are conceptually indistinguishable but they may have unitarily inequivalent Hilbert-space realizations. Theories having non-*-isomorphic algebras of observables employ different patterns of explanation, prediction and retrodiction; they make use of different kinds of basic concepts and rely upon different modes of confirmation.

The most important single algebraic characterization of an algebra refers to its centre. The centre $Z(A)$ of an algebra A is defined as the set of those operators in A that commute with every operator in A,

$$Z(A) \overset{\text{def}}{=} \{Z \mid Z \in A, \quad ZA=AZ \text{ for every } A \in A\} .$$

The centre $Z(A)$ of a W*-algebra is a commutative W*-algebra. Since two projectors are compatible if and only if they commute,

$$F \leftrightarrow G \quad \text{iff} \quad FG=GF \ , \quad F,G \in P(A) \ ,$$

the centre $Z(A)$ of the algebra is generated by the centre $Z(P)$ of the projection lattice $P=P(A)$. The propositions in $Z(P)$ represent the classical properties, and on that account the observables in $Z(A)$ are called classical observables.

A W*-algebra is called a factor if its centre is trivial (i.e. if the centre consists of scalar multiples of the identity operator only). Every W*-algebra with separable predual can be expressed uniquely as a direct sum or a direct integral of factors. This so-called central decomposition reduces the study of W*-algebras to the study of their centres (i.e. commutative W*-algebras) and the study of factors.

The classification of W*-algebras is based on the properties of their projection lattices. In the projection lattice $P(A)$ of a W*-algebra A there is a natural equivalence relation: two projections F,G are called equivalent (written $F\sim G$) if there exists in A an operator T such that $T^*T=F$ and $TT^*=G$. A projection F is said to be finite if $G\sim F$ and $G\leq F$ imply $G=F$. If F is not finite, it is called infinite. A W*-algebra with the unit 1 is called

(i) finite if for every $F\in P(A)$, $F\sim 1$ implies $F=1$.

(ii) infinite if A is not finite.

Recall that an element $F\in P(A)$ is called an atom if $F\neq 0$ and if for every $G\in P(A)$ the relation $G\leq F$ implies either $G=0$ or $G=F$. The factors of W*-algebras are classified into three major groups, those of type I, II and III. A factor is said to be

(i) of type I if it contains an atom,

(ii) of type II if it is atom-free and if it contains some nonzero finite projection,

(iii) of type III if it does not contain any nonzero finite projection.

Clearly, the type of a factor is preserved under *-isomorphisms. A factor of type I is said to be of type I_n $(n=1,2,\ldots)$ if it is *-isomorphic to the W*-algebra of $n\times n$-matrices, and of type I_∞ if it is *-isomorphic to the algebra $B(H)$ of all bounded operators acting on some infinite-dimensional separable Hilbert space. A factor of type II is said to be of type II_1 if all its projections are finite; it is called type II_∞ if it contains also infinite projections. Every type II_∞ factor can be realized as the W*-tensor product of a type II_1 factor and a type I_∞ factor.

With this, the classification of type I factors is complete. However, there are uncountably infinitely many nonisomorphic factors of type II_1, II_∞ and III whose complete classification is altogether out of reach. This is an indication that the category of W*-algebras with separable preduals is much too large and not necessary for the theoretical description of nature. There is, however, an important special class of W*-algebras which is physically well motivated and whose full classification is essentially known: the class of approximately finite-dimensional W*-algebras. A W*-algebra A with a separable predual A_* is called approximately finite-dimensional (synonym: hyperfinite) when it is generated by an increasing sequence of finite-dimensional subalgebras of A . To the best of our knowledge, there is no physically relevant W*-system whose algebra of observables is not approximately finite-dimensional. The class of approximately finite-dimensional W*-algebras is closed under direct sums, and contains all commutative W*-algebras and all factors of type I. Murray and von Neumann (1936, 1943) showed that there is (up to *-isomorphisms) exactly one approximately finite-dimensional factor of type II , and forty years later Connes (1976) proved that there is exactly one approximately finite-dimensional factor of type II_∞. The Tomita-Takesaki theory of modular automorphisms (reviewed in Takesaki, 1970) gave a major breakthrough in the classification of type III. Connes (1973) introduced in his thesis a one-parameter classification of type III factors into factors of type III_λ with $0 \leq \lambda \leq 1$, and showed later (Connes, 1976) that there is (up to isomorphisms) only one approximately finite-dimensional factor of type III_λ for each λ with $0<\lambda<1$. Approximately finite-dimensional factors of type III_0 are classified exactly in terms of so-called Krieger factors. In case III_1, the only known approximately infinite-dimensional factor is the factor R_∞ of Araki and Woods (1968) which is the relevant algebra of observables for an infinite free boson gas above the critical temperature (Araki and Woods, 1963). The question of the unicity of the approximately finite-dimensional factor of type III_1 is of great physical interest; however, this important mathematical problem is still open. Note that Golodec's (1977) proof of the unicity is wrong! (compare Golodec's withdrawal added in the English translation by the American Mathematical Society).

4.6 The States of a W*-System

The ontic states of a W*-system are related to the pure states on the algebra A of observables (recall that the concept of a pure state is a mathematical notion, it is a synonym for an extremal normalized positive linear functional on the algebra of observables). In order to show that every pure state $\rho \varepsilon A^*$ generates an ontic state, we define the definite set $\mathcal{D}_\rho$ of ρ as the set of all self-adjoint elements $A \varepsilon A$ that fulfil $\rho(A^2)=\{\rho(A)\}^2$. The subset consisting of all projections in $\mathcal{D}_\rho$ will be denoted by T_ρ,

$$T_\rho = \{F | F \varepsilon P(A) \quad , \quad \rho(F)\rho(1-F)=0\} \ .$$

Clearly, $F \varepsilon T_\rho$ iff $F^\perp \varepsilon T_\rho$. Let $F, G \varepsilon T_\rho$, then the relation $F \wedge G = \inf(F,G)$ implies either $\rho(F \wedge G)=1$, or $\rho(F \wedge G)=0$, so that $F \wedge G \varepsilon T_\rho$. Accordingly, T_ρ is a lattice; it is easy to show that T_ρ is even a maximal assessable orthomodular sublattice of $L=P(A)$. With the definition given in sect.4.3, it follows that the restriction of every pure state ρ to T_ρ is an ontic state of the system.

The ontic states generated by the pure states on the algebra of observables have a conceptually important continuity property. We say that two properties **f** and **g** represented by the propositions F and G are similar if the angle $\phi(F,G)$ between F and G is small (Primas, 1975), where

$$\sin \phi(F,G) \overset{\text{def}}{=} \|F-G\| \quad , \quad F,G\in P(A)$$

In the ontic interpretation the relation $\rho(F)=1$ implies that in the ontic state ρ the system has the property **f**. It would be rather artificial not to attribute a certain degree of "**f**-ness" to an ontic state ρ for which $\rho(G)=1$ and $\|F-G\|$ is very small. The possibility of such an extended interpretation is feasible if and only if the functional ρ is norm-continuous (recall that a linear functional on a Banach algebra A is norm-continuous if for every sequence $\{A_n\}$, $A_n\in A$ converging uniformly to an element $A\in A$, $\rho(A_n)$ converges towards $\rho(A)$). Since every positive linear functional on a C*-algebra is norm-continuous, every ontic state coming from a pure state is norm-continuous, hence allows an extended ontic interpretation.

Does every continuous ontic state arise from a pure state on the algebra of observables? This is a mathematically highly non-trivial question. The main stumbling block is whether there exists an extension of every continuous state to a linear functional on the algebra, and whether this extension is given by a unique pure state. For approximately finite-dimensional algebras of observables, a theorem by Aarnes (1970) implies that every continuous ontic state ρ has a linear extension from T_ρ to D_ρ. Since D_ρ is a maximal definite set in the dense of Kadison and Singer (1959), a theorem by Størmer (1968) can be used to prove the existence of a unique extension to a pure state on A. We have some evidence that this line of arguments can be extended to a proof (G. Raggio). In the following we will assume that there is a one-to-one correspondence between the continuous ontic states of a W*-system and the pure states on the algebra of observables. (A disproof of our conjecture does not imply a disaster, all we have to do is to add some inelegant conditions.)

Remark: Which ontic states are accessible by experiments?
There are no normal pure states on atom-free commutative W*-algebras, and on factors of type II and type III. Since every epistemic state is normal, in these cases the ontic states are experimentally inaccessible. On the other hand, every factor of type I (or more generally, every W*-algebra generated by its atoms) has normal pure states. Every factor of type I is *-isomorphic to the algebra $B(H)$ of all bounded operators acting on some Hilbert space H, its predual $B(H)_*$

can be identified with the Banach space $\mathcal{B}_1(\mathcal{H})$ of all bounded nuclear operators on $\mathcal{H}$. The pure normal states on $\mathcal{B}(\mathcal{H})$ correspond to the one-dimensional projections in $\mathcal{B}_1(\mathcal{H})$. Note that in the factor of type I_∞ not all pure states are normal.

4.7 Dynamical W*-Systems

As discussed in sect.3.6, the dynamics of a classical W*-system is given by a σ-weakly continuous semigroup $\{\alpha_t | t \geq 0\}$ of positive, identity-preserving normal maps α_t of the algebra of observables into itself. In the case of general W*-systems we have to add stronger positivity assumptions.

Recall that a linear map α of one W*-algebra $\mathcal{A}_1$ into another W*-algebra $\mathcal{A}_2$ is called positive if α(A) is positive for all positive $A \in \mathcal{A}_1$. Let M_n denote the W*-algebra of all complex n×n-matrices. A linear map $\alpha : \mathcal{A}_1 \to \mathcal{A}_2$ of one W*-algebra $\mathcal{A}_1$ into another W*-algebra $\mathcal{A}_2$ is said to be n-positive if the linear map $\phi \otimes \mathbf{id} : \mathcal{A}_1 \otimes \mathcal{M}_n \to \mathcal{A}_2 \otimes \mathcal{M}_n$ is positive, where **id** denotes the identity map of $\mathcal{M}_n$. The map α is called completely positive if it is n-positive for every n=1,2,... Note that there exist n-positive maps that fail to be (n+1)-positive (Choi, 1972). Two important classes of completely positive maps are the homomorphisms and the projections of norm one (for a review, compare Størmer, 1974).

Completely positive linear maps are the natural generalization of positive maps of commutative algebras. Stinespring (1955) has shown that on a commutative C*-algebra every positive linear map is completely positive. In the noncommutative case, complete positivity is an essential restriction. For example, every *-anti-automorphism of a noncommutative W*-algebra is positive, but not 2-positive (Gorini et al., 1976).

Complete positivity of operations is a necessary consistency requirement for composite systems (Kraus, 1971). Consider two noninteracting systems whose algebra of observables is given by the W*-tensor product $\mathcal{A}_1 \bar{\otimes} \mathcal{A}_2$. Since the tensor product of two positive maps is not necessarily positive, positivity is not a sufficient requirement. On the other hand, Stinespring (1955) has shown that the tensor product of completely positive maps is again completely positive. Accordingly, we arrive at the following definition of a dynamical semigroup on an abritrary W*-system:

> *Dynamical semigroup on a W*-system*
> A dynamical semigroup $\{\alpha_t | t \geq 0\}$ on a W*-algebra $\mathcal{A}$ is a σ-weakly continuous one-parameter semigroup of completely positive, identity-preserving normal maps α_t of $\mathcal{A}$ into itself.

The dynamical semigroup $\alpha_t:A \to A$ corresponds to the Heisenberg picture for the time evolution. The physically more relevant Schrödinger picture is given by the preadjoint semigroup $\alpha_{*t}:A_* \to A_*$, defined by $\rho_t(A)=\rho(A_t)$ for all $\rho \varepsilon A_*$ and all $A \varepsilon A$, where $\rho_t \overset{def}{=} \alpha_{*t}(\rho)$ and $A_t \overset{def}{=} \alpha_t(A)$. If $t \to \alpha_t$ is σ-weakly continuous, then $t \to \alpha_{*t}$ is strongly continuous. Note that the dynamics is also well-defined on the dual A^* but the time-evolution of non-normal states are as a rule not blessed with nice continuity properties.

Invertible time evolutions are given by symmetries, i.e. by Jordan automorphisms of the algebra of observables. Since Jordan automorphisms of A that are not *-automorphisms of A are not completely positive, an invertible dynamics of a W*-system is necessarily given by a *-automorphism. So we have:

> *Dynamical group on a W*-system*
> A closed and invertible dynamical W*-system is a pair $(A, t \to \alpha_t)$ consisting of a W*-algebra A having a separable predual A_*, and a σ-weakly continuous one-parameter group $\{\alpha_t | t \varepsilon \mathbb{R}\}$ of automorphisms $\alpha_t \varepsilon \mathbf{Aut}(A)$, called the dynamical group.

Remark: Continuity of flows
The assumption that the dynamical group $t \to \alpha_t$ is σ-weakly continuous is fulfilled for every reasonable system since it follows from the measurability of the map $t \to \rho\{\alpha_t(A)\}$ for every normal state ρ and for every $A \varepsilon A$. Moreover, $t \to \alpha_t$ is σ-weakly continuous (i.e. continuous in the $\mathbf{s}(A,A_*)$-topology). Any stronger continuity assumption is physically not defensible since a flow $t \to \alpha_t$ on a W*-algebra is weakly continuous if and only if it is uniformly continuous. Uniformly continuous flows have bounded generators; however, we know that most physically interesting systems have unbounded generators.

The complete positivity of a dynamical semigroup is a necessary condition that it can be obtained by the restriction of an automorphic dynamics on a larger system via a projection of norm one. However, the conjecture that every dynamical semigroup on a W*-algebra arises this way does not seem settled yet. In the two important cases of a commutative W*-algebra and a factor of type I the conjecture is known to be true (it follows in both cases from a representation theorem by Stinespring, 1955; compare also Evans and Lewis, 1976).

4.8 Conclusions

Chemistry deals with objects that are characterized by the fact that they have both classical and quantum-mechanical properties.

Such systems can be described consistently neither by classical theories nor by quantum mechanics in its original form but by a generalized quantum theory called the theory of W*-systems. The ontic interpretation assumes that natural objects have properties which are well-determined and which are possessed independently of our knowledge of them. In the ontic interpretation, the logical structure of W*-systems is non-Boolean what implies a radical change in the classical attitude toward nature. The philosophical merit of quantum logic lies in the fact that it accounts for the incompatibility of complementary properties in a fundamental way.

The characterization of W*-systems can be summarized as follows:

(i) A closed and invertible dynamical W*-system is a pair $(\mathcal{A}, t\rightarrow\alpha_t)$ consisting of a W*-algebra $\mathcal{A}$ (called the algebra of observables) having a separable predual $\mathcal{A}_*$, and a σ-weakly continuous group $\{\alpha_t | t\in\mathbb{R}\}$ of automorphisms $\alpha_t \in Aut(\mathcal{A})$, called the dynamical group. There are good reasons to believe that for every system of chemical interest the algebra $\mathcal{A}$ can be chosen to be approximately finite-dimensional.

(ii) The referent of the theory is a single system (in contradistinction to a virtual Gibbsian ensemble) together with its objective properties.

(iii) The potential properties of the system are represented by the elements of the σ-complete orthomodular projection lattice $\mathcal{L}=\mathcal{P}(\mathcal{A})$. The elements of the centre $\mathcal{Z}(\mathcal{L})$ represent those objective properties of the system which are actualized at every instant.

(iv) The time evolution $\rho\rightarrow\rho_t$ of a state ρ in the dual $\mathcal{A}^*$ is given by

$$\rho_t(A) \overset{\text{def}}{=} \rho\{\alpha_t(A)\} \text{ for every } A\in\mathcal{A}.$$

(v) The pure states in the dual $\mathcal{A}^*$ represent the ontic states of the system; they describe the state in which the system can be.

(vi) The normal states in the predual $\mathcal{A}_*$ represent the result of experimental determination of states. Two different normal states are in principle distinguishable by a countable number of experiments having finite accuracy.

(vii) Non-normal ontic states exist, they describe the ultimate truth not accessible by experiments.

(viii) The self-adjoint operators in $\mathcal{A}$ are called observables. Every observable determines a projection-valued spectral measure, hence a family of potential properties, called a potential feature of the system.

(ix) Every observable $A=A^*\varepsilon\mathcal{A}$ that is dispersion-free with respect to an ontic state $\rho_t, \rho_t(A^2)=\rho_t(A)^2$, has at the instant t a definite value in an objective and realistic sense. This value is given by $\rho_t(A)$.

(x) The centre of the W*-algebra $\mathcal{A}$ consists of all those elements of $\mathcal{A}$ that commute with every element of $\mathcal{A}$. The centre of $\mathcal{Z}$ is a commutative W*-algebra; the corresponding W*-system is called the classical part of the system. The self-adjoint elements of $\mathcal{Z}$ are called classical observables.

(xi) Every classical observable $Z=Z^*\varepsilon\mathcal{Z}(\mathcal{A})$ has at every instant t the definite value $\rho_t(Z)$ in an objective and realistic sense.

(xii) Every state can be decomposed into a mixture of states which are dispersion-free on the centre (central decomposition). Two states which are dispersion-free on the centre but assume different values for some classical observable, are said to be separated by a <u>superselection rule</u>. A superselection rule invalidates the universal validity of the quantum mechanical <u>superposition principle</u>. The superposition principle holds only in a superselection sector.

5. CHEMISTRY AND W*-SYSTEMS

5.1 <u>Chemistry is More Than The Study of Molecules</u>

Chemistry has never been well-treated by fundamental physical theories. Some contemporary theoreticians have attempted to narrow down the scope of scientific inquiry by requiring operational definitions and reductionistic first-principle underpinnings for all concepts. In theoretical chemistry, there is a distinct tendency to throw out typically chemical variables, admitting that they have served a noble purpose in the past but that now they are obsolete. However, the task of theoretical chemistry is the sharpening and explanation of chemical concepts and not the rejection of a whole area of inquiry.

Current quantum chemistry tends to give the impression that the major difficulties in applying quantum mechanics to complex molecular systems are computational. Yet *the main stumbling block for the development of a genuine theoretical chemistry is not computational but conceptual.* The concept of an isolated object is fundamental for every scientific inquiry but by no means unproblematic; in fact it is in direct contradiction to some formulations of pioneer quantum mechanics. Our ability to describe objects cannot go farther than our theoretical ability to isolate them. Theoretical isolation means abstraction, and abstraction generates the relevant patterns of a scientific inquiry. The creation of the physical reality by abstractions is a deep conceptual problem which cannot be discussed separately from more general philosophical problems. We have to understand what a phenomenon is, how we recognise it, and how we communicate about it.

It is well-known from elementary quantum mechanics that composite systems are in general not in product states. That even non-interacting systems are in general not in product states is truly surprising and has been noted for the first time by Einstein, Podolsky and Rosen in 1935. The holistic nature of these Einstein-Podolsky-Rosen correlations puts severe limitations on the classical concept of an "object" as something having individuality and having properties. We will sketch a way to render precise the notion of an object in the framework of W*-systems, adopting an interpretation close to scientific realism. We consider the concept "object" as mind-independent. That is not to say that objects exist in an absolute sense, independent of any abstraction, since nothing can be said about nature unless some abstractions have been made. Objects exist only in virtue of abstractions. *The notion "object" is abstraction-dependent but can be taken as mind-independent.*

Objects are created by abstractions from the Einstein-Podolsky-Rosen correlations with their environment. The W*-algebra $\mathcal{A}$ of the observables of an object has to be chosen in accordance with these abstractions. Different abstractions lead in general to nonisomorphic W*-algebras, hence to conceptually inequivalent theories. The set of all chemically relevant subtheories cannot be totally ordered (Primas, 1977), so that it is a pure illusion to think that learning all about molecules is the distinguished way to chemical and biological knowledge.

5.2 Objects and Their Environments

Consider two independent invertible dynamical W*-systems $(\mathcal{A}_1, t\to\alpha_{1t})$ and $(\mathcal{A}_2, t\to\alpha_{2t})$ and let ρ_1 and ρ_2 be the respective ontic states at time $t=0$. The dynamics transforms ontic states ρ_ν into ontic states $\rho_{\nu t}=\alpha_t^*(\rho_\nu)$, $\nu=1,2$, where the Schrödinger group $t\to\alpha_{\nu t}^*$ is defined in terms of the Heisenberg group $t\to\alpha_{\nu t}$ by

$$\alpha_{\nu t}^*\{\rho(A)\} = \rho\{\alpha_{\nu t}(A)\} \quad , \quad A\in\mathcal{A}_\nu,\ \rho\in\mathcal{A}_\nu^* \quad .$$

The joint system of $(\mathcal{A}_1, t\to\alpha_{1t})$ and $(\mathcal{A}_2, t\to\alpha_{2t})$ is again a W*-system $(\mathcal{A}, t\to\alpha_t)$, where the algebra of observables is given by the W*-tensor product $A = A_1 \bar{\otimes} A_2$. If $\rho_{\nu t}$ is an ontic state of the W*-system $(A_\nu, t \to \alpha_{\nu t})$, $\nu = 1,2$, then $\rho_{1t} \otimes \rho_{2t}$ is an ontic state of the joint system $(A, t \to \alpha_t)$. However, the reverse is in general not true, in general the joint system has not only product states as ontic (i.e. pure) states. If the two subsystems of the joint system are noninteracting, the Heisenberg dynamics of the joint system is given by $\alpha_t = \alpha_{1t} \otimes \alpha_{2t}$, so that the time evolution transforms product states $\rho_1 \otimes \rho_2$ into product states $\rho_{1t} \otimes \rho_{2t}$. If the product states are not the only pure state on A, the slightest interaction between the two subsystems changes the situation dramatically. In this case, an invertible dynamics transforms pure states into pure states, but not product states into product states.

There exists a distinguished class of factorizations $A = A_1 \bar{\otimes} A_2$ not showing any Einstein-Podolsky-Rosen correlations between the subsystems $A_1 \bar{\otimes} \{C\}$ and $\{C\} \bar{\otimes} A_2$ (where $\{C\}$ denotes the trivial algebra consisting of the multiples of the identity only). We call such a factorization an object factorization and define:

Object factorization
A factorization $A = A_1 \bar{\otimes} A_2$ of the algebra A of observables of a W*-system is called an object factorization, if

(i) neither A_1 nor A_2 is the trival factor of type I_1,
(ii) the restriction of every pure state on A to A_ν is a pure state on the algebra A_ν, $\nu = 1,2$.

In an object factorization $A_1 \bar{\otimes} A_2$ we call one subsystem (say $A_1 \bar{\otimes} \{C\}$) the <u>object system</u> and the other subsystem (say $\{C\} \bar{\otimes} A_2$) its <u>environment</u>. In this situation, every ontic state of the joint system is the product state of an ontic state of the object and an ontic state of its environment. If not all factorizations of a W*-system are object factorizations, then either the object or its environment show holistic features, so that we speak of holistic systems. If a W*-system allows no factorizations at all, it is called a prime system.

Examples:

(i) Every nontrivial factor is holistic.
(ii) The factors of type I_n, with n a prime number, are prime.
(iii) The factors of type I_∞, II_∞ and III are not prime
(iv) Commutative W*-systems are not holistic.

Since the concept of an object entails individuality, and individuality necessitates the existence of ontic states, we define objects as those subsystems of the universe of discourse that are at any instant in an ontic state. Equivalently, *we can characterize an object as a W*-subsystem that cannot have Einstein-Podolsky-Rosen correlations with its environment.*

Objects exist in object factorizations of the universe of the universe of discourse only. The following theorem, due to Guido Raggio (1978) gives a complete control over the feasible factorizations: *a factorization* $A_1 \bar{\otimes} A_2$ *is an object factorization if and only if either* A_1*, or* A_2*, or* A_1 *and* A_2 *are commutative W*-algebras.* Accordingly, we have only three possibilities:

(i) a nonclassical object in a classical environment,

(ii) a classical object in a nonclassical environment,

(iii) a classical object in a classical environment.

To speak of a nonclassical object in interaction with its nonclassical environment is a logical incongruity.

5.3 Elementary Objects

The concept of a so-called "elementary particle" is inherently related to the structure of space and time. The structure of space-time is characterized by an abstract kinematical group. For a Galilei-relativistic theory the kinematical group is given by the Galilei group (a 10-parameter Lie group which is equal to the semidirect product of the 6-parameter Euclidean group with the 4-parameter translation group of Galilean space-time).

Quantum logic is prior to mechanics. Mechanical systems are defined within the framework of quantum logic by the appropriate kinematical group which has to be represented as a subgroup of the logical symmetries of the theory. A W*-system is called a Galilei system (A,G) if the Galilei group is faithfully realized as a group G of automorphisms of the algebra A of observables of the system. An automorphism group $G \subset \mathrm{Aut}(A)$ of a W*-algebra A is called ergodic on A if the only A in A which satisfies $g(A)=A$ for every $g \in G$ is a multiple of the identity. A Galilei system (A,G) is called elementary if the Galilei group G acts ergodically on the algebra A of observables. It can be shown (Amann, 1978) that this definition is the proper generalization of the elementarity concept used in Newtonian mechanics (transitive action of the Galilei group on a symplectic manifold) and in pioneer quantum mechanics (irreducible projective unitary representation of the Galilei group).

If the algebra A of observables is chosen to be a commutative and atom-free W*-algebra, then the Galilei group creates Newtonian mechanics. If A is a factor of type I_∞, then the Galilei group creates pioneer quantum mechanics. According to Bargmann's (1954) representation theory of the Galilei group, every Galilean object is characterized by a pair (m,s), $m\in\mathbb{R}$, $2s+1\in\mathbb{N}$, where the parameter m is called the mass and the parameter s is called the spin of the elementary Galilean object. Their exists a logical symmetry (given by an antiautomorphism of the algebra A) transforming the Galilean object (m,s) into the object (m,-s), so that these two systems are logically equivalent (i.e., they cannot be distinguished in isolation). By convention, we call an elementary Galilean object with positive mass a Galileon. To every Galileon (m,s), there is a twin system (-m,s), called anti-Galileon. In contradistinction to Einsteinian relativity, elementary Galilean objects with negative mass seem to play no role in science.

A classical Galileon with s=0 is called a particle, it is punctually localized in the physical 3-dimensional space and has all the properties of a Newtonian mass point. A quantal Galileon has very different properties so that the usual name "elementary particle" is bad and very misleading. The prototype of a quantal Galileon is the electron, characterized as a Galileon associated with a factor of type I , and having the parameters $m=9,109 \ldots \cdot 10^{-31}$ kg and $s=\frac{1}{2}$. The electron is neither a particle nor a wave, it has a nature of its own. Note that with every quantal Galileon having nonvanishing mass there is associated an ergodic Weyl system whose generators P and Q fulfil the famous Heisenberg canonical commutation relation. Accordingly, the origin of the canonical commutation relations of pioneer quantum mechanics is the assumption of Galilean space-time. Moreover, Bargmann's (1954) superselection rule (intrinsically associated with the representation theory of the Galilei group) says that it is impossible to have in pioneer quantum mechanics ontic states which are superpositions of states describing Galileons of different mass. Bargmann's superselection rule ensures the temporal stability of Galileons and guarantees the strict conservation of mass in any mechanics associated with Galilean space-time.

5.4 Are There Molecules?

This question is theoretically highly nontrivial and mathematically difficult. There exist no objects in total isolation (with the possible exception of the whole cosmological universe). We can interact the Galileons and more complicated molecular systems only via one of the two existing long-range forces: electromagnetism and gravitation. Both electromagnetism and gravitation are interactions which cannot be turned off, so that we have to require that an object does not lose its identity by including such interactions. If we want to isolate a molecular system as an object, we have to investigate the stability of such an isolation procedure. This is an exceedingly difficult mathematical problem, since it has to be proved by a stability analysis that the weak coupling to the environment (which is a system having infinitely many degrees of freedom!) the qualitative behavior of the molecular system remains unchanged. If a stability analysis validates the behaviour expected by ignoring the environment, we call the molecular system a robust one (Primas, 1975). A thorough analysis shows that there are both robust as well as nonrobust molecular systems. The qualitatively new behaviour of nonrobust systems can be described in terms of inherited classical properties. These new properties are in fact properties of the environment but it is possible and consistent to assign them to the molecular object itself. Equivalently, we speak of molecular superselection rules induced by the environment.

Illustration:
A nontechnical example may be helpful. The author of this article is a Swiss since he is a native of Switzerland. However, to be a Swiss is not an intrinsic property of a person, it is a relational property which makes sense only in an appropriate environment. Yet, if we presuppose a reasonable environmental stability, it is consistent to attribute a nationality to a person.

In particular, large molecular systems are as a rule strongly coupled with their environment (heuristically, the reason is that they have many almost degenerate eigenvalues), so that as a rule large molecular systems have more classical properties than small ones. Nevertheless, already rather small molecules can acquire classical properties by the interaction with their environment. As a paradigmatic example we may mention chiral molecules whose left- and right-handed ontic states are separated by a strict superselection rule. As shown by Pfeifer (1979), chirality is a classical property, inherited from the infrared-singular part of the electromagnetic radiation field associated with the molecule.

The analysis by Pfeifer (1979) indicates that (at least in a reasonable approximation) the factorization into a molecular and a radiation system is as a rule indeed an object factorization. If we suppose this to be true, we can simplify the analysis tremendously. If molecular objects exist at all, they are non-classical, hence they have a classical environment. We choose the usual factorization that describes the radiation field by a transversal vector potential and attributes the longitudinal field to the molecular system. In this way, we get the usual molecular Hamiltonian with Coulomb interactions between the Galileons. *The only consistent ontic description of a molecule as an object having individuality in spite of its interaction with the radiation field is to describe the transversal electromagnetic field classically.* Here, classically means a description in terms of a Boolean propositional system, or equivalently, with a commutative W*-system. It does not mean that we just take over the standard methods of classical electrodynamics. The proper description can be obtained by using the variational principle of pioneer quantum mechanics and restricting the admissible states to product states (i.e. to tensor products of arbitrary molecular states and arbitrary field states). It turns out that the radiation field always is a generalized coherent state, no superpositions of coherent states arise. Such a coherent electromagnetic field is fully equivalent to a Gaussian classical stochastic field, fulfilling the classical Maxwell equations but which is not vanishing when the molecular sources vanish. Nonrobust systems are characterized by the emergence of disjoint coherent states of the radiation field, or in the terminology of classical stochastic fields, by the emergence of stochastic fields having mutually singular probability measures. The emergence of new classical properties in nonrobust systems is again due to a singular behaviour of the electromagnetic radiation field of extremely long wavelength. Note that the radiation responsible for the infrared singularities cannot be screened, it finds its way through any screen or container and couples every molecular system with any other system.

The upshot of this analysis is that molecules exist in spite of their interactions with the radiation field. This interaction is a low-energy effect but it can generate superselection sectors for the joint system, which in turn can be rationalized by a purely molecular type I W*-system whose algebra of observables has a nontrivial centre. Such dynamically generated classical observables are very common in chemical and biological systems.

5.5 Does a Molecule Have a Shape?

Woolley (1978a,b) has reminded us convincingly that the concept of molecular structure is a classical idea, foreign to the principles of pioneer quantum mechanics. In fact, pioneer quantum mechanics neither gives a correct nor a consistent description of molecules. This is a hard fact, well-known (to those who know quantum mechanics well) but is usually put out of our mind merely because pioneer quantum mechanics gives perfect predictions for all spectroscopic experiments. However, chemistry is not spectroscopy. I agree with Woolley that the notion of a molecular structure is a classical concept, and I agree with the chemists that without this concept there is no chemistry.

That pioneer quantum mechanics cannot consistently explain molecular structures is almost trivial. Pioneer quantum mechanics uses as algebra of observables a factor of type I, hence an algebra with a trivial centre only. Accordingly, there are no classical observables in pioneer quantum mechanics, and no consistent approximation staying within the framework of pioneer quantum mechanics can produce classical properties.

In pioneer quantum mechanics, molecular structures have been discussed by the time-honoured Born-Oppenheimer method. According to Born and Oppenheimer one can - to a good approximation - discuss a molecular system by holding the slow nuclear variables fixed, solving first for the energy eigenvalues of the fast electronic variables, and then using these eigenvalues as an effective potential for the motion of the slow nuclear variables. In spite of its apparent descriptiveness, this description of the Born-Oppenheimer method misses the crucial point. Since in the algebra of observables there are no classical observables, it remains a mystery why the Born-Oppenheimer description should lead to a notion of molecular structure which can be interpreted classically.

A more fitting appreciation of the Born-Oppenheimer method stresses its singular nature: it is an expansion about the singular point of infinite nuclear masses. It is well-known that often it is the limiting behaviour that presents regularities. The Born-Oppenheimer limit, like the limit of vanishing Planck's constant, is highly singular and has to be discussed very carefully. For both limits, the traditional representation of pioneer quantum mechanics in terms of an irreducible algebra of operators acting on some Hilbert space is mathematically extremely inconvenient. The fact that the symmetries of W*-systems are *-automorphisms and not spatial unitary equivalences allows us to use a much more convenient Hilbert-space representation of pioneer quantum mechanics: the so-called standard representation. The standard representation of a factor of type I_∞ is highly reducible and has many most convenient mathematical properties. Above all, the limit of infinite mass can be discussed easily and rigorously in the standard representation. The result is that in the Born-Oppenheimer limit the algebra of nuclear observables degenerates into a commutative W*-algebra, whereby the nuclear position and momentum observables become classical observables. That is, in the Born-Oppenheimer limit of infinite nuclear masses, the notion of molecular structure becomes a truly classical concept which can be described in terms of asymptotically induced classical observables.

This result is gratifying since from another point of view it is inevitable. If we want to consider either the electronic or the nuclear part of a molecule as an object having individuality, one of these systems has to be classical as a necessity. Whether such a description is useful depends both on the molecule and on our questions. In the case of the molecule H_2^+ we know the Schmidt expansion of the electronic-nuclear factorization of the nonadiabatic ground state. Bishop and Cheung (1979) give for the population of the first product state n_1=0.995 722... (where the n_i are normalized such that $\Sigma_i n_i=1$), so the product approximation associated with any object description can be considered as reasonable.

I would like to stress that every method whatsoever (e.g., the adiabatic approximation, the generator coordinate method), which is intended to give a description of a molecule in terms of electronic structure and nuclear framework cannot avoid to use one commutative algebra of observables (probably always for the nuclei).

5.6 Conclusions

Matter admits many modes of description, all equally valid and real. Every description is a caricature. In every observation and in every experiment we have to leave out of consideration an unlimited number of effects as irrelevant. All phenomena are created by abstractions alone and do not otherwise exist. Different viewpoints lead to inequivalent descriptions of nature, represented by nonisomorphic W*-systems.

The qualities characterizing Lorentzian elementary objects, Galileons, molecules and substances are not the same. For example, substances may have a temperature, single molecules do not. Besides the Galilean concepts like momentum, energy, force and power, nonmechanical concepts like temperature, entropy, organization, specifity, function, memory, adaptiveness, growth or purpose are important for the scientific inquiry. These patterns can only be understood in a holistic relationship of objects with their environment. Which features of a system we regard as essential or interesting does not depend on the structure of the system alone but also on our mode of interaction with this system.

ACKNOWLEDGEMENT

Much of the material appearing in these notes was first presented in lectures I have given at ETH in Zürich, and has been further developed in the course discussions held with students. It is a pleasure to thank my coworkers Anton Amann, Wolfgang Gasche, Werner Gans, Eberhart Müller, Dr. Ulrich Müller-Herold, Peter Pfeifer and Guido Raggio for many critical discussions and substantial improvements. Particular thanks are due to Guido Raggio whose cooperation has been indispensable. Last but not least I thank cordially Mrs. H. Rohrer for providing invaluable editorial assistance.

REFERENCES

Aarnes, J.F., 1970. "Quasi-states on C*-algebras". Trans.Amer. Math.Soc., 149, 601-625.

Abraham, R, 1967. "Foundations of mechanics". Benjamin, New York; second edition (with J.E. Marsden), 1978.

Akcoglu, M.A., 1975. "Positive contractions of L_1-spaces". Math.Z. 143, 5-13.

Alfsen, E.M. and Shultz, F.W., 1976. "Non-commutative spectral theory for affine function spaces on convex sets". Memoirs of the Amer. Math. Soc., number 172.

Amann, A., 1978. "Eine Verallgemeinerung des quantenmechanischen Elementaritätsbegriffs". Diplomarbeit an der Abteilung für Chemie der ETH Zürich. Unpublished.

Araki, H. and Woods, E.J., 1963. "Representations of the canonical commutation relations describing a nonrelativistic infinite free Bose gas". J.Math.Phys. (N.Y.) 4, 637-662.

Araki, H. and Woods, E.J., 1968. "A classification of factors". Publ.Res.Inst.Math.Sci. (Kyoto) A 4, 51-130.

Arnold, V.I., 1962. "The classical theory of perturbations and the problem of stability of planetary systems". Sovjet Math. Dokl.3, 1008-1012. (Russian original: Dokl.Akad.Nauk SSSR 145, 487-490 (1962).)

Arnold, V.I., 1963a. "Proof of a theorem of A.N. Kolmogorov on the invariance of quasi-periodic motions under small perturbations of the Hamiltonian". Russian Math. Surveys 18, No.5, 9-36. (Russian original: Uspehi Mat.Nauk 18, No.5, 13-40 (1963).)

Arnold, V.I., 1963b. "Small denominators and problems of stability of motion in classical and celestial mechanics". Russian Math. Surveys 18, No.6, 85-191. (Russian original: Uspehi Mat.Nauk 18, No.6, 91-192 (1963).)

Arnold, V.I. and Avez, A., 1968. "Ergodic problems of classical mechanics". Benjamin, New York, 1968.

Banach, S., 1932. "Théorie des opérations linéaires". Warsaw.

Bargmann, V., 1954. "On unitary ray representations of continuous groups". Annals of Mathematics 59, 1-46.

Bartle, R.G., 1966. "The elements of integration", Wiley, New York.

Birkhoff, G., 1940. "Lattice theory". American Mathematical Society, Providence, Rhode Island; second revised edition, 1948; third new edition, 1967.

Birkhoff, G. and Bartee, T.C., 1970. "Modern applied algebra", McGraw-Hill, New York.

Bishop, D.M. and Cheung, L.M., 1979. "Natural orbital analysis of non-adiabatic H_2^+ wave functions". Int.J. Quantum Chem. 15, 517-532.

Blau, U., 1973. "Zur 3-wertigen Logik ner natürlichen Sprache". Papiere zur Linguistik 4, 20-96.

Blau, U., 1978. "Die dreiwertige Logik der Sprache". De Gruyter, Berlin.

Bohr, N., 1928. "The quantum postulate and the recent development of atomic theory". Nature (London) 121, 580-590.

Bohr, N., 1948. "On the notions of causality and complementarity". Dialectica 2, 312-319. (Reprinted in: Science 111, 51-54, 1950).

Bohr, N., 1949. "Discussion with Einstein on epistemological problems in atomic physics". In: "Albert Einstein: Philosopher-Scientist", ed. by P.A. Schilpp; Library of Living Philosophers, Evanston, Illinois, pp.199-241.

Boole, G., 1847. "The mathematical analysis of logic". Cambridge.
Boole, G., 1854. "An investigation of the laws of thought". Macmillan, London. Reprinted by Dover, New York, 1958.
Borel, E., 1914. "Introduction géométrique à quelques théories physiques". Gauthier-Villars, Paris.
Bourbaki, N., 1963. "Eléments de mathématiques. Livre VI. Intégration". Hermann, Paris.
Brown, A., 1974. "A version of multiplicity theory". In: "Topics in operator theory"; ed. by C. Pearcy; Mathematical surveys, number 13; American Mathematical Society, Providence, Rhode Island, pp.129-160.
Brown, A. and Pearcy, C., 1977. "Introduction to operator theory I. Elements of functional analysis". Springer, New York.
Brown, J.R., 1976. "Ergodic theory and topological dynamics". Academic Press, New York.
Bruns, H., 1884. "Bemerkungen zur Theorie der allgemeinen Störungen". Astron. Nachr. 109, 215-222.
Choi, M.D., 1972. "Positive linear maps on C*-algebras". Canad.J. Math. 24, 520-529.
Chong, K.M., 1976. "Doubly stochastic operators and rearrangement theorems". J.Math.Analysis and Applications 56, 309-316.
Clauser, J.F. and Shimony, A., 1978. "Bell's theorem: experimental tests and implications". Rep.Prog.Phys. 41, 1881-1927.
Connes, A., 1973. "Une classification des facteurs de type III". Annales Scientifiques de l'Ecole Normal Supérieur 6, 133-253.
Connes, A., 1976. "Classification of injective factors". Annals of Mathematics 104, 73-115.
Czelakowski, J., 1974. "Logics based on partial Boolean σ-algebras.1". Studia Logica 33, 371-396.
Dell'Antonio, G.F., 1967. "On the limits of sequences of normal states". Commun.Pure Appl.Math. 20, 413-429.
Dixmier, J., 1957. "Les algèbres d'opérateurs dans l'espace Hilbertien. (Algèbres de von Neumann)". Gauthier-Villars, Paris, premièr édition, 1957; deuxième édition, revue et augmentée, 1969.
Dixmier, J., 1964. "Les C*-algèbres et leurs représentations". Gauthier-Villars, Paris; première édition, 1964; deuxième édition 1969. (English translation: "C*-algebras", North-Holland, Amsterdam 1977).
Domotor, Z., 1974. "The probability structure of quantum-mechanical systems". Synthese 29, 155-185.
Doob, J.L., 1953. "Stochastic processes". Wiley, New York.
Douglas, R.G., 1972. "Banach algebra techniques in operator theory". Academic Press, New York.
Dye, H.A., 1955. "On the geometry of projections in certain operator algebras". Annals of Mathematics 61, 73-89.
Einstein, A., Podolsky, B. and Rosen, N., 1935. "Can quantum-mechanical description of physical reality be considered complete?" Phys.Rev.47, 777-780.

Erber, T., Schweizer, B. and Sklar, A., 1973. "Mixing transformations on metric spaces". Commun.Math.Phys. 29, 311-317.

Evans, D.E. and Lewis, J.T., 1976. "Dilations of dynamical semigroups". Commun.Math.Phys. 50, 219-227.

Finch, P.D., 1969. "On the structure of quantum logic". J.Symbolic Logic 34, 275-282.

Golodec, V.Ja., 1977. "On automorphisms of von Neumann algebras". Dokl.Akad.Nauk SSSR 237, 770-772. (English translation incorporating corrections made by the author: Sov.Math.Dokl. 18, 1477-1480 (1977).)

Gorini, V., Kossakowski, A. and Sudarshan, E.C.G., 1976. "Completely positive dynamical semigroups of N-level systems". J.Math.Phys. (N.Y.) 17, 821-825.

Greechie, R.J., 1971. "Orthomodular lattices admitting no states". J.Combinatorial Theory 10, 119-132.

Halmos, P.R., 1950. "Measure theory". Van Nostrand Reinhold, New York.

Halmos, P.R. and Neumann, J. von, 1942. "Operator methods in classical mechanics, II". Annals of Mathematics 43, 332-350.

Holland, S.S., 1964. "Distributivity and perspectivity in orthomodular lattices". Trans.Amer.Math.Soc. 112, 330-343.

Holland, S.S., 1970a. "An m-orthocomplete orthomodular lattice is m-complete". Proc.Amer.Math.Soc. 24, 716-718.

Holland, S.S., 1970b. "The current interest in orthomodular lattices". In: "Trends in lattice theory"; ed. by J.C. Abbott; van Nostrand-Reinhold, New York, pp.41-126.

Hooker, C.A., 1975. "The logico-algebraic approach to quantum mechanics. Volume 1. Historical evolution". Reidel, Dordrecht-Holland.

Hopf, E., 1934. "On causality, statistics and probability". J. Mathematics and Physics (Cambridge) 13, 51-102.

Huntington, E.V., 1933. "New sets of independent postulates for the algebra of logic, with special reference to Whitehead and Russell's Principia Mathematica". Trans.Amer.Math.Soc. 35, 274-304, 557-558, 971.

Jammer, M., 1974. "The philosophy of quantum mechanics". Wiley, New York.

Jauch, J.M., 1968. "Foundations of quantum mechanics". Addison-Wesley, Reading, Massachusetts.

Kadison, R.V., 1976. "Normal states and unitary equivalence of von Neumann algebras". In "C*-algebras and their applications to statistical mechanics and quantum field theory", ed. by D. Kastler; Proceedings of the international school of physics "Enrico Fermi", Course 60, North-Holland, Amsterdam, pp.1-18.

Kadison, R.V. and Singer, I.M., 1959. "Extensions of pure states". Amer.J.Math. 81, 383-400.

Kaplansky, I., 1951. "Projections in Banach algebras". Annals of Mathematics 53, 235-249.

Kochen, S. and Specker, E.P., 1965a. "Logical structures arising in quantum theory". In: "The theory of models"; ed. by J. Addison, L. Henkin and A. Tarski; North-Holland, Amsterdam, pp.177-189.

Kochen, S. and Specker, E.P., 1965b. "The calculus of partial propositional functions". In: "Logic, methodology and philosophy of science"; ed. by Y. Bar-Hillel; North-Holland, Amsterdam; pp.45-57.

Kochen, S. and Specker, E.P., 1967. "The problems of hidden variables in quantum mechanics". J. Mathematics and Mechanics 17, 59-88.

Kolmogorov, A.N., 1933. "Grundbegriffe der Wahrscheinlichkeitsrechnung". Springer, Berlin. (English translation: "Foundations of the theory of probability"; Chelsea, New York, 1950.)

Kolmogorov, A.N., 1948. "Algèbres de Boole métriques complètes". VI. Zjazd Matematykow Polskich. Appendix to Ann.Soc.Pol.Math. 20, 21-30.

Kolmogorov, A.N., 1957. "General theory of dynamical systems and classical mechanics" (in Russian), Proc. 1954 Intern.Congr. Math.; North-Holland, Amsterdam; pp.315-333. (English translation in: R. Abraham, "Foundations of mechanics"; Benjamin, New York, 1967, pp.263-279.)

Kolmogorov, A.N., 1958. "A new metric invariant of transient dynamical systems and automorphisms in Lebesgue spaces". (Russian). Dokl.Akad.Nauk SSSR 119, 861-864.

Koopmann, B.O., 1931. "Hamiltonian systems and linear transformations in Hilbert space". Proc.Nat.Acad.Sci.U.S. 17, 315-318.

Kraus, K., 1971. "General state changes in quantum theory". Annals of Physics 64, 311-335.

Lamperti, J., 1958. "On the isometries of certain function spaces". Pacific J.Math. 8, 459-466.

Loomis, L.H., 1947. "On the representation of σ-complete Boolean algebras". Bull.Amer.Math.Soc. 53, 757-760.

MacLane, S. and Birkhoff, G., 1967. "Algebra". Macmillan, New York.

Maeda, F. and Maeda, S., 1970. "Theory of symmetric lattices". Springer, Berlin.

Misra, B., 1967. "When can hidden variables be excluded in quantum mechanics?" Nuovo Cimento A 47, 841-859.

Moser, J., 1962. "On invariant curves of area-preserving mappings of an annulus". Nachr.Akad.Wiss. Göttingen, Math.-Phys.Kl. 1962, 1-20.

Moser, J., 1963. "Stable and random motions in dynamical systems". Princeton University Press, Princeton, New Jersey.

Murray, F.J. and Neumann, J. von, 1936. "On rings of operators". Annals of Mathematics 37, 116-229.

Murray, F.J. and Neumann, J. von, 1943. "On rings of operators, IV". Annals of Mathematics 44, 716-808.

Neumann, J. von, 1929. "Zur Algebra der Funktionaloperatoren und Theorie der normalen Operatoren". Mathematische Annalen 102, 370-427.

Neumann, J. von, 1932a. "Mathematische Grundlagen der Quantenmechanik". Springer, Berlin. (English translation: Mathematical foundations of quantum mechanics". Princeton University Press, Princeton, New Jersey, 1955.)

Neumann, J. von, 1932b. "Zur Operatorenmethode in der klassischen Mechanik". Annals of Mathematics 33, 587-642, 789-791.

Ornstein, D.S., 1974. "Ergodic theory, randomness and dynamical systems". Yale University Press, New Haven.

Parthasarathy, K.P., 1977. "Introduction to probability and measure". Macmillan, Delhi and London.

Pfeifer, P., 1979. "Chiral molecules - a classical observable induced by the radiation field". Thesis, ETH, Zürich. Preprint.

Plymen, R.J., 1968. "Dispersion-free normal states". Nuovo Cimento A 54, 862-870.

Poincaré, H., 1892. "Les méthodes nouvelles de la mécanique céleste". Gauthier-Villars, Paris.

Primas, H., 1975. "Pattern recognition in molecular quantum mechanics". Theoret.Chim.Acta 39, 127-148.

Primas, H., 1977. "Theory reduction and non-Boolean theories". J.Math.Biology 4, 281-301.

Primas, H., 1980. "Chemistry, quantum mechanics and reductionism. Perspectives in theoretical chemistry". Lecture Notes in Chemistry, Springer, Berlin; in preparation.

Raggio, G.A., 1978. "Dispersion-free states on the centre of C*- and W*-algebras". Internal progress report, Lab.Phys.Chem. ETH-Zürich, October 1978. Unpublished.

Royden, H.L., 1963. "Real analysis". Macmillan, New York; second edition 1968.

Sakai, S., 1971. "C*-algebras and W*-algebras". Springer, Berlin.

Sato, H. and Oka, Y., 1974. "A characterization of unitary operators induced by nonsingular transformations and its applications" Nagoya Math.J. 53, 189-198.

Segal, I.E., 1947a. "Irreducible representations of operator algebras". Bull.Amer.Math.Soc. 53, 73-88.

Segal, I.E., 1947b. "Postulates for general quantum mechanics". Annals of Mathematics 48, 930-948.

Sikorski, R., 1949. "On the inducing of homomorphisms by mappings". Fund.Math. 36, 7-22.

Sikorski, R., 1960. "Boolean algebras". Springer, Berlin; second revised edition, 1962; third edition (corrected reprint), 1968.

Sinai, Ya., 1963. "On the foundations of the ergodic hypothesis for a dynamical system of statistical mechanics". Sov.Math.Dokl. 4, 1818-1822. (Russian original: Dokl.Akad.Nauk SSSR 153, 1261-1264 (1963).)

Sinai, Ya., 1976. "Introduction to ergodic theory". Princeton University Press, Princeton.

Smoluchowski, M. von, 1918. "Ueber den Begriff des Zufalls und den Ursprung der Wahrscheinlichkeitsgesetze in der Physik". Naturwissenschaften 6, 253-263.

Specker, E., 1960. "Die Logik nicht gleichzeitig entscheidbarer Aussagen". Dialectica 14, 239-246.

Stinespring, W.F., 1955. "Positive functions of C*-algebras*. Proc.Amer.Math.Soc. 6, 211-216.

Stone, M.H., 1936. "The theory of representations for Boolean algebras". Trans.Amer.Math.Soc. 40, 37-111.

Størmer, E., 1968. "A characterization of pure states of C*-algebras" Proc.Amer.Math.Soc. 19, 1100-1102.

Størmer, E, 1974. "Positive linear maps of C*-algebras". In: "Foundations of quantum mechanics and ordered linear spaces"; ed. by A. Hartkämper and H. Neumann; Lecture Notes in Physics, vol.29; Springer, Berlin; pp.85-106.

Stroescu, E., 1973. "Isometric dilations of contractions on Banach spaces". Pacific J.Math. 47, 257-262.

Takesaki, M., 1970. "Tomita's theory of modular Hilbert algebra and its applications". Lecture Notes in Mathematics, vol.128, Springer, Berlin.

Topping, D.M., 1967. "Asymptoticity and semimodularity in projection lattices". Pacific J.Math. 20, 317-325.

Varadarajan, V.S., 1968. "Geometry of quantum theory. Volume 1". Van Nostrand, Princeton.

Watanabe, S., 1969. "Knowing and guessing". Wiley, New York.

Whittaker, E.T., 1943. "Chance, freewill and necessity in the scientific conception of the universe". Proc.Phys.Soc.55, 459-471.

Woolley, R.G., 1978a. "Must a molecule have a shape?" J.Amer.Chem. Soc. 100, 1073-1078.

Woolley, R.G., 1978b. "Further remarks on molecular structure in quantum theory". Chem.Phys.Lett. 55, 443-446.

Yosida, K., 1965. "Functional analysis". Springer, Berlin; fifth edition 1978.

SUPERSONIC MOLECULAR BEAMS AND VAN DER WAALS MOLECULES

Donald H. Levy

The James Franck Institute and The Department of Chemistry, The University of Chicago
Chicago, Illinois 60637, USA

These lecture notes will concern themselves with van der Waals molecules, molecules that are bound in part by forces that are two or three orders of magnitude weaker than the forces responsible for ordinary chemical bonds. The interest in these species arises in part from a desire to understand the van der Waals forces themselves and in part from the fact that knowledge of the structura. and dynamic features of these molecules may shed some light on analogous features in more complicated chemically bound molecules.

Because van der Waals bonds are very weak, they are unstable with respect to binary collisions in ordinary laboratory environments, and therefore the study of these molecules first requires their synthesis in an environment where they are stable. The development of supersonic molecular beams and the relatively recent application of supersonic beam techniques to chemical physics problems represents a great breakthrough in our ability to prepare van der Waals molecules. Moreover these techniques are so powerful that they will find application in attacking a wide variety of problems of concern to this Institute. If indeed experimentalists are about to offer a new challenge to theorists, supersonic molecular beams are likely to be one of their more powerful weapons.

I have organized these lectures into three parts. In the first part I shall describe supersonic molecular beams, the principles which govern experimental design and the properties of the beams themselves. In the second part I shall discuss the structure of van der Waals molecules although limitations of space require that this topic be covered only in the lectures and not in these notes. Finally, in the third part I shall discuss the photochemistry of van der Waals molecules.

1. SUPERSONIC MOLECULAR BEAMS AND JETS

A supersonic free jet is formed when a high pressure gas initially contained in a reservoir is expanded through an orifice or nozzle into a vacuum or low pressure region. If the flow pattern of the free jet expansion is collimated by aperatures placed downstream of the nozzle, and if the pressure in the region downstream of the collimating aperatures is low enough so that there are no collisions, the result is a supersonic molecular beam.

In the older and perhaps more familiar effusive molecular beams (Ramsey, 1963) the pressure in the gas reservoir was low enough or the orifice was small enough so that the mean free path (Λ_0) of gas molecules in the reservoir was very much larger than the diameter (D) of the orifice. In this case there are no binary collisions as molecules leave the orifice, and therefore there is no exchange of information in the expansion process and no change of properties as the gas leaves the reservoir. Therefore an effusive orifice is simply a sampling device for transferring molecules from a reservoir where they suffer collisions to a low pressure region where they do not, but the sampling is performed without affecting the properties of the gas.

The supersonic expansion is essentially a device for converting random thermal energy in a static gas into energy associated with directed mass flow in the expanding jet. In a supersonic expansion either the reservoir pressure or the orifice diameter is large enough so that $\Lambda_0 << D$ and there are binary collisions in and downstream of the nozzle. This means that unlike an effusive beam where the flow is molecular, in a supersonic expansion the flow is hydrodynamic and can be described using thermodynamic principles. We can consider a small segment of gas that is large enough to contain enough molecules so that continuum thermodynamics can be applied, yet small enough so that the intensive thermodynamic variables are constant throughout the segment. We can then describe the flow by describing how the thermodynamic properties of the segment change as the gas expands and flows downstream.

If in the post nozzle expansion there are no sources or sinks of entropy such as viscous forces, heat conductivity, shock waves, or chemical reaction, then the system remains isentropic. In actual practice this isentropic flow can in fact be maintained. This being the case it is easy to show from Maxwell's relationships that

$$\left(\frac{\partial T}{\partial V}\right)_S < 0 \qquad (1)$$

and that therefore the gas is cooled as it expands.

It is physically reasonable that this should be the case. As the expansion proceeds the flow velocity increases and the enthalpy associated with the directed flow must come from the random thermal enthalpy originally present in the static gas. It is possible to use the principle of conservation of enthalpy (Liepmann and Roshko, 1957) to obtain a relationship between the downstream temperature and the flow velocity. If h is the enthalpy per unit mass at some point downstream of the nozzle, h_0 is the enthalpy per unit mass in the reservoir, and u is the flow velocity, then

$$h_0 = h + u^2/2. \tag{2}$$

For a calorically ideal gas

$$h_0 - h = C_p(T_0 - T) = [\gamma/(\gamma-1)\ r(T_0 - T)] \tag{3}$$

where $\gamma = C_p/C_v$, r = the gas constant per unit mass = $(C_p - C_v)$, and C_p and C_v are the heat capacities per unit mass at constant pressure and constant volume, respectively. The quantity T is the downstream temperature and T_0 is the reservoir temperature. For an ideal gas the local speed of sound, a, is related to the local temperature, T, by

$$a = (\gamma r T)^{1/2} \tag{4}$$

and therefore we can combine (2)-(4) to give

$$T/T_0 = \left(1 + (\tfrac{\gamma-1}{2})M^2\right)^{-1} \tag{5}$$

where M = the Mach number ≡ u/a. For an isentropic process in an ideal gas P = const. ρ^γ and this relationship plus the ideal gas law may be used to provide the relationship between pressure and density and the Mach number. These relationships are

$$\frac{\dot{P}}{P_0} = \left(1 + (\tfrac{\gamma-1}{2})M^2\right)^{\frac{\gamma}{1-\gamma}} \tag{6}$$

$$\frac{\rho}{\rho_0} = \left(1 + (\tfrac{\gamma-1}{2})M^2\right)^{\frac{1}{1-\gamma}} \tag{7}$$

As an example of the use of these relationships we can evaluate the pressure, density, and temperature at the throat of the nozzle where M ≡ 1. Using *'s to indicate quantities at the throat, and sub-zero's to indicate quantities in the reservoir (stagnation quantities) these quantities are

$$\frac{T^*}{T_0} = \frac{2}{\gamma+1} = \begin{cases} 0.75 \text{ monatomic gas} \\ 0.83 \text{ air} \end{cases} \quad (8)$$

$$\frac{P^*}{P_0} = \left(\frac{2}{\gamma+1}\right)^{\frac{\gamma}{\gamma-1}} = \begin{cases} 0.49 \text{ monatomic gas} \\ 0.53 \text{ air} \end{cases} \quad (9)$$

$$\frac{\rho^*}{\rho_0} = \left(\frac{2}{\gamma+1}\right)^{\frac{1}{\gamma-1}} = \begin{cases} 0.65 \text{ monatomic gas} \\ 0.63 \text{ air} \end{cases} \quad (10)$$

where the quantities have been evaluated for a monatomic gas ($\gamma = 5/3$) and for air ($\gamma = 7/5$).

In order to calculate the temperature, pressure, or density profiles as a function of position downstream of the nozzle we must know the Mach number as a function of position. The solution to this problem requires solution of the hydrodynamic equations of flow and cannot be obtained from thermodynamics. Ashkenkas and Sherman (1966) have shown that for a continuous gas and for distances greater than a few nozzle diameter from throat, the Mach number profile is given by

$$M = A(X/D)^{\gamma-1} \quad (11)$$

where X is the distance downstream of the nozzle, D is the nozzle diameter, and A is a constant that depends on γ and is 3.26 for a monatomic gas. In Fig. 1 the Mach number and the temperature, density, and pressure ratios for a monatomic gas are plotted as a function of the downstream distance.

The implication of Eq. (11) and Fig. 1 is that any arbitrarily large Mach number or arbitrarily low temperature can be reached at some point sufficiently far downstream of the nozzle. This is not the case because at some point the density becomes low enough that the assumption that the gas is continuous breaks down and Eq. (11) is no longer valid. When the density is sufficiently low so that there are no binary collisions, then the individual gas atoms can no longer exchange information and the temperature of the expanding gas cannot change. This sets an upper limit on the Mach number, and this asymptotic value is known as the terminal Mach number, M_T. Anderson and Fenn (1965) have predicted that the terminal Mach number should be a function of the product (P_0D) and they deduced an expression

$$M_T = \varepsilon(D/\Lambda_0)^{(\gamma-1)/\gamma} \quad (12)$$

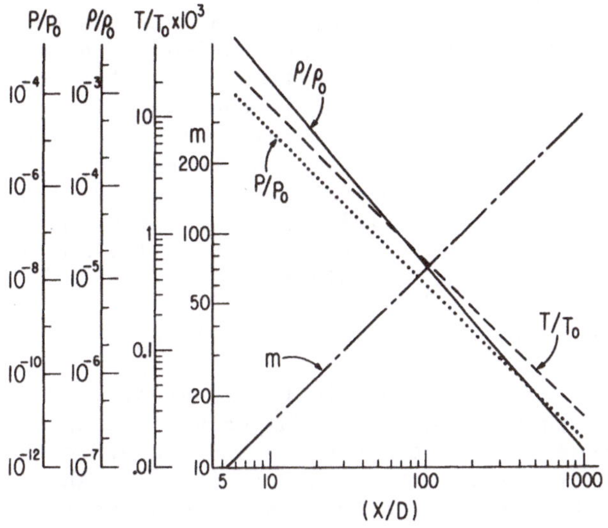

Fig. 1. Pressure, density, temperature, and Mach number for a supersonic expansion of a monatomic gas as a function of the downstream distance.

where Λ_0 is the mean free path in the reservoir and ε is a constant characteristic of the particular gas in question. This expression was found experimentally to be valid for most gases and for argon the experimental value of ε led to

$$M_T = 133(P_0D)^{0.4} \tag{13}$$

where P_0 is the stagnation pressure in atmospheres and D is the nozzle diameter in cm.

The most significant exception to Eq. (13) is the case of helium at high Mach numbers. Miller et al. (1974) have shown that because of quantum effects, the collisional cross section increases rapidly with decreasing energy of the gas, and therefore at low temperatures helium continues to collide at distances far beyond what would have been the low density limit for a classical gas. This allows one to obtain far higher terminal Mach numbers and far lower temperatures in helium expansions than would be predicted by Eq. (12). Mach numbers of up to 250 corresponding to a temperature of 0.015°K (assuming T_0 = 300°K) have been achieved in helium expansions (Campargue, 1975), and Mach numbers of up to 770 corresponding to a temperature of 0.0015°K have been predicted theoretically (Toennies and Winkelmann, 1977).

At this point in the discussion it would be well to mention that these very large Mach numbers do not imply a particularly large flow velocity. As previously mentioned, the enthalpy associated with the directed mass flow is supplied by the random thermal enthalpy that was present in the gas prior to expansion. Since a gas at finite temperature only contains a finite amount of enthalpy, even if in the expansion the temperature is reduced to zero and all of the random enthalpy is converted into direct flow, this still only produces a finite flow velocity and not a particularly large one. As may be seen from Eqs. (2) and (3), this maximum flow velocity, u_M, is given by

$$u_M^2/2 = h_0 = [\gamma/(\gamma-1)]r\ T_0 = [1/(\gamma-1)]a_0^2 \tag{14}$$

where a_0 is the speed of sound in the reservoir. For a monatomic gas this means that $u_M/a_0 = \sqrt{3}$. The very large Mach numbers are achieved by lowering the temperature and therefore the local speed of sound, not by raising the flow velocity. It should also be noted that although the flow velocity does increase as the distance from the nozzle is increased, it very rapidly approaches its asymptotic limit and remains nearly constant for the rest of the expansion. For a monatomic gas the flow velocity reaches 99% of its terminal value at M = 12 which happens at ∿7 nozzle diameters downstream.

An interesting physical picture of the cooling mechanism of a supersonic expansion has been provided by Toennies and Winkelmann (1977). To understand their point of view we must consider a bit more carefully just what is meant by temperature in the expanding gas. For the time being let us confine our attention to a monatomic gas. In this case the gas has no significant internal degrees of freedom and the only temperature is the translational temperature. In a static gas the translational temperature is defined by the width of the Maxwell-Boltzmann velocity distribution. As seen in Fig. 2, in a static gas the distribution of velocities in one direction is peaked at $v_x=0$ and has a width that increases with increasing temperature. In the supersonic expansion two changes take place, the centre of the distribution shifts to a non-zero value because of the non-zero flow velocity, and the distribution narrows because of the cooling. The downstream temperature is defined by the width of the distribution, not by its most probable velocity. This is physically reasonable since from the point of view of a molecule flowing downstream, the temperature is determined by the relative velocity of its collision partners, and the fact that the entire mass of gas is moving at some finite flow velocity is irrelevant.

Along with this concept of temperature we must introduce the idea of two different temperatures, a parallel temperature and a perpendicular temperature. Since the expansion is anisotropic, the different velocity components are not equivalent, and we must

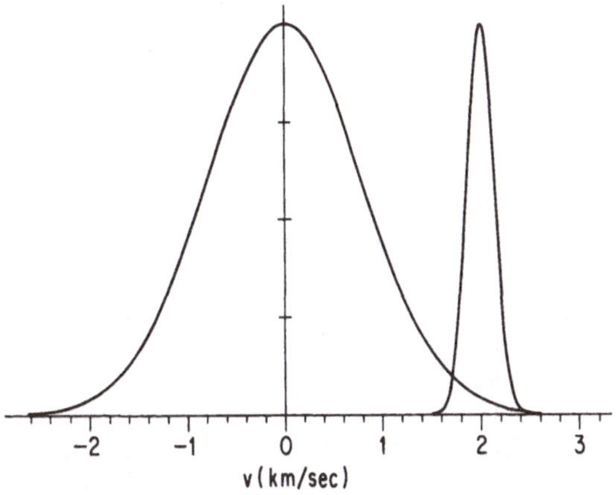

Fig. 2. Distribution of parallel velocities for helium in a static gas and in a Mach 10 supersonic expansion

individually examine their distribution. Hence the need for two temperatures. In regions of high density near the nozzle there will be collisions to equilibrate the different velocity components, and therefore the parallel and perpendicular temperatures will be equal. However, far downstream where there are no collisions, these two temperatures are not the same.

The parallel temperature which defines the velocity distribution parallel to the expansion axis will indeed reach an asymptotic value and then remain constant from that point on downstream. However, the perpendicular temperature will continue to drop due to a phenomenon known as geometric cooling. Even in the absence of collisions molecules with smaller perpendicular velocities will tend to cluster near the centreline of the expansion while molecules of larger perpendicular velocity will move farther off the centreline as the expansion proceeds. This means that the farther downstream one goes, the more nearly uniform the perpendicular velocities will be in a given volume element, and this is a cooling effect which does not require collisions. Of course pure geometric cooling is, by itself, not a particularly interesting phenomenon inasmuch as in the absence of collisions there is no way for the very low perpendicular temperatures to effect any physical process of interest.

In Toennies and Winkelmann's (1977) view of supersonic cooling, geometric cooling is the primary process that lowers the temperature of at least the perpendicular degrees of freedom. Geometric cooling is uninteresting in the absence of collisions; however, it takes place throughout the expansion including the higher density regions where collisions can equilibrate the parallel and perpendicular temperatures. Thus supersonic cooling requires two processes, expansion to provide the geometric cooling and collisions to provide equilibration between the perpendicular velocity and other degrees of freedom.

Thus far in our discussion we have neglected one important practical detail, pumps. One cannot produce cooling if the expanding gas is allowed to scatter from background gas which is at ambient temperature since this scattering only reheats the supersonic jet. Therefore one must pump away the gas that is discharged through the nozzle and this requirement sets the practical limit on the degree of cooling. We have already noted that the terminal Mach number can be increased and the terminal temperature can be decreased by increasing the quantity P_0D. However, the gas discharge through the nozzle and the required pumping speed increase as P_0D^2, and therefore the brute force approach has some natural limit.

For many years the brute force approach was more or less the only common technique. However, fairly recently there have been two new methods which have greatly improved the performance of supersonic molecular beams. The first method is the use of the pulsed valve (Bier and Hagena, 1966; Gentry and Giese, 1978). If the supersonic jet is pulsed, one can use very large values of P_0D which produce very large peak gas loads. However, if the duty cycle is low, the average gas load will be small and can be handled by reasonably-sized pumps. If the expansion chamber is large enough and if the valve is fast enough, it may be possible to perform the experiment of interest between the time that isentropic flow is established and the time when the first gas molecules hit a wall of the apparatus. If this is the case then the pump need only be large enough to evacuate the expansion chamber before the next pulse of the valve. Therefore increased pump size provides increased repetition rate, but if a low duty cycle can be tolerated, very large expansions can be carried out with pumps of moderate size. For many experiments such as those involving pulsed lasers, the experimental duty cycle is limited by other parts of the apparatus. An example of the state of current pulsed valve technology is the molecular beam apparatus at the University of Minnesota (Gentry and Giese, 1978). This has a pulsed valve with 10 μsec·pulse width, 2 atm stagnation pressure, 0.6 mm orifice diameter, 1 Hz repetition rate, and 10^{21} molecules sr^{-1} sec^{-1} beam flux. Liverman et al. (1979) have made a modified version of the Gentry-Giese valve which is slower (200 μsec) but has a higher repetition rate (5 Hz), stagnation pressure (18 atm), and beam flux (10^{23} molecules sr^{-1} sec^{-1}).

An ingenious solution to the pumping speed problem was developed by Campargue (1964, 1970), and his principle has been heavily used in my own laboratory. The traditional approach to eliminating background scattering was to reduce the ambient pressure in the expansion chamber to the point where the mean free path was larger than the dimensions of the apparatus. This meant that the pump that evacuated the bulk of the nozzle discharge worked in the 10^{-3} -10^{-5} torr pressure region. Operation at this low pressure required very large volumetric pumping speeds to remove any substantial mass of gas.

Campargue's idea was to raise the pressure in the expansion chamber and thus the operating pressure of the pump to allow the removal of a given mass of gas with pumps of much smaller volumetric pumping speed. This approach makes use of the fact that if a supersonic expansion takes place in a chamber of finite pressure, that shock waves are formed around the cold isentropic core of the expansion (Rothe, 1966) as shown in Fig. 3. Campargue realized that the shock structure would shield the cold core from the background gas forming what has been referred to as a "zone of silence". The shock waves forming the side walls of the zone of silence are called the barrel shock and the shock wave transverse to the expansion axis is called the Mach disc. The position of the Mach disc, X_M, is given by

$$X_M/D = 0.67(P_0/P_1)^{1/2} \tag{15}$$

where P_1 is the background pressure. As long as one is prepared to either perform the experiment of interest before the molecules hit the shock or to skim the jet, Campargue's technique allows one to produce very high Mach number jets with moderate-size pumps. Operation at P_1 = 0.1–1 torr is possible allowing the use of a fixed P_0D using pumps with two to five orders of magnitude less volumetric pumping speed than those required in conventional low pressure operation.

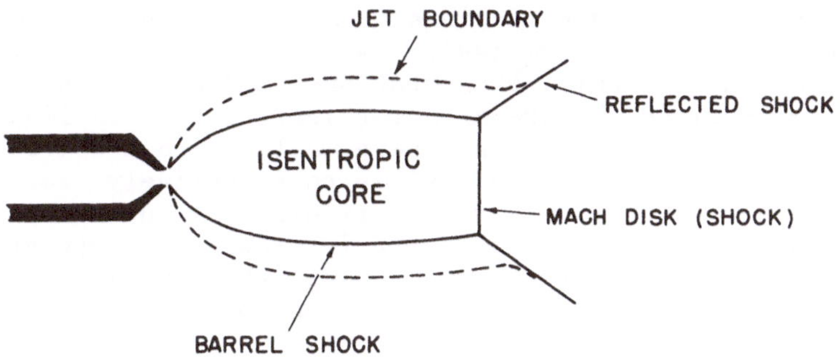

Fig. 3. Shock structure surrounding the expanding gas of a free jet.

Thus far we have only considered free jet expansions where there is no attempt to collimate the downstream flow. To produce a supersonic molecular beam a skimmer must be used in a differentially pumped system. A skimmer is a collimating orifice placed downstream of the nozzle which extracts the portion of the expansion near the centreline and transfers it to a region of low pressure. Most of the gas discharged through the nozzle is rejected by the skimmer and therefore the highest capacity pump in the system should be that which evacuates the region between the nozzle and skimmer. Since only a small fraction of the total discharge passes through the skimmer, the post skimmer pump may be of modest capacity and operating at low pressure. It is not uncommon to have additional beam defining aperatures and additional stages of differential pumping downstream of the skimmer.

Although the skimmer is conceptually simple, there are practical problems involved with the actual design and fabrication. Unless a proper design is used the skimmer will interfere with the post nozzle flow and destroy the isentropic expansion. A particular problem arises in trying to skim high density Campargue type expansions where the skimmer opening must be placed through the Mach disc. Any imperfections in the skimmer will interfere with the shock structure, detach a shock wave, and reheat the beam. Campargue's success in producing very high Mach number molecular beams was due both to a good idea and also to his ability to fabricate very high quality skimmers.

The design and fabrication of skimmers is partly science and partly black art, and the skimmer literature while extensive is obscure. One of the best sources is Campargue's thesis (1970) but this is not readily available. Gentry and Giese (1975) have developed a technique for fabricating high quality skimmers which are capable of high Mach number operation, and skimmers made by this technique are commercially available.

In the discussion thus far the emphasis has been on translational cooling. For applications to molecular spectroscopy, my own particular interest, the desire is to cool the internal degrees of freedom of the molecule, the rotations, and vibrations. If only cooling were required the use of a refrigerant such as liquid helium would be satisfactory and there would be no need for a supersonic expansion. However, the goal is to selectively cool the internal degrees of freedom without bringing the entire system into thermodynamic equilibrium at the internal temperatures. At the temperatures we have been considering, the equilibrium state for most materials is a solid, and we wish to study internally cooled gas phase molecules.

The utility of a supersonic expansion is that it is a system very far out of equilibrium and one can therefore accomplish selective cooling. If a small amount of the molecule of interest is seeded into a monatomic carrier gas, in the post nozzle region, the molecule finds itself in a translationally cold bath. Given enough time all degrees of freedom would equilibrate with the translational bath, but in fact the molecules are only in contact with the bath for a limited period of time. As already mentioned, at some point downstream of the nozzle the density becomes too low to provide binary collisions, and relaxation between the seed molecule and the bath can only take place through collisions. Therefore the state of the molecules in the collision free region is determined more by kinetics than by thermodynamics; those degrees of freedom that equilibrate rapidly with translations are cooled and those that do not are not cooled.

Both translation-translation and translation-rotation equilibration are fast, and therefore there is extensive cooling of both the translational and rotational degrees of freedom of the seed molecule. Translation-vibration equilibration is a slower process and the actual rates depend very much on which molecule and which vibrational mode of a particular molecule is being considered. In general we observe vibrational cooling, particularly of those low frequency modes whose excited states are initially populated, but the vibrational cooling is not nearly as complete as rotational cooling.

Phase equilibration is an extremely slow process requiring multi-body collision to initiate it, and for this reason one can produce internally cold molecules which enter the collision free region before they have had time to condense. The point at which condensation becomes a problem depends on the gas mixture and on the expansion conditions. Using gases with weak intermolecular forces tends to suppress condensation, and this is the reason that it is common to use a rare gas carrier, particularly helium, to achieve very high Mach numbers and low temperatures. Of course, sometimes a small amount of condensation is desirable, and the study of van der Waals molecules which is the subject of my remaining lectures is in fact the study of the first stages of condensation in a mixed gas expansion.

Let me conclude this section with a brief description of our own apparatus to provide some feel for the operating parameters of at least one working apparatus. None of these parameters are the limit of what can be done, and they have all been exceeded in other laboratories. We typically use a 0.025 mm diameter circular pinhole for a nozzle and expand gas mixtures of up to a few percent molecule in a helium carrier gas at a total pressure of 100 atm. Our pumping plant consists of three 4" ring jet booster pumps operating in parallel with a compined pumping speed of 200-400 ℓ/sec at a background pressure of 0.1 torr. The total gas discharge through the nozzle is $\sim$100 STP liters/hr, most of this being helium.

Under these conditions the flow velocity is $\sim 2\times 10^5$ cm/sec and the Mach disc is 1.5 cm downstream of the nozzle which corresponds to $\sim$600 nozzle diameters. At a point just upstream of the Mach disc the calculated Mach number is $\sim$225 assuming no Mach number freezing, and the calculated terminal Mach number for a classical gas is 76. We commonly observe the spectra of molecules $\sim$5 mm from the nozzle where the calculated Mach number is 110 corresponding to a calculated translational temperature of 0.07°K. We have no way of experimentally measuring the translational temperature of the jet, but from spectroscopic intensities we have measured rotational temperatures of $\sim$0.05°K and vibrational temperatures in the range 20-50°K.

2. THE PHOTOCHEMISTRY OF VAN DER WAALS MOLECULES

It is clear that photochemistry is a proper concern of this Institute in that this certainly is an area where experimentalists have recently been able to produce information far more detailed than that which had formerly been available. Technical advances in lasers, molecular beams, picosecond sources and detectors, etc., are beginning to allow us to probe the details of a photochemical reaction in ways that demand qualitatively new levels of theoretical understanding. Some state-to-state dynamical information is already available, and it is clear that much more is in the offing. We are rapidly developing the ability to measure very fast rates of elementary processes, and it is therefore profitable to consider theoretically what physical phenomena are responsible for making these rates what they are. With precise experimental data to guide the development of our theoretical understanding, it is likely that we are on the verge of understanding the details of how energy flows in a photochemical reaction. It is reasonable to suppose that with this understanding will come the ability to predict the outcome of a photochemical reaction in a given system, or even the ability to control and alter the reaction toward some desired end product.

In this lecture I would like to consider a particular class of photochemical reactions, those of van der Waals molecules. Naturally we would really like to understand the details of the breaking of a true chemical bond under the influence of light, but it may be that, at present, van der Waals photochemistry is a more suitable meeting ground for theory and experiment. Until recently van der Waals molecules have been more difficult to study experimentally than chemically bound molecules because their instability in ordinary laboratory environments has made them difficult to prepare. However, the advances in supersonic molecular beams which were described in part 1 have made the study of these species a good bit easier than was the case several years ago. If it is possible to obtain detailed experimental information about these weakly bound species, the theoretical understanding of their photochemistry is likely to be easier than that of chemically bound molecules, and this would seem to be a fruitful area for interplay between theory and experiment.

Because chemical bonds are relatively strong, the energy required to break them is a major perturbation of the system. Photochemical reactions of chemically bound systems of necessity take place on a portion of the potential surface far above the zero point level, and this is a region where little is known of the surface. In this region vibrational modes are very anharmonic and very tightly coupled, and the density of states is very high. The repulsive electronic states responsible for many photochemical reactions are little known, and the couplings between electronic states that dominate the course of the photochemistry can only be guessed at.

In contrast, the study of the photochemistry of van der Waals molecules bypasses many of these problems. The dissociation energy of van der Waals molecules is small, and therefore the reaction takes place on a part of the potential surface about which, at least in principle, much may be known. Many of the vibrational modes of the molecule, the vibrations of the chemically bound part, remain reasonably harmonic and reasonably isolated throughout the reaction. The reaction can take place on a single electronic surface and therefore problems of the coupling between electronic states need not arise. For these reasons we may be able to progress more rapidly in the study of van der Waals photochemistry than in the study of ordinary photochemistry, while the lessons to be learned may provide some insight into the more difficult problem.

In this lecture I would like to describe a number of experimental observations that have recently been made on photodissociating van der Waals molecules. The specific set of van der Waals molecules that we will consider (at least in these written notes) will all be complexes of one or more rare gas atoms van der Waals bound to a molecular iodine core. The process that we will consider is alternatively called vibrational predissociation, type II predissociation, or unimolecular decomposition (Herzberg, 1966).

The process is illustrated in its most simple form in Fig.4 for the van der Waals molecule I_2-Ne. The molecule is formed in its ground electronic and vibrational state and in a narrow distribution of rotational levels near J=0 in a supersonic free jet by the process

$$I_2 + Ne + (\text{third body?}) \rightarrow I_2Ne. \qquad (16)$$

This ground vibronic state complex is then laser excited to the $B^3\Pi_{0_u^+}$ electronic state of the complex and to a vibrational state

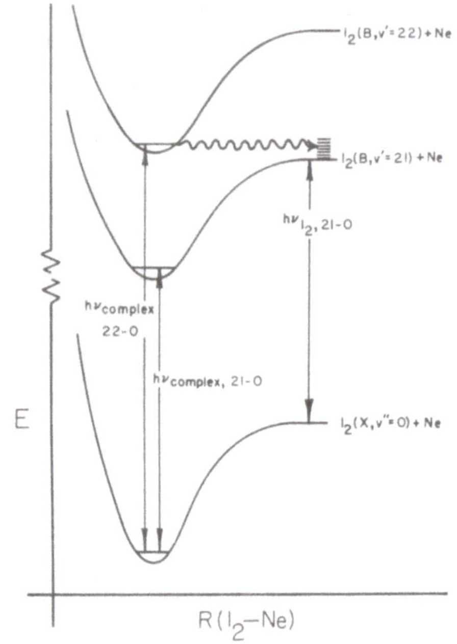

Fig.4. Potential energy of the van der Waals molecule I_2Ne as a function of the I_2-Ne bond length. Vibrational predissociation from the bound v'=22 state to the continuum v'=21 state is indicated by wavy line. The fluorescence excitation transition of the complex and of uncomplexed I_2 are indicated by solid arrows.

containing v' vibrational quanta in the iodine stretching mode (in the figure v'=22) and perhaps one or more quanta in the van der Waals stretching mode by the process

$$I_2Ne(v''=0) + h\nu \rightarrow I_2Ne^* (v'). \quad (17)$$

The combination of laser and molecular beam technology allows very precise excitation of the photochemically reactive state. In the excitation step the optical selection rules guarantee that, even with relatively broad band excitation, the rotational distribution in the excited state will be similar to that of the ground state. With narrow band excitation it is possible in some cases to excite a single known rovibronic state of the complex.

Because of the weakness of the van der Waals bond, the energy originally deposited in the chemically bound mode, the I_2 stretch, is more than sufficient to break the van der Waals bond, and therefore after some time the energy migrates from the original storage mode to the van der Waals stretch and the molecule dissociates

$$I_2Ne*(v') \rightarrow I_2*(v'-z) + Ne. \qquad (18)$$

In this process the iodine stretching mode loses z vibrational quanta (in the figure z=1) and the energy in these quanta is distributed between the bond dissociation energy, rotational energy of the product I_2*, and relative kinetic energy of the recoiling fragments.

Finally, after roughly a 1 μsec radiative lifetime, the electronically excited product I_2* emits a photon

$$I_2*(v'-z) \rightarrow I_2(v'') + h\nu. \qquad (19)$$

If there is little or no relaxation between the time of dissociation and the time of emission of the product, the emission spectrum contains information on the product state distribution of the dissociation reaction.

There are several questions which we are now in a position to attempt to answer both experimentally and theoretically. What is the time for energy redistribution from the storage mode to the dissociating mode? What is the final distribution of the energy initially deposited in the storage mode? What other processes, if any, compete with photodissociation? How do the redistribution times, the final energy distribution, and the competing processes depend on the composition of the complex and on the choice of initially excited state? We will now take up these questions individually as space and time permit.

A. Lifetimes

Our knowledge of the photodissociation lifetime of van der Waals molecules comes from measurements in the frequency domain, i.e. measurements of the spectral line broadening produced by the finite lifetime of the initially excited state. As is frequently the case in this type of measurement, some care must be taken to assure that the linewidth is produced only by the lifetime broadening or, at least, that other sources of broadening are understood and that appropriate corrections have been made.

In the case of the complex I_2He, the spectroscopy is rather well understood (Smalley et al., 1978), and it was possible to correct for heterogeneous broadening. The effect of lifetime broadening on the fluorescence excitation spectrum of the complex is illustrated in Fig.5, and the measured values of the lifetimes in the range v'=12-25 are shown in Fig.6 (Johnson et al., 1978). The lifetimes are seen to lie in the range of tens to hundreds of picoseconds and the predissociation rate (reciprocal lifetime)

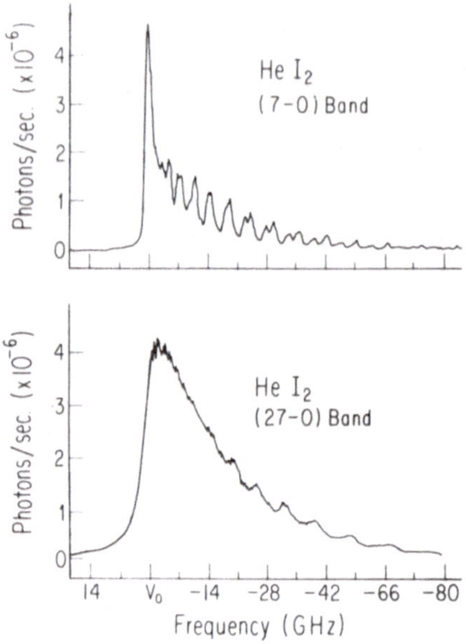

Fig. 5. The fluorescence spectrum of I_2He excited in the $B(v'=7) \leftarrow X(v''=0)$ and $B(v'=27) \leftarrow X(v''=0)$ bands.

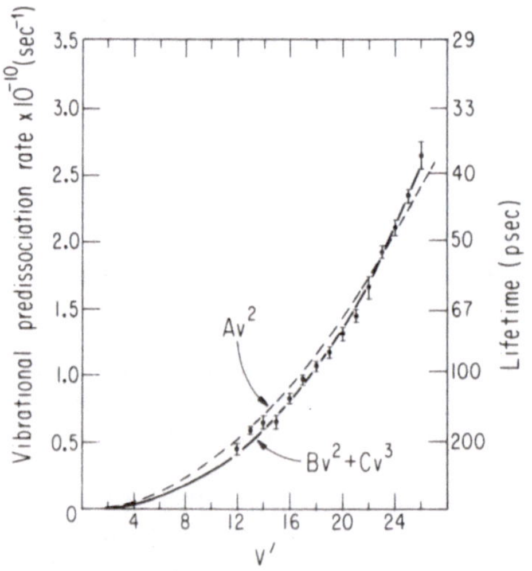

Fig. 6. Vibrational predissociation rate (and lifetime) of I_2He as a function of the vibrational state of the I_2 stretching mode that was excited. Points are experimental measurements, and dashed and solid curves are the best least squares fits of functional form $A\ v^2$ and $B\ v^2 + C\ v^3$.

increases nearly quadratically with increasing vibrational quantum number. Since the dissociation rate is so much faster than the radiative rate, we may be sure that all of the observed emission comes from the product I_2^* and not from the complex itself.

Beswick and Jortner (1977; Beswick et al., 1979) have developed a theory of vibrational predissociation in van der Waals molecules and have applied it to the case of I_2He. The theory involves a close coupling calculation where the interaction potential is assumed to be the sum of Morse pair potentials between the helium atom and the two iodine atoms plus the known I_2 B state potential. They have treated the case of both linear and T-shaped geometries, and have assumed both a harmonic and an anharmonic potential for the I_2 stretch. They have developed an energy gap law which predicts that the rate increases as the energy gap between the storage mode frequency and the dissociating mode frequency decreases. For the case of a harmonic storage mode vibration they find that the rate increases linearly with v', but the effect of anharmonicity is to reduce the energy gap and make the rate increase superlinearly with v'.

The numerical results of the theory are very sensitive to the assumed values of the van der Waals potential parameters. Nonetheless, with reasonable choice of parameters they get qualitative agreement with the experimental measurements, and in particular they calculate a near quadratic v' dependence for the case of the anharmonic oscillator. Certainly more work can be done to improve the quantitative fit, one particular area of interest being the inclusion of the other van der Waals mode, the I_2-He bend, in the calculation. Nonetheless, it appears that even in its present state the theory provides at least a qualitative picture of dissociation in triatomic van der Waals molecules, and it allows us to think about the interpretation of new experiments on this class of molecules.

B. Product State Distributions

Since a van der Waals bond is weak, the energy originally stored in the I_2 stretching mode of an iodine-rare gas van der Waals complex may be far more than enough to break all van der Waals bonds, and it is of some interest to determine how the excess energy is distributed in the products after the reaction is completed. Since the spectroscopy of I_2 itself has been so heavily studied, it is possible to deduce the product state distributions of the fragment I_2^* by observing its emission spectrum. Not only are the energy levels of I_2 very well known (Wei and Tellinghuisen, 1974) so that one can identify an observed emission line with a particular product state, but the Franck-Condon factors are also known (Tellinghuisen, 1978) so that one can derive a cross section for dissociation into an observed product state channel from the intensity of the spectral line.

One final requirement for the experiment is that there be no collisional relaxation of the excited state population between the time it is formed in the photodissociation reaction and the time it is observed roughly 1 μsec later. If collisional relaxation cross sections were the same in the supersonic free jet as they are in the static gas, the low density of the jet would assure collision-free conditions. In fact, as will be discussed in other lectures, the vibrational relaxation cross section is observed to increase as the relative kinetic energy of the collision partners decreases, and at the very low temperature of the supersonic jet collisions are extremely effective. We do observe some collisional relaxation, but fortunately it is a minor effect and the raw data can be corrected to account for it.

An example of emission spectra from product fragments is shown in Fig.7 (Sharfin et al., 1979). In these two spectra the van der Waals molecules I_2He and I_2He_2 were laser excited to the B(v'=22) vibronic state. Although the raw intensities shown in the figure must be corrected by both Franck-Condon factors and collisional relaxation to obtain quantitative branching ratios, the qualitative picture may be seen in the spectrum itself. In the case of I_2He, the strongest emission lines are from the B(v'=21) state, and it is clear that the one quantum dissociation channel dominates the photodissociation process. This means that most of the excess energy stays in the storage mode which was originally excited, and most of the time only one quantum is used to break the van der

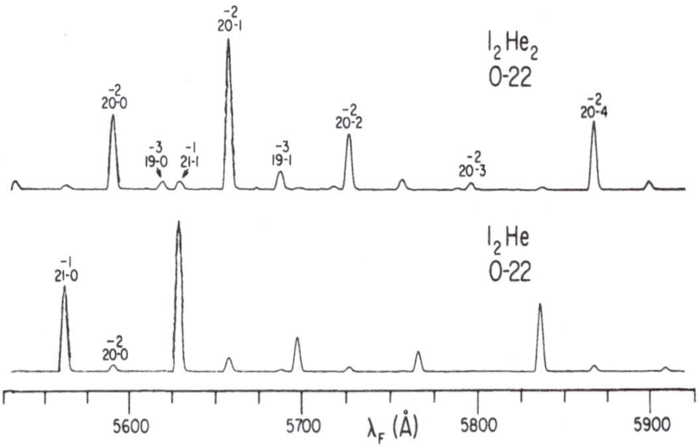

Fig. 7. Dispersed fluorescence spectra of the product I_2^* formed upon dissociation of the van der Waals complexes I_2He and I_2He_2. In both cases the complex was originally excited to the B(v'=22) state.

Waals bond. A two quantum channel is observed, but its cross section is only a few percent that of the major channel.

In the case of I_2He_2, the strongest emission is from B(v'=20), and the two quantum channel has the largest cross section. This is an interesting observation since we know (Blazy et al., 1979) that one iodine stretching quantum has enough energy to break two and even three I_2-He van der Waals bonds. This means that the one quantum process is energetically allowed but does not seem to be experimentally observed. It appears that in the dissociation of I_2He_2, the breaking of the two van der Waals bonds takes place sequentially, and that the two dissociation steps cannot share one I_2 stretching quantum. Moreover, to the extent that we can measure it, the weak three quantum process has twice the relative cross section that the two quantum process had in I_2He. This implies that the two steps in the dissociation are independent, i.e., that the branching ratios for the two steps

$$I_2He_2 \rightarrow I_2He + He \tag{20}$$

$$I_2He \rightarrow I_2 + He \tag{21}$$

are the same.

We have also observed the emission spectrum from the product I_2^* formed when I_2He_3 is dissociated. The spectrum is very weak and we can only obtain qualitative information. Only the three quantum channel is observed which suggests that the dissociation in I_2He_3 may be like that in I_2He_2, sequential and independent.

When a mixture of I_2, He, and Ne is expanded through a supersonic nozzle, a large number of different van der Waals molecules of the type $I_2He_aNe_b$ are formed (Kenny et al., 1979). In Fig. 8 a portion of the fluorescence excitation spectrum (similar to the absorption spectrum) is shown, and bands from the various van der Waals molecules are indicated. Not only is the chemistry of the I_2, Ne, He system more complicated than that of the I_2, He system, but the dynamics of photodissociation is also much more complex. By fixing the exciting laser on the absorption band of a particular series, it is possible to dissociate it and study the product state distribution of the fragment I_2^*.

In Fig. 9 is shown the emission spectra of the product I_2^* produced when the van der Waals complexes I_2Ne, I_2Ne_2, and I_2Ne_3 were excited to the B(v'=22) state. The spectrum produced in the case of I_2Ne is qualitatively similar to that produced by I_2He, a strong one quantum channel and a very much weaker two quantum channel. However, the spectrum produced by the dissociation of I_2Ne_2 begins to show some new effects. As in the case of I_2He_2, the first observed channel is the two quantum channel, the one

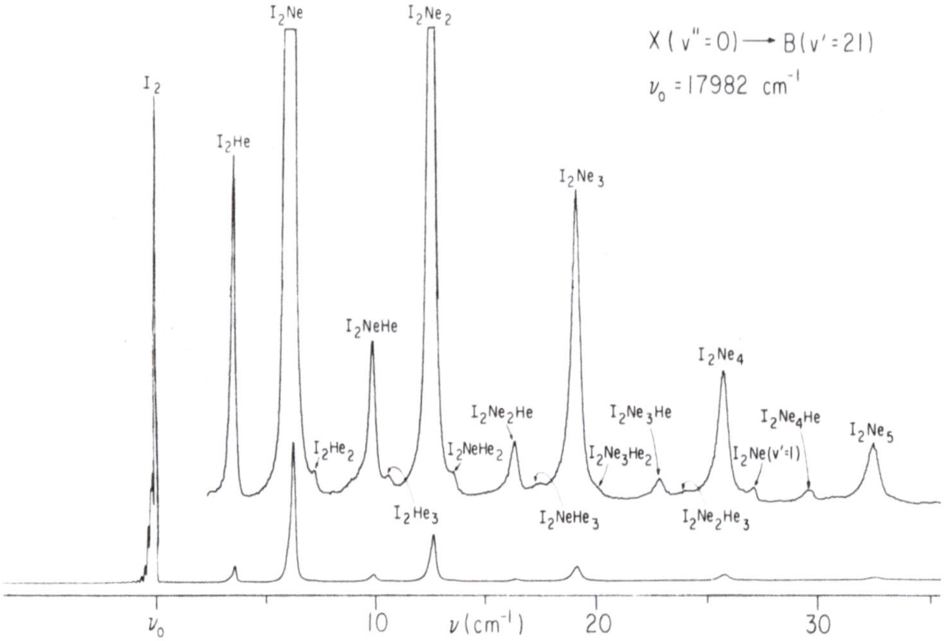

Fig. 8. The fluorescence excitation spectrum of a supersonic expansion of iodine in a mixture of neon and helium. Assignment of the various van der Waals species formed in the expansion is shown.

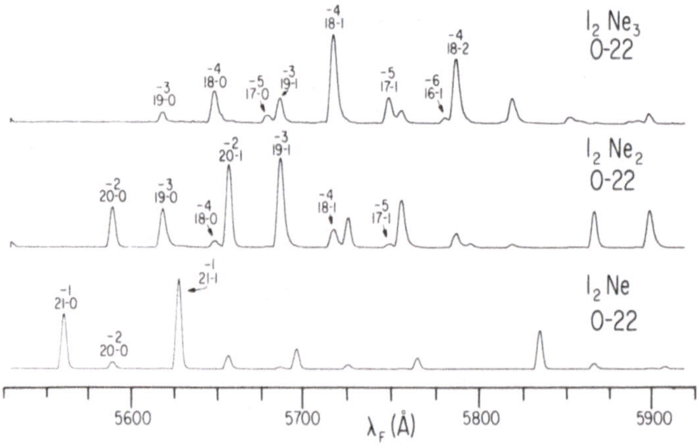

Fig. 9. Dispersed fluorescence spectra of the product I_2^* formed upon dissociation of the van der Waals complexes I_2Ne, I_2Ne_2, and I_2Ne_3. In all cases the complex was originally excited to the $B(v'=22)$ state.

quantum channel in this case being energetically closed. However, now the next open channel, the three quantum channel, is essentially as big as the two quantum channel, and the higher four and five quantum channels are clearly observed. It is still possible that the complex dissociates in two sequential steps, but now the two steps are by no means independent. That is to say, the overall product state distribution cannot be made up by convoluting the product state distribution of I_2Ne with itself.

The spectrum produced by the dissociation of I_2Ne_3 continues the trend observed in I_2Ne_2. The three quantum channel is the first observed, but the four quantum channel is now much stronger and even the five quantum channel is somewhat stronger. This general behaviour is also observed in larger van der Waals complexes as shown in Table 1. Here we list the branching ratio between the first two observed channels for a number of van der Waals complexes. The second column of the table identifies the first observed channel, and the third column gives the branching ratios.

It can be seen from this table that as the complex gets bigger, the dissociation process favours removal of a larger number of quanta per particle from the storage mode. In I_2Ne_4, the four quantum channel is very much weaker than the major five quantum channel, whereas in I_2Ne_5, the five quantum channel is not even observed. In I_2Ne_6 the six quantum channel is not observed, and the process with the largest cross section is the eight quantum photodissociation.

The qualitative lesson to be learned from these data is that as the complex gets larger, the dissociation process gets less efficient. In the small complex I_2Ne, the dominant process was the transfer of one quantum, the minimum necessary, from the storage mode to the dissociating mode. In the larger complexes the transfer was not nearly so direct with the transfer of additional quanta from the storage mode being necessary.

A qualitative explanation of these observations would be that as the complex gets bigger we are going from a small molecule limit where the dynamics are governed by the strength of the intramolecular couplings to a statistical limit where the dynamics are governed by the density of states. In the small molecule limit one may invoke the energy gap law to predict the dynamical behaviour. In I_2Ne we have three vibrational modes, the chemically bound storage mode and the two van der Waals modes, the van der Waals stretch which is the dissociating mode and the van der Waals bend. Certainly the highest frequency mode must be the chemically bound I_2 stretch, and the highest frequency van der Waals mode is likely to be the van der Waals stretch. The energy gap law would then predict that the strongest coupling would involve the modes where the energy gap was least, and that energy would be preferentially

Table 1. State distribution of product I_2^* (v') produced by photodissociation of rare gas iodine van der Waals complexes. The first observed channel $[XI_2(v_i) \rightarrow X + I_2(v_i-z)]$ involves the loss of z quanta from the iodine stretching vibration. The quantity k_{z+1}/k_z is the branching ratio between the second and first observed channels

Species	z	k_{z+1}/k_z
I_2He	1	0.04
I_2Ne	1	0.07
I_2Ar	3	0.21
I_2He_2	2	0.05
I_2HeNe	2	
I_2Ne_2	2	0.75
I_2HeAr	3	1.69
I_2Ar_2	6	1.73
I_2Ne_2He	3	0.54
I_2Ne_3	3	3.73
I_2Ne_3He	4	1.16
I_2Ne_4	4	4.8
I_2Ne_5	6	0.74
I_2Ne_6	7	1.16

channelled from the storage mode to the dissociating mode leading to an efficient photodissociation.

However, as the complex becomes larger, the number of van der Waals modes increases, and the density of vibrational states begins to grow. The strongest coupling will still be between the chemically bound storage mode and the van der Waals stretch, but now there are a large number of bending and torsional states, and excitation of these does not lead to dissociation. As the density of non-dissociating states gets larger, they will begin to accept energy from the storage mode, and the dissociation process becomes less efficient. In the large molecule limit all modes must be heated statistically, and there will no longer be preferential transfer of energy to the dissociating modes. In going from I_2Ne to I_2Ne_6, we seem to be seeing a change from the small molecule limit to the statistical limit.

Product state distributions have also been measured for the photodissociation of complexes of argon, helium, and iodine. When a mixture of argon, helium, and iodine is expanded, the fluorescence excitation spectra of the complexes I_2Ar, I_2Ar_2, and I_2Ar_3 may be observed. In addition to these argon complexes, various mixed complexes of argon and helium are also observed. Although we do not observe any van der Waals molecules as large as the complex I_2Ne_6 that was observed in the neon expansion, it seems likely that the problem may be observation and not formation.

The binding energy of argon to iodine is larger than that of neon to iodine, and there seems to be no reason why the larger complexes do not form. A more reasonable interpretation is that the larger complexes do form but that neither they nor their dissociation products fluoresce due to competition from another decay mechanism, induced electronic predissociation of the I_2 core. It is known that in iodine itself electronic predissociation to atoms can be induced by collisions, and that the cross section for this process is larger with an argon collision partner than with either helium or neon. In the case of the smallest argon complex, I_2Ar, the intensity of the fluorescence excitation spectrum was found to vary greatly depending on the vibrational state of the I_2 stretch that was initially excited (Kubiak et al., 1978). The interpretation of this effect was that the radiationless electronic predissociation

$$I_2Ar^* \rightarrow 2I + Ar \tag{22}$$

competes with the vibrational predissociation process

$$I_2Ar^*(v') \rightarrow I_2^*(v'-z) + Ar \tag{23}$$

and that the vibrational state dependence of the intensity was due to the vibrational state dependence of the cross sections for these two processes. Extending this interpretation to the larger argon-iodine van der Waals complexes, we assume that the effect of adding more argon atoms is to increase the electronic predissociation cross section to the point where fluorescence is no longer observed.

The product state distribution of the smaller complexes that we do observe is illustrated in Fig. 10 and seems to follow the same qualitative trend that was observed in neon, i.e., that the photo dissociation process takes more quanta per particle as the complex

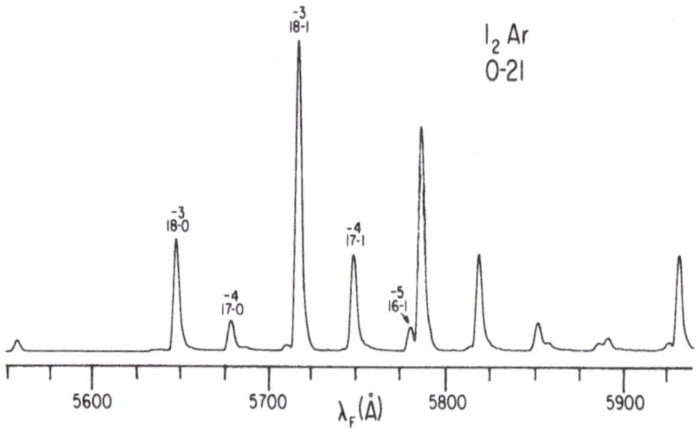

Fig. 10. Dispersed fluorescence spectrum of the product I_2^* produced upon dissociation of the van der Waals complex I_2Ar. The complex was originally excited to the B(v'=21) state, and the feature near 5560A is scattered light from the exciting laser.

gets bigger. In the case of the mixed complex I_2ArHe, we observe a new type of behaviour, the two particles are able to share an I_2 stretching quantum in the dissociation process. When excited to the B(v'=22) state, the complex I_2Ar requires three vibrational quanta to dissociate while I_2He requires one. If the two steps in the dissociation of I_2ArHe were unable to share a vibrational quantum, the first observed channel should be the four quantum channel. As seen in Table 1, although the four quantum channel is the largest, the three quantum channel is observed. This means that the energy originally contained in one quantum of the I_2 storage mode is shared in breaking the two van der Waals bonds.

The energy gap model may be used to understand why this quantum sharing is observed in argon complexes but not in neon complexes. In the case of neon complexes, one quantum of the iodine stretching vibration in the region of v'=22 contains more than the dissociation energy of the I_2-Ne bond. Therefore, if any energy flows from the storage mode to the I_2-Ne van der Waals bond, the bond breaks. This is not the case in the argon complexes. Since the first observed dissociation channel is the three quantum channel, it takes more than two iodine stretching quanta to break the I_2-Ar van der Waals bond. Therefore, energy can flow from the storage mode to the I_2-Ar stretch and produce a bound state.

The I_2Ar stretching vibration has a higher frequency than the I_2He stretch, and therefore the energy gap law would predict that energy would flow preferentially from the storage mode to the argon stretch. However, since this produces a bound state, energy could then flow from the I_2-Ar stretch to the I_2-He stretch. In other words, the I_2-Ar stretch could serve as a secondary storage mode for I_2-He dissociation, and what was originally one I_2 stretching quantum could be shared between the two van der Waals bonds.

The final topic which should be discussed is the other degree of freedom that is available to accept some of the energy that is initially stored in the iodine stretching mode, namely the molecular rotation of the fragment I_2^* that is formed in the photodissociation reaction. In principle one can measure the product rotational state distribution in the same way that one measures the product vibrational state distribution. It should be realised that there is the practical experimental problem that one must have much higher spectral resolution to analyse the emitted light since the rotational structure is more compressed than the vibrational structure. The requirement of increased spectral resolution reduces the intensity of the observed emission, and the experiment is rather more difficult than those where only vibrational resolution is required. Nonetheless some experiments to measure the rotational state distribution have been performed and we can expect more in the future. Resolution of individual rotational lines has not been achieved, but band contours have been measured. It should be remembered that the rotational constant of iodine is very small, $\sim 0.03\ cm^{-1}$, and this series of iodine van der Waals molecules is a particularly difficult case.

In Fig.11 we may see, at least qualitatively, the rotational state distribution in the product I_2 that is produced when I_2Ar is photodissociated. In the lower figure the complex had been excited to the B(v'=29) state and, as was previously the case, the first strong observed channel is the three quantum channel. In the upper trace, the complex had been excited to the B(v'=30) state, and in this case the four quantum channel is the first to be observed. Because of the anharmonicity in the iodine stretching vibration, the amount of energy contained in a vibrational quantum decreases as the quantum number increases. Therefore, at some point three stretching quanta will have insufficient energy to break the van der Waals bond and a fourth quantum becomes necessary. The fact that this occurs between v'=29 and v'=30 allows us to bracket the argon-iodine van der Waals binding energy as $220\ cm^{-1} \leq D_0 \leq 226\ cm^{-1}$.

One other consequence of this switch is that in the case of v'=29 excitation, the three quantum dissociation channel requires essentially all of the available energy to break the van der Waals bond, and little is left over to be used in rotational excitation. This may be observed qualitatively in Fig.11, where the four quantum band is clearly broader than the three quantum band due to the broader rotational distribution produced by the extra quantum of excess energy.

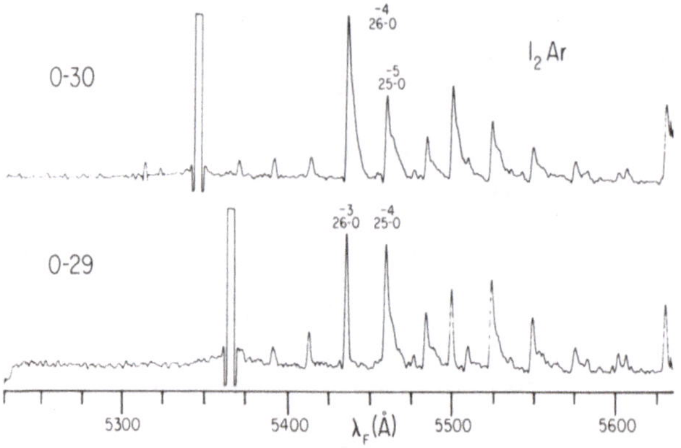

Fig.11. Dispersed fluorescence spectra of the product I_2^* produced upon dissociation of the van der Waals complex I_2Ar. In the upper spectrum the complex was originally excited to the B(v'=30) state, and in the lower spectrum the complex was excited to the B(v'=29) state. The offscale features to the short wavelength side of each spectrum are from scattered laser light.

One final comment should be made regarding the role of excess energy in the rotational excitation of the fragment. The more important limitation on rotational excitation is angular momentum conservation, not energy conservation. As mentioned previously, the original rotational state distribution in the cold jet is very narrow, and the restrictive optical selection rules prevent it from being significantly broadened in the process of optical excitation. If additional angular momentum is to be deposited in the product iodine to produce a broadened rotational state distribution, conservation of total angular momentum requires that an equal amount of angular momentum must be carried off as orbital angular momentum of the recoiling argon atom. Classically this orbital angular momentum is given by μvb where μ is the reduced mass of the dissociating pair, v is the relative velocity of recoil, and b is the impact parameter. The reduced mass is, of course, fixed, and the range of available impact parameters is limited by the geometry of the bound complex. Therefore, the amount of angular momentum available for I_2 rotation can only be increased by increasing the velocity and thereby the kinetic energy. Because the rotational constant of I_2 is so small, relatively little energy is required directly to produce several h of rotational excitation, and the amount of kinetic energy required to conserve angular momentum is likely to be very much greater.

This material is based upon work supported by the National Science Foundation under Grant CH78-25555, and by the Louis Block Fund of The University of Chicago.

REFERENCES

Anderson, J.B., and Fenn, J.B., 1965, Velocity distributions in molecular beams from nozzle sources, Phys. Fluids, 8: 780.

Ashkenhas, H. and Sherman, F.S., 1966, The structure and utilisation of supersonic free jets in low density wind tunnels, in "Rarefied Gas Dynamics", Fourth Symposium, Vol.II, J.H. de Leeuw, ed., Academic Press, New York.

Beswick, J.A., Delgado-Barrio, G. and Jortner, J., 1979, Vibrational predissociation lifetime of the van der Waals molecule HeI_2, J.Chem.Phys., 70: 3895.

Beswick, J.A. and Jortner, J., 1977, Model for vibrational predissociation of van der Waals molecules, Chem.Phys.Lett., 49: 13.

Beswick, J.A. and Jortner, J., Vibrational predissociation of triatomic van der Waals molecules, J.Chem.Phys., 68: 2277, 2525.

Bier, K. and Hagena, O., 1966, in "Rarefied Gas Dynamics", Fourth Symposium, Vol.II: 260, Academic Press, New York.

Blazy, J.A., DeKoven, B.M., Russell, T., and Levy, D.H., The Binding Energy to Iodine-Rare Gas van der Waals molecules, J. Chem. Phys. 72, in press.

Campargue, R., 1964, High intensity supersonic molecular beam apparatus, Rev. Sci.Instr., 35: 111.

Campargue, R., 1970, Aerodynamic separation effect on gas and isotope mixtures induced by invasion of the free jet shock wave structure, J.Chem.Phys., 52: 1795.

Campargue, R., 1970, Etude, par simple et double extraction de jets supersoniques purs ou dopes, des effets intervenant dans la formation d'un faisceau moleculaire de haute intensité et d'energie comprise entra 0 et 25 eV, Ph.D. Thesis, University of Paris.

Campargue, R., Lebehot, A., Lemonnier, J.C., Marette, D. and Pebay, J., 1975, Generateur de jet moleculaire supersonique functionant en monochromateur dans le domaine de 10^{-2} à 40 eV, in "Abstracts of Vth Symposium International sur Jets Moleculaires, Nice.

Gentry, W.R. and Giese, C.F., 1975, High precision skimmers for supersonic molecular beams, Rev.Sci.Instr., 46: 104.

Gentry, W.R. and Giese, C.F., 1978, Ten-microsecond pulsed molecular beam source and a fast ionization detector, Rev.Sci.Instr., 49: 595.

Herzberg, G., 1966, "Molecular Spectra and Molecular Structure", Vol.3: 469, Van Nostrand, New York.

Johnson, K.E., Wharton, L. and Levy, D.H., 1978. The photodissociation lifetime of the van der Waals molecule I_2He, J.Chem.Phys., 69: 2719.

Kenny, J.E., Johnson, K.E., Sharfin, W.F. and Levy D.H., The photodissociation of van der Waals molecules: complexes of iodine, neon and helium, J. Chem. Phys. 72, in press, January 15, 1980.

Kubiak, G., Fitch, P.S.H., Wharton, L. and Levy, D.H., 1978, The fluorescence excitation spectrum of the ArI_2 van der Waals complex, J.Chem.Phys., 68: 4477.

Liepmann, H.W. and Roshko, A., 1957, "Elements of Gas Dynamics", Wiley, New York.

Liverman, M.G., Beck, S.M., Monts, D.L. and Smalley, R.E., 1979, Fluorescence excitation spectrum of the 1Au(nΠ*) ← 1Ag(0-0) band of oxalyl fluoride in a pulsed supersonic free jet, J.Chem.Phys., 70: 192.

Miller, D.R., Toennies, J.P. and Winkelmann, K., 1974, Quantum effects in highly expanded helium nozzle beams, in "XIth Symposium of Rarefied Gas Dynamics", M. Becker and M. Fiebig, eds., DFVLR Press, Porz-Wahn.

Ramsey, N.F., 1963, "Molecular Beams", Clarendon Press, Oxford.

Rothe, D.E., 1966, Electron beam studies of the diffusive separation of helium-argon mixtures, Phys. Fluids, 9: 1643.

Sharfin, W., Johnson, K.E., Wharton, L., and Levy, D.H., 1979, Energy distribution in the photodissociation products of van der Waals molecules: iodine-helium complexes, J.Chem.Phys., 71, 1292 (1979).

Smalley, R.E., Wharton, L. and Levy, D.H., 1978, The structure of the He I_2 van der Waals molecule, J.Chem.Phys., 68: 671.

Tellinghuisen, J., 1978, Intensity factors for the I_2 B↔X band system, J.Quant.Spec.Rad.Trans., 19: 149.

Toennies, J.P. and Winkelmann, K., 1977, Theoretical studies of highly expanded free jets: influence of quantum effects and a realistic intermolecular potential, J.Chem.Phys., 66: 3965.

Wei, J. and Tellinghuisen, J., 1974, Parameterizing diatomic spectra: "best" spectroscopic constants for the I_2 B↔X transition, J.Mol.Spectry., 50: 317.

A GENERAL PHENOMENOLOGY FOR SMALL CLUSTERS, HOWEVER FLOPPY

R. Stephen Berry

Department of Chemistry and the James Franck Institute
The University of Chicago
Chicago, Illinois 60637, U.S.A.

1. BACKGROUND: NONRIGID MOLECULES

The central topic of this Advanced Study Institute, "The Quantum Dynamics of Molecules", is a new incarnation of a subject very old in molecular physics, the description of quantum states of molecules that do not conform strictly to traditional rigid-structure models. While the general three-body problem has been attacked successfully without reliance on this model (Kolos, Roothaan and Sack, 1960; Smith, 1960; Mayer, 1975), virtually all of the interpretation of molecular vibrational and rotational spectra has been carried out within the context of the model of rigid structures.

The classical theory of molecular structure grew from a tradition that relied on the permanence and rigidity of each molecular species for its stereochemistry and its chemical identity. The notion of chemical isomers, in particular, is predicated on the rigidity of molecular structures: we observe and work with compounds having the same composition but altogether different properties, including those properties from which we infer <u>structures</u>.

Probes of the quantum states of matter long ago began to show us some of the limitations of this concept. Ammonia was the first example of a molecule for which classical concepts of structure are inadequate (Dennison and Uhlenbeck, 1932; Morse and Rosen, 1932). The heat capacity of cyclopentane showed an anomaly that implied the existence of a degree of freedom available at low temperatures that could not be accounted for in terms of a rigid structure (Kilpatrick, Pitzer and Spitzer, 1947). This molecule has the structure shown in Fig. 1, a pentagon with one carbon vertex out

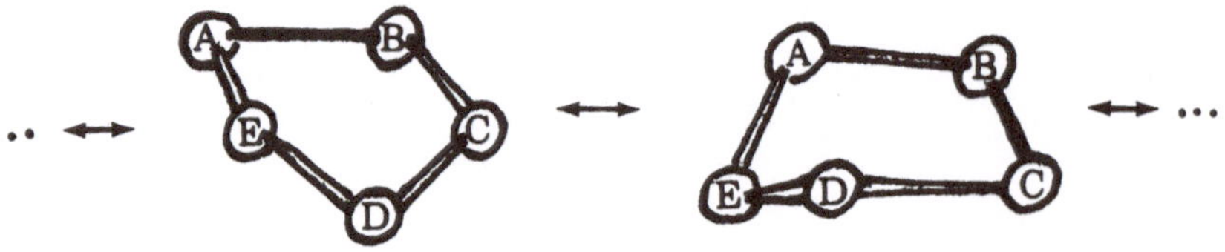

Fig. 1. Pseudorotation of cyclopentane. In the left-hand form, vertex (carbon atom) D lies below the plane of the other four carbons; in the right-hand figure, vertex E is below the plane of A, B, C and D. Hydrogen atoms are not shown.

of the plane of the other four. The molecule exhibits its special degree of freedom when the out-of-plane carbon exchanges roles with one of the four in-plane carbons. This process is accomplished by a large-amplitude concerted motion; it is equivalent to a permutation of the two carbons and their attached hydrogens that exchange roles, and a rotation through $2n\pi/5$ about the axis perpendicular to the plane of the four coplanar carbons; n is the number of steps of $2\pi/5$ required to bring the molecule in its first configuration into register with the second configuration. Such a process, a product of a permutation of identical particles and a rotation, is called a pseudorotation.

Cyclopentane is a simple example of a molecule in which we can observe the establishment of equivalence among identical nuclei occupying chemically inequivalent sites. By exhibiting such a process, this molecule makes us acutely aware of the apparent paradox presented to us by most molecules. The fundamental property of the indistinguishability of identical particles in any stationary state seems at odds with the obvious fact of everyday life upon which most of chemistry is built. The behaviour of all but the simplest molecules is predicated on the concept of structures, in which there may be any number of chemically and physically distinguishable sites available to atoms of a single kind.

This apparent paradox is easily resolved by rejecting the notion that we observe a stationary state when we assign distinguishable sites to identical nuclei (Berry, 1960a; Woolley, 1976). We know perfectly well that the stationary states of a molecule must exhibit the permutation symmetry - totally symmetric or totally antisymmetric - appropriate to all the nuclei of any given isotopic species (Longuet-Higgins, 1963). Thus the wave function for cyclopentane's ground state is not only a superposition of the five terms in which each CH_2 occupies the out-of-plane position; it must also contain symmetrically all the other terms that represent the $5!/2\times5$ permutations of the five identical C^{12} atoms among the sites on the ring. It must also contain in antisymmetric form the corresponding 10!/2x5 permutations among the hydrogen sites,[1] just as the electronic wave function is fully antisymmetrized. And the wave function must also be odd or even with respect to inversion so that if it does not have a centre of symmetry, we must double the number of configurations by making symmetric and antisymmetric superpositions of left-handed and right-handed forms. Strictly, this last point would only be true if parity were a conserved property; the nonconservation of parity in weak interactions generates a small inequivalence of left-handed and right-handed optical isomers. Such inequivalence will be discussed later in this series of lectures by R. Harris.

In cyclopentane, the establishment of chemical equivalence among the CH_2 groups is quite a separate matter from the establishment of permutational symmetry, in the sense that the conventional pseudorotation of C_5H_{10} (Fig. 1), leaves invariant the serial positions of all the atoms. In other species, such as a triatomic alkali molecule or a 5-coordinate phosphorus compound, the pseudorotation process generates the full set of permutations of the identical atoms (see Fig. 2). Thus, in some molecules, the observable dynamic process that makes identical nuclei chemically equivalent generates the entire permutation group; in other molecules, that process generates some subgroup of the permutation group. We shall return to this point when we address the problem of choosing the appropriate symmetry group for a molecule.

Based on the existence of isomers and the methods from which we infer structures, especially the various methods of diffraction, we have constructed an elaborate theoretical model for representing the behaviour of molecules. At the heart of this model is the Born-Oppenheimer approximation, its immediate consequent, the rigid motor, small-amplitude oscillator representation of nuclear motion, and the extension of these ideas to the concept of a potential surface for each electronic state. Let us pause to examine just what lies behind our ideas of molecular structure. First, we recall that the separation of nuclear and electronic motion is established by writing a perturbation series in a parameter κ that is not very small, the <u>fourth root</u> of the ratio of the electron mass to a typical nuclear mass (Born and Oppenheimer, 1927; Born and Huang, 1962). Born and Oppenheimer showed that the zeroth order eigenvalues in this series corresponds to the electronic energy in a system of infinitely heavy nuclei. The zeroth order

---

1. If one wishes to be very sticky about "ultimate" symmetries, one can argue that the wave function should exhibit the correct permutational symmetry of its <u>elementary</u> particles, not merely the permutational symmetry of its identical nuclei and electrons.

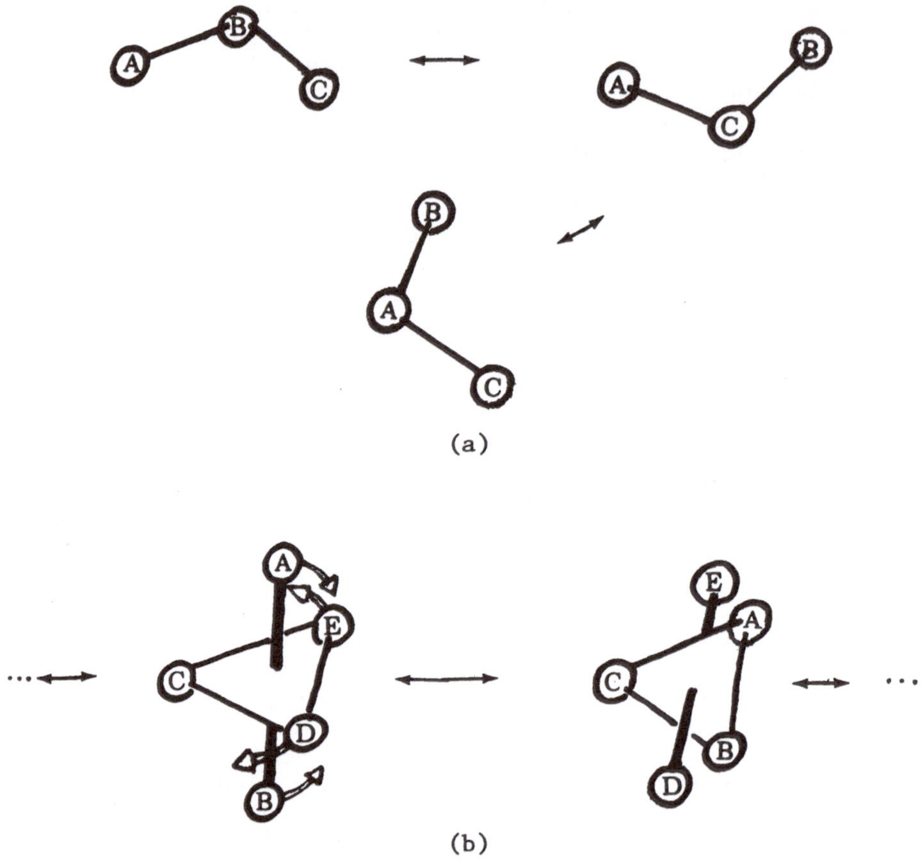

Fig. 2. Pseudorotation of a) a homonuclear triatomic molecule with isosceles triangular geometry at equilibrium, such as ozone or Na_3; b) a trigonal bipyramidal molecule such as PF_5. The fluorines are indicated by atoms A-E, and the phosphorus, at the centre of the polyhedron, is not shown.

Hamiltonian H_0 contains the electronic kinetic energy and all the terms of the potential of the total $(N_e + N_n)$-body Hamiltonian of N_e electrons and N_n nuclei. Moreover the wave function, to zeroth order in the energy, is separable into an electronic factor and a nuclear factor. Now we postulate that a minimum exists for the zeroth order eigenvalue, for some particular set of internuclear coordinates $\{R\} = \{R_e\}$. This assumption accompanies a subtle transition in the way we treat the nuclear coordinates in the zeroth order Schroedinger equation. Up to this point, those coordinates have been considered dynamical variables represented by operators R, with corresponding conjugate momenta $(\hbar/i)\partial/\partial R$ in a coordinate representation. With these nuclear momentum operators absent from H_0 we begin to treat the set {R} as multiplication parameters in the equation

$$H_0 \Psi_0(\{r\},\{R\}) = E_0 \Psi_0(\{r\},\{R\}) , \quad (1)$$

rather than as operators. Then we do the obvious, natural thing: we expand the potential of the Hamiltonian around $\{R_e\}$. This expansion has the particular attributes that the first-order and third-order terms in the eigenvalue series vanish, and that the second-order Schroedinger equation, in which the nuclear coordinates are dynamical variables, becomes the equation for a collection of harmonic oscillators. Moreover expansions around $\{R_e\}$ lets us continue to use the simple product wave function

$$\Psi(\{r\},\{R\}) = \psi(\{r\};\{R\}) \chi(\{R\}) \quad (2)$$

through second order in the total energy. The functions ψ are the eigenfunctions of $H_0(\{r\};\{R\})$ for whatever values one chooses for the parameters {R}. The zeroth order eigenvalues $E_0(\{R\})$ enter as numerical functions in the Schroedinger equation for the vibrational energy and wave functions

$$H_2(\{R\}) \chi(\{R\}) + E_0(\{R\}) \chi(\{R\}) = E_2 \chi(\{R\}) \quad (3)$$

The second-order part of the Hamiltonian, H_2, contains the momenta associated with vibrations - i.e. with changes in internuclear coordinates - which are of order κ^2 in the full expansion, in the mass-weighted coordinates of Born and Oppenheimer, but it does not contain the terms in total angular momentum, which are of order κ^4. The kinetic energy thus contains operators of order κ^0, κ^2 and higher; the effective potential for $\chi(\{R\})$ provies the harmonic potential when we expand it around $\{R_e\}$. Terms beyond the quadratic are relegated to the next nonvanishing orders. The derivation of Eq. (3) requires that terms from the action of H_2 on $\psi(\{r\};\{R\})$ also be relegated to higher-order equations, in other words that

$$H_2 \psi \chi = \psi H_2 \chi \quad (4)$$

The derivation also requires that there be no degeneracy in the electronic state. If there is, then terms involving both nuclear and electronic motion may invalidate the model by introducing terms that lower the total energy linearly with nuclear displacements away from $\{R_e\}$ - the Jahn-Teller effect (Jahn and Teller, 1937; Liehr, 1963a) or quadratically - the Renner effect (Renner, 1934) for linear molecules or the "pseudo-Jahn-Teller" effect (Jahn, 1938; Liehr, 1963b,c) for nonlinear polyatomic molecules.

Equations (3) and (4) and the assumptions that get us these equations are the essence of the Born-Oppenheimer approximation. They are the justification most often used for the concept of an effective potential surface for describing the motions of the nuclei of a molecule. We shall be examining the Born-Oppenheimer approximation in considerable detail in this Institute, so I shall not launch into a critique of it now. A relatively rigorous derivation of a second-order equation better than (3) was carried out by Bunker and Moss (1977) for diatomic molecules and applied to the vibration-rotation spectra of H_2 and D_2 (Bunker, McLarnon and Moss, 1977). Nevertheless this method still retains the spirit of the Born-Oppenheimer approximation, beginning with a zero-order Hamiltonian in which the nuclear coordinates are treated as parameters, thereby leading to an effective potential surface for the motion of the nuclei. Suppose one accepts the notion of an effective potential, either in the form of $E_o(\{R\})$ or, as derived by Bunker and Moss, E_o plus the <u>diagonal</u> nonadiabatic terms in a representation in which the off-diagonal elements of the first-order term H_1 in the total Hamiltonian are zero. Then Liehr (1963c) has shown that the full electronic potential energy surface of any molecule must be invariant under all permutations of identical nuclei, whether or not the permutations are operations of the molecular point group. His demonstration, intended originally with $E_o(\{R\})$ as the effective potential, is equally valid for a Bunker-Moss effective potential.

We must be careful to distinguish between the symmetry of the full electronic potential energy surface and the <u>molecular symmetry group</u> as this term is used by Longuet-Higgins (1963) and Bunker (1976). The latter is defined operationally by Bunker as the group with which we can give symmetry labels to the rotation-vibration levels "as much as necessary but not as much as possible". This definition carries the tacit assumption that we can specify the time scale or spectral resolution of our observations, in order to give precise meaning to what we mean by an energy level (c.f. Berry, 1960a). For example the molecular symmetry group of ammonia, NH_3, would be C_{3v} if we were unable to resolve the inversion doubling, but we need the higher symmetry of D_3 to classify the rotation-vibration levels when the resolving power of our experiments allows us to distinguish the doublets. The relationship between the time scale and resolving power is of course the energy-time form of the uncertainty principle.

A very important relationship exists between the appropriate symmetry group for a molecule - or any other system - and its Hamiltonian. First, we must recognize the distinction between the true Hamiltonian H with its symmetry group G, and whatever approximate Hamiltonian we want to use, H_0 with its symmetry group G_0. For example if a molecular Hamiltonian H_0 is written in terms of displacements of nuclei from an equilibrium geometry, then G_0

may be a point group or a direct product of a point group and the inversion-rotation group 0(3). The true Hamiltonian by contrast, contains the coordinates and momenta of all the particles as operators, and its symmetry group is at least as large as the direct product of 0(3) with the permutation groups of all the sets of identical particles, $S_n^{(e)}$ for the n electrons and the permutation group $S_{N_J}^{(J)}$ for each Jth isotopic species of nucleus. Clearly G_0 is a subgroup of G. Suppose G_j^o represents an element of G_o, G_j represents an element of G, and G_k' is an element of G not in G_o. The invariances of H and H_o require that

$$[H,G_j] = 0 \tag{5}$$

and

$$[H_o,G_j^o] = 0. \tag{6}$$

However, we cannot expect H_o to commute with elements G_k'. Specifically, we may construct

$$[H_o,G_k'] = \Delta E_k' \tag{7}$$

for all elements G_k'. Let

$$\Delta E' = \max \{\Delta E_k'\} . \tag{8}$$

This quantity provides us with the formal criterion as to whether H_0 is adequate to represent the spectrum of interest. If ΔE_{obs} is the maximum resolving power of our experiment ($\Delta E_{obs} = \hbar/2\tau$, where τ is the time inverval of each observation), then so long as $\Delta E' < \Delta E_{obs}$, the symmetry of H_0 is high enough to let us label all the observable spectral levels according to G_0. But if $\Delta E' > \Delta E_{obs}$, we'd better use a Hamiltonian with symmetry higher than that of H_o, if we want to label the quantum states with enough precision to extract selection rules from analysis of their symmetry. Put another way, if H_o is the most approximate Hamiltonian for which

$$[H_o,G_k'] < \Delta E_{obs} \tag{9}$$

then G_0 is the best choice of the molecular symmetry group. Any larger group would be superfluous and any smaller group would be inadequate to the resolving power of the experiment.

2. OBSERVABLE EXAMPLES

We have already seen one of the first examples of a dynamical process, the pseudorotation of cyclopentane. In this case, the physical property crucial to the identification of the process is the heat capacity (Kilpatrick, Pitzer and Spitzer, 1977). Most of the experimental history of such dynamical processes in molecules has relied on magnetic resonance, most frequently nuclear magnetic resonance (nmr) as its primary tool. The observation that the nmr spectrum of the trigonal bipyramidal molecule PF_5 shows a single fluorine resonance (Gutowsky, McCall and Slichter, 1953) was interpreted by this writer as due to the mechanism shown in Fig. 2b. It was proposed that the pseudorotation of PF_5 occurs rapidly enough to make all five fluorines equivalent on the scale of time or energy resolution of a nuclear magnetic resonance observation. The same mechanism had been introduced earlier in an entirely different context: Teller and Wheeler (1938) had considered the Ne^{20} nucleus as a trigonal bipyramid of five α particles, in order to get a picture of its low-lying rotational and pseudorotational levels.

Possible pathways for intramolecular rearrangements became a subject of interest when it was shown that they could be categorized and that many of the possible mechanisms could be tested by suitable contributions of substitution and nuclear magnetic resonance spectroscopy, especially over a range of temperatures (Muetterties, 1965, 1967, 1969a,b; 1970a,b). The pathway of Fig. 2 was established by Whitesides and Mitchell (1969) by an ingenious choice of species for study by nuclear magnetic resonance. They examined the temperature dependence of the ^{31}P nuclear resonance in the molecule $(CH_3)_2NPF_4$. The dimethylamine group, $(CH_3)_2N-$, is permanently locked to an equatorial position in the trigonal bipyramid about the P atom. The fluorines can exchange between equatorial and apical positions. Whitesides and Mitchell found that at "high" temperatures (-50°C), the fluorines are magnetically equivalent according to the splittings they produce in the ^{31}P resonance. At lower temperatures (-100°C), the splittings of the phosphorus resonance indicate that there are two inequivalent kinds of fluorine atoms. Between the two extremes, some lines remain sharp but others broaden. By assigning the components of the split ^{31}P resonance to particular fluorine spin functions, Whitesides and Mitchell showed that the mechanism of Fig. 2 would transfer magnetization among only a few of these components while other mechanisms would cause coupling among several more. The observed broadening was consistent with only the former situation, demonstrating that the fluorines exchange two pairs at a time, rather than two atoms at a time as the most general mechanisms would have. The result is consistent with phenomenological and semiempirical calculations (Florey and Cusachs, 1970; Russegger and Brickmann, 1975a,b).

This elegant experiment determined, in effect, what coordinate changes occur when a five-coordinate phosphorus compound establishes chemical equivalence by exchange of identical nuclei. If one wishes to use the language of potential surfaces, one can say the Whitesides-Mitchell experiment showed the relationship on the potential surface between the locations of the equivalent minimum and the region of the lowest-energy saddle connecting them.

Inorganic and organometallic molecules were the first to be explored extensively (c.f. Cotton 1968, 1975) with regard to their large-amplitude motions and internal rearrangements. Particularly great interest arose with the recognition that the hydrolysis of phosphate esters, a crucial reaction in many biological processes, involves intermediates that undergo the rearrangement of Fig. 2 (c.f. Westheimer, 1968). A dramatic example among pure organic species, which has much aesthetic appeal, is the interval rearrangement of the hydrocarbon bullvalene, shown in Fig. 3 (Doering and Roth, 1963). This substance shows equivalent protons in its high-temperature nuclear magnetic resonance spectrum but four different kinds of protons at low temperatures. The theoretical interpretation of the relationship between the nuclear resonance spectrum and the internal rearrangements of a molecule is based on the work of Redfield (1957) and Sack (1958). It was first applied to the non-rigid molecule $(C_5H_5)_2Fe(CO)_2$, in which each ring exhibits a single proton line at room temperature (Bennett et al., 1966; Cotton, 1968, 1975).

Among six-coordinate species, XeF_6 shows the characteristic of a molecule that doesn't quite have a structure: its <u>average</u> geometry is that of a regular octohedron, but the most probable geometry is not that at all. The potential has minima, probably, at the three-fold symmetric distorted octahedral configurations, near enough to the octahedral structure that the zero-point vibrational level may well have an energy above the local maximum at the octahedral geometry. The fluorines of this molecule appear to move about, over distances large compared to ordinary vibrations but small relative to the F - F distances. Hence the molecule is non-rigid and the fluorines are equivalent chemically but do not permute among themselves (Bartell and Gavin, 1968). The reason for the large amplitude motions in XeF_6 can be interpreted on a phenomenological basis as a pseudo-Jahn-Teller effect (Liehr, 1963b, c; Nicholson and Longuet-Higgins, 1965), or at a microscopic level in terms of the repulsive interactions of a lone pair of electrons

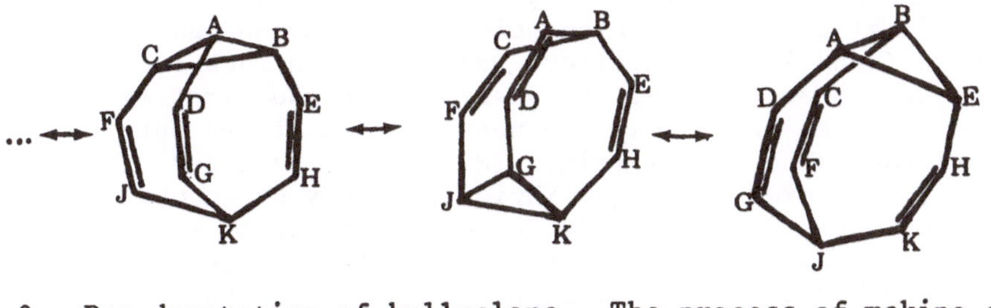

Fig. 3. Pseudorotation of bullvalene. The process of making and breaking bonds involves only small shifts of nuclear positions but the molecule appears to rotate nearly 180° with each pseudorotation operation.

on xenon with the six fluorine atoms (Bartell and Gavin, 1968). Other, more complex six-coordinate species exhibit the sort of large amplitude motions associated with permutations of substituents (Muetterties, 1968; Meakin, 1971). With sufficient vibrational excitation, species such as SF_6 take on the character of nonrigid molecules; this phenomenon will be discussed later, in the context of spectroscopic studies.

The simplest system in which equivalent nuclei may occupy nonequivalent sites is the homonuclear isosceles triangle. Ozone was long thought to be the easiest example to study, among such species. Its analysis was carried out (Berry, 1960a), but the barrier for permutational exchange seemed far too high to make the process observable. Now, it appears that ozone may have an excited state with a potential minimum not far above that of the ground state (Peyerimhoff and Buenker, 1967; Hay, Dunning and Goddard, 1975, and refs. therein). Electron energy loss spectroscopy indicates that such a state or states do exist (Swanson and Celotta, 1975) in the range expected from at least some of the calculations (Hay, Dunning and Goddard, 1975). However none of the predicted excited states has the same symmetry as the ground state, so some electronic-vibrational energy exchange would be required to permit ozone to go from one isosceles geometry through an excited electronic state with an equilateral geometry into another isosceles configuration. Such processes might be produced by excitation of O_3 with radiation between about 1.4 and 2.0 ev. The natural way to carry out the experiments would be to search for isotopic exchange between the two kinds of oxygen positions. [A much cruder isotope effect, a difference in rates of exchange of chlorine atoms with other molecules, was used to distinguish two kinds of chlorine in PCl_5 (Downs and Johnson, 1955).]

Recently a far more striking group of nonrigid triatomic molecules has come to light, the trimeric alkali molecules such as Li_3 and Na_3. A diatomics-in-molecules calculation (Companion, Steible and Starshak, 1968) indicated that the equilibrium isosceles geometry of Li_3 lay only about 5 kcal/mole or about 1750 cm^{-1} below the saddle at the equilateral geometry. This corresponds to about six vibrational quanta. A calculation based on configuration interaction (Kendrich and Hillier, 1977) gives a much smaller barrier, only about 1.2 kcal, suggestive that the first excited vibrational states of Li_3 may lie near the top of the barrier for pseudorotation. The sodium trimer (Martin and Davidson, 1978) is similar; its predicted equilibrium geometry of an obtuse isosceles triangle is 1.6 kcal/mole below the lowest energy for an equilateral triangle. These triatomic alkalis are probably the most nonrigid molecules we know, in the sense that they violate the small-amplitude vibrator model even in their ground states. The spectroscopic investigation of the alkali trimers is being conducted via resonant two-photon ionization followed by mass analysis of the product ions (Hermann et al., 1978; Gerber and Schumacher, 1979). If these notions are born out, the computed thermodynamic properties of Li_3 will warrant reexamination (Wu, 1976). Another dramatic example of a molecule whose zero-point energy is greater than the barrier between configurations is UF_5, according to Wadt and Hay (1979).

A large class of molecules that may well exhibit considerable nonrigidity consists of the van der Waals molecules. Among the small species studies thus far, the nonrigidity takes the form of some internal rotation. For example the N_2Ar molecule has a T-shaped structure in its lowest two librational states but the N_2 is more or less free to rotate in higher levels (Henderson and Ewing, 1974; Ewing, 1976).

The complexes of rare gas atoms with HCl, for example, seem to be very nonrigid. This can be inferred from their spectra (Novick et al., 1973, 1976; Boom, Frenkel and van der Elsken, 1977) and from various phenomenological and model potential calculations (Holmgren, Waldman and Klemperer, 1977; Dunker and Gordon, 1976; Detrich and Conn, 1976). In particular these molecules offer very little resistance to the "bending" movement of the proton, which lies off the rare gas-halogen axis but between the two heavy atoms.

Molecular dimers such as $(H_2)_2$, $(HF)_2$, $(H_2O)_2$ and $(O_2)_2$ have all proved to be interesting nonrigid species. The infrared spectrum of the H_2 dimer was studied in successively greater detail (McKellar and Welsh, 1974, and refs. therein). One calculation (Jaszunski, Kochanski and Siegbahn, 1977) based on conventional configuration mixing puts the energy of the most stable form, a T-shaped structure, only 22cm^{-1} below the lowest-energy rectangle and 25cm^{-1} below the energy of a linear structure. Another calculation, based on mixing of nonorthogonal configurations (Gallup, 1977) also gives energy differences in this range for the various spherical harmaric contributions to the effective H_2-H_2 potential and thus to the various geometric structures. Hence we can assume that the $(H_2)_2$ molecule is extremely nonrigid even while retaining the integrity of the individual H_2 molecules, and that its vibration-rotation spectrum will be interpretable only with this fact taken into account. That the two hydrogen molecules are nearly free to rotate within the dimmer was recognized early (Watanabe and Welsh, 1964; Gordon and Cashion, 1966). The determination of the molecular symmetry group and its selection rules has been a recent effort by Bunker (1979), toward a full assignment of the $(H_2)_2$ spectrum. We shall later examine the symmetry of this species in the general context of the energy levels of a four-body system.

Nonrigidity associated with tunneling of hydrogen atoms has been reported both for $(NF)_2$ (Dyke, Howard and Klemperer, 1972) and $(H_2O)_2$ (Dyke, Mack and Muenter, 1976; Dyke, 1976). In these cases, the hydrogens presumably move rapidly enough to influence the overall molecular rotation slightly but observably. The barrier for exchange of hydrogens in the nonlinear HFHF is about 1 kcal/mole or 350 cm^{-1}, low for proton tunneling (Curtiss and Pople, 1976). It has been suggested from time to time that the hydrogens in such systems could be treated by a sort of Born-Oppenheimer separation; this approach, ironically, is quite the opposite of what one would choose according to the discussion of Woolley (1976; 1978a,b; 1979; Woolley and Sutcliffe, 1977).

We shall not dwell in more detail on van der Waals molecules. They and their spectra are discussed in much more detail in the preceding chapter by Donald Levy. But before leaving the subject of empirical information about nonrigid molecules, we should take note of two related kinds of systems which nonrigidity is important. Both are molecules in highly excited vibrational or rotational-vibrational states: one consists of the diatomics called by Stwalley "long-range molecules"; the other consists of polyhedral molecules such as SF_6 excited by absorption of many vibrational quanta.

Long-range molecules include ordinary diatomics in very high vibrational states and especially such molecules in high rotational states as well, so that centrifugal forces keep the nuclei from coming close to one another (LeRoy, 1973; Stwalley, 1977). But they also include molecules in states whose classical distances of closest approach are of order 30Å or more, states such as the 0_g^- and 1_u^- states of the alkali diatomics that correlate with $M(^2P_{3/2}) + M(X^2S_{1/2})$ (Niemax and Pichler, 1975; Movre and Pichler, 1977; Stwalley, Uang and Pichler, 1978). A challenging question in the context of these lectures is whether we shall find similar states among triatomic or larger molecules. Polyatomic molecules with much "open space" would be the molecular analogues of nuclei, and would epitomize molecular nonrigidity.

Polyatomic molecules, especially simple polyhedral molecules, exhibit centrifugal distortions that couple rigid-molecule rotation-vibration levels. The subject has a long history (Jahn, 1938; Childs and Jahn, 1939; Hecht, 1960; Amat and Nielsen, 1962, for example), but moved ahead rapidly with Watson's recasting of the centrifugal distortion Hamiltonian (Watson, 1967, 1968a,b) and the application of this method to the interpretation of splittings and clusterings of rotational lines of CH_4 (Dorney and Watson, 1972). The physical picture inherent in their interpretation is a locking of the axis of molecular rotation to a symmetry axis of the molecular polyhedron. This model was developed more extensively, with application to octahedral molecules such as SF_6, by Harter, Patterson and their collaborators (Harter and Patterson, 1976, 1977a,b; Patterson and Harter, 1976; Harter, Patterson and Paixao, 1978), and especially for the interesting case of rotational coupling with degenerate vibrational modes (Gray and Robiette, 1976; Robiette, Gray and Birss, 1976; Gray, Robiette and Johns, 1977; Michelot, 1976; Harter, Galbraith and Patterson, 1978; Harter, Patterson and Galbraith, 1978). These approaches have become particularly important for experimentalists with the advent of very high resolution infrared spectroscopy and the concern with isotope effects and isotope separation by laser excitation of molecules. The dynamical problem in such systems is one of interpreting the degree to which a centrifugal distortion is locked to a particular molecular axis or can hop from one axis to an axis that would be equivalent in the undistorted polyhedron.

3. TAXONOMY OF NONRIGID SPECIES

In Section 1, we looked quickly at some of the history of the analysis of energy levels of nonrigid molecules. Now we return to this topic, to see how the spectroscopic problem was attacked from the viewpoint that nonrigidity is a deviation from nearly rigid behaviour. This approach has its roots in the Born-Oppenheimer approximation and is based on the concepts of a potential surface for motion of the atoms within the molecule, and the rigid rotor, small-amplitude vibrator representation of the molecule. Moreover the rigid rotor picture inherent in the method made it natural to use the methods of finite groups to classify levels and construct selection rules. Then, in the larger part of this section, we shall turn to a method that avoids the restrictions of a Born-Oppenheimer approximation, an explicit potential surface and a rigid rotor model, but, in exchange for greater generality, requires a) that one artificially idealize the symmetry of one's system before one falls back to using the true symmetry of that system, and b) that one uses the methods of continuous groups rather than the simpler tools available with finite groups.

What should one try to extract from an analysis of nonrigid molecules or other nonrigid species? In classical terms, we would like to know the nature - meaning trajectories of moving particles - and degree - meaning frequencies or velocities of motion - of nonrigidity of the species. We would of course like to know how these depend on the internal energy of the species and how the species exchanges energy with its surroundings. These ideas translate into quantum mechanical ideas when we replace specifications of trajectories and velocities with a labeling of the energy levels and spectral lines that is as complete as our spectral resolution allows. The exchange of energy is replaced, of course, by selection rules and line strengths. These may be either simple electric or magnetic dipole, or, for more general kinds of exchange, the generalized oscillator strength based on the operator exp $(i\underset{\sim}{k}\cdot\underset{\sim}{r})$. Later we shall see that one can extract more than these data from the analysis.

Coriolis Coupling

Before examining the group theoretic approaches that are the main subject of this section, a few words are in order about the dynamics of nonrigid species. Rigid molecules, to a large degree, have vibrational motions that are separable from their rigid body rotations. However at least two effects spoil this separability: the variation of moments of inertia with vibrational amplitude and, more important, Coriolis interactions that couple vibrational motion with rigid body rotations. The latter is the effect usually responsible for breakdown of the rigid rotor, small-amplitude vibrator model, particularly when the molecule of interest has degenerate vibrational modes which carry angular momentum of their own. When the Coriolis interactions are small, the rotational quantum number $\tilde{J}$ associated with rigid body rotation and, for symmetric tops, the quantum number $\tilde{K}$ for rotation about the axis of the unique principal inertial axis, remain good quantum numbers but the degeneracies of the rotational levels are split, due to the different ways of coupling the rigid body rotation to the vibrational angular momentum (Teller, 1934; Johnson and Dennison, 1935; Boyd and Longuet-Higgins, 1952). For example Boyd and Longuet-Higgins showed that if a symmetric top has one quantum in a doubly-degenerate vibration, each $\tilde{K}$-level is split by an amount

$$\Delta E_{Cor} = \frac{4\hbar^2}{I_A} \tilde{K} \zeta$$

$$= \frac{4\hbar^2}{I_A} \tilde{K} \sum_{\substack{\text{atoms}\\ j}} \frac{1}{m_j} \left(\frac{\partial X}{\partial x_j}\frac{\partial Y}{\partial y_j} - \frac{\partial X}{\partial y_j}\frac{\partial Y}{\partial x_j}\right) \quad (10)$$

where X and Y are the coordinates of the doubly-degenerate vibration; x_j and y_j are the Cartesian coordinates of atom i in a plane perpendicular to the unique inertial axis, z, m_i is the mass of that atom and I_A is the moment of inertia about the z-axis. If the Coriolis coupling becomes large, corresponding to the Coriolis splitting of the rotational levels becoming comparable to the separation of the states of different $\tilde{J}$ or $\tilde{K}$, then a simple perturbation approach is inadequate and one may have to diagonalize a Hamiltonian matrix containing several $\tilde{J}$-levels in order to reproduce an observed vibration-rotation spectrum. But to preserve total angular momentum, states entering such a matrix with different $\tilde{J}$-values must also have different amounts of vibrational angular momentum.

Naturally, the process of diagonalizing a matrix of several rigid-rotor angular momenta and different vibrational angular momenta is the mathematical counterpart of a strong coupling of these two angular momenta, and to a flow of angular momentum between the two trial degrees of freedom. In the extreme conceptual rigid-rotor limit, they would be entirely separable. In the traditional Coriolis-coupled case (for a symmetric top) they are coupled only enough to spoil the $\tilde{K}$-degeneracy; that is, the rigid-rotor component of angular momentum along the unique inertial axis and the corresponding component of the vibrational angular momentum do not independently maintain good quantum numbers $\tilde{K}$ and $\tilde{\ell}_r$. Although $|\tilde{K}|$ is a good quantum number at this level of approximation, $\tilde{K}$ itself is not. The situation is analogous to the Russell-Saunders coupling of electrons in an atom: there, principal quantum numbers n_j and angular momentum quantum numbers ℓ_j are good, but the coupling that makes L and M good quantum numbers for total angular momentum and its z-component, respectively, spoils the orientation of the individual electron "orbits" and thus spoils the m_j's. In the extreme case of strong coupling, the rigid rotor angular momentum itself ceases to be a good quantum number and only the quantum numbers of <u>total</u> angular momentum, J, (without tilde) and its z-component, M (in a space-fixed coordinate system), are good quantum numbers.

The breakdown of the rigid rotor model through Coriolis coupling is one of the most important ways to think if the transition away from rigidity. Another is in terms of a change of symmetry, which we examine next. A third, in principle, is through the breakdown of the concept of a potential surface, or its reverse, through the construction of a potential surface from a more general approach to the coordinates of the many-body problem (Griffin and Wheeler, 1957; Griffin, 1957; Lathouwers, van Leuven and Bouten, 1977; Lathouwers and van Leuven, 1978). This third method seeks to arrive at the concepts of a molecular geometry and an effective potential by deduction from the full molecular dynamics, rather than from a prior assumption of the Born-Oppenheimer approximation. In effect, the <u>generator coordinate method</u>, as it is called, begins with no assumption of rigidity, but presumably often yields the nearly-rigid picture as the solution to the Schrödinger equation of the full many-body problem. This method will be discussed in detail in other lectures here, so we shall not discuss it in any more detail here (see the following two chapters).

Nonrigidity as a Change of Symmetry

The vibration-rotation levels of a free molecule are commonly taken to be determined by a Hamiltonian of the near-rigid-rotor, small-amplitude vibrator sort, whose potential is determined by the Born-Oppenheimer approximation. The symmetry of the nonrelativistic molecular Hamiltonian is, strictly, the permutation group of the electrons and the permutation-inversion group of the nuclei (Bunker, 1979), and the interior product of that group with the direct product of the unitary unimodular groups $SU_j(n)$ describing the spins of all the particles j with spins n (Wybourne, 1970, p.69;

Flurry and Siddall, 1978). However, the symmetry of the Hamiltonian is generally taken to be the direct product of the point group G^o of the potential and the frame rotation group ${}^FO(3)$. If the molecule is a spherical top, this group may conveniently be taken as a broken symmetry of the direct product of two rotation groups:

$$G^o \times {}^FO(3) \subset {}^mO(3) \times {}^FO(3) \tag{11}$$

where ${}^mO(3)$ is an idealized spherically symmetric molecular field (Hilico, Berger and Loete, 1976). Such an approach offers a mathematical and physical rationale for the phenomenological description of rotational states given by Harter et al. (Harter and Patterson, 1976, 1977a,b; Patterson and Harter, 1976): The centrifugally distorted octahedral or tetrahedral molecule may still be considered a spherical top in zeroth order, and its distortion to a symmetric top becomes a perturbation. Later in this section, we shall see another application of approximate, artificially high symmetries as devices to organize and classify energy levels.

The appropriate symmetry group for a molecule has been a subject of extensive discussion and occasionally of controversy. That it should include the full permutation group - if the resolving power of the observation justifies it - was pointed out by this writer (Berry, 1960a). Following the logic developed for finding the symmetry groups for rigid molecules (Hougen, 1962, 1963), Longuet-Higgins (1963) showed how to generate an appropriate group for describing vibrational structure, provided one knows enough about the dynamics of the molecule to know what constitutes a "feasible process" for rearrangement. The problem of relating "the" molecular symmetry group to the permutation group was considered by Hougen (1962, 1963), who made it clear that the appropriate symmetry group could well be something lower than the direct product of the permutation groups of the various sets of identical nuclei. Longuet-Higgins (1963) worked out the appropriate groups and character tables for the vibrational states of several molecules: CH_3BF_2, C_2H_6, N_2H_4 and $B(CH_3)_3$. Bunker carried out similar analyses for ferrocene (1964) and $CH_3C{\equiv}CSiH_3$, methylsilylacetylene (1965).

Dalton (1966a) pointed out the need and then Dalton and Nicholson (1975) extended the arguments of Longuet-Higgins to show how one can apply the concept of a "group of feasible transformations" when more than one kind of geometric configuration can be attained by the molecule.

Let us review briefly the construction that Longuet-Higgins made. Let G^o be an element of the molecular point group G^o. Suppose there is a feasible transformation G' that permutes identical nuclei in some manner not attainable by any G^o in G^o. We imagine labels on the nuclei and arbitrarily set one equilibrium configuration of these nuclei as a reference point. Then we can apply the transformation G' to the molecule and its point group remains G^o; we may apply G' again, over and over. Because there are only a finite number of identical nuclei in the molecule, there is some n such that eventually $(G')^n$ takes the molecule back to some configuration that it had exhibited previously. The first of these repeated configurations must be the original reference configuration (Exercise: why?) Therefore the set G', G'^2, ..., $(G')^{n-1}$ is a closed set. Mechanics assures that $(G')^{-1}$ exists, and it must be $(G')^{n-1}$. Therefore G' and its powers form an Abelian group. Moreover G' and G^o generate the cosets of a group G' larger than G^o. If G' and its powers represent all the feasible transformations, corresponding to other generators G'', ..., then the appropriate group is generated by G^o and the cosets of G^o with G', G'', ... etc. Each individual coset corresponds to the molecule transformed, e.g. by a tunneling process, to some configuration other than the reference configuration. The vibronic states of the molecule then correspond to irreducible representations of the group G rather than of group G^o.

Within the context of many of the chemical species of spectroscopic interest, it is sufficient for vibrational analysis to suppose that all the available configurations are geometrically equivalent, as they are for ethane, cyclopentane and phosphorus pentafluoride. However a molecule such as 1,2-dichloroethane has three potential minima, as indicated in Fig. 4. If this species is observed at a temperature high enough to exhibit two or three of these minima, the symmetry group must reflect the attainability of these configurations. This situation becomes particularly important to consider when one tries to describe molecular vibrations and rotations without making the Born-Oppenheimer approximation. It will be an amusing challenge to solve the problem of finding the symmetry groups of molecular Hamiltonians by methods of generator coordinates.

Still within the limitation that rigid-body rotation is a constant of the motion, several extensions and alternative formulations have been developed. Dalton (1971) showed how to calculate the splittings and relative line intensities of individual rovibrational spectral lines for a molecule whose appropriate symmetry group G (which Dalton calls the Q-group) is larger than the point group G^o of the equilibrium geometry. The vehicle for this calculation was the PF_5 molecule, for which G is the full permutation group S_5, times the rotation group $\mathbf{O}(3)$. The treatment is based on first finding the splittings and weights of each (2J + 1)-degenerate level with definite values of J and K, due to vibration-rotation interaction, and then, the further splittings of these levels to nonrigid motions from the geometry of one equilibrium configuration to another, e.g., by tunneling. The assumption that made possible the latter step was the assumption of a specific

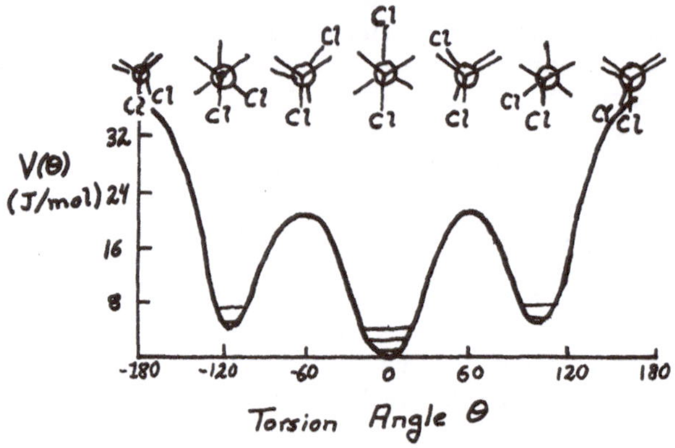

Fig. 4. The effective potential of 1,2-dichloroethane as a function of the torsion angle; θ = 0 is taken at the minimum energy, with the chlorine atoms trans to one another. The curve illustrates two inequivalent kinds of potential minima - that at 0° and those at $\pm$ 120°, respectively - in a nonrigid molecule.

mechanism corresponding to nonrigid behaviour; for the example of PF_5, Dalton used the mechanism of Fig. 2b. Each configuration can be the precursor of three other equivalent configurations, if the molecule obeys these dynamics. Dalton supposed that each configuration persists long compared with the rotational period. Each configuration can be thought of as possessing its own D_{3h} point group and 0(3) rotation group, and its own set of rovibrational levels; together they constitute G^o. From this viewpoint, the levels of one configuration are exactly degenerate with the corresponding rovibrational levels of the nineteen other equivalent configurations. By requiring the molecular rovibrational wave functions to transform as irreducible representations, not of $G^o = D_{3h} \otimes 0(3)$, but of the group G generated from G^o by the operation of Fig. 2b, call it G', Dalton classified the vibration-rotation states of the trigonal bipyramid and counted their degeneracies. He also carried out perturbation calculations using as a basis configurational wave functions (small-amplitude, localized vibrational wave functions). With this basis, and the assumption that the only significant matrix elements are those connecting configuration connected by one transformation of Type G', Dalton computed the relative splittings of the rovibronic energy levels and intensities of the optical transitions among these levels, due to the interaction among different configurations.

This calculation is parallel to the better known calculations of the splittings of degenerate atomic states of the same angular momentum J by a crystal or ligand field. However there is one often-overlooked but significant difference between the two cases. In crystal field problems, one begins with a system of high degeneracy, in which this degeneracy is a consequence of high symmetry - the rotational symmetry of the free atom. There one splits the degeneracy by breaking the symmetry, and determines how the energy levels and states split by finding how the large irreducible representations of the group of high symmetry split into smaller (subduced) representations of the point group appropriate to the crystal field. In the approach of Longuet-Higgins, Dalton and others, one begins with a highly degenerate set of levels, but with no group to whose irreducible representations they correspond, no group analogous to the O(3) starting point of crystal field theory. The rigid molecule limit, in this picture, has levels each of whose degeneracy is the degeneracy of the jth point group level g_i^o, times c^0, the number of equivalent configurations with the point group G^o. Therefore the zero-order picture is one of $g_j^o c^o$ degenerate states for the jth level, of which only g_j^o at a time are accounted for by symmetry. The rest of the degeneracy must be considered accidental, which presents a situation that at very least offends one's aesthetic sensibilities. More important, this apparent high accidental degeneracy suggests that we look for a group whose irreducible representations correspond to the states, and from which the splittings, such as those derived by Dalton, can be derived as consequences of symmetry breaking, rather than by inducing a higher symmetry group G from a lower symmetry group G^o and a generator G'. The induction procedure was that of Longuet-Higgins (1963), Altmann (1967, 1971, 1978), Dalton (1971), Chevon and Bordé (1974), Dalton and Nicholson (1975), and Bunker (1976, 1979).

There is no way yet to describe nonrigid molecules in a manner precisely analogous to the elegant symmetry-breaking picture provided by crystal field theory. We simply have no way yet to write a Hamiltonian to describe nuclear motion that retains the real high symmetry that presumably underlies the degeneracy $g_j^o c^o$. One approach to resolving this problem may come through the use of generator coordinates. Another, to be discussed shortly, starts with an artificial high symmetry. A third, developed by Günthard and his collaborators, is based on the small-amplitude oscillator picture, but with the symmetry group derived from the permutations and substitutions of structural coordinates (internuclear coordinates of the molecule in its equilibrium geometry) and the molecular point group (Bauder, Meyer and Günthard, 1974; Gut, Bauder and Günthard, 1975; Frei and Günthard, 1976; Frei et al., 1976; Bossert et al., 1978; Gut et al., 1978; Frei et al., 1978). This approach is conceptually different from that of Longuet-Higgins, Altmann, et al., but is quite close to the induced symmetry methods in practice, as the unifying approach of Ezra shows (Ezra, 1979).

The line of reasoning that begins with construction of the level splittings continues naturally to the inference of the dynamics of nonrigid motions from molecular spectra. If one can construct energy levels and a hypothetical spectrum for a molecule undergoing a particular process, as Dalton did for pseudorotation in PF_5, then one could do the same for several proposed processes, predict the spectrum for each and infer the mechanism from the appearance of the spectrum. This entails the computation of the splittings, the statistical weights and the intensities of the resolvable transitions in each model one wishes to test. This was done for PF_5 (Brocas and Fastenakel, 1975; Dalton, Brocas and Fastenakel, 1976). The mechanisms of rearrangements (Muetterties, 1967, 1969; Ugi et al., 1970; Hasselbarth and Ruch, 1973; Brickmann, 1971; Russegger and Brickmann, 1975a,b; Fastenakel and Brocas, 1975) were enumerated and their spectroscopic implications constructed. In this particular case, of five proposed mechanisms, each gives a distinctive pattern if both $A_1'' \leftrightarrow A_1'$ and $E'' \leftrightarrow E'$ transitions are observable. The results of this analysis are shown in Fig. 5. Note that the signs of the terms that determine the splittings are not known, so that mechanisms 1 and 4, and mechanisms 2 and 3 give indistinguishable $A_1'' \leftrightarrow A_1'$ patterns, but differ in the predictions for their $E'' \leftrightarrow E'$ transitions. Furthermore the mechanism of Fig. 2b and the so-called turnstile mechanism (Ugi et al., 1970) are both equivalent to the permutation designations of mechanism 1; to distinguish between them one must use a still more sensitive spectroscopic probe, such as that previously mentioned, of Whitesides and Mitchell (1969), which detects pairwise correlations.

To close this section, we should point out that there were a few pieces of work in which the interaction of large-amplitude vibrations with rotations were considered, but only in lowest order. Watson (1965) made a first step but still did not include the rotation group as part of the total molecular symmetry. Hougen (1965) moved the problem a step ahead by examining with great care how to relate the symmetry operations of the molecular symmetry group to the internal and laboratory coordinates of the nuclei within the molecule. The problem was again considered by Meyer and Günthard (1968) and by Pickett (1972) from the view of extending the methods of choosing axes from near-rigid molecules to more general methods for nonrigid molecules. Other discussions of the choices of frames and of internal coordinates are the review by Makusklin and Ulenikov (1977), the rederivation of the small-amplitude oscillator Hamiltonian (Louck, 1976) and the review of the Eckhart problem, the choice of the appropriate references frame for a rotating small-amplitude vibrator (Louck and Galbraith, 1976). The latter, as well as the methods of Meyer and Günthard and of Pickett, still leave unresolved the problem of how to define a set of rotating axes for a nonrigid molecule, in a quantum-mechanical description of molecular motion (c.f. Sutcliffe's chapter).

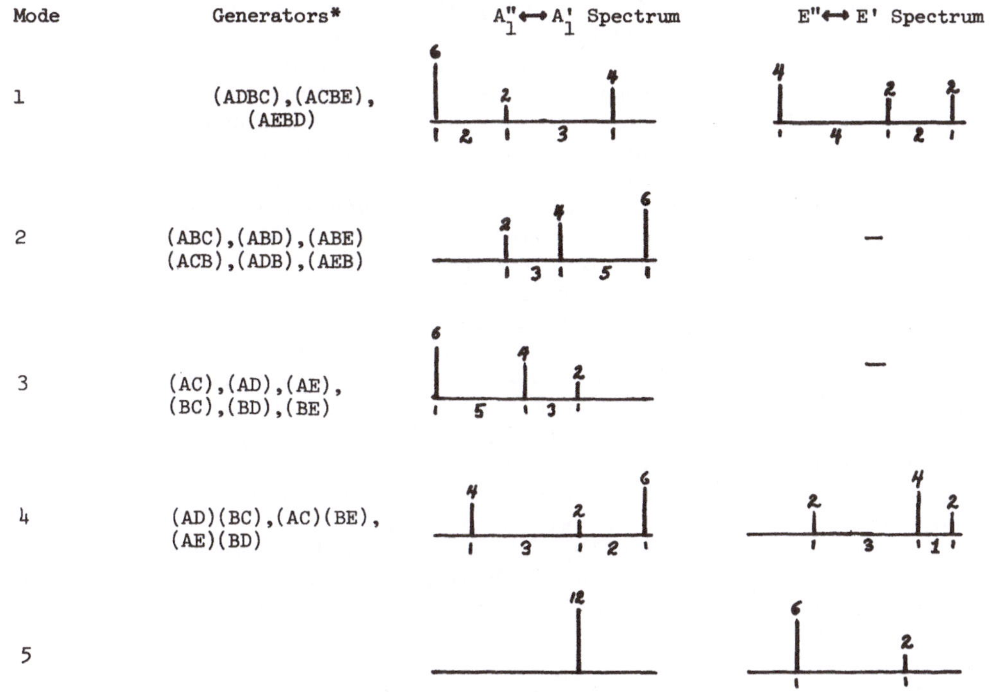

Fig. 5. Predicted vibration-rotation spectra of PF_5 corresponding to five different modes of large-amplitude motion (based on Dalton, Brocas and Fastenakel, 1976).
*Permutations are labled according to the left structure of Fig. 2b.

The Renner effect (Renner, 1934) is a phenomenon in which bonding of a triatomic and its vibrational angular momentum is the inherent large-amplitude degree of freedom. Yet even here, the treatments to date seem to be restricted to cases which, admittedly very important, preserve the rigid body rotational quantum numbers rather well (Pople and Longuet-Higgins, 1958; Pople, 1960; Aarts, 1978).

5. RELATING RIGID AND NONRIGID STRUCTURES BY SYMMETRY

Thus far, the approaches we have considered dealt with the energy levels and symmetry of nonrigid species in a frame of reference based on deviations from a rigid model. Even methods such as Günthard's, that seek as general a group as possible, begin with the concept of a nearly rigid equilibrium structure and the symmetry properties of parameters characteristic of that structure. The approach of Altmann (1967, 1971, 1978) was clearly stimulated by a desire to make the mathematical representation of a molecule somehow reflect its fundamental symmetries even when the nuclei move off their equilibrium positions. This method required the construction of a group that is only a symmetry group of the molecule when the nuclear positions correspond to equilibrium. Altmann's picture represents the extreme use of the Born-Oppenheimer approximation, in that the construction of the "isodynamic group" in fact portrays the nuclei as absolutely static; the nuclear coordinates are strictly parameters, not quantum mechanical operators. This in turn means that the Hamiltonian as treated by Altmann is always examined in terms of its symmetry with respect to those parameters. One consequence of Altmann's method of treating nuclear displacements was the need to construct some rather awkward operations, which led to misunderstandings in the relation between Altmann's method and Longuet-Higgins' (Watson, 1971; Altmann, 1971). The same limitations regarding the Born-Oppenheimer approximation apply to Longuet-Higgins' method (1963) and its subsequent extensions, but Longuet-Higgins, Dalton (1971), Dalton and Nicholson (1975) and Bunker (1976, 1979) did not deal with the symmetry of the molecule displaced from equilibrium, so the dilemma of reconciling displacements with the molecular symmetry group did not arise in their work.

Apart from the approach via generator coordinates and the correlation approach described below, there have been some other inroads towards treating the problem of molecular structure outside the Born-Oppenheimer approximation. Flurry and Siddall (1978) have given the detailed group structure of the many-body Hamiltonian, which we discussed in Section 3. The formally exact integral equations of Fadeev (1961) for the three-body problem have been applied to the analysis of the helium trimer (Lim, 1977).

The closest thing to a direct confrontation of the role of rotation-vibration coupling that starts from the rigid molecule picture is probably the work cited in Section 3 on the way Coriolis interactions split rotational levels in symmetric and spherical top molecules.

An approach rather different in spirit from those we have discussed is the one introduced by Kellman and this writer (Kellman and Berry, 1976), and developed by Kellman, Amar and myself (Amar, Kellman and Berry, 1979; Kellman, Amar and Berry, 1979-80). One can introduce the viewpoint with a question: "How could one construct a correlation diagram for the energy levels of a cluster of N identical particles, if one limit of the diagram corresponds to some specific rigid (or otherwise constrained) structure and the other represents a completely nonrigid but still cohesive structure?" We interpret the former limit, which we shall call the rigid limit, to be that of a rigid rotor, small-amplitude vibrator for most cases of interest. However there are some cases, such as the dimers of simple molecules, for which one may wish to consider limiting cases in which the component molecules are free to rotate. And, as we shall see, there are other limiting cases of interest that are not strictly nearly rigid rotors. As for the nonrigid limit, a significant part of the problem of answering the question consists of deciding what the nonrigid limit should be. We shall come to this point very quickly.

It may be of interest to give some historical perspective by pointing out the stimuli for the question regarding correlation diagrams. By the end of the 1960's, there was a division of opinion among chemists studying nonrigid molecules. One group argued that the important consideration in their interpretation is finding the number of equivalent configurations that can be reached by a feasible process from any reference configuration (Muetterties, 1969a,b, 1970a,b). In effect, this group argued that one must understand the topology of the potential surface. (Naturally all the dialogue concerning this subject was carried out in terms of potential surfaces or of classical trajectories of the rearranging nuclei.) The other viewpoint (Ugi, 1971; Ugi et al., 1970) was that the specific configuration of the molecule at and near the saddle of the potential between two stable configurations is the most important thing to consider. This group argued, in effect, that one must understand the geometry of the potential surface, especially near the highest points relevant to the large-amplitude motions of the nuclei.

Both of these are of course correct in part, but not in full. The topology of the potential surface determines the form of the splittings of the levels that would be "accidentally" degenerate if there are no large-amplitude motions. That is, the degeneracies and relative separations of levels of a real nonrigid molecule are determined by the topology of the potential surface. But the magnitudes of these splittings are determined by the geometry of the potential surface through which the localized levels communicate. In a one-dimensional representation, where we can use a WKB treatment of tunneling, the rates of passage and the level splittings are given by an exponential in the square root of the area of the classically forbidden region of potential above the energy of the states in question. Hence if one wants to understand the quantum states of nonrigid molecules in terms of potential surfaces, one must know both the geometries and topologies of the surfaces.

The quantum states, rather than the topologies or trajectories of nonrigid molecules, are of course their meaningful physical characteristics. The debate, which may have seemed at the time to hold up progress, in fact served to stimulate a renewed focus on the energy levels and quantum states. Specifically, the debate made one ask whether there might not be some alternative way to inquire directly about the energy levels, in a manner that would give information about the nonrigidity of a molecule. Could one find a way to infer from a high-resolution vibration-rotation spectrum the nature and degree of nonrigidity of a molecule, however nonrigid it may be? One knows to expect spectra of very nonrigid molecules to be very different from those of nearly rigid rotating vibrators. The rotational "constants" will appear to change drastically from band to band or perhaps even from one end of a band to the other, for example. We expect such problems to appear in the spectra of the alkali triatomics, in UF_5, and perhaps in XeF_6, the most severe cases now known to us. This problem is far from solved, but the approach through correlation diagrams opens a possible path more deeply based on phenomenology and less dependent on microscopics than the methods described above.

The essence of the method is a) the idealization of both rigid and nonrigid limits to make the corresponding Hamiltonians analyzable as much as possible in terms of symmetry - but the symmetry must be one we know how to break, to obtain realistic representations even of the limiting cases, and b) the connection of the symmetry-derived states from the two limits by use of the good conserved quantities of the system however nonrigid it may be - meaning total angular momentum, parity and the character of its permutation symmetry of the identical particles - and the usual noncrossing rule.

The direct outcomes of the method thus far have been: the provision of a one-parameter characterization of the degree of nonrigidity of any molecule (one-parameter, but dependent on the choice of the rigid limit model); a method to construct approximate density-of-state functions for rigid, solid-like and nonrigid, liquid-like clusters of arbitrary numbers of identical particles, and the beginnings of a pattern-identification method to infer from high-resolution vibration-rotation spectra the nature and type of nonrigidity in a molecule.

Details of the approach are given elsewhere (Kellman and Berry, 1976; Amar, Kellman and Berry, 1979; Kellman, 1977; Amar, 1979; Berry, 1979; Kellman, Amar and Berry, 1979-80). Here, I shall describe the physics and merely outline the mathematics.

First we consider the nonrigid limit. We want a model of a cluster of N identical particles that lets us construct the spectrum with as few parameters as possible, that is based on forces consistent with interatomic interactions and that has a symmetry high enough to contain all the symmetries that must apply for an arbitrary degree of nonrigidity. The natural choice for this limit is very much like that used for the shell model of nuclear physics (Gartenhouse and Schwartz, 1957) and of three-quark baryons (Faiman and Hendry, 1968). We take the interparticle potential energies to be simple harmonic potentials,

$$V_{ij} = \frac{1}{2} K(\underset{\sim}{r}_i - \underset{\sim}{r}_j)^2. \tag{12}$$

The N-particle Hamiltonian is then

$$H = \sum_i \frac{p_i^2}{2m} + \frac{1}{2} K \sum_{i<j} (\underset{\sim}{r}_i - \underset{\sim}{r}_j)^2.$$

The properties of this Hamiltonian are quite well-known (Louck, 1965; Kramer and Moshinsky, 1965, 1968; Moshinsky, 1969). However this Hamiltonian contains the centre-of-mass motion as well as the relative motion of all the particles. To remove the centre-of-mass motion, we first make a transformation to Jacobi vectors

$$\begin{aligned}
\underset{\sim}{R} &= N^{-1/2} (\underset{\sim}{r}_1 + \underset{\sim}{r}_2 + \cdots \underset{\sim}{r}_N), \\
\underset{\sim}{\lambda}_1 &= 2^{-1/2} (\underset{\sim}{r}_1 - \underset{\sim}{r}_2) \\
\underset{\sim}{\lambda}_2 &= 6^{-1/2} (\underset{\sim}{r}_1 + \underset{\sim}{r}_2 - 2\underset{\sim}{r}_3), \\
&\vdots \\
\underset{\sim}{\lambda}_{N-2} &= [(N-1)(N-2)]^{-1/2} [\sum_{j=1}^{N-2} \underset{\sim}{r}_j - (N-2)\underset{\sim}{r}_{N-1}] \\
\underset{\sim}{\lambda}_{N-1} &= [N(N-1)]^{-1/2} [\sum_{j=1}^{N-1} \underset{\sim}{r}_j - (N-1)\underset{\sim}{r}_N].
\end{aligned} \tag{13}$$

The Hamiltonian in the new coordinates becomes

$$H = \frac{P_R^2}{2Nm} + H_{int}, \tag{14}$$

where

$$H_{int} = \frac{1}{2m} \sum_{i=1}^{N-1} \underset{\sim}{P}_{\lambda i}^2 + \frac{NK}{2} \sum_{i=1}^{N-1} \underset{\sim}{\lambda}_i^2 \tag{15}$$

and the $P_{\lambda i}$'s are momenta conjugate to the Jacobi coordinates, $\underset{\sim}{P}_\lambda = m\dot{\underset{\sim}{\lambda}}$. This is similar to the nuclear shell model, but without the artificial confining potential that gives rise to "spurious states" in the shell model. The Hamiltonian of interest to us is of course H_{int}; we do not need to concern ourselves with translational motion of the centre of mass. Moreover H_{int} is clearly separable in Jacobi coordinates, so they correspond to independent normal modes.

The Hamiltonian (15) is that of N-1 identical 3-dimensional harmonic oscillators or, in more abstract terms, of an isotropic harmonic oscillator in 3N-3 dimensions. This Hamiltonian is invariant under unitary transformations (including inversion) among all its 3N-3 degrees of freedom, so its symmetry group is the unitary group U(3N-3). The states of the molecule in this limit correspond to all those irreducible representations of U(3N-3) that correspond to assignments of 0,1,2,... quanta to the 3N-3 oscillators. Since the quanta are bosons, the representations must be totally symmetric. The energy levels are of course equally spaced because the oscillators are harmonic; let us call the spacing $\hbar\omega$.

The dimensionality of the irreducible representation of U(n) corresponding to quanta is g_ν, the degeneracy of the νth-energy level. For n = 3N-3,

$$g_\nu = \frac{[(3N-3) + \nu - 1]!}{\nu!(3N-4)!}, \tag{16}$$

the number of ways of assigning ν indistinguishable particles among 3N-3 compartments. From this information, we can construct the level pattern and the density of states distribution for the nonrigid limit.

Like the united atom limits of the correlation diagrams for diatomic molecules, we cannot expect more than a very few (if any) real molecules to approach this limit. Nevertheless, as with the diagrams for diatomic molecules, the existence of a well-defined limit is itself useful, because it allows us to set up a closed pattern for the behaviour of all the vibration-rotation levels. The one class of molecules that might correspond closely to the nonrigid limit would be polyatomic analogues of the long-range molecules described in Section 2. As yet, there is no evidence for such species.

The nonrigid limit, with its neglect of hard core repulsions between pairs of atoms, carries a tacit implication that so much of the cluster is "open space" that pairwise collisions occur almost always at impact parameters greater than the hard core diameter. This assumption is often appropriate for nuclei. The ratio of typical nuclear densities to the densities of the corresponding nucleons close-packed is in the range 0.01 to 0.1. The same ratio for a cluster of argon atoms is in the range 0.7-0.9, based on the Ar-Ar pair potential. In short, clusters of atoms are mostly "filled up" so cannot conform closely to the nonrigid limit.

Now consider the rigid limit. The obvious first choice in a polyhedron of high symmetry, such as a tetrahedron for four particles or an octahedron for six particles. These are of course among the rigid-limit models we choose, but in order to associate a vibration-rotation spectrum with a structure and a mechanism for large-amplitude motion, we must compare the experimental data with predicted spectra based on all the reasonable alternatives, in the same spirit as the predictions of Dalton, Brocas and Fastenakel.

Among the rigid-limit cases, one in fact does hold a special position, and much of our discussion will be expressed in terms of that case. It consists of a spherical top, meaning that all three principal moments of inertia are equal, with 3N-6 vibrational modes all harmonic and degenerate. This model, artificial as it is, has a symmetry high enough to allow us to construct a level pattern just as we did for the nonrigid limit. Moreover the artificial degeneracy of the 3N-6 oscillators can be broken easily to correspond to the level pattern appropriate to the molecular symmetry group of the polyhedron.

The symmetry of this rigid limit, the spherical top with 3N-6 degenerate harmonic oscillators, is the direct product of the $O(3) \times O(3)$ symmetry of the top and the $U(3N-6)$ symmetry of the oscillators. One $O(3)$, call it $O(3)^{\ell}$, refers to the laboratory frame; the other, call it $O(3)^{f}$, refers to body frame rotations. (Naturally we neglect all vibration-rotation interactions in this limit.) The level pattern of this model is a set of rotationless vibrational levels with equal spacing $\hbar\omega$ and degeneracy

$$g_{\nu} = \frac{[3N-6 + \nu - 1]!}{\nu!(3N-7)!} \tag{17}$$

upon each of which is built a ladder of rotational levels with rotational energy $\hbar BJ(J+1)$ and degeneracy $(2J+1)^2$. Thus this rigid limit is characterized by two parameters, ω and B; the nonrigid limit is characterized by the single parameter w.

It is useful to consider other limits akin to the extreme rigid model just described. For example for four identical particles, while the tetrahedron is the obvious choice for the high-symmetry limit, the square, the double diatomic (with two freely-rotating diatomics loosely bound to each other) and the ammonia-like triangular pyramid are all important cases if one is trying to establish a correspondence between hypothetical level patterns and real spectra. Then, even in the tetrahedral limit, one may break the symmetry of $O(3)^{\ell} \times O(3)^{f} \times U(6)$ to that of the spinning tetrahedron $O(3)^{\ell} \times O(3)^{f} \times G$ in which G, the molecular symmetry group, is a subgroup of the group of 48 elements, the permutation-inversion group $S_4 \times (E,E^*)$. Then G may be the 24-element group T_d, or the 12-element group of rotations of the tetrahedron. This choice depends on what operations one considers "feasible" and on the energy resolution of the observations. The most probable case for high resolution spectroscopy corresponds to taking G as T_d.

For the square, the appropriate molecular symmetry group is the 16-element subgroup of $S_4 \times (E,E^*)$ isomorphic to the point group D_{4h}. The double diatomic has two limits of its own, depending on the degree to which rotations of the individual diatomics are hindered. One is isomorphic with either the group of the square or of the tetrahedron. The other is the product of the rotation groups of the two free internal rotors and the rotation group of the entire four-body system.

Now we turn to the task of correlating the rigid limit with the nonrigid limit. The procedure, in essence, consists of breaking the symmetries of the two limiting cases, splitting up their irreducible representations to correspond to a lower symmetry in which the only constants of the motion are those we can assure ourselves are universally valid: total angular momentum, parity and the character of the permutational symmetry.[1] The group for this intermediate situation is the kinematic group $0(3) \times S_4$ or $SO(3) \times (E,E^*) \times S_4$. More generally, the kinematic group $0(3) \times S_N$ is the appropriate group for the intermediate region that connects the rigid limit of $0(3) \times 0(3) \times U(3N-6)$ with the nonrigid limit of $U(3N-3)$. The problem of correlation is now the mathematical problem of finding the irreducible representations of $0(3) \times S_N$ in the physically symmetric representations of the limiting cases, that is, the totally symmetric representations of $U(3N-3)$ and the representation of $0(3)^{\ell} \times 0(3)^{f} \times U(3N-6)$ that are totally symmetric in their $U(3N-6)$ parts.

The rotational part of the correlation at the rigid limit is easy. The $0(3)^{\ell}$ of the rigid limit, corresponds to the rotation of the laboratory frame, and thus to the isotropy of space; this $0(3)^{\ell}$ is the same as the $0(3)$ of the kinematic group. The rest of the correlation at the rigid limit is a bit more subtle. The reasoning here follows the lines of Hougen (1962, 1963) and Bunker

---

1. If spin were to be treated from the outset, one would also include the groups $[SU(2s+1)]^N$, taking the appropriate representations corresponding to states of total spin S. This is formally correct and is probably the most efficient way to handle the analysis of systems with more than four or five particles: however for smaller systems, it is easy to put in the restrictions of spin after the spatial analysis is complete.

(1976). The permutations are not to be identified solely with elements of the group of rotations of body-fixed axes or with elements of the point group or molecular symmetry group. Rather, they are in general equivalent to products of one element from each of these groups. Figure 6 illustrates this, with the permutation (123) shown as the product of the point group operation C_3 (about the C-axis) and the rotation $R(2\pi/3)$ or R_{3c}, for a three-body system drawn with an equilateral triangle as the reference configuration. The permutation changes only the indexing; the body-fixed frame, however, is defined with respect to the indexing. The point group operation (clockwise) interchanges the local displacement coordinates but leaves the indexing the same and therefore does not change the body-fixed frame. The rotation (counterclockwise) moves atoms and displacement coordinates. Thus $(123) = C_{3c} \times R_{3c}$.

v	π	g_v	J=0	J=1	J=2	J=3	J=4	J=5	J=6	J=7	J=8
0	+	1	1e 1,0,0								
1	-	6		2o 0,0,1							
2	+	21	3e 1,0,1	1e 0,1,0	3e 1,0,1						
3	-	56		6o 1,1,2	2o 0,0,1	4o 1,1,1					
4	+	126	6e 2,0,2	3e 0,1,1	9e 2,1,3	3e 0,1,1	5e 1,0,2				
5	-	252		12o 2,2,4	6o 1,1,2	12o 2,2,4	4o 1,1,1	6o 1,1,2			
6	+	462	10e 3,1,3	6e 0,2,2	18e 4,2,6	9e 1,2,3	15e 3,2,5	5e 0,1,2	7e 2,1,2		
7	-	792		20o 3,3,7	12o 2,2,4	24o 4,4,8	12o 2,2,4	18o 3,3,6	6o 1,1,2	8o 1,1,3	
8	+	1287	15e 4,1,5	10e 1,3,3	30e 6,4,10	18e 2,4,6	30e 6,4,10	15e 2,3,5	21e 4,3,7	7e 1,2,2	9e 2,1,3

$\alpha=4$ $\alpha=3$ $\alpha=2$ $\alpha=1$ $\alpha=0$

Fig. 6. Decomposition of U(6) nonrigid limit states according to parity, (e or O), O(3) and S_3. Large entries in the square table are the number of times each J appears; under each is the triplet of numbers giving the number of times the S, A and M representations occur. The index α corresponds to the number of vibrational quanta the state has at the rigid limit.

By expressing all the elements of the permutation group S_N as products of elements of $\mathcal{D}$ and elements of the rotation group $O(3)^f$, one can find the decomposition chain that allows the classification of irreducible representations of the rigid limit group according to the representations of the kinematic group. For example the three-body case breaks down as follows: $O(3)^f$ breaks to $O(2)^f$ when one makes a symmetric top (equilateral triangle) from the hypothetical spherical top, which breaks further to give R_3, the finite rotations $\pm 2\pi/3$; U(3) breaks to give the point group D_3 or C_{3v} via the chain

$$U(3) \supset U(2) \supset O(2) \supset C_{3v} \text{ or } D_3. \tag{18}$$

The result for this example is

$$O(3)^\ell \times O(3)^f \times U(3) \supset O(3)^\ell \times [O(2)^f \times U(2)] \supset O(3)^\ell \times [O(2)^f \times O(2)] \times O(3)^\ell \times [R_3 \times D_3] \supset O(3)^\ell \times S_3. \tag{19}$$

The parity classification is readily derived in any of several ways, e.g., by direct examination of the form of the wavefunctions of the kinematic group.

The decomposition of the nonrigid limit U(3N-3) to $O(3)^\ell \times S_N$ is an extension of the problem of decomposing a unitary group to an orthogonal group. The decomposition of U(3) and SU(3) have been studied particularly carefully (Elliott, 1958a,b; Bargmann and Moshinsky, 1960, 1961; Karl and Obryk, 1968; Hecht and Braunschweig, 1975; McKay, Patera and Sharp, 1975; Moshinsky et al., 1975) but several general methods are available for performing this reduction. Some are very straightforward but awkward for large N or for high quantum states (Hamermesh, 1962, esp. Chap. 10; Wybourne, 1974; Amar, Kellman and Berry, 1979; Kellman, Amar and Berry, 1979-80), and others are more mathematically elaborate but are well suited for machine computation of the reduction of representations of large size (Louck, 1965; Butler and Wybourne, 1971; Wybourne, 1973; Braunschweig and Hecht, 1978; Mikhailov, 1978).

The most natural path for decomposing U(3N-3) follows the chain

$$U(3N-3) \supset U(N-1) \times U(3) \supset \underset{\displaystyle \underset{\displaystyle S_N}{\cup}}{O(N-1)} \times O(3)^\ell. \tag{20}$$

This chain becomes clear when we write the Hamiltonian in terms of ladder operators doubly labeled for the N-1 three-dimensional oscillators and for the three space dimensions of each:

$$H = \sum_{j=1}^{N-1} \frac{p_{\lambda_j}^2}{2m} + \frac{K}{2} \lambda_j^2$$

$$= \sum_{s,t=1}^{N-1} \sum_{j,k=1}^{3} [\delta_{st}\, \delta_{jk}\, C_{jk}^{st} + \frac{3N-3}{2}] \tag{21}$$

where

$$C_{jk}^{st} = (a^+)_j^s\, a_k^t \tag{22}$$

and $(a^+)_j^s$ and a_k^t are the usual ladder (creation and annihilation) operators

$$a_i^s = (2m\hbar w)^{-1/2}(mw\lambda_i^{(s)} + ip_{\lambda i}^{(s)}),$$

$$(a^+)_j^t = (2m\hbar w)^{-1/2}(mw\lambda_j^{(t)} - ip_j^{(t)}) \tag{23}$$

with the usual commutation rules

$$[a_i^s, a_j^t] = 0, \; [(a^+)_i^s, (a^+)_j^t] = 0, \; [a_i^s, (a^+)_j^t] = \delta_{st}\delta_{ij} \tag{24}$$

The operators C_{jk}^{st} and their adjoints, $C_{kj}^{ts} = (C_{jk}^{st})^+$, are generators for the Lie algebra of U(3N-3) (Jauch and Hill, 1940; Kramer and Moshinsky, 1968).

Contracting the operators of (22)

$$C^{st} = \sum_{j=1}^{3} C_{jj}^{st}$$

and

$$C_{jk} = \sum_{s=1}^{N-1} C_{jk}^{ss}, \tag{25}$$

we obtain two new, smaller sets of operators, 9 of the C_{jk}'s and $(N-1)^2$ of the C^{st}'s. These each satisfy the commutator combination rule that defines Lie multiplication:

$$[C_{ij}, C_{k\ell}] = \delta_{jk} C_{i} - \delta_{i\ell} C_{kj}$$

and

$$[C^{qr}, C^{st}] = \delta^{rs} C^{qt} - \delta^{qt} C^{sr}. \tag{26}$$

The operators C_{jk} generate the Lie algebra and the associated Lie algebra and the associated Lie group U(3). The operators C^{st}, similarly, generate U(N-1). Among all the unitary transformations associated with the three space dimensions are the proper and improper rotations. Hence U(3) contains O(3) as a subgroup. The group U(N-1) contains the symmetric group S_N as a proper subgroup (Kramer and Moshinsky, 1968). In physical terms, constructing the C_{jk}'s is constructing the spatial properties of the system averaged over all the N-1 oscillators; constructing the C^{st}'s is constructing the properties of the system of N-1 oscillators averaged over all orientations. This can be done because we can transform the Hamiltonian into its expressions in Jacobi vectors without losing its separability into x,y and z-components in the laboratory frame.

We shall not give a general method here for determining the O(3) content of the U(3) representations or the S_N content of U(N-1). Rather, we illustrate with the simple example of the 3-body problem (Kellman, Amar and Berry, 1979-80). Define the Jacobi coordinates

$$\underset{\sim}{\rho} = 2^{-1/2}(\underset{\sim}{r}_1 - \underset{\sim}{r}_2)$$

and

$$\underset{\sim}{\lambda} = 6^{-1/2}(\underset{\sim}{r}_1 + \underset{\sim}{r}_2 - 2\underset{\sim}{r}_3). \tag{27}$$

The ν quanta, corresponding to the νth symmetric irreducible representation of U(3), may be assigned to the $\underset{\sim}{\rho}$ and $\underset{\sim}{\lambda}$-degrees of freedom and catalogued $\nu = \nu_p + \nu_\lambda$. The possible assignments of three quanta are

$$[\nu_\rho,\nu_\lambda] = [3,0],[2,1],[1,2] \text{ and } [0,3] \tag{28}$$

Each assignment ν_ρ or ν_λ has the rotational state content of a three-dimensional harmonic oscillator with ν_ρ or ν_λ quanta. For example the one-quantum state of the three-dimensional oscillator contains the triply-degenerate J = 1 state once and no other states. The three-quantum state contains J = 1 and J = 3 each once, for a total degeneracy of (2×1+1) + (2×3+1) = 10. The compound state with $\nu_\rho + \nu_\lambda$ quanta has a J-content corresponding to the combinations that can be generated by addition of the J_ρ's with the J_λ's. Hence the [2,1] partition of ν = 3 corresponds to the combinations of J_ρ = 0 and 2 with J_λ = 1, giving J = 1,1,2 and 3. We have denoted this as

$$\{J_1,J_2,...\}[\nu_\rho,\nu_\lambda],$$

so the J-content of [2,1] is

$$\{1,1,2,3\}[2,1]$$

and the J-content of [2,2] is

$$\{0,0,1,2,2,2,3,4\}[2,2].$$

The total J-content of $\nu = 3$ which we denote $\langle J(3)\rangle$, consists of

$$\langle J(3)\rangle = \{1,3\}[3,0] + \{1,1,2,3\}[2,1] + \{1,1,2,3\}[1,2] + \{1,3\}[0,3]. \tag{29}$$

with a total degeneracy of 56. The parity is given by $(-1)^{\nu}$.

A similar partitioning procedure can be used to find the S_3 content of the $U(2) \times O(2)^{r}$ appropriate to the 3-body problem. Instead of partitioning according to ρ and λ, we now partition into three groups according to the value of J_Z, i.e. (ρ_1,λ_1), (ρ_2,λ_2) and (ρ_3,λ_3), defined by (27). It is easy to construct the short catalogue of the action of the permutation operators of S_3 - E, (12), (13) (23), (123) and (132) - on the components of ρ and λ. For example if we call ρ_x as defined by (27) $\rho_x(12)$, then

$$(123)\rho_x(12) = \rho_x(31) = 2^{-1/2}(\underset{\sim}{r}_{3x} - \underset{\sim}{r}_{1x}). \tag{30}$$

These permutations are a subset of the unitary transformations of ρ_x and λ_x, and similarly of the other pairs ρ_y,λ_y and ρ_z,λ_z. They are in fact a subset of these unitary transformations that constitute the rotations in two dimensional spaces defined by ρ_i,λ_i. In short, S_3 is a subgroup of O(2) which is itself a subgroup of U(2). The same properties hold whether we use the Cartesian ρ_x,λ_x and ρ_y,λ_y or the chiral ρ_+,λ_+, and ρ_-,λ_-; $\rho_z,\lambda_z = \rho_o,\lambda_o$ of course.

Now we partition ν, the number of quanta, into $[\nu_1,\nu_0,\nu_{-1}]$; all partitions of ν into these groups are allowed. Hence for $\nu = 3$, we have [3,0,0], [0,3,0], [0,0,3], [2,1,0], [1,2,0], [0,1,2], [0,2,1], [1,0,2] and [2,0,1], and [1,1,1]. The S_3 states themselves can be classified according to the three kinds of irreducible representations of S_3, which in turn correspond to the three classes of S_3, the symmetric (S), the antisymmetric (A) and the other, which we call mixed (M). The first two are one-dimensional; M is two-dimensional. For $\nu = 0$, only the nondegenerate S representation occurs. For $\nu = 1$, one obtains the M representation once, for $\nu = 2$, one has S + M with $g_\nu = 3$. With three quanta, $g_\nu = 4$ and one has representations S, A and M, and so forth. The multiplication of representations of S_3 is $S \times S = A \times A = S$, $S \times M = A \times M = M$, $S \times A = A$, and $M \times M = S + A + M$. Thus, for example, the partition [1,2,0] has content

$$(M) \times (S+M) \times (S) = M + (A+S+M) = S + A + 2M. \tag{31}$$

The total S_3 content of the $\nu = 3$ level is (S+A+2M) six times, plus (S+M) three times, plus M×M×M = S+A+3M, or 10S + 10A + 18M, 56 states in all as required.

Finally, we must classify the states according to the direct product group $S_N \times O(3)$. This requires finding the S_N content for each specific J value. The straightforward method used for the three-body case illustrates what is involved (Kellman, Amar and Berry, 1979-80). One looks first at the S_N content of J_{max} within the states of a given ν - here, S_3 content. From (29) we know that

$$\langle J(3)\rangle = 4(J=3) + 2(J=2) + 6(J=1) \tag{32}$$

The four states J = 3, J_z = 3 correspond to the partition [3,0,0], which, in turn has the S_3 content S+A+M because all three quanta are assigned to ρ_+,λ_+. But the other components of J = 3 with J_z = 2,1,0,-1,-2 and -3 must also have the same S_3 composition, so, in all, the set of states with J = 3 contains 7(S+A+M). The [2,1,0] partition contains all the J_z = 2 states, the S+A+M of J = 3 and whatever is left to belong to J = 2. But [2,1,0] contains S+A+2M so J = 2, J_z = 2, can contain only M. There are necessarily 5M's contained in each of the manifolds of the two J = 2 states. Thus the J = 2 and J = 3 states together contain S+A+2M for J_z = 1,0 and -1. The partitions [1,2,0] and [2,0,1] together contribute 2S + 2A + 4M, of which we have just accounted for S+A+2M. Hence the other S+A+2M of these belong to the J = 1, J_z = 1 states. The total S_3 content of $\nu = 3$ is thus

$$7(S+A+M) + 5(M) + 3(S+A+2M) = 10S + 10A + 18M, \tag{33}$$

with total degeneracy of 56 as required for $\nu = 3$.

Figure 6 gives the decompositions of the U(6) nonrigid limit states under O(3) , S_3 and parity, for states through the eight-quantum level. This table was derived by Karl and Obryk (1968) in the context of constructing a concrete three-quark model of a hadron, well before we rediscovered it independently. One point not noticed by Karl and Obryk (and probably not relevant to their analysis) makes the three-body case especially neat: the staircases of heavy lines separate the states according to the number of rigid-limit vibrational quanta, denoted by α in the Figure. In other words, for this case the nonrigid levels fall into an especially simple pattern with respect to their rigid-limit counterparts, a pattern that has an immediate physical interpretation.

We can now construct the correlation diagrams by starting with the ground state, connecting limiting cases with the same J, permutation group symmetry and parity, relying on the noncrossing rule to give us the ordering. The first example is Figure 7, the

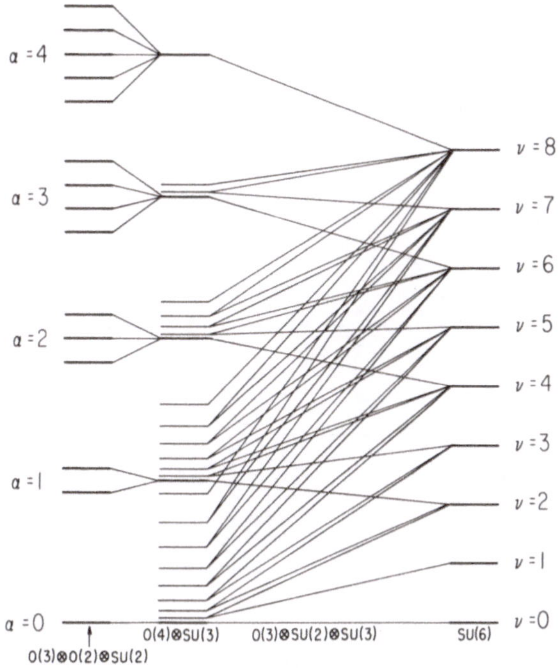

Fig. 7. The correlation diagram for the three-body problem; the rigid limit of an equilateral triangle is at the left and the nonrigid limit, at the right. The index α is the number of vibrational quanta in the rigid limit. At the far left we indicate very schematically how the successive levels with no rigid-rotor rotations split when the symmetry is reduced to that inherent in the equilateral triangle, without the requirement that the molecule be a spherical top. The O(3) × O(3) of the top is indicated at bottom by the isomorphic group O(4).

three-body case. The second, Figure 8, shows the correlation of levels for the 4-body problem with the tetrahedron as a rigid limit. Figure 9 is a second 4-body case, in which the "rigid" limit is the double diatomic. The ratio ω/w of rigid to nonrigid oscillator frequencies, is taken to be ca. 2. The ratio of ω to the rotational constant B is based on the model of $(H_2)_2$.

6. THE PARAMETRIZATION OF NONRIGIDITY

One can extract from the correlation diagrams a parameter to characterize the degree of nonrigidity of a molecule. The concept is related to the ratio B_e/ω_e of rotational constant to vibrational frequency of a diatomic molecule; this ratio measures the degree

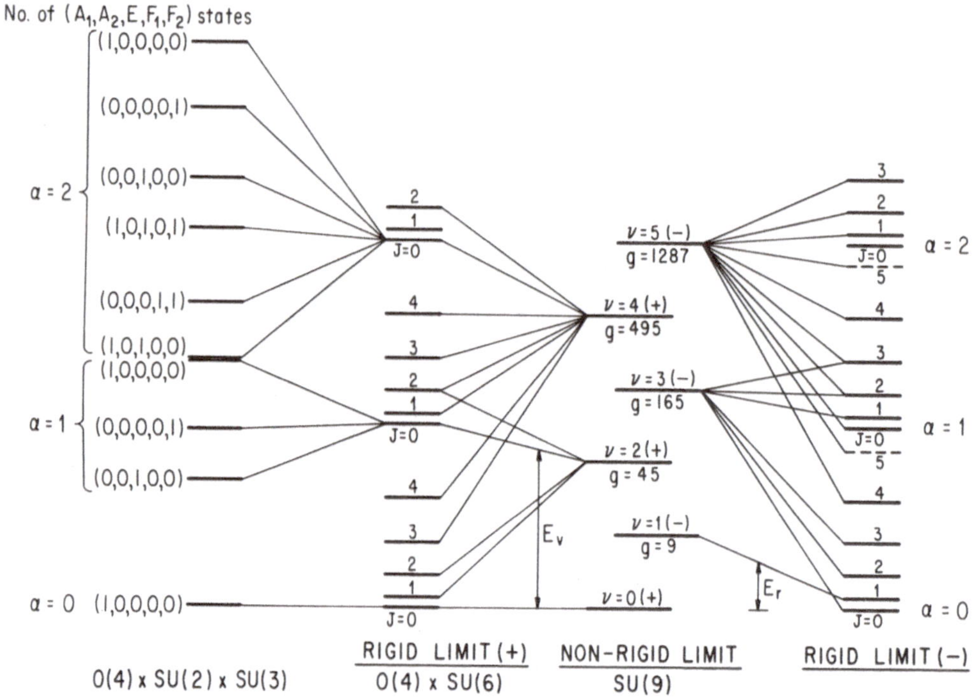

Fig. 8. The correlation diagram for the 4-body problem with a tetrahedron as the rigid limit. For clarity, the even and odd sets of levels are put on opposite sides of the nonrigid case. At far left are the splittings of some of the J = 0 states when the requirements of degeneracy are reduced from the artificial U(6) to the degeneracies inherent in the oscillations of a regular tetrahedron. The O(3) × O(3) of the spherical top is again given as the isomorphic O(4). (Taken from Amar, Kellman and Berry, 1979).

to which vibrations and rotations are separable in the simplest of molecules. The parameter we use is even more closely related to the parameter introduced by Yamada and Winnewisser (1976) to describe the degree of deviation from linearity of chain-like molecules. Their parameter is

$$\gamma_{yw} = 1-4[\Delta E_{min}(K \text{ or } \ell=1)/\Delta E_{min}(K \text{ or } \ell=0)] \tag{34}$$

the ratio of the excitation energy to the lowest state with K or ℓ=1 to the excitation energy to the lowest state with the same quantum number, K or ℓ, equal to zero. This ratio falls between -1 for most linear molecules and +1 for well-bent molecules, but may be less than -1.

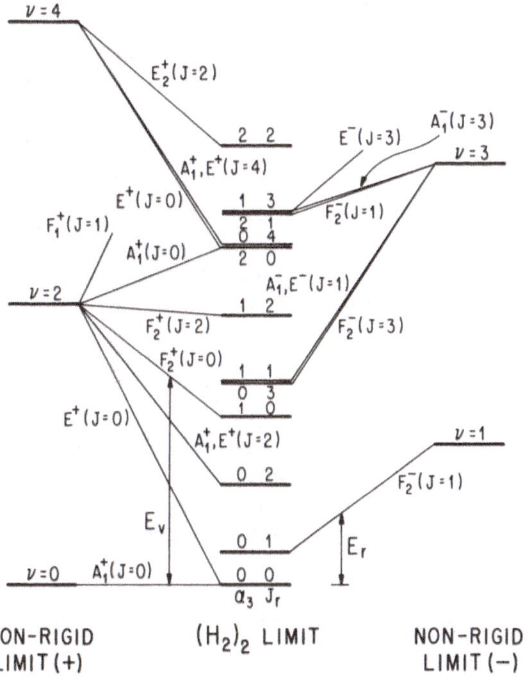

Fig. 9. The correlation diagram for the 4-body problem with $(H_2)_2$ as the model for the "rigid" limit. Here the even and odd levels of the nonrigid limit are put at the left and right of the diagram. The vibrational quantum of the H_2-H_2 oscillation is α_3; the individual H_2 frequencies are too large for their first excited states to appear in this diagram.

Our parameter is based on the correlation diagrams such as those of Figures 7, 8 and 9 and on the vibration-rotation spectrum of the molecule of interest. The former makes it dependent on the choice of the model of the rigid limit, however in most cases, we expect one would use the $O(3) \times O(3) \times U(3N-6)$ spherical top with degenerate normal modes.

The parameter is defined as the ratio of two excitation energies of the idealized model or, for any real molecule, the ratio of two <u>average</u> excitation energies. Let E_r be the excitation energy from the ground state to the first state that correlates with the level $\alpha = 0$, $J = 1$ of the rigid limit, i.e., to the first rotational, nonvibrational state of that limit. Let E_v be the excitation energy to the first state of the rigid limit for which $\alpha = 1$, $J = 0$, i.e., to the first nonrotating vibrationally excited state of the rigid limit.

Note that the state defining E_r is an odd-parity state that correlates with states of the level $\nu = 1$ of the nonrigid limit. Similarly, the excited state that defines E_v is an even-parity state (it always includes the totally symmetric vibrations) that correlates with the level $\nu = 2$ of the nonrigid limit. Because the nonrigid limit is harmonic, the excitation energy to $\nu = 2$ is twice that to $\nu = 1$. Hence we introduce a factor of 2 to define our parameter

$$\gamma = 2E_r/E_v \tag{35}$$

So defined, $0 < \gamma < 1$; if the vibrational frequency were allowed to become arbitrarily high, corresponding to the molecule becoming more and more rigid, γ would approach zero. At the other limit, γ becomes 1 as the distinction between rotations and vibrations becomes completely lost.

For any real molecule, one must compute γ by finding all those states that correlate with the state $\alpha = 0$, $J = 1$ and with $\alpha = 1$, $J = 0$, and computing the mean energy of each set. These means correspond to what E_r and E_v would be if the molecule had the high symmetry of the correlation diagrams. Table 1 contains values of γ for several triatomic (Kellman, Amar and Berry, 1979-80) and tetratomic molecules (Amar, Kellman and Berry, 1980). Just as we expect virtually no molecules to correspond closely to the nonrigid limit, we expect to find most values of γ near zero, not near unity. If one were to compute γ for <u>nuclei</u>, on the other hand, we would expect most values to lie between 0.1 and 1, reflecting their openness and proximity to the nonrigid limit (Elliott, 1962; Chacon and Moshinsky, 1977).

Note that γ is a phenomenological parameter that depends in no way on any construction of a potential surface. This is especially important because it releases us from the dilemma created by the self-consistent, <u>effective</u> nature of any potential surface one constructs for motion of atoms within a molecule. We normally work with effective potentials based on the Born-Oppenheimer approximation, but higher approximations make the effective potential reflect deviations from this limit (Kolos and Wolniewicz, 1964, 1965, 1968, 1969). In particular this means that the effective potential will, in general, depend on the velocities of the nuclei and not only on their positions; more broadly, it means that strictly the effective potential must depend on the degree of nonrigidity of the molecule. We can only expect a single

Table 1. The nonrigidity parameter γ for various triatomic and tetratomic molecules. Values of γ are based on observed spectra except for Ar_3, Ar_4 and $(H_2)_2$.
(See Amar, Kellman and Berry, 1979, and Kellman, Amar and Berry, 1979-80, for details.)

Molecule	γ
H_2O	0.023
Ar_3	0.013
NO_2	0.0082
O_3	0.0061
SO_2	0.0035
ClO_2	0.0033
$(H_2)_2$	0.40
NH_3	0.0336 (inversion only) or 0.0133 (all fundamentals)
CH_4	0.0127
Ar_4	0.00772
P_4	0.00096

effective potential to be a very approximate tool, at best, across the full range of the correlation diagram. It will be interesting in coming years to see the differences between effective potentials calculated for clusters in rigid and nonrigid limits. These calculations would correspond to the evaluation of the effective potentials for solid-like and liquid-like clusters with the same number of particles. We shall return to this subject later.

7. THE INTERPRETATION OF NONRIGIDITY

We have just seen that it is possible to parametrize nonrigidity. We have a clear mechanical notion of its meaning in the limit of small deviations from the nearly-rigid oscillator model; there, we can think in terms of passage along a well-defined path of changing molecular configurations, in the manner that one attributes "mechanisms" to chemical reactions. But what can we use for a conceptual picture of a species far from the rigid limit? How can we visualize the particle dynamics in a cluster such as a nucleus, an atom or a very nonrigid molecule such as Li_3? Having good methods of obtaining the eigenfunctions and eigenvalues for such systems may be necessary for understanding their internal dynamics but it is certainly not sufficient.

In this section, we shall outline briefly how one may approach this question, and how we have used one of these approaches to interpret nonrigidity in atoms, for which we can solve Schroedinger equations. The method appears to be a powerful interpretive tool whenever one has wavefunctions that describe the system's rovibronic states (or their counterparts). The discussion points to the need for methods of calculating such wavefunctions for nonrigid molecules and clusters comparable to the methods we have for treating electrons in atoms, including their correlation.

Concepts of spatial correlation in atoms have been sought from the time it was apparent that we could find wavefunctions that could represent correlation reasonably accurately. The notions of Coulomb and Fermi holes, regions around one electron depleted in the probability density of other electrons, have been variously defined and computed (Maslen, 1956; Löwdin, 1959; Coulson and Neilson, 1961; Curl and Coulson, 1965; Sinanoglu and Brueckner, 1970). Coulson and Neilson define the Coulomb hole of electrons 1 and 2 as the difference between the distribution functions of the distance r_{12} calculated with a correlated wavefunction and with a Hartree-Fock wavefunction. As one would expect, this "hole" in the helium atom ground state is deepest when the two electrons are close, about 0.7-0.8 bohr, and the total negative part of the "hole" contains only about one-twentieth of an electron. Moreover for He and isoelectronic positive ions, the shape of this density difference seems to have a single form. Doggett (1977) extended the use of this kind of correlation study to diatomic molecules by defining longitudinal and transverse "holes".

Extensive computational studies have been made of other distribution functions. Joint radial distributions, $f(r_1,r_2)$, of the distances of two electrons from their common nucleus have been used by Dickens and Linnett (1957), Macek (1967, 1968), Fano and Lin (1974) and Fano (1976). The distribution function of the angle $\theta_{12} = \underline{r}_1 \cdot \underline{r}_2 / |\underline{r}_1||\underline{r}_2|$ or its cosine has also been used (Banyard and Ellis, 1972, 1975; Tatum, 1976; Rehmus, Kellman and Berry, 1978). Previously Banyard and Baker (1969) had used the mean of $\cos\theta_{12}$. All these approaches were predicated on the existence of previous "good" calculations of wavefunctions that include correlation.

A completely different philosophy underlay the approach of Herrick and Sinanoglu (Sinanoglu and Herrick, 1975, 1976; Herrick and Sinanoglu, 1975) and Wulfman (1973, 1976; Wulfman and Kumei, 1973). Here, the authors began with the picture that electrons in a helium atom are, first of all, much like hydrogenic electrons but deviate from hydrogenic behaviour because of the electron-electron repulsion, $-e^2/r_{12}$. This in turn implies that a reasonable first approximation for describing two-electron or even many-electron atoms would be a representation based on the O(4) symmetry of the bound states of the hydrogen atom. Recall that the

n^2-degeneracy of the nth level of the hydrogen atom arises from the invariance of its Runge-Lenz vector

$$\underset{\sim}{A} = \frac{1}{2m}[\underset{\sim}{p} \times \underset{\sim}{L} - \underset{\sim}{L} \times \underset{\sim}{p}] - e^2 \frac{\underset{\sim}{r}}{r}$$

(Pauli, 1926; Fock, 1935; Wulfman, 1971). Wulfman showed that an approximate way of diagonalizing the interaction potential r_{12}^{-1} is to diagonalize the square of the difference ot the two one-electron Runge-Lenz vectors, $A = |\underset{\sim}{A}_1-\underset{\sim}{A}_2|^2$. Herrick and Sinanoglu showed independently that if one diagonalizes this operator with a very small basis set, that one obtains remarkably good representations of doubly-excited states of helium. Specifically, they used only the states of given L and S having the same assignments of principal quantum numbers, such as the ^1S states based on $2s^2$ and $2p^2$ configurations. Physically, the Runge-Lenz vector is essentially the vector of the semimajor axis of the classical Kepler ellipse. Hence being approximately diagonal is equivalent to the length of the vector connecting the Kepler ellipses of the two electrons being a constant. In other words the Herrick-Sinanoglu-Wulfman model implies that the Kepler orbits of the two electrons of a helium atom precess together. This is quite a different sort of information than one gets from an analysis of distribution functions. The method is essentially one of finding approximate invariants for an $O(4) \times O(4)$ symmetry broken by r_{12}^{-1}. It works well when $n_1 \cong n_2$, but a different decomposition of $O(4) \times O(4)$ must be used when the two principal quantum numbers are very different (Nikitin and Ostrovsky, 1976).

Stimulated by the striking accuracy of the Herrick-Sinanoglu-Wulfman picture, and by its use of broken Lie group symmetry, we sought to find a more precise spatial picture of what made those functions so successful. An analytic expression for two-electron atoms was derived (Rehmus, Kellman and Berry, 1978) giving the distribution function $\rho(r_1,r_2,\theta_{12})$ in the three variables of electron-nuclear distances and electron-electron angle. From this function one can readily construct the two-electron <u>conditional</u> probability density

$$\rho(r_2, \theta_{12}|r_1=\alpha) = \frac{\rho(r_1=\alpha,r_2,\theta_{12})}{[\int dr_2 \int d\theta_{12}\, \rho(r_1,r_2,\theta_{12})]_{r_1=\alpha}} \qquad (36)$$

This function is particularly useful as an interpretive tool because it can be computed readily, it shows how angular correlations depend on the radial coordinates of the electrons, and it can be displayed graphically. Long before an analytical representation of $\rho(r_1,r_2,\theta_{12})$ was found, Munschy and Pluvinage (1963) constructed a single contour diagram giving essentially $\rho(r_2,\theta_{12}|r_1)$ for a single value of r_1, for the ground state of helium. More recently, we have computed many diagrams of conditional probability densities for Herrick-Sinanoglu-Wulfman functions for doubly-excited states of helium (Rehmus, Kellman and Berry, 1978), for a number of approximate and accurate representations of the ground states of H^- and helium (Rehmus, Roothaan and Berry, 1978), and for accurate representations of several S-states - $1s2s\,^1S$ and ^3S and ($2s^2$ and $2p^2$) doubly-excited states, as well as the ground state - of helium (Rehmus and Berry, 1979). An example is given in Figure 10, to show the way the conditional probability distribution varies as the distance r_1 changes.[1]

The use of broken O(4) symmetry has been extended to many-electron atoms (Herrick and Kellman, 1978), where one can obtain particularly clear information regarding the pair angular distribution $\rho(\theta_{12})$. For two electron atoms, Herrick (1977) showed that the orbital basis that diagonalizes A^2 is very much like a tetrahedral hybrid orbital basis, for the ($2s^2$, $2p^2$) S-states of helium. The most recent results to come from this line of investigation has been the recognition of sets of rigid-rotor-like states among the doubly-excited states of helium (Kellman and Herrick, 1978) that can be classified as supermultiplets (Herrick, Kellman and Poliak, 1979-80) analogous to the supermultiplet structure of nuclei (e.g. Kramer and Moshinsky, 1968).

7. CONCLUSION. THE PROPERTIES OF CLUSTERS

This final section is intended as a brief overview of the application of the analysis of small, nonrigid molecules to the theory of properties, especially melting properties, of small clusters and of nucleation. The ideas coming from the previous discussion come from the recognition that our models of rigid and nonrigid molecules give us their energy levels and their degeneracies; that from these, we can construct approximate density-of-state functions, from which we can construct partition functions and thermodynamic functions; and from these, we can try to predict the melting behaviour of clusters of fixed size and the dependence of the stability of a cluster on its number of component particles. This approach completely avoids the problems of defining bulk and surface properties that have plagued classical nucleation theory for decades.

McGinty (1971) used two forms of a harmonic oscillator, rigid rotor model to compute partition functions and free energies of clusters containing as few as 2 or 3 particles and as many as 100.

---

1. This is particularly vivid when it is displayed as a motion picture; the film, produced by Paul Rehmus, David Campbell and R. Stephen Berry, will be released soon.

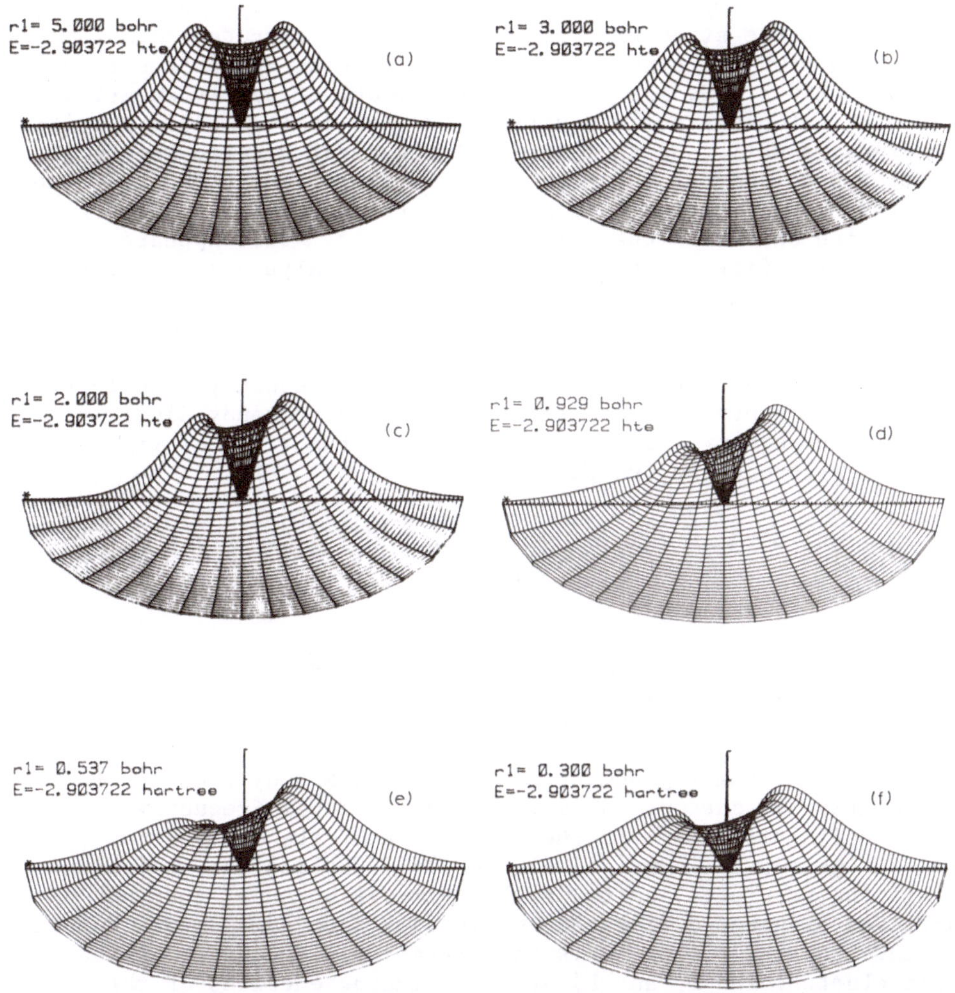

Fig. 10. Conditional probability densities; $\rho(r_2\theta_{12}|r_1 = \alpha)$, for the helium ground state demonstrating how the "core" density of electron 2 shifts under the influence of a second electron which is at various distances from the nucleus. The maximum radial distances in each graph is 2.5 bohr and the interelectronic angle, θ_{12}, sweeps from 0 to π radians. The asterisk always denotes $\theta_{12} = 0$ radians. The chosen distances of r_1 in (a) and (b) are off the scale. The shift toward $\theta_{12} = \pi$ in core density evident from (a) to (c) is the polarization effect exerted by electron 1. The Coulomb hole is most evident in (d) where r_1 has its expected value, and in (e) where r_1 has its most probable value. The densities are computed from a 26-term Hylleraas-Kinoshita function. (From Rehmus and Berry, 1979).

He allowed the computer to move atoms in the cluster until the classical stable structure was achieved, then computed normal mode frequencies for the structure and used these to obtain their partition functions. He also used another model of nearly spherical clusters, not necessarily equilibrated, and derived their partition functions as well. He then used the classical kinetic theory of nucleation to compute the rate of formation of the cluster of the size that gives it the maximum free energy per particle of formation, which should be the rate of nucleation. This appears to be one of the first statistical-mechanical analyses of cluster behaviour based on modeling the energy spectra of the clusters, and is, in that sense, very close to the spirit of our discussion.

If, instead of the general normal mode frequency distribution assumed by McGinty, we used the same single-frequency Einstein model we invoked previously, the partition functions of rigid and nonrigid clusters are, respectively,

$$q_{rigid} = \frac{\pi^{1/2}}{\sigma} \left(\frac{T}{B_e}\right)^{3/2} \left(\frac{e^{-\omega/2T}}{1-e^{-\omega/T}}\right)^{3N-6} e^{D_e/T} \tag{37}$$

and

$$q_{nonrigid} = f(N) \left(\frac{e^{-w/2T}}{1-e^{-w/T}}\right)^{3N-3} e^{-D_e'/T} \tag{38}$$

Here σ is the symmetry number for the rigid N-body cluster, B_e is its rotational constant and ω is its vibrational frequency. The function f(N) is $(N!)^{-1}$ if the time scale of the observation is long enough for the nonrigid cluster to establish the equivalence of all the particles, unity if that time is so short that no particles exchange during the observation, and some smooth function for intermediate times; w is the vibrational frequency of the nonrigid cluster, and D_e and D_e' are the binding energies of the rigid and nonrigid clusters, respectively. The supposition that we can give physical significance to both rigid and nonrigid clusters of N identical particles of a given kind is equivalent to the assumption that such a cluster may take on either solid-like or liquid-like behaviour. It may be that we must invoke the previously-mentioned idea that the effective potentials of the rigid and nonrigid limits may be different, in order to make this notion hold at realistic temperatures and for realistic cluster sizes; this is only a caveat, not a fact, but it should be kept in mind.

If clusters can melt, one should be able to observe such behaviour. In fact, melting of small clusters has been observed in computer simulations. These simulations have been done by both Monte Carlo and molecular dynamics methods (McGinty, 1973; Lee, Barker and Abraham, 1973; Briant and Burton, 1975; Kaelberer and Etters, 1977). Both Briant and Burton's dynamics calculations and Kaelberer and Etters' Monte Carlo work implied that very small clusters of argon atoms should exhibit melting.

We look for melting by finding whether there is a temperature for which the Helmholtz free energy, $A = -kT\ln q$, of "solid" and "liquid" clusters are equal. Figure 11 shows the values of A for several hypothetical clusters, all based on argon. This means that r(Ar-Ar) = 3.4 Å, and B_e is 0.412 $N^{-5/3}$ (°K). Figure 12 shows three examples of the behaviour of the melting temperature as a function of N: $\sigma = 1$ and $f(N) = 1$ and $D_e = D_e'$; σ assuming a maximum value for the polyhedrons of each N and $f(N) = 1/N!$ and $D_e = D_e'$; and $\sigma = 1$, $f(N) = 1$ but $D_e - D_e' = 0.2\ D_e$. The conclusion is that a set of assumptions can be found that will reproduce the classical calculations of the computer simulations (crosses of Figure 12c) for all but very small N, where the Einstein model is suspect and the classical calculations are not so trustworthy either, especially at the melting temperatures for the low values of w.

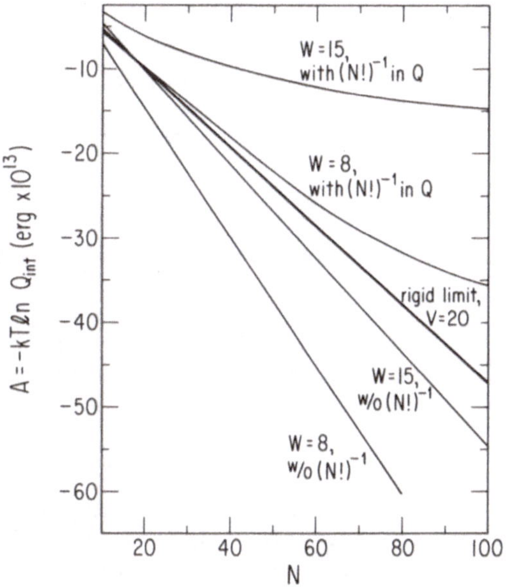

Fig. 11 The Helmholtz free energy A for rigid argon clusters and for several models of nonrigid argon clusters, as functions of the number of atoms N in the cluster.

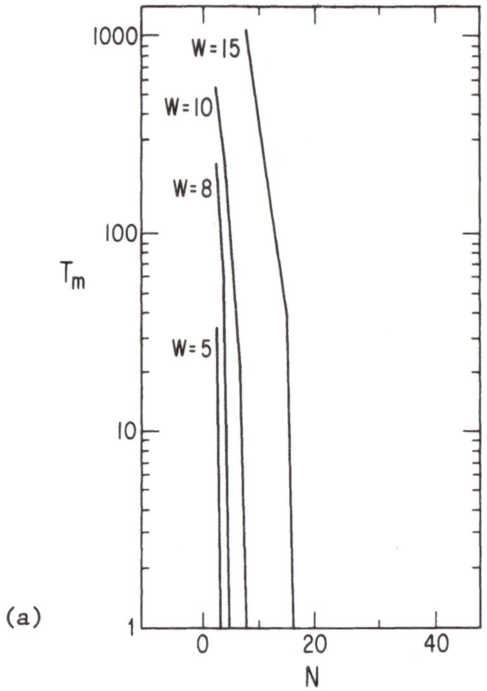

Fig. 12. Three examples of the behaviour of the "melting temperature" T_m of small argon clusters as functions of the number of atoms N in the cluster.

a) σ = 1, f(N) = 1 and $D_e D_e'$;

b) σ has its maximum for each N, f(N) = $(N!)^{-1}$ and $D_e = D_e'$;

c) $D_e - D_e'$ = 142 K, σ = 1 and F(N) = 1; crosses are points from Briant and Burton's (1975) calculations.

The surface free energy of a cluster may be evaluated from its free energy per added particle (Nishioka, 1977). Amar (1979) has shown recently how the free energy computed from a model as apparently naive as the Einstein oscillator we have used does in fact give a nonzero surface tension. Moreover its value is within about 30% of the semiempirical results given by Nishioka, for N above about 100.

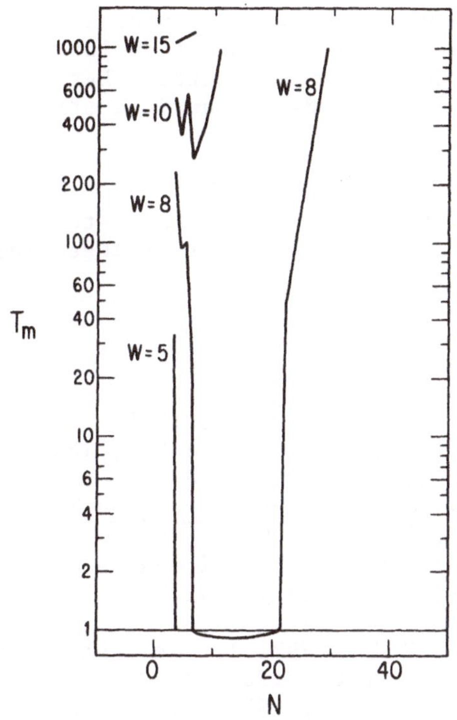

Fig. 12. (b)

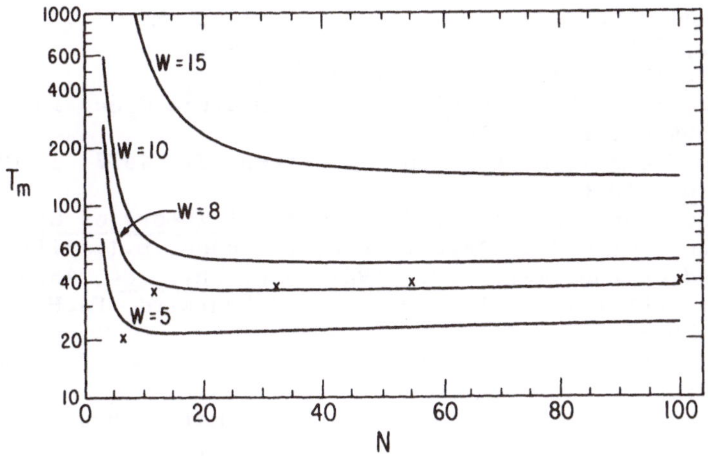

Fig. 12. (c)

We cite this result as a suggestion of the power inherent in methods based on statistical-mechanical evaluation of properties from a knowledge of their energy spectra. Let it emphasize the potential of pursuing the study of nonrigid structures from that viewpoint.

Acknowledgements: The author would like to thank the Aspen Center for Physics for its hospitality; most of these lecture notes were prepared there.

REFERENCES

Aarts, J.F.M., 1978, Molec. Phys. 35, 1785.
Altmann, S.L., 1967, Proc. Roy. Soc. A 298, 1841.
Altmann, S.L., 1971, Mol. Phys. 21, 587.
Altmann, S.L., 1978, Induced Representations in Crystals and Molecules, Academic Press, New York.
Amar, F., 1979, Ph.D. thesis, The University of Chicago.
Amar, F., Kellman, M.E. and Berry, R.S., 1979, J. Chem. Phys. 70, 1973.
Amat, G. and Nielsen, H.H., 1962, J. Chem. Phys. 36, 1859.
Banyard, K.E. and Baker, C.C., 1969, J. Chem. Phys. 51, 2680.
Banyard, K.E. and Ellis, J.D., 1972, Molec. Phys. 24, 1291.
Banyard, K.E. and Ellis, J.D., 1975, J. Phys. B. 14, 2311.
Bargmann, V. and Moshinsky, M., 1960, Nucl. Phys. 18, 697.
Bargmann, V. and Moshinsky, M., 1961, Nucl. Phys. 23, 177.
Bartell, L.S. and Gavin, R.M., Jr., 1968, J. Chem. Phys. 48, 2466.
Bauder, A., Meyer, R. and Günthard, Hs.H., 1974, Molec. Phys. 28, 1974. (erratum, 1976, Molec. Phys. 31, 647)
Bennett, M.J., Cotton, F.A., Davison, A., Faller, J.W., Lippard, S.J. and Morehouse, S.M., 1966, J. Am. Chem. Soc. 88, 4371.
Berry, R.S., 1960a, Revs. Mod. Phys. 32, 447.
Berry, R.S., 1960b, J. Chem. Phys. 32, 933.
Berry, R.S., 1978, Proc. Conf. on The Symmetric Group in Chemistry and Physics, Bielefeld.
Boom, E.W., Frenkel, D. and van der Elsken, J., 1977, J. Chem. Phys. 68, 1826.
Born, M. and Huang, K., 1962, The Dynamical Theory of Crystal Lattices, Oxford University Press, London, p.166-171, 402-405.
Born, M. and Oppenheimer, J.R., 1927, Ann. Phys. 84, 457.
Bossert, W., Ekkers, J., Bauder, A., and Günthard, Hs.H., 1978, Chem. Phys. 27, 433.
Boyd, D.R.J. and Longuet-Higgins, H.C., 1952, Proc. Roy. Soc. A213, 55.
Braunschweig, D. and Hecht, K.T., 1978, J. Math. Phys. 19, 720.
Briant, C.L. and Burton, J.J., 1975, J. Chem. Phys. 63, 2045.
Brickmann, J., 1971, Zeit. fur Electrochem. 75, 747.
Brocas J. and Fastenakel, D., 1975, Molec. Phys. 30, 193.

Bunker, P.R., 1964, Molec. Phys. 8, 81.
Bunker, P.R., 1965, Molec. Phys. 9, 257.
Bunker, P.R., 1971, J. Mol. Spectrosc. 68, 367.
Bunker, P.R., 1976, Chapter 1 in Vibrational Spectra and Structure, J.R. Durig, ed., Marcel Dekker, New York.
Bunker, P.R., 1979 (in press), Can. J. Phys.
Bunker, P.R., 1979, Molecular Symmetry and Spectroscopy, Academic Press, New York.
Bunker, P.R. and Landsberg, B.M., 1977, J. Mol. Spec. 67, 374.
Bunker, P.R., Landsberg, B.M. and Winnewisser, B.P., 1979, J. Molec. Spectrosc. 74, 9.
Bunker, P.R., McLarnon, C.J. and Moss, R.E., 1977, Molec. Phys. 33, 425.
Bunker, P.R., and Moss R.E., 1977, Mol. Phys. 33, 417.
Butler, P.H. and Wybourne, B.G., 1971, Atomic Data 3, 133.
Chacon, E. and Moshinsky, M., 1977, J. Math. Phys. 18, 870.
Cheron, M. and Bordé, J., 1974, Le Journal de Physique 35, 641.
Childs, W.H.J. and Jahn, H.A., 1939, Proc. Roy. Soc. (London) A169, 451.
Cotton, F.A., 1968, Acc. Chem. Res. 1, 257.
Cotton, F.A., 1975, J. Organomet. Chem. 100, 29.
Coulson, C.A. and Neilson, A.H., 1961, Proc. Phys. Soc. 78, 831.
Curl, R.F., Jr. and Coulson, C.A., 1965, Proc. Phys. Soc. 85, 647.
Curtiss, L.A. and Pople, J.A., 1976, J. Mol. Spec. 61, 1.
Dalton, B.J., 1966, Molec. Phys. 11, 265.
Dalton, B.J., 1971, J. Chem. Phys. 54, 9745.
Dalton, B.J., Brocas J. and Fastenakel, D., 1976, Molec. Phys. 31, 1887.
Dalton, B.J. and Nicholson, P.D., 1975, Int. J. Quantum Chem. 9, 325.
Dennison, D.M. and Ehlenbeck, G.E., 1932, Phys. Rev. 41, 313.
Detrich, J. and Conn, R.W., 1976, J. Chem. Phys. 64, 3091.
Dickens, P.G. and Linnett, J.W., 1957, Quart. Revs. 11, 291.
Doering, W. and Roth, W.R., 1963, Angew. Chem. Intern. Ed. Engl. 2, 115.
Doggett, G., 1977, Mol. Phys. 34, 1739.
Dorney, A.J. and Watson, J.K.G., 1972, J. Molec. Spectrosc. 42, 1.
Downs, J.J. and Johnson, R.E., 1955, J. Am. Chej. Soc. 77, 2098.
Dunker, A.M. and Gordon, R.G., 1976, J. Chem. Phys. 64, 354.
Dyke, T.R., 1977, J. Chem. Phys. 66, 492.
Dyke, T.R., Mack, K.M. and Muenter, J.S., 1977, J. Chem. Phys. 66, 498.
Elliott, J.P., 1958a, Proc. Roy. Soc. London A 245, 128.
Elliott, J.P., 1958b, Proc. Roy. Soc. London A 245, 562.
Elliott, J.P. and Harvey, M., 1962, Proc. Roy. Soc. London A 272, 557.
Ewing, G.E., 1976, Can. J. Phys. 54, 487.
Ezra, G.S., 1979 (in press), Molec. Phys.
Fadeev, L.D., 1961, Soviet Phys. JETP 12, 1014.
Faiman D. and Hendry, A.W., 1968, Phys. Rev. 173, 1720.

Fano, U., 1976, Physics Today 29(9), 32.
Fano, U. and Lin, C.D., 1974, Invited Talk for the Fourth International Conference on Atomic Physics, Heidelberg, July.
Fastenakel, D. and Brocas, J., 1975, Bull. Soc. Chim. Belg. 84, 1093.
Florey, J.B. and Cusachs, L.C., 1972, J. Am. Chem. Soc. 94, 3040.
Flurry, R.L. and Siddall, T.H., III, 1978, Molec. Phys. 36, 1309.
Fock, V., 1935, Zeit. Phys. 98, 145.
Frei, H., Grover, P., Bauder, A. and Günthard, Hs.H., 1978, Molec. Phys. 36, 1467.
Frei, H. and Günthard, Hs.H., 1976, Chem. 15, 155.
Frei, H., Meyer, R., Bauder, A. and Günthard, Hs.H., 1976, Molec. Phys. 32, 443 (Erratum: 1978, Molec. Phys. 34, 1198.)
Gallup, G.A., 1977, Molec. Phys. 33, 943.
Gartenhaus, S. and Schwartz, C., 1957, Phys. Rev. 108, 482.
Gerber, W.H. and Schumacher, E., 1978, J. Chem. Phys. 69, 1692.
Gordon, R.G. and Cashion, J.K., 1966, J. Chem. Phys. 44, 1190.
Gray, D.L. and Robiette, A.G., 1976, Mol. Phys. 32, 1609.
Griffin, J.J., 1957, Phys. Rev., 108, 328.
Griffin, J.J. and Wheeler, J.A., 1957, Phys. Rev. 108, 311.
Gut, M., Bauder, A. and Günthard, Hs.H., 1975, Chem. Phys. 8, 252.
Gut, M., Meyer, R., Bauder, A. and Günthard, Hs.H., 1978, Chem. Phys. 31, 433.
Gutowsky, H.S. McCall, D.W. and Slichter, C.P., 1956, J. Chem. Phys. 21, 279.
Hamermesh, M., 1962, Group Theory and Its Application to Physical Problems, Addison-Wesley, Reading, Mass.
Harter, W.G. and Patterson, C.W., 1977, Phys. Rev. Lett. 38, 224.
Harter, W.G., Galbraith, H.W. and Patterson, C.W., 1978, J. Chem. Phys. 69, 4888.
Harter, W.G., Patterson, C.W. and Galbraith, H.W., 1978, J. Chem. Phys. 69, 4896.
Harter, W.G., Patterson, C.W. and de Paixao, F.J., 1978, Revs. Mod. Phys. 50, 37.
Hasselbarth, W. and Ruch, E., 1973, Theor. Chim. Acta 29, 259.
Hay, P.J., Dunning, T.H. and Goddard, W.A., 1975, J. Chem. Phys. 62, 3912.
Hecht, K.T., 1960, J. Molec. Spectrosc. 5, 355, 390.
Hecht, K.T., and Braunschweig, D., 1975, Nucl. Phys. A244, 365.
Henderson, G. and Ewing, G.E., 1974, Molec. Phys. 27, 903.
Hermann, A., Leutwyler, S., Schumacher, E. and Wöste, L., 1978, Helv. Chim. Act. 61, 453.
Herrick, D.R., 1977, J. Chem. Phys. 67, 5406.
Herrick, D.R. and Kellman, M.E., 1978, Phys. Rev. A 18, 1770.
Herrick, D.R., Kellman, M.E. and Poliak, R.D., to be published.
Herrick, D.R. and Sinanoglu, O., 1975, Phys. Rev. A 11, 97.
Hilico, J-C., Berter, H. and Loete, M., 1976, Can. J. Phys. 54, 1702.
Holmgren, S.L., Waldman, M. and Klemperer, W., 1977, J. Chem. Phys. 67, 4414.

Hougen, J.T., 1962, J. Chem. Phys. 37, 1433.
Hougen, J.T., 1963, J. Chem. Phys. 39, 358.
Hougen, J.T., 1965, Invited Lectures presented at the VIIIth European Congress on Molecular Spectroscopy, Copenhagen, Denmark, August 14-20, p.481.
Jahn, H.A., 1938a, Proc. Roy. Soc. (London) A 164, 117.
Jahn, H.A., 1938b, Proc. Roy. Soc. A 168, 469, 495.
Jahn, H.A. and Teller, E., 1937, Proc. Roy. Soc. (London) A 161, 220.
Jaszunski, M., Kochanski, E. and Siegbahn, P., 1977, Mol. Phys. 33, 139.
Jauch, J.M. and Hill, E.L., 1940, Phys. Rev. 57, 641.
Johnson, M. and Dennison, D.M., 1935, Phys. Rev. 48, 868.
Kaelberer, J.B. and Etters, R.D., 1977, J. Chem. Phys. 66, 3233.
Karl, G. and Obryk, E., 1968, Nucl. Phys. B8, 609.
Kellman, M.E., 1977, Ph.D. thesis, The University of Chicago.
Kellman, M.E., Amar, F., and Berry, R.S., "Correlation diagrams for rigid and nonrigid 3-body systems", submitted for publication.
Kellman, M.E. and Berry, R.S., 1976, Chem. Phys. Lett. 42, 327.
Kellman, M.E. and Herrick, D.R., 1978, B. 11, L755.
Kendrick, J. and Hillier, I.H., 1977, Mol. Phys. 33, 635.
Kilpatrick, J.E., Pitzer, K.S. and Spitzer, R., 1947, J. Am. Chem. Soc. 69, 2483.
Kolos, W., Roothaan, C.C.J. and Sack, R.A., 1960, Revs. Moc. Phys. 32, 178.
Kolos, W. and Wolniewicz, L., 1964, J. Chem. Phys. 41, 3663.
Kolos, W. and Wolniewicz, L., 1965, J. Chem. Phys. 43, 2429.
Kolos, W. and Wolniewicz, L., 1968, J. Chem. Phys. 49, 404.
Kolos, W. and Wolniewicz, L., 1969, J. Chem. Phys. 51, 1417.
Kramer, P. and Moshinsky, M., 1966, Nucl. Phys. 82, 241.
Kramer, P. and Moshinsky, M., 1968, in Group Theory and Its Applications, E. Loebl, ed., Academic Press, New York, pp.340ff.
Lathouwers, L. and van Leuven, P., 1978, J. Quantum Chem. 12, 371.
Lathouwers, L., van Leuven, P. and Bouten, M., 1977, Chem. Phys. Lett. 52, 3.
Lee, J.K., Barker, J.A. and Abraham, F.F., 1973, J. Chem. Phys. 58, 3166.
LeRoy, R.J., 1973, in Molecular Spectroscopy, R.F. Barrow, D.A. Long and D.J. Millin, eds., Chem. Soc. Spec. Periodical, London.
Liehr, A.D., 1963, J. Phys. Chem. 67, 389.
Liehr, A.D., 1963b, J. Phys. Chem. 67, 471.
Liehr, A.D., 1963c, Prog. Inorg. Chem. 5, 385.
Lim, T.K., 1977, Molec. Phys. 33, 373.
Longuet-Higgins, H.C., 1963, Mol. Phys. 6, 445.
Louck, J.D., 1965, J. Math. Phys. 6, 1786.
Louck, J.D., 1976, J. Mol. Spec. 61, 107.

Louck, J.D., and Galbraith, H.W., 1976, Rev. Mod. Phys. 48, 69.
Lowdin, P.O., 1959, Adv. Chem. Phys. 2, 207.
Macek, J.H., 1967, Phys. Rev. 160, 170.
Macek, J.H., 1968, J. Phys. B 2, 831.
Martin, R.L. and Davidson, E.R., 1978, Mol. Phys. 35, 1713.
Maslen, V.W., 1956, Proc. Phys. Soc. A 69, 734.
Mayer, H., 1975, J. Phys. A 8, 1562.
McGinty, D.J., 1971, J. Chem. Phys. 55, 580.
McGinty, D.J., 1973, J. Chem. Phys. 58, 4733.
McKay, W., Patera, J. and Sharp, R.T., 1975, Computer Phys. Commun. 10, 1.
McKellar, A.R.W. and Welsh, H.L., 1974, Can. J. Phys. 52, 1082.
Meakin, P., Muetterties, E.L., Tebbe, F.N. and Jesson, J.P., 1971, J. Am. Chem. Soc. 93, 4701.
Meyer, R. and Günthard, Hs.H., 1968, J. Chem. Phys. 49, 1510.
Michelot, F., 1976, J. Mol. Spec. 63, 227.
Mikhailov, V.V., 1978, J. Phys. A 11, 443.
Moshinsky, M., 1969, The Harmonic Oscillator in Modern Physics: From Atoms to Quarks, Gordon and Breach, New York.
Moshinsky, M., Patera, J., Sharp, R.T. and Winternitz, P., 1975, Annals of Phys. 95, 139.
Movre, M. and Pichler, G., 1977, J. Phys. B 10, 2631.
Muetterties, E.L., 1965, Inorg. Chem. 4, 769.
Muetterties, E.L., 1967, Inorg. Chem. 6, 635.
Muetterties, E.L., 1968, J. Am. Chem. Soc. 90, 5097.
Muetterties, E.L., 1969a, J. Am. Chem. Soc. 91, 1636.
Muetterties, E.L., 1969b, J. Am. Chem. Soc. 91, 4115.
Muetterties, E.L., 1970a, Accts. Chem. Res. 3, 266.
Muetterties, E.L., 1970b, Rec. Chem. Prog. 31, 51.
Munschy, G. and Pluvinage, P., 1963, Revs. Mod. Phys. 35, 494.
Nicholson, B.J. and Longuet-Higgins, H.C., 1965, Mol. Phys. 9, 461.
Niemax, K. and Pichler, G., 1975, J. Phys. B 8, 179.
Nikitin, S.I. and Ostrovsky, V.N., 1976, B 9, 3141.
Nishioka, K., 1977, Phys. Rev. A 16, 2143.
Novick, S.E., Davies, P., Harris, S.J. and Klemperer, W., 1973, J. Chem. Phys. 59, 2273.
Novick, S.E., Janda, K.C., Holmgren, S.L., Waldman, M. and Klemperer, K., 1976, J. Chem. Phys. 65, 1114.
Patterson, C.W. and Harter, W.G., 1977, J. Chem. Phys. 66, 4886.
Pauli, W., 1926, Zeit. Phys. 36, 336.
Peyerimhoff, S.D. and Buenker, R.J., 1967, J. Chem. Phys. 47, 1953.
Pickett, H., 1972, J. Chem. Phys. 56, 1715.
Pople, J.A., 1960, Molec. Phys. 3, 16.
Pople, J.A. and Longuet-Higgins, H.C., 1958, Molec. Phys. 1, 372.
Redfield, A.G., 1957, J. Res. and Develop. 1, 19.
Rehmus, P. and Berry, R.S., 1979, Chem. Phys. 38, 257.
Rehmus, P., Kellman, K. and Berry, R.S., 1978, Chem. Phys. 31, 239.
Rehmus, P., Roothaan, C.C.J. and Berry, R.S., 1978, Chem. Phys. Lett. 58, 321.

Renner, R., 1934, Z. Physik 92, 172.
Robiette, A.G., Gray, D.L. and Birss, F.W., 1976, Mol. Phys. 32, 1591.
Rosen, N., and Morse, P.M., 1932, Phys. Rev. 42, 10.
Russegger, P. and Brickmann, J., 1975, J. Chem. Phys. 62, 1086.
Russegger, P. and Brickmann, J., 1975, Chem. Phys. Lett. 30, 276.
Sack, R.A., 1958, Mol. Phys. 1, 163.
Sinanoglu, O. and Brueckner, K., 1970, Three approaches to electron correlation in atoms, Yale, New Haven.
Sinanoglu, O. and Herrick, D.R., 1975, J. Chem. Phys. 62, 886.
Sinanoglu, O. and Herrick, D.R., 1976, J. Chem. Phys. 65, 850.
Smith, F.T., 1960, Phys. Rev. 120, 1058.
Stwalley, W.C., 1978, Contemp. Phys. 19, 65.
Stwalley, W.C., Uang, Y.-H. and Richler, G., 1978, Phys. Rev. Lett. 41, 1164.
Swanson, N. and Celotta, R.J., 1975, Phys. Rev. Lett. 35, 783.
Tatum, J.P., 1976, Int'l. J. of Quant. Chem. X, 967.
Teller, E., 1934, Handb. Jb. Chem. Phys. 9, II, 43.
Teller, E. and Wheeler, J.A., 1938, Phys. Rev. 53, 778.
Ugi, I., 1971, Angew. Chem. Int'l. Ed. 10, 637.
Ugi, I., Marquarding, D., Klusacek, H., Gokel, G. and Gillespie, P., 1970, Angew. Chem. Int. Ed. 9, 703.
Wadt, W.R. and Hay, P.J., 1979, J. Am. Chem. Soc.
Watanabe, A. and Welsh, H.L., 1964, Phys. Rev. Lett. 13, 810.
Watson, J.K.G., 1965, Canad. J. Phys. 43, 1996.
Watson, J.K.G., 1967, J. Chem. Phys. 46, 1935.
Watson, J.K.G., 1968, J. Chem. Phys. 48, 181.
Watson, J.K.G., 1968, J. Chem. Phys. 48, 4517.
Watson, J.K.G., 1971, Mol. Phys. 21, 577.
Westheimer, F.H., 1968, Accts. Chem. Res. 1, 70.
Whitesides, G.M. and Mitchell, H.L., 1969, J. Am. Chem. Soc. 91, 5384.
Woolley, R.G., 1976, Adv. in Phys. 25, 27.
Woolley, R.G., 1978a, Chem. Phys. Lett. 55, 443.
Woolley, R.G., 1978b, J. Am. Chem. Soc. 100, 1073.
Woolley, R.G., 1979, Israel J. Chem. 18.
Woolley, R.G. and Sutcliffe, B.T., 1977, Chem. Phys. Lett. 45, 393.
Wu, C.H., 1976, J. Chem. Phys. 65, 3181.
Wulfman, C., 1973, Chem. Phys. Lett. 23, 367.
Wulfman, C.E., 1971, in Group Theory and Its Applications, Vol. II, E.M. Loebl, ed., Academic Press, New York.
Wulfman, C.E., 1976, Chem. Phys. Lett. 40, 139.
Wulfman, C. and Kumei, S., 1973, Chem. Phys. Lett. 23, 367.
Wybourne, B.G., 1970, Symmetry Principles and Atomic Spectroscopy, Wiley Interscience, New York.
Wybourne, B.G., 1973, Int. J. Quant. Chem. 7, 1117.
Wybourne, B.G., 1974, The Classical Groups for Physicists, Wiley Interscience, New York.
Yamada, K. and Winnewisser, M., 1976, Z. Naturforsch, 31a, 139.

THE GENERATOR COORDINATE METHOD IN MOLECULAR PHYSICS

P. Van Leuven and L. Lathouwers

Dienst Teoretische en Wiskundige Natuurkunde
Rijksuniversitair Centrum
University of Antwerp, Belgium

1. INTRODUCTION

There are several reasons why a new method for the study of molecular properties might be proposed. According to some authors (Woolley, 1976; Woolley and Sutcliffe, 1977) there are inherent defects in the semiclassical theory resulting from the Born-Oppenheimer approximation. In the traditional method nuclei and electrons are treated unsymmetrically; the nuclei are considered as classical point charges in the initial stage of the theory. A number of concepts such as the energy surface and the molecular shape are introduced, the precise physical meaning and the theoretical consistency of which may be severely criticized. Although the complete adiabatic method yields an exact result in principle, in practice the nature of the convergence of the Born-Huang series and the complexity of the system of coupled differential equations to which it leads, seem to inhibit fruitful applications. Both of these aspects of the traditional molecular theory will be discussed elsewhere in this course.

From the experimental point of view, more and more precise measurements lead to spectroscopic data which reveal non-adiabatic effects. There seems to be a growing interest in regions of strongly interacting energy surfaces (avoided crossings) where the adiabatic approximation does not hold. We may also mention that in the case of mesic molecules, which can now be studied, a drastic departure from the adiabatic situation occurs. For all these reasons a practical and systematic procedure for handling non-adiabatic effects is called for.

In these lectures we shall propose a new method for solving the molecular Schrödinger equation which is entirely quantal. The formulation is put in a non-adiabatic form: it contains the coupling between nuclear and electronic motion. As it contains the adiabatic approximation as a limiting case, the theory conserves all the useful concepts we know from the traditional theory but relegates all semi-classical concepts (energy surface, molecular structure ...) to a space of fictitious coordinates which serve as a mathematical auxiliary and do not require physical interpretation.

The Generator Coordinate Method (GCM), as we will call this new method, has only reached an initial state of development in molecular physics. We shall present the fundamental ideas and some of the preliminary applications. The GCM originated in nuclear physics where it was introduced for the ab-initio description of collective nuclear motion, where non-adiabatic effects are very large. The conjecture we wish to put forward here is that the GCM will prove significant as a means of describing in a rigorous and compact way non-adiabatic effects in molecular physics.

2. OUTLINE OF THE GENERATOR COORDINATE METHOD

2.1 The GCM trial function

The GCM is a variational approximation to the eigenfunctions of a many particle system. The trial function $\Psi(x)$, where x stands for all dynamical variables, is chosen in the form

$$\Psi(x) = \int f(\alpha)\, \Phi(x|\alpha)d\alpha \tag{1}$$

Here α is a set of parameters (generator coordinates) whose number is equal to or less than the total number of degrees of freedom of the system. Φ, the intrinsic function, is a chosen function and $f(\alpha)$ is the unknown "weight function" which has to be varied such as to make the energy stationary. Mathematical properties and assumptions about all quantities occuring in the theory will be discussed at the appropriate place.

There are different ways in which to interpret the ansatz (1). First there is the physical picture in which the choices of α and Φ are based on physical insight. We shall see that according to the nature of the motion one wants to concentrate upon, α can be chosen correspondingly; if one desires to describe, e.g., the translational motion of the system as a whole one should choose α as the translational increment of the coordinates and $\Phi(x|\alpha)=\Phi(x-\alpha)$. $f(\alpha)$ can loosely speaking be interpreted as a weight factor. Care must be taken, however, not to interpret f as a probability amplitude.

In a mathematical context (1) would be interpreted as an integral transform between the physical wave function $\Psi(x)$ and its transform $f(\alpha)$, the kernel of the transformation being $\Phi(x|\alpha)$. This aspect of the method will be exploited in the method of natural states discussed below. In the case of translations the integration (1) represents a convolution in the ordinary sense.

The form of Ψ can also be looked upon as a linear superposition of basis functions Φ. These basis functions are labelled by a continuous index α. As $\Phi(x|\alpha)$ for different values α are not meant to be orthogonal this is an expansion in a non-orthogonal basis. In this sense $f(\alpha)$ is a representative of the state vector Ψ in the α-space just as $\Psi(x)$ is its representative in x-space.

Many approximate wave functions devised in the literature are in fact special cases of the GCM trial functions. Hence GCM is a good starting point for the comparison of different methods.

2.2 The GCM-subspace

For a fixed choice of Φ, all functions Ψ which can be written in the form (1) will span a certain linear space. If definite restrictions on the amplitude $f(\alpha)$ are introduced so as to make the integral in (1) meaningful, e.g., so as to make Ψ square integrable, i.e.,

$$(\Psi|\Psi) = \iint f^*(\alpha)\Delta(\alpha,\beta)f(\beta)d\alpha d\beta = 1 \quad (2)$$

$$\Delta(\alpha,\beta) = \int \Phi^*(x|\alpha)\Phi(x|\beta)dx \quad (3)$$

then the functions Ψ span a subspace H_Φ of the total Hilbert space $L^2(x)$. There is a theoretical difficulty involved in considering limits of the form

$$\Psi_n = \int f_n(\alpha)\Phi(x|\alpha)d\alpha \rightarrow \Psi \quad (4)$$

As the limiting procedure does not necessarily commute with the integration, even if we know that $\Psi_n \rightarrow \Psi$ we do not know whether $f_n \rightarrow f$ and what the nature of this limit is. We have already mentioned the interpretation of Φ as a continuously labelled basis set: due to the separability of $L^2(x) \supset H_\Phi$ there exists a countable subset $\Phi(x|\alpha_i)$ $i \in \mathbb{N}$ which spans the space H_Φ. So, in fact, the ansatz

$$\Psi(x) = \sum_{i=0}^{\infty} c_i \, \Phi(x|\alpha_i) \quad (5)$$

is equivalent to (1). The continuous family $\Phi(x|\alpha)$ constitutes an overcomplete set and can be replaced by a countable family $\Phi(x|\alpha_i)$. Hence $f(\alpha)$ is not unique and there is no <u>a priori</u> relation between

$f(\alpha_i)$ and c_i. This discretisation procedure is used to construct practical algorithms for the GCM.

2.3 The GCM-equation

In the spirit of the Variational Principle, the expression (1) is used as a trial function. If the GCM-subspace is not the entire Hilbert space this variational procedure will lead to an approximate wave function for the ground state of the system. Ψ is determined in the usual way by the requirement

$$\delta \frac{(\Psi H \Psi)}{(\Psi\Psi)} = 0 \tag{6}$$

where H is the total Hamiltonian of the system and δ stands for the first order variation with respect to the quantity $f(\alpha)$. The expectation value of H occuring in (6) can be written

$$(\Psi H \Psi) = \iint f^*(\alpha) H(\alpha,\beta) f(\beta) d\alpha d\beta \tag{7}$$

$$H(\alpha,\beta) = \int \Phi^*(x|\alpha) H \Phi(x|\beta) dx \tag{8}$$

Both the Hamiltonian kernel $H(\alpha,\beta)$ and the overlap kernel $\Delta(\alpha,\beta)$ are Hermitian in the sense

$$H(\beta,\alpha) = H^*(\alpha,\beta), \qquad \Delta(\beta,\alpha) = \Delta^*(\alpha,\beta) \tag{9}$$

The condition (6) leads to an equation for the optimal $f(\alpha)$

$$\int [H(\alpha,\beta) - E\Delta(\alpha,\beta)] \, f(\beta) d\beta = 0 \tag{10}$$

This integral equation, usually called the Hill-Wheeler (HW) equation, (Hill and Wheeler, 1953) together with (2) constitutes an eigenvalue problem whose solutions are denoted f_n. It is equivalent to the Schrödinger equation projected on the GCM-subspace. Its lowest eigenvalue E_0 is an approximation for the ground state energy and a variational upper-bound. The other eigenvalues E_n are considered approximations to excitation energies. It must be remembered that in choosing the generator coordinates we are concentrating on a particular kind of motion and hence the excited states to which the eigenfunctions Ψ_n (belonging to f_n) refer, must be interpreted as special kinds of excitations, e.g., vibrations or rotations.

2.4 Symmetry Properties

If the exact Hamiltonian commutes with a unitary operator S, the exact eigenfunctions can be taken as eigenfunctions of S:

$$S^+ = S^{-1} \tag{11}$$

$$[H,S] = 0 \tag{12}$$

The approximate eigenfunctions given by the GCM are in general not eigenfunctions of the conserved quantity S. If we assume that the intrinsic function $\Phi(x|\alpha)$ has the property

$$S\Phi(x|\alpha) = \Phi(x|\sigma(\alpha)) \tag{13}$$

which means that there corresponds to S a transformation σ in the GC's, then the GC-trial space is stable under S,

$$S\Psi = \int f(\alpha)\Phi(x|\sigma(\alpha))d\alpha = \int g(\alpha')\Phi(x|\alpha')d\alpha' \tag{14}$$

where

$$g(\alpha') = f(\sigma^{-1}(\alpha'))/J(\alpha') \tag{15}$$

and J is the Jacobian of the transformation $\alpha'=\sigma(\alpha)$. So $S\Psi$ may be written in the form (1) and hence belongs to $\mathcal{H}_\Phi$. Furthermore it immediately follows from (11) and (12) that the kernels have the property

$$\Delta(\sigma(\alpha),\sigma(\beta)) = \Delta(\alpha,\beta) \tag{16}$$

$$H(\sigma(\alpha),\sigma(\beta)) = H(\alpha,\beta) \tag{17}$$

Then also the HW-equation (10) is invariant under S. Therefore f and g are both solutions of (10) with same eigenvalue and (if there is no degeneracy) they are proportional, so Ψ is an eigenfunction of S. If we consider a group G of symmetry operators the same arguments as above are valid but the conclusion now is that the solutions of the HW-equation form an irreducible representation of G. We thus have shown that condition (13) is sufficient for the GCM-wave functions to be symmetry adapted.

2.5 Exactly soluble problems in GCM

2.5.1 Translations. Suppose we take as generator coordinate the translation vactor $\underset{\sim}{\alpha}$. Then any L^2-function $\Phi(x)$ can be used as an intrinsic function by defining (Wiener, 1933)

$$\Phi(x|\alpha) = \Phi(\underset{\sim}{x}-\underset{\sim}{\alpha})$$

This function is stable under translations $T_a\Phi(\underset{\sim}{x}|\underset{\sim}{\alpha})\equiv\Phi(\underset{\sim}{x}-\underset{\sim}{a}|\underset{\sim}{\alpha}) = \Phi(\underset{\sim}{x}|\underset{\sim}{\alpha}+\underset{\sim}{a})$. Because of translational invariance of the exact Hamiltonian H, the kernels are functions of $\underset{\sim}{\alpha}-\underset{\sim}{\beta}=\underset{\sim}{\delta}$ only. From the above arguments we know that $f(\underset{\sim}{\alpha})$ and $f(\underset{\sim}{\alpha}+\underset{\sim}{a})$ are proportional, so

$$f(\underset{\sim}{\alpha}) \sim \exp[i\underset{\sim}{k}.\underset{\sim}{\alpha}] \tag{18}$$

This is indeed a representation of the irreps of the translation group. The solution of the HW-equation is explicitly known, the (continuous) eigenvalues are

$$E(\underset{\sim}{k}) = \int H(\underset{\sim}{\delta})\exp[i\underset{\sim}{k}.\underset{\sim}{\delta}]d\underset{\sim}{\delta} / \int \Delta(\underset{\sim}{\delta})\exp[i\underset{\sim}{k}.\underset{\sim}{\delta}]d\underset{\sim}{\delta}$$

The wave function is of the form

$$\Psi_{\underset{\sim}{k}} = \int \exp[i\underset{\sim}{k}.\underset{\sim}{\alpha}]\Phi(\underset{\sim}{x}-\underset{\sim}{\alpha})d\underset{\sim}{\alpha} \quad (19)$$

It is an eigenfunction of total linear momentum with eigenvalue $\underset{\sim}{k}$. In the form (19) we can recognize the "projection operator"

$$P_{\underset{\sim}{k}} = \int \exp[i\underset{\sim}{k}.\underset{\sim}{\alpha}] \; T_{\underset{\sim}{\alpha}} \; d\underset{\sim}{\alpha} \quad (20)$$

We see that in this case the GCM is equivalent to the projection of total linear momentum.

2.5.2 Rotations about an axis. Analogous expressions exist for the rotations about an axis. Suppose we take as a GC the rotation angle ϕ about the z-axis of the reference system, and

$$\Phi(xy|\alpha) \equiv \Phi(x \cos\phi + y \sin\phi, -x \sin\phi + y \cos\phi)$$

This intrinsic state is stable with respect to rotations R_a

$$\begin{aligned} R_a\Phi(xy|\phi) &= \Phi(x \cos a + y \sin a, -x \sin a + y \cos a|\phi) \\ &= \Phi(xy|\phi+a) \end{aligned}$$

Again it is possible to determine $f(\phi)$ by quite general arguments; we find

$$f_m(\phi) = \exp(i\, m\, \phi) \qquad m \in Z \quad (21)$$

These are the irreducible representations of R(2) and again the GCM is equivalent to the projection of the z-component m of angular momentum.

2.5.3 Three dimensional rotations. In this case we have to deal with a non-abelian symmetry group. It is well-known that the irreducible representations may be given by the rotation matrices $D^J_{MK}(\Omega)$ where Ω stands for a set of Euler angles characterizing the rotation $R(\Omega)$. Let us take Ω as our GC. It can be shown that a solution of the HW equation can be obtained in the form

$$f_{JM}(\Omega) = \sum_K C^{JM}_K D^J_{MK}(\Omega) \quad (23)$$

where the coefficients C_K are determined by the secular equation

$$\sum_{K'} (H^J_{KK'} - E\Delta^J_{KK'}) C^J_{K'} \;=\; 0 \tag{24}$$

with

$$\begin{aligned} \Delta^J_{KK'} &= \int D^J_{KK'}(\Omega)\Delta(\Omega)d\Omega \\ H^J_{KK'} &= \int D^J_{KK'}(\Omega)H(\Omega)d\Omega \end{aligned} \tag{25}$$

We note here that due to the rotational invariance of H the HW kernels depend on one set of Euler angles Ω only. The GC wave function is then of the form

$$\psi_{LM}(x) \;=\; \sum_K C^J_K \int D^J_{MK}(\Omega)\Phi(x|\Omega)d\Omega \tag{26}$$

Each term in (26) can be considered as a projection of Φ on a state of good angular momentum. In fact, the genuine projection operators that enter here are

$$P_{JK} \;=\; \frac{2J+1}{8\pi^2}\int D^J_{KK}(\Omega)\; R(\Omega)d\Omega \tag{27}$$

They project Φ on a state with angular momentum J and z-component K. A state with component M can be reached by subsequent raising operators J_+ or lowering operators J_-.

2.5.4 <u>The Harmonic Oscillator</u>. In the above examples a solution of the HW-equation could be obtained due to the symmetry properties of H. The GCM did nothing more than restore the symmetry that was missing in the intrinsic function. In the case of a particle moving in a harmonic oscillator potential

$$H \;=\; -\frac{1}{2m}\,\frac{d^2}{dx^2} + \frac{k}{2}\,x^2 \tag{28}$$

the HW-equation can be solved exactly, if we take the displacement α as a GC and use an intrinsic function of the form

$$\Phi(x|\alpha) \;=\; \left(\frac{4s}{\pi}\right)^{1/4} \exp\,[-2s(x-\alpha)^2] \tag{29}$$

We note that in this case H is not invariant under translations. The kernels can be calculated easily: we get

$$\Delta(\alpha,\beta) \;=\; \exp\,[-s(\alpha-\beta)^2] \tag{30}$$

$$H(\alpha,\beta) \;=\; [K(0) + \tfrac{1}{2}(B\alpha^2+2A\alpha\beta+B\beta^2)]\Delta(\alpha,\beta) \tag{31}$$

where

$$A = \frac{k}{4} + \frac{4s^2}{m} \quad , \quad B = \frac{k}{4} - \frac{4s^2}{m} \tag{32}$$

$$K(0) = \frac{k}{16s} + \frac{s}{m} \tag{33}$$

The HW-equation can be solved through the ansatz

$$f_v(\alpha) = P_v(\alpha) \exp\,[-\tfrac{1}{2}\alpha^2/c^2] \tag{34}$$

By substituting (34) in the HW-equation and requiring that the result should vanish for all α one can determine c and the polynomials P_v. The solution is

$$c = [(2s - \frac{m\omega}{2} /2m\omega s]^{1/2} \tag{35}$$

and $P_v(\alpha)$ are Hermite polynomials $H_v(\alpha/c')$ with a different scaling (Lathouwers, 1975). The corresponding energy eigenvalues are

$$E_v = (v + \frac{1}{2})\omega \tag{36}$$

We obtain the exact solution of the oscillator problem, thanks to the fact that the translated Gaussians form a complete set of functions in $L^2(x)$ and due to the simple form of the HW kernels. It must be emphasized, however, that the choice of scale factor s is crucial. From (35) it can be seen that only $s > m\omega/4$ yields solutions $f_v(\alpha) \in L^2(\alpha)$.

2.5.5 The Hydrogen Atom; scaling parameters as GC. In the case where a scaling parameter is used as a GC there does not correspond a conserved quantity to the transformation. It can be proved (Broeckhove, 1979) that the scaled Gaussians $\Phi(x|\alpha)=(2\alpha|\pi)^{1/4}\exp(-\alpha x^2)$ form a complete set for positive parity functions in $L^2(x)$. An example which comes close to this is the Hydrogen atom where the ground state can be represented as a GC-wave function. In this case the Hamiltonian (for the s-states) is

$$H = -\frac{1}{2m} r^2 \frac{d}{dr} r^2 \frac{d}{dr} - \frac{1}{r} \tag{37}$$

Starting from the intrinsic function

$$\Phi(x|\alpha) = \alpha^{3/2}\,\pi^{-3/4} \exp(-\alpha^2 r^2/2) \tag{38}$$

one can easily calculate the kernels of the HW equation

$$\Delta(\alpha,\beta) = (\frac{2\alpha\beta}{\alpha^2+\beta^2})^{3/2} \tag{39}$$

$$H(\alpha,\beta) = [\frac{3}{2}(\frac{\alpha^2\beta^2}{\alpha^2+\beta^2}) - \sqrt{\frac{2}{\pi}}(\alpha^2+\beta^2)^{1/2}]\Delta(\alpha,\beta) \tag{40}$$

Although a straightforward analytical solution of this HW-equation has not been given (Hurley, 1967) the solution can be inferred from the known integral transformation

$$\exp(-r) = \sqrt{\frac{2}{\pi}} \int_0^\infty \alpha^{-2} \exp(-1/2\alpha^2)\exp(-\alpha^2r^2/2)d\alpha \tag{41}$$

which shows that the ground state wave function of the Hydrogen atom can be exactly represented as a GC-wave function. This kind of approach has been extended to the Helium atom, where an approximate numerical solution has been given (see further).

2.6 Solution Methods of the GC Equation

The HW equation not being a classical type of integral equation, no direct methods of solution are known. One way of attack is to transform the equation into a Fredholm integral equation, containing only one Hermitian kernel. We therefore define a transformation on the GC-function f (Griffin and Wheeler, 1957)

$$g(\xi) = \int B(\xi,\alpha)f(\alpha)d\alpha \tag{42}$$

and its inverse

$$f(\alpha) = \int B^{-1}(\alpha,\xi)g(\xi)d\xi \tag{43}$$

with

$$\int B^{-1}(\alpha,\xi)B(\xi,\beta)d\xi = \delta(\alpha-\beta) \tag{44}$$

If then B is chosen such that

$$\int B(\xi,\alpha)\Delta(\alpha,\beta)B^{-1}(\beta,\eta)d\alpha d\beta = \delta(\xi-\eta) \tag{45}$$

and if we denote

$$\int B(\xi,\alpha)H(\alpha,\beta)B^{-1}(\beta,\eta)d\alpha d\beta = F(\xi,\eta) \tag{46}$$

The HW-equation can be put in the form

$$\int F(\xi,\eta)g(\eta)d\eta = Eg(\xi) \tag{47}$$

which is of the classical type. In the particular case of the harmonic oscillator as described in 2.5.4 we can determine these kernels:

$$B(\xi,\alpha) \sim \exp[-2s(\xi-\alpha)^2] \tag{48}$$

$$B^{-1}(\alpha,\xi) \sim \int dq \exp[q^2/8s+iq(\alpha-\xi)] \tag{49}$$

From the expression for B^{-1} we see that $f(\alpha)$ is an integral of the Fourier transform of g times $\exp(q^2/8s)$; this will not converge if the Fourier transform of g does not fall off fast enough. These difficulties are connected with those mentioned in connection with (35).

Another way to reduce the HW-equation to a Fredholm type is to consider the expression $H(\alpha,\beta)-E\Delta(\alpha,\beta)$ as a new, E-dependent kernel $L(\alpha,\beta,E)$ and consider the "supersecular equation" (Lathouwers, 1976)

$$\int L(\alpha,\beta,E)h(\beta,E)d\beta = \lambda(E)h(\alpha,E) \tag{50}$$

The HW eigenvalues are then determined by the roots E_n of the equation

$$\lambda(E) = 0 \tag{51}$$

and

$$f_n(\alpha) = h(\alpha,E_n) \tag{52}$$

The above methods transform the HW-equation into another integral equation of a more common type. Attempts have been made to transform the HW-equation into a Schrödinger equation, but this attempt has succeeded so far only in some special cases, like the harmonic approximation which will be discussed in the next section. Another idea is to develop the relevant functions on a well-chosen discrete basis and reduce the HW-equation to a matrix equation. One possibility in this direction is the so-called "natural state formalism" (Lathouwers, 1975). Considering the left- and right-iterated kernels

$$A(\alpha,\alpha') = \int \Phi^*(x|\alpha)\Phi(x|\alpha')dx \tag{53}$$

$$X(x,x') = \int \Phi^*(x|\alpha)\Phi(x'|\alpha)d\alpha \tag{54}$$

one can prove the following mathematical results: considered as operator kernels in $L^2(\alpha)$ and $L^2(x)$ respectively, their eigenvalues λ_n are equal, their eigenfunctions $b_n(\alpha)$ and $y_n(x)$ form orthonormal sets and are related by an integral transform using $\Phi(x|\alpha)$ as a kernel

$$b_n(\alpha) = \lambda_n^{-1/2} \int \Phi^*(x|\alpha)y_n(x)dx \tag{55}$$

$$y_n(x) = \lambda_n^{-1/2} \int \Phi(x|\alpha)b_n(\alpha)d\alpha \tag{56}$$

and of all possible N-term sums of the form $\Phi_N=\sum_0^N c_n b_n'(\alpha)y_n'(x)$ the one that minimizes $\int |\Phi_N-\Phi|^2 dxd\alpha$ is just the "natural expansion"

$$\Phi \sim \sum^{N} \lambda_n^{1/2} \, b_n^*(\alpha) y_n(x) \tag{57}$$

For compact proofs of these theorems we refer to Vincent (1973). Now the HW-equation can be replaced by the matrix equation

$$\sum_n H_{mn} c_n = E c_m \tag{58}$$

where the kernel H_{mn} can be calculated either from $H(\alpha,\beta)$ or directly from H. The GC wave function acquires the form

$$\Psi = \sum_n c_n y_n(x) \tag{59}$$

or

$$f(\alpha) = \sum_n f_n b_n(\alpha) \tag{60}$$

with

$$f_n = c_n \lambda_n^{-1/2} \tag{61}$$

None of the above mentioned methods, however, has been used in practical problems. The usual way to solve the HW-equation is through discretization of the GC. One chooses a set of parameter values $\{\alpha_i\}$ and solves the algebraic set

$$\sum_{j=1}^{N} [H(\alpha_i,\alpha_j)-E\Delta(\alpha_i,\alpha_j)]f_j = 0 \quad (i=1...N) \tag{62}$$

In most practical applications the values of α_i are chosen arbitrarily or guided by the shape of the "energy surface" $E(\alpha,\alpha)=H(\alpha,\alpha)/\Delta(\alpha,\alpha)$. The most consistent procedure used (Caurier, 1975) is a variational determination of the values α_i. One chooses the first value α_1 such that $E(\alpha_1,\alpha_1)$ is minimum. The second point α_2 is determined such that the lowest eigenvalue of the matrix equation (62) with N=2 is minimum etc... One difficulty arising in this method is the approximate linear dependence of two basis vectors $\Phi(x|\alpha_i)$ and $\Phi(x|\alpha_j)$. This circumstance is reflected in the appearance of very small eigenvalues of the overlap matrix Δ which, in a numerical treatment, cannot be distinguished from zero. The overlap matrix is interpreted as singular and the numerical diagonalisation stops. An approximation procedure consists in cutting off the matrix equation when an eigenvalue of Δ occurs below a certain limit. The experience in nuclear physics problems shows that in the variational choice of meshpoints a good convergence can be obtained before the above mentioned trouble arises.

2.7 The Helium atom, a numerical application

Thakkar and Smith (1976) have made a numerical application of the GCM for the ground and lowest excited states of the He-like ions through Mg^{10+}. The S-states of a two electron ion can be described by taking an intrinsic function of essentially the form

$$\Phi(r_1 r_2 r_{12}|\alpha\beta\gamma) \sim \exp(-\alpha r_1 - \beta r_2 - \gamma r_{12})$$

Due to the exponential form of Φ all kernels can be calculated analytically. The HW-equation is solved using the discretization technique. The meshpoints are chosen to be the abscissa of Monte-Carlo or number theoretic quadrature formulae inside a parallelopipedum in $\alpha\beta\gamma$-space the sides of which have been restricted such that all integrals occuring in the kernels exist. The calculation has been pushed to 66 terms, it yields a ground state energy of -2.903724363 a.u. as compared with the 1078-term wave function of Pekeris which yields an energy of -2.903724376 a.u... Besides the energy, various quantities of the form $\langle r^n \rangle$ and $\langle r_{12}^n \rangle$ have also been calculated in the GC-ground state with remarkable success. After comparison with other calculations in the literature, the authors conclude that the GC function is accurate and more compact than any other function available (with the exception of those containing logarithmic terms) and certainly the most easy to manipulate.

2.8 The Harmonic Approximation

Whenever the overlap kernel $\Delta(\alpha,\beta)$ is expected to be a sharply peaked function of $\alpha-\beta$, it can be approximated by a Gaussian

$$\Delta(\alpha,\beta) \sim \exp[-s(\alpha-\beta)^2] \tag{63}$$

If furthermore also the Hamiltonian kernel is expected to be peaked around $\alpha=\beta$ one can represent it in the form

$$H(\alpha,\beta) \sim \Delta(\alpha,\beta)K(\alpha,\beta) \tag{64}$$

and develop $K(\alpha,\beta)$ up to and including quadratic terms. If we assume that the GC are defined such that $K(\alpha,\alpha)$ has a minimum at $\alpha=0$, then no linear terms would occur in the expansion of $K(\alpha,\beta)$. Hence

$$K(\alpha,\beta) = K(0) + \tfrac{1}{2}(B\alpha^2 + 2A\alpha\beta + B\beta^2) \tag{65}$$

The harmonic approximation thus consists of two assumptions: the Gaussian overlap approximation and the quadratic approximation for the energy kernel. It may be noted that we have the same problem here as for the harmonic oscillator 2.5.4. The constants occuring in the kernels can easily be derived from the intrinsic function Φ in the following way:

$$s = \frac{1}{2}\left(\frac{\partial\Phi}{\partial\alpha}\Big|\frac{\partial\Phi}{\partial\alpha}\right)_{\alpha=0} \tag{66}$$

$$K(0) = (\Phi|H\Phi)_{\alpha=0}, \quad A = \left(\frac{\partial\Phi}{\partial\alpha}\Big|(H-K(0))\frac{\partial\Phi}{\partial\alpha}\right)_{\alpha=0},$$

$$B = \left(\frac{\partial^2\Phi}{\partial\alpha^2}\Big|(H-K(0))\Phi\right)_{\alpha=0} \tag{67}$$

The HW-equation can now be solved analytically and one gets, as before, for the ground state

$$f_0(\alpha) = N_0 \exp(-\alpha^2/2a^2) \tag{68}$$

with

$$a^2 = \frac{1}{2s}\left[\left(\frac{A-B}{A+B}\right)^{1/2} - 1\right] \tag{69}$$

From this expression it is found that the solution f_0 will only be $L^2(\alpha)$ if $B < 0$. The eigenvalue of the ground state is

$$E_0 = K(0) + \frac{\omega}{2} - \frac{A}{2s} \tag{70}$$

and the oscillator frequency ω is obtained as

$$\omega = \frac{1}{s}\sqrt{(A^2 - B^2)} \tag{71}$$

The effective mass of the oscillator can be expressed as

$$m = 2s^2/(A-B) \tag{72}$$

It is possible to generalize the formalism for the case of a n-dimensional GC parameter space $\{\alpha_i\}$. Supposing all GC to be real parameters the Gaussian overlap and quadratic energy kernel may now be written in the analogous form:

(eq. (73) is obtained after renormalizing the intrinsic state Φ)

$$\Delta(\alpha_i,\beta_j) = \exp\left[+2s\sum_i \alpha_i\beta_i\right] \tag{73}$$

$$K(\alpha_i,\beta_j) = K(0)+\frac{1}{2}\sum_{i,j}(B_{ij}\alpha_i\alpha_j+2A_{ij}\alpha_i\beta_j+B^*_{ij}\beta_i\beta_j) \tag{74}$$

Here B and A are n x n matrices (A is Hermitian and B is symmetric). The HW-equation now describes a set of coupled oscillators, and can be solved by a method described in Brink and Weiguny (1968).

3. A GENERATOR COORDINATE THEORY FOR MOLECULAR STRUCTURE AND DYNAMICS

3.1 The Molecular Generator Coordinates

Although, in the traditional description the molecule is considered to be a system of electrons and nuclei, as a starting assumption, the latter are clamped at fixed positions $\underset{\sim}{R}_i$, which form the nuclear "equilibrium form" or nuclear "skeleton". In a subsequent stage of the theory these parameters $\underset{\sim}{R}_i$ are turned into dynamical variables of the nuclear wave function. In the Generator Coordinate description we shall try to treat electrons and nuclei on the same footing, both as dynamical coordinates, but we shall introduce another set of parameters $\underset{\sim}{\alpha}_i$ (equal in number to the $\underset{\sim}{R}_i$) which we shall consider to represent "nuclear skeleton configurations". These parameters will be treated as generator coordinates. From the outset we can suppose the $\underset{\sim}{\alpha}_i$ to consist of a centre-of-mass $\underset{\sim}{A}$, the 3 Euler angles $\{\Omega\}$ which determine the orientation of the skeleton with respect to the laboratory and a set of 3N-6 vibrational coordinates η_i describing the intrinsic structure of the skeleton.

3.2 The Molecular GC-wave function

The intrinsic wave function will depend on all electronic variables x, all nuclear coordinates X, and all GC's α

$$\chi(x,X|\alpha)$$

As we aim at an approximate description which will incorporate non-adiabatic corrections, we shall at a first stage take the intrinsic function χ to be of the form

$$\chi = \phi(x|\alpha)\Phi(X|\alpha) \tag{1}$$

At the "intrinsic level" this corresponds to a decoupling between electrons and nuclei, but both sets of particles are coupled to the generator coordinates. The full GC-wave function will then be

$$\Psi(x,X) = \int f(\alpha)\phi(x|\alpha)\Phi(X|\alpha)d\alpha \tag{2}$$

Ψ will not be of the product type: through the GC superposition the motion of electrons and nuclei will be coupled. This coupling entails a number of non-adiabatic effects, the nature of which will be investigated in this chapter.

3.3 The Molecular GC-kernels

Let us start from the exact molecular Hamiltonian

$$H = T+t+V \tag{3}$$

where T and t are the nuclear and electronic kinetic energy-operators respectively and V is the potential energy. The overlap kernel factorizes as

$$(\phi(\alpha)|\alpha(\beta))(\Phi(\alpha)|\Phi(\beta)) = n(\alpha,\beta)N(\alpha,\beta) \tag{4}$$

The Hamiltonian kernel is

$$\begin{aligned} H(\alpha,\beta) &= n(\alpha,\beta)(\Phi(\alpha)T\Phi(\beta))+N(\alpha,\beta)(\phi(\alpha)t\phi(\beta)) + \\ &\qquad (\phi(\alpha)\Phi(\alpha)V\phi(\beta)\Phi(\beta)) \\ &= n(\alpha,\beta)T(\alpha,\beta)+N(\alpha,\beta)t(\alpha,\beta)+V(\alpha,\beta) \end{aligned} \tag{5}$$

The diagonal kernels define a "GC energy-surface" (we assume henceforth ϕ and Φ to be normalized for all α)

$$E(\alpha) = T(\alpha\alpha)+t(\alpha\alpha)+V(\alpha\alpha) \tag{6}$$

The absolute minimum of this quantity defines the nuclear "equilibrium configuration" α_0 in generator space.

3.4 The adiabatic limit of the GCM

In the adiabatic approximation one starts from the extreme situation of infinite nuclear mass. Then the Hamiltonian reduces to the electronic Hamiltonian $H_0=t+V$. Its eigenfunctions $\phi_n(x,X)$ and eigenvalues $U_n(X)$ depend parametrically on X. The next step in the theory is to develop the unknown molecular wave function in the set ϕ_n:

$$\Psi(x,X) = \sum_n f_n(X)\phi_n(x,X) \tag{7}$$

Substituting this in the full Schrödinger equation

$$(H-E)\Psi = 0$$

and after multiplication with $\phi_n^*(x,X)$ and integration over x one is led to a set of coupled differential equations

$$(T+U_n(X)-E)f_n(X) = \sum_m \Lambda_{mn}f_m(X) \tag{8}$$

with

$$\Lambda_{mn} = -(\phi_m T \phi_n) + \sum_i \frac{1}{M_i} (\phi_m \underset{\sim}{\nabla}_i \phi_n)\cdot\underset{\sim}{\nabla}_i \tag{9}$$

The above procedure is in principle exact. The adiabatic approximation is introduced if one decouples the set (8) by taking only the diagonal terms in the r.h.s. Because ϕ_n is a real function satisfying $(\phi_n|\phi_n)=1$ for all X, the Λ_{nn} is simply

$$\Lambda_{nn} = -(\phi_n T \phi_n) = -C_n(X) \tag{10}$$

which adds a correction term to the potential $U_n(X)$. The nuclear factor f(X) is then finally determined from the equation

$$[T+U_n(X)+C_n(X)]f_n(X) = E\, f_n(X) \tag{11}$$

with eigenvalues E_{nm} and eigenfunctions $f_{nm}(X)$.

Since the adiabatic wave function is of the form

$$\Psi^A(x,X) = f(X)\phi(x,X) \tag{12}$$

it can be considered as a special case of (2) if one rewrites it as

$$\Psi^A = \int f_{nm}(\alpha)\ \phi_n(x,\alpha)\delta(X-\alpha)d\alpha \tag{13}$$

The adiabatic equation (11) can also be considered as a special case of the HW-equation which is obtained by replacing in the kernels (4) and (5), $\Phi(X|\alpha)$ by $\delta(X-\alpha)$ and carrying out the integration over X first. Indeed we then find

$$\Delta(\alpha,\beta) \rightarrow \delta(\alpha-\beta) \tag{14}$$

and

$$H(\alpha,\beta) \rightarrow -\sum_i \frac{1}{2M_i}\Delta_i\ \delta(\alpha-\beta) + [U(\alpha)+C(\alpha)]\delta(\alpha-\beta) \tag{15}$$

From the previous discussion we learn that if $\Phi(X|\alpha)$ is taken as the delta function and $\phi(x|\alpha)$ is taken to be an electronic wave function $\phi_n(x,X)$ where we replace X by α, the GC wave function is just the adiabatic wave function, so

$$\lim_{\Phi\to\delta} f_{GCM}(\alpha) = f^A(\alpha)$$

$$\lim_{\Phi\to\delta} E_{GCM} = E^A$$

The energy kernel $H(\alpha,\beta)/\Delta(\alpha,\beta)$ for $M_i \to \infty$ is nothing but the potential energy surface U+C. The GCM contains as a limiting case the adiabatic approximation. We also notice that this limit automatically includes the correction term C. The physical meaning of all this is clear. The intrinsic function $\Phi(X|\alpha)$ describes the nuclear motion about the skeleton $\{\alpha\}$. In the extreme limit where the nuclei do not move, the nuclei are localized on the skeleton.

3.5 Non-adiabatic Effects

The main purpose of the GC theory is to try and incorporate in a compact and easy way, the non-adiabatic corrections. We have just seen that the adiabatic effects are contained in the theory. If we choose the intrinsic nuclear function Φ different from $\delta(X-\alpha)$, the GC space will be different; a priori no comparison with the adiabatic eigenvalues is possible. Both descriptions being variational, they both will yield upper bounds to the exact energies, but which will yield the best cannot be decided. However, it is expected that the non-product type of the GC wave function will account for some non-adiabaticity. In fact, as is shown on a model problem (see further), if one uses the freedom one has to choose Φ appropriately, a significant improvement both on ground state energy and excitation energy is obtained. From now on we will choose $\Phi(X|\alpha)$ to be a function that localizes the nuclei near the skeleton but allows a certain fluctuation in their motion. In other words $\Phi(X|\alpha)$ will be chosen as $\Phi(X-\alpha)$, where $\Phi(t)$ is a sharply peaked even function about $t=0$. In the realistic case, care must be taken to treat the translational and rotational motion first by the projection techniques mentioned in chapter 2. The discussion which follows is intended for vibrational motion of semi-rigid molecules. Other types of motion can also be described by the GCM, but the approximations we are going to consider do not apply there. It should be mentioned that the same idea can be used to describe non-adiabatic coupling between any pair of "slow" and "rapid" motions whatever their nature.

The procedure to introduce vibrational modes is in fact analogous to the expansion of a potential energy surface around its equilibrium configuration. We know from the discussion in chapter 2 that harmonic vibrations can be treated in GCM by assuming a Gaussian overlap kernel and expanding K up to second order around the minimum configuration $\{\alpha_0\}$. When the intrinsic function is very sharply peaked the overlap $\Delta(\alpha,\beta)$ will also be a sharply peaked function around $\alpha=\beta$ and it will be a good approximation to assume

$$\Delta(\alpha,\beta) = \exp[-(\alpha-\beta)S(\alpha-\beta)] \tag{16}$$

and also

$$K(\alpha,\beta) = E(\alpha_0)+\frac{1}{2}[(\alpha-\alpha_0)B(\alpha-\alpha_0)+2(\alpha-\alpha_0)A(\beta-\beta_0) + (\beta-\beta_0)B(\beta-\beta_0)] \qquad (17)$$

where S, B and A are 3Nx3N matrices. The HW-equation corresponding to these kernels is equivalent to a set of coupled harmonic oscillators. Provided the space of GC trial functions is stable under the symmetry group of the Hamiltonian, normal generator coordinates η can be introduced making use of the point group G_0 which leaves the equilibrium structure α_0 invariant. This transformation $\eta=L\alpha$ is identical to the one which gives normal coordinates from the displacements $\delta R=R-\alpha_0$. As a result of this change of variables the oscillators decouple. The six frequencies associated with normal coordinates of rotation and translation vanish. The remaining ones are the 3N-6 positive, non-zero eigenvalues of the matrix eigenvalue problem (Brink and Weiguny, 1968)

$$\begin{pmatrix} A & B \\ B & A \end{pmatrix}\begin{pmatrix} Y \\ Z \end{pmatrix} = \omega \begin{pmatrix} 1 & 0 \\ 0 & -1 \end{pmatrix}\begin{pmatrix} Y \\ Z \end{pmatrix} \qquad (18)$$

and correspond to the normal vibrational modes of the system. As a consequence of the decoupling scheme the weight functions in the harmonic approximation factorise

$$f(\eta) = f^{trans}(\underset{\sim}{A})\, f^{rot}(\Omega) \prod_{k=1}^{3N-6} f^{vib}_{n_k}(\eta_k) \qquad (19)$$

and the approximate eigenvalues are written as

$$E = E(\alpha_0) + E_{trans} + E_{rot} + E_{vib} \qquad (20)$$

This separation is here obtained as a result of a mathematical approximation scheme in the space of GC's rather than from semiclassical arguments involving dynamical degrees of freedom. It is noteworthy that the resulting molecular wave functions are not of the product form (19).

Deviations from the harmonic approximation can be taken into account by considering the difference between the exact and harmonic kernels as perturbations (Lathouwers and Lozes, 1977). They give rise to anharmonic corrections, rotation-vibration coupling, etc. The rotational and vibrational energies in (20) consist of a constant correction to the adiabatic ground state energy $E(\alpha_0)$

$$E_0 = E(\alpha_0) - \Delta E_{rot} - \Delta E_{vib}$$

and of excitation energies $\sum_i (v_i+1/2)\,\omega_i$ and (for the linear molecules) $(J(J+1)-\Lambda^2)/2I$. The inertial moment I and the frequencies ω_i of the normal modes, are expressed in terms of the intrinsic function χ and cannot be considered as adaptable parameters. Explicit examples of these statements are given in the next paper.

3.6 Vibronic Couplings

Until here we considered only one electronic state ϕ, consequently the nuclear motions described above refer to rotational and vibrational bands built on one electronic configuration. In order to incorporate different electronic states and so be able to describe vibronic couplings we shall generalize the GC wave function to

$$\Psi = \sum_n \int f_n(\alpha)\phi_n(x|\alpha)\Phi(X|\alpha)d\alpha \tag{21}$$

Variation of the weight functions $f_n(\alpha)$ leads to a set of coupled HW equations

$$\int [H_{nn}(\alpha,\beta)-E\Delta_{nn}(\alpha,\beta)]\, f_n(\beta)d\beta = \sum_m [H_{nm}(\alpha,\beta)-E\Delta_{nm}(\alpha,\beta)] f_m(\beta)d\beta \tag{22}$$

where we have introduced the generalized kernels

$$\Delta_{nm}(\alpha,\beta) = (\phi_n(\alpha)|\phi_m(\beta))(\Phi(\alpha)|\Phi(\beta)) \tag{23}$$

$$H_{nm}(\alpha,\beta) = (\phi_n(\alpha)\Phi(\alpha)|H\phi_m(\beta)\Phi(\beta)) \tag{24}$$

In the delta function limit (21) gives the Born and Huang series (7) whereas (22) goes over into the coupled differential equations (8). We thus arrive at the GCM analogue of the Born theory.

The importance of the coupling terms in (22) depends mainly upon the off-diagonal overlap kernels

$$\Delta_{nm}(\alpha,\beta) = (\phi_n(\alpha)|\phi_m(\beta))(\Phi(\alpha)|\Phi(\beta)) \tag{25}$$

which are usually very small. Indeed since $(\Phi(\alpha)|\Phi(\beta)) \sim \delta(\alpha-\beta)$ and $(\phi_n(\alpha)|\phi_m(\alpha))=0$ we see that $\Delta_{nm}(\alpha,\beta)$ is a product of two functions one of which vanishes whenever the other is significantly different from zero. The only possibility for Δ_{nm} to be large is when the electronic state m and n are degenerate at some point α_0. It is well known that in this case the $\phi_n(x|\alpha)$ vary rapidly with α in the crossing region. Therefore $(\phi_n(\alpha_0)|\phi_m(\alpha_0+\Delta\alpha))$ can be large when $(\Phi(\alpha_0)|\Phi(\alpha_0+\Delta\alpha)$ is not very small. The same reasoning applies to the Hamiltonian kernels except for the term $(\Phi(\alpha)|T\Phi(\beta))(\phi_n(\alpha)|\phi_m(\beta))$ where T is the nuclear kinetic energy.

If one were to use translated Gaussians as nuclear basis states a typical contribution would look like $[3-2c(\alpha-\beta)^2]\exp[-c(\alpha-\beta)^2]$ where c, in view of the strong localisation of the nuclei, is a large number (Shavitt, 1962). In contrast to the other terms these do not vanish in the delta function limit. Instead they give rise to the familiar couplings $(\phi_n|\partial^2/\partial^2X^2\phi_m)$ and $(\phi_n|\partial/\partial X|\phi_m)\partial/\partial X$.

3.7 A Soluble Model: coupled oscillators as a molecular analogue

As an illustration we consider the following model Hamiltonian of two interacting one-dimensional oscillators

$$H = h_0 + H_0 + V$$

$$h_0 = -\frac{1}{2m}\frac{\partial^2}{\partial r^2} + \frac{kr^2}{2} \qquad H_0 = -\frac{1}{2M}\frac{\partial^2}{\partial R^2} + \frac{KR^2}{2}$$

$$V = -\lambda rR \tag{26}$$

The eigenvalue problem for H is exactly soluble by scaling away the masses and rotating the reference frame over an angle $\theta=-1/2\ \mathrm{arctg}[\mu/(\omega_0^2-\Omega_0^2)]$, where $\omega_0=(k/m)^{1/2}$, $\Omega_0=(K/M)^{1/2}$ are the frequencies of the uncoupled ($\lambda=0$) oscillators and $\mu=2\lambda/(mM)^{1/2}$. In the resulting coordinate system (26) is equivalent to two non interacting oscillators with modified frequencies ω and Ω given by

$$\begin{matrix}\omega^2\\ \Omega^2\end{matrix} = (\omega_0^2+\Omega_0^2)/2 \pm (1/2)[(\omega_0^2-\Omega_0^2)^2+\mu^2]^{1/2} \tag{27}$$

These results are valid for coupling constants λ in the range $\pm(kK)^{1/2}$. We would like to interpret the energy spectrum

$$E_{nN} = (n+\tfrac{1}{2})\omega + (N+\tfrac{1}{2})\Omega \tag{28}$$

as a set of "vibrational bands" built on "electronic levels". For this purpose we choose $m \ll M$ and $k=K$ such that $\Omega_0=\kappa^2\omega_0$ ($\kappa=(m/M)^{1/4}$). If furthermore λ is not close to $\pm k$, the exact frequencies will not deviate too much from ω_0 and Ω_0 such that $\Omega\cong\kappa^2\omega$ (counterpart of $E_{vib}\cong\kappa^2E_{el}$). Thus, under these conditions, ω plays the role of electronic level spacings while Ω is the analogue of a nuclear vibration frequency. In the following we apply the adiabatic approximation and the GCM to the ground state vibrational band $E_{0N}\equiv E_N$.

The adiabatic approximation is easily worked out analytically. Both the "electronic" and "nuclear" eigenvalue problems reduce to one dimensional harmonic oscillator equations. We find

$$E_N^{AD} = \omega_0/2 + \Delta E_{AD} + (N+1/2)\Omega_{AD}$$
$$\Omega_{AD} = \Omega_0(1-c^2)^{1/2} \qquad \Delta E_{AD} = \omega_0 c^2\kappa^4/4 \tag{29}$$

where $c=\lambda/k=\lambda/K$. According to Messiah (1961) the term ΔE_{AD}, arising from averaging the "nuclear kinetic energy" over the "electronic ground state", should be of the order $E_{rot}\cong\kappa^4 E_{el}$. Although there are obviously no rotational type energy levels in our model (29) confirms this result.

For the GC treatment we supplement the "electronic ground state" $\phi_0(r|\alpha)$ with a Gaussian in R space to form the total intrinsic state (unnormalised)

$$\chi(r,R|\alpha) = \exp[-(r-c\alpha)^2/w^2]\exp[-(R-\alpha)^2/W^2] \tag{30}$$

where a width parameter W is introduced and $w=\sqrt{2/m\omega_0}$ is the analogous quantity in ϕ_0. The variational principle applied to the trial function leads to the HW equation:

$$\int [H(\alpha,\beta)-E\Delta(\alpha,\beta)]\, f(\beta)d\beta = 0$$
$$H(\alpha,\beta) = \langle\chi(\alpha)|H|\chi(\beta)\rangle \qquad \Delta(\alpha,\beta) = \langle\chi(\alpha)|\chi(\beta)\rangle \tag{31}$$

The Hamiltonian and overlap kernels, H and Δ, can easily by calculated and put in the familiar form

$$\Delta(\alpha,\beta) = e^{-S(\alpha-\beta)^2}$$

$$H(\alpha,\beta)/\Delta(\alpha,\beta) = E_g + \frac{1}{2} M_{GC}[2S-4S^2(\alpha-\beta)^2] + \frac{M_{GC}\Omega_{GC}^2}{2}\left(\frac{\alpha+\beta}{2}\right)^2 \tag{32}$$

known as the case of harmonic kernels.

The calculation of the kernels from (26) and (30) is straightforward. For normalised intrinsic states one obtains

$$S = (1/2)[(1/W)+(c^2/w)] :$$

introducing the variable $x=(w/W)^2$ we can write the effective frequency Ω_{GC} and effective mass M_{GC} of the generator coordinate oscillator as

$$\Omega_{GC} = \Omega_{AD}[F(x)]^{1/2} \qquad M_{GC} = M/F(x)$$

$$F(x) = (x^2+c^2/\kappa^4)/(x+c^2)^2$$

The constant E_g entering the ground state energy is given by

$$E_g+S/M_{GC} = (1/2)[(1/mw^2)+(1/MW^2)] + (1/8)[m\omega_0^2 w+M\Omega_0^2 W]$$

The integral equation for the kernels (32) is equivalent to the differential equation for a harmonic oscillator with mass M_{GC} and frequency Ω_{GC} (Griffin and Wheeler, 1957). The GC spectrum is given by

$$E_N^{GC} = E_g-M_{GC}\Omega_{GC}^2/16S + (N+1/2)\Omega_{GC} \tag{33}$$

and is a function of the width parameter W. In Figs. 1 and 2 we have plotted E_0^{GC} and Ω_{GC} versus 1/W. The adiabatic limit, i.e., $\dot{W}\to 0$,

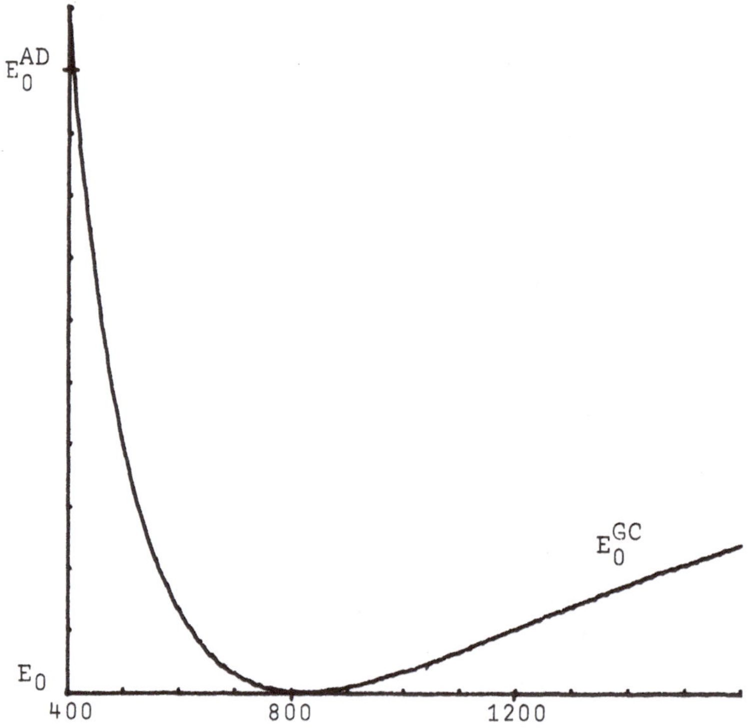

Fig. 1. E_0^{GC} as a function of $1/W^2$ (parameters: k=K=1, m=1, M=1600, c=.5)

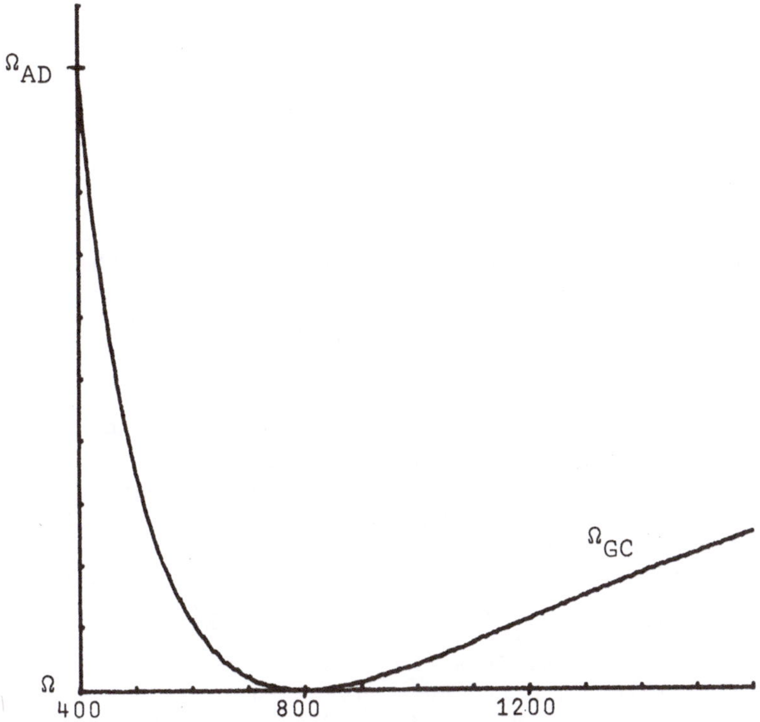

Fig. 2. Ω_{GC} as a function of $1/W^2$ (parameters: k=K=1, m=1, M=1600, c=.5)

can be performed analytically and confirms the tendency of the figures

$$\lim_{W\to 0} E_0^{GC} = E_0^{AD} \qquad \lim_{W\to 0} \Omega_{GC} = \Omega_{AD} \tag{34}$$

The minima of both quantities do in fact not coincide with the exact results, as the figure might suggest, the scale of the drawing is too large to show the deviation. For the vibrational frequency the optimum width can be computed exactly and equals $W_0=\kappa^2 w$. The behaviour of E_0^{GC} as a function of W is dominated by the zero point vibrational contribution $\Omega^{GC}/2$ such that its minimum occurs very close to W_0. We have therefore chosen to analyse the results further for $W=W_0$.

In order to compare the adiabatic and optimum ($W=W_0$) GCM results with the exact ones we expanded all quantities as a Born-Oppenheimer series i.e., in the parameter $\kappa=(m/M)^{1/4}$. The table below gives the order of magnitude of the error in the ground state and the ground state vibrational frequency.

	E_0	Ω
adiabatic	$\kappa^6\omega_0$	$\kappa^6\omega_0$
GCM	$\kappa^8\omega_0$	$\kappa^{10}\omega_0$

The adiabatic energy levels are off by $\kappa^6\omega_0=\kappa^6 E_{el}$ as predicted by Messiah. The GC ground state agrees to order κ^8, the GC frequency even to κ^{10}, with the exact values.

REFERENCES

Brink, D.M. and Weiguny, A., 1968, Nucl. Phys. A120: 59.

Broeckhove, J., 1979, private communication.

Caurier, E., 1975, in "Proceedings of the 2nd International Seminar on the GCM for nuclear Bound states and Reactions", M. Bouten and P. Van Leuven, eds., S.C.K., Mol.

Griffin, J.J. and Wheeler, J.A., 1957, Phys. Rev. 108: 311.

Hill, D.L. and Wheeler, J.A., 1953, Phys. Rev. 89: 1106.

Hurley, A.C., 1967, Int. J. Quant. Chem., 1S: 677.

Lathouwers, L., 1975, Ann. Phys. 102: 347.

Lathouwers, L., 1976, J. Math. Phys. 17: 1274

Lathouwers, L. and Lozes, R., 1977, J. Phys. A10: 1465.

Messiah, A., 1961, Quantum Mechanics, North Holland Publ., Amsterdam.

Shavitt, I., 1962, Methods in Computational Physics 2, Ac. Press, New York.

Thakkar, A.J. and Smith, V.H., Jr., 1976, Phys. Rev. A15: 15.

Vincent, C.M., 1973, Phys. Rev. C8: 929.

Wiener, N., 1933, The Fourier Integral and some of its Applications, Cambridge University Press, Cambridge.

Woolley, R.G., 1976, Adv. Phys. 25: 27.

Woolley, R.G. and Sutcliffe, B.T., 1977, Chem. Phys. Letters, 45: 393.

GENERATOR COORDINATE THEORY OF DIATOMIC MOLECULES

L. Lathouwers and P. Van Leuven

Dienst Teoretische en Wiskundige Natuurkunde
Rijksuniversitair Centrum
University of Antwerp, Belgium

1. INTRODUCTION

In the first part of these lectures (hereafter referred to as I) we introduced the mathematical framework of the Generator Coordinate Method (GCM) and outlined how nuclear positions can be used as generator coordinates to construct a non-adiabatic theory of molecules (Van Leuven and Lathouwers, this volume).

Here we will demonstrate in some detail how the GCM can be applied to diatomic systems. In particular we will work out the angular momentum projection technique and derive a set of coupled radial integral equations as a counterpart of the Born-Huang series (BH-series) defined in a rotating or body-fixed (BF) frame.

For bound state diatomic systems we will rederive the symmetry labeling of the rotation vibration levels in a new way and demonstrate that the GCM can explain the separation of the eigen-energies into electronic, vibrational and rotational parts without having to actually decouple the respective dynamical motions. A simple formula explaining the qualitative behaviour of the Λ-doubling effect, absent in the adiabatic approximation (AA) is given.

Asymptotic GC states and their application to inelastic atom-atom scattering are studied in the last sections. A set of inhomogeneous coupled integral equations is derived. Its solutions yields the transition matrix in the helicity representation without having to transform any operators to the BF-frame.

These notes are based on a recent series of papers (Lathouwers, 1977-1979).

2. DIATOMIC SYSTEMS

2.1 Notations and Definitions

After separating off the centre of mass motion the Hamiltonian for a system of two nuclei with charges and masses (Z_1,M_1), (Z_2,M_2) and a number of electrons can be written as

$$H = -(1/2\mu)\Delta_{\underset{\sim}{R}}-(1/2)\sum_i \Delta_{\underset{\sim}{r}_i} + (Z_1Z_2)/R-Z_1\sum_i (1/|\underset{\sim}{r}_i-\underset{\sim}{R}_1|)$$

$$- Z_2\sum_i (1/|\underset{\sim}{r}_i-\underset{\sim}{R}_2|) + \sum_{i<j} (1/|\underset{\sim}{r}_i-\underset{\sim}{r}_j|) \qquad (1)$$

Here μ is the reduced nuclear mass $(M_1M_2)/M$, $\underset{\sim}{r}$ the position of the ith electron measured from the centre of mass of the nuclei, $\underset{\sim}{R}$ the vector connecting the two nuclei and $\underset{\sim}{R}_1=-(M_2\underset{\sim}{R})/M$, $\underset{\sim}{R}_2=+(M_1\underset{\sim}{R})/M$ $(M=M_1+M_2)$. Before going any further one should notice that in writing down the Hamiltonian as in (1) we have neglected relativistic and mass-polarisation effects and assumed the coupling between the electron spin and orbital angular momenta to be negligible.

H commutes with the total angular momentum

$$\underset{\sim}{J} = \underset{\sim}{R} \times \underset{\sim}{P} + \sum_i \underset{\sim}{r}_i \times \underset{\sim}{p}_i = \underset{\sim}{L} + \underset{\sim}{\ell} \qquad (2)$$

which is the sum of a nuclear $\underset{\sim}{L}$ and an electronic $\underset{\sim}{\ell}$ component. From the above expression it is clear that H depends upon the internuclear vector $\underset{\sim}{R}$ (not on $\underset{\sim}{R}_1$ and $\underset{\sim}{R}_2$ separately) such that we can write Schrödinger's equation as

$$H(r,\underset{\sim}{R})\ \Psi_{JM}(r,\underset{\sim}{R}) = E_J\ \Psi_{JM}(r,\underset{\sim}{R}) \qquad (3)$$

The Born-Oppenheimer and adiabatic approximations have made it customary to decompose H as the sum of the kinetic energy of the nuclear relative motion and the so-called electronic Hamiltonian

$$H(r,\underset{\sim}{R}) = -(1/2\mu)\Delta_{\underset{\sim}{R}} + H_e(r,\underset{\sim}{R}) \qquad (4)$$

As an operator acting on the space of functions square integrable in the electron coordinates H_e is clearly invariant under rotations about $\underset{\sim}{R}$ and reflexions through planes containing $\underset{\sim}{R}$. Therefore the solutions to the electronic Schrödinger equation can be labelled as follows (Landau-Lifschitz, 1965)

$$H_e(r,\underset{\sim}{R})\ \phi_{n\Lambda}(r|\underset{\sim}{R}) = U_{n\Lambda}(R)\ \phi_{n\Lambda}(r|\underset{\sim}{R}) \qquad (5)$$

Here Λ denotes the absolute value of the electronic angular momentum along the internuclear axis, i.e. $\underset{\sim}{\ell}.\underset{\sim}{R}/R$, while n stands for all other quantum numbers. The fact that H_e commutes with $\underset{\sim}{R}$ itself is demonstrated by the parametric dependence of the electronic eigenvalues and eigenfunctions upon this quantity. By writing $U_{n\Lambda}=U_{n\Lambda}(R)$ we have anticipated the well known result (to be derived later) that the electronic eigenvalues depend upon the relative distance between nuclei only.

The GC form of the wavefunction for a diatomic system described by the above Hamiltonian is

$$\psi(r,\underset{\sim}{R}) = \sum_{n\Lambda} \int F_{n\Lambda}(\underset{\sim}{\alpha})\ \phi_{n\Lambda}(r|\underset{\sim}{\alpha})\ \Phi(|\underset{\sim}{R}-\underset{\sim}{\alpha}|)d\alpha \tag{6}$$

It should be stressed that the completeness of the set $\{\phi_{n\Lambda}(\underset{\sim}{\alpha})\}$ and the closure of translations (Wiener, 1933) imply that (6) is an exact form of the wavefunction. In the delta function limit (see I) one gets the BH-series

$$\psi(r,\underset{\sim}{R}) = \sum_{n\Lambda} F_{n\Lambda}(\underset{\sim}{R})\ \phi_{n\Lambda}(r|\underset{\sim}{R}) \tag{7}$$

Since the Dirac δ-function is even we have chosen Φ to depend on $|\underset{\sim}{R}-\underset{\sim}{\alpha}|$. There are other reasons for this choice which will become clear later. We will often restrict ourselves to a one state approximation of (6,7) appropriate if the PE curve $U_{n\Lambda}(R)$ is well separated from the rest of the electronic spectrum. With the notation $\chi_n(r,\underset{\sim}{R}|\underset{\sim}{\alpha})=\phi_{n\Lambda}(r|\underset{\sim}{\alpha})\Phi(|\underset{\sim}{R}-\underset{\sim}{\alpha}|)$ for the full intrinsic state a single term in (6) is written as $\int F_{n\Lambda}(\underset{\sim}{\alpha})\chi_{n\Lambda}(\underset{\sim}{\alpha})d\underset{\sim}{\alpha}$. It should be noticed that the $\chi_n(\underset{\sim}{\alpha})$ describe electrons moving in the average, rather than instantaneous, field of the nuclei. Indeed one has

$$\langle\chi_{n\Lambda}(\underset{\sim}{\alpha})|\underset{\sim}{R}|\chi_{n\Lambda}(\underset{\sim}{\alpha})\rangle = \underset{\sim}{\alpha} \tag{8}$$

2.2 Conservation of Angular Momentum

As outlined in I total angular momentum of a many-body system can be projected out by considering Euler angles as GC's. This is common in nuclear physics. However, in molecular theory it is customary to account for rotational invariance by introducing a body-fixed (BF) or rotating reference frame (Wigner, 1959; Wigner and Hirschfelder, 1935). In practice a projection technique follows an already approximate expansion of the wavefunction whereas in the BF-frame approach the transformation law of the exact states is used, after which a basis for the internal states is introduced. This interchange of a technique to obtain good angular momentum quantum numbers and the expansion of the wavefunction produces some differences between the adiabatic and GC treatment of diatomic systems. An obvious one is the use of active rotation operators in projection as opposed to passive rotation operators in a BF-frame transformation.

A convenient definition of the rotating reference frame for a diatomic system is to choose the z'-axis along the internuclear vector, i.e., the Euler angles α,β are taken as the polar angles $\omega_R=(\phi_R,\theta_R)$ of $\underset{\sim}{R}$. The third angle, say γ_R, then serves to fix the x' and y' axes. It can be seen that this angle is redundant in the diatomic case and it is therefore common (Goodisman, 1973) to put $\gamma_R=0$. This corresponds to always taking y' perpendicular to the z,z' plane. The transformation law of the exact wavefunction then takes the form

$$\psi_{JM}(r,\underset{\sim}{R}) = \sum_{\Lambda} D^{J^*}_{M\Lambda}(\phi_R,\theta_R,0)\psi_{J\Lambda}(r',R\underset{\sim}{e}_{z'}) \qquad (9)$$

$$\equiv \sum_{\Lambda} D^{J^*}_{M\Lambda}(\omega_R)\ \psi_{J\Lambda}(r'|R)$$

The new electron coordinates are functions of the polar angles of $\underset{\sim}{R}$, i.e. $r'=r'(\omega_R)$. The transformation of a position vector $\underset{\sim}{r}_i$ to $\underset{\sim}{r}_i'$ is given by a (passive) rotation matrix with $(\alpha,\beta,\gamma)=(\phi_R,\theta_R,0)$. Clearly one has

$$(\underset{\sim}{J}.\underset{\sim}{R})/R = J_{z'} = \ell_z' = (\underset{\sim}{\ell}.\underset{\sim}{R})/R \qquad (10)$$

i.e. there is no nuclear angular momentum about the body fixed axis. We have therefore used the notation Λ, common for the electronic angular momentum $\ell_{z'}$, to specify the $J_{z'}$ components in (9).

An appropriate basis to expand the $\psi_{J\Lambda}(r'|R)$ are the electronic states $\phi_{n\Lambda}(r'|R)$ defined in the BF-frame. For each Λ and R they form a complete set such that the full wave function can be written as

$$\psi_{JM}(r,\underset{\sim}{R}) = \sum_{n\Lambda} g^J_{n\Lambda}(R)D^{J^*}_{M\Lambda}(\omega_R)\ \phi_{n\Lambda}(r'|R) \qquad (11)$$

in which the radial functions $g^J_{M\Lambda}(R)$ remain to be determined. This is the Born-Huang (BH) series (7) (Born, 1951) adapted to rotational invariance of the diatomic system.

Since (6) is an exact representation of the wavefunction, the variational principle applied to the weight functions will automatically lead to angular momentum eigenstates. The question is whether one can also determine part of the $F_{n\Lambda}(\underset{\sim}{\alpha})$ from first principles in analogy to the separation of the Wigner D-functions (9). For this purpose we will use the following property of the electronic Hamiltonian: the parametric dependence of H_e on the generating vector $\underset{\sim}{\alpha}$ is such that

$$H_e(\underset{\sim}{\alpha}) = \mathcal{R}_e(\omega_\alpha)H_e(\alpha\underset{\sim}{e}_z)\mathcal{R}_e^{-1}(\omega_\alpha) \qquad (12)$$

where $R_e(\omega_\alpha)$ is the (active) rotation operator in the electron space. This relation is a direct consequence of the fact that Coulomb interactions depend upon the mutual distance between the particles only. Equation (12) states that $H_e(\underset{\sim}{\alpha})$ and $H_e(\alpha\underset{\sim}{e}_z)$ are unitary equivalent. This implies that the spectra of these operators are identical, i.e., $U_{n\Lambda}=U_{n\Lambda}(\alpha)=U_{n\Lambda}(\underset{\sim}{\alpha})$, a result we used earlier (5). Furthermore the associated eigenfunctions are transformed into each other through the unitary operator $R_e(\omega_\alpha)$, i.e.,

$$\phi_{n\Lambda}(r|\underset{\sim}{\alpha}) = R_e(\omega_\alpha)\,\phi_{n\Lambda}(r|\alpha\underset{\sim}{e}_z) \tag{13}$$

As for the nuclear part of the basis functions it is clear that

$$\Phi(|\underset{\sim}{R}-\underset{\sim}{\alpha}|) = R_n(\omega_\alpha)\Phi(|\underset{\sim}{R}-\alpha\underset{\sim}{e}_z|) \tag{14}$$

$R_n(\omega_\alpha)$ indicating the rotation operator in nuclear space. Combining (13) and (14) we can state that for the full intrinsic states one has

$$\chi_{n\Lambda}(r,\underset{\sim}{R}|\underset{\sim}{\alpha}) = R(\omega_\alpha)\chi_{n\Lambda}(r,\underset{\sim}{R}|\alpha\underset{\sim}{e}_z) \tag{15}$$

It is also easily seen that $\chi_{n\Lambda}(\alpha\underset{\sim}{e}_z)$ is an eigenstate of J_z with eigenvalue Λ

$$J_z\,\chi_{n\Lambda}(\alpha\underset{\sim}{e}_z) = \Lambda\chi_{n\Lambda}(\alpha\underset{\sim}{e}_z) \tag{16}$$

The last two equations can now be used to partially determine the weight function $F_{n\Lambda}(\underset{\sim}{\alpha})$. Using (15) we can write (6) as

$$\psi(r,\underset{\sim}{R}) = \sum_{n\Lambda} \int F_{n\Lambda}(\underset{\sim}{\alpha})R(\omega_\alpha)\chi_{n\Lambda}(r,\underset{\sim}{R}|\alpha\underset{\sim}{e}_z)d\underset{\sim}{\alpha} \tag{17}$$

In this expression one starts to recognize the form of the angular momentum projection operators (see part I). Firstly, one can easily build in the third, redundant, Euler angle, say γ_α, by inserting $e^{i\gamma_\alpha\Lambda}\,e^{-i\gamma_\alpha J_z}$ in front of $\chi_{n\Lambda}(\alpha\underset{\sim}{e}_z)$. Secondly, knowing that P^J_{MK} working on an eigenfunction of J_z will give zero unless the associated eigenvalue equals K and taking into account the variational procedure outlined in I the angular dependence of the weight functions must be such that

$$F^{JM}_n(\underset{\sim}{\alpha}) = f^J_{n\Lambda}(\alpha)D^{J^*}_{M\Lambda}(\omega_\alpha) \tag{18}$$

With this choice the GCM wavefunctions will be the angular momentum eigenstates

$$\begin{aligned}\psi_{JM}(r,\underset{\sim}{R}) &= \sum_{n\Lambda} \int f^J_{n\Lambda}(\alpha) D^{J^*}_{M\Lambda}(\omega_\alpha) \chi_{n\Lambda}(r,\underset{\sim}{R}|\underset{\sim}{\alpha}) d\underset{\sim}{\alpha} \\ &= \sum_{n\Lambda} \int f^J_{n\Lambda}(\alpha) D^{J^*}_{M\Lambda}(\Omega_\alpha) R(\Omega_\alpha) \chi_{n\Lambda}(r,\underset{\sim}{R}|\alpha \underset{\sim}{e}_z) d\underset{\sim}{\alpha} \\ &= \sum_{n\Lambda} \int f^J_{n\Lambda}(\alpha) P^J_{M\Lambda} \chi_{n\Lambda}(r,\underset{\sim}{R}|\alpha \underset{\sim}{e}_z) \alpha^2 d\alpha\end{aligned} \tag{19}$$

where the radial weight functions are the only unknowns left.

2.3 Coupled Equations

The coupled differential equations and coupled integral equations derived in I can now be written down for each value J of the total angular momentum.

In substituting the BH-series into Schrödinger's equation one must take into account the implicit dependence of the electron coordinates on the polar angles ω_R. This means that differentiations with respect to ϕ_R and θ_R are to be replaced by

$$\begin{aligned}\frac{\partial}{\partial\theta_R} + \sum_i \frac{\partial x_i'}{\partial\theta_R}\frac{\partial}{\partial x_i'} + \ldots &= \frac{\partial}{\partial\theta_R} - i\ell_{y'} \\ \frac{\partial}{\partial\phi_R} + \sum_i \frac{\partial x_i'}{\partial\phi_R}\frac{\partial}{\partial x_i'} + \ldots &= \frac{\partial}{\partial\phi_R} - i\cos\theta_R \ell_{z'} \\ &\quad + i\sin\theta_R\,\ell_{x'}\end{aligned} \tag{20}$$

where the partial derivatives indicate differentiation with respect to explicit occurence of θ_R and ϕ_R. Making this substitution amounts to transforming the Hamiltonian from the space-fixed (SF-frame) to the BF-frame). The extra terms arising from the nuclear kinetic energy can be combined to give the total angular momentum squared

$$J^2 = -[\Delta(\omega_R) - 2i\frac{\cot g\,\theta_R}{\sin\theta_R}\ell_{z'}\frac{\partial}{\partial\phi_R} - \frac{1}{\sin^2\theta_R}\ell^2_{z'}] \tag{21}$$

and a number of coupling terms. Inserting (11) into the transformed Schrödinger equation then gives a set of coupled differential equations for the radial functions $g^J_{n\Lambda}(R)$

$$[-\frac{1}{2\mu}\frac{d^2}{dR^2} + \frac{J(J+1)-\Lambda^2}{2\mu R^2} + U_{n\Lambda}(R)]\ (Rg^J_{n\Lambda}(R)) =$$

$$\sum_{n'\Lambda'} C_{n\Lambda,n'\Lambda'}(R,\frac{d}{dR})\ (Rg^J_{n'\Lambda'}(R)) \tag{22}$$

The coupling terms contained in $C^J_{n\Lambda,n'\Lambda'}$ are proportional to matrix elements of the following form

$$<\phi_{n\Lambda}(R)|\frac{d}{dR}|\phi_{n'\Lambda}(R)>\frac{d}{dR}, \quad <\phi_{n\Lambda}(R)|\frac{d^2}{dR^2}|\phi_{n'\Lambda}(R)> \tag{23}$$

$$<\phi_{n\Lambda}(R)|\ell'_{\pm}|\phi_{n'\Lambda\pm 1}(R)>, \ <\phi_{n\Lambda}(R)|\ell^2_{x'}+\ell^2_{y'}|\phi_{n'\Lambda}(R)> \tag{24}$$

The two groups are referred to as radial and angular couplings respectively. The explicit expression for the full set of equations is rather lengthy and will not be given here. It can be found in Van Vleck (1951).

The optimal radial weight function obey a set of coupled integral equations which can readily be derived using (19) and the projection operator properties

$$\int K^J_{n\Lambda,n\Lambda}(\alpha,\beta|E)f^J_{n\Lambda}(\beta)\beta^2 d\beta = \sum_{n'\Lambda'} \int K^J_{n\Lambda,n'\Lambda'}(\alpha,\beta|E)$$
$$f^J_{n'\Lambda'}(\beta)\beta^2 d\beta \tag{25}$$

$$K^J_{n\Lambda,n'\Lambda'}(\alpha,\beta|E) = H^J_{n\Lambda,n'\Lambda'}(\alpha,\beta)-E\Delta^J_{n\Lambda,n'\Lambda'}(\alpha,\beta)$$

$$H^J_{n\Lambda,n'\Lambda'}(\alpha,\beta) = <\chi_{n\Lambda}(\alpha\underset{\sim}{e}_z)|HP^J_{\Lambda\Lambda'}|\chi_{n'\Lambda'}(\beta\underset{\sim}{e}_z)> \tag{26}$$

$$\Delta^J_{n\Lambda,n'\Lambda'}(\alpha,\beta) = <\chi_{n\Lambda}(\alpha\underset{\sim}{e}_z)|P^J_{\Lambda\Lambda'}|\chi_{n'\Lambda'}(\beta\underset{\sim}{e}_z)>$$

The expressions for the kernels can further be simplified inserting the explicit form of the projection operators and using properties (15) and (16) of the intrinsic states. For example, the most complicated matrix elements to be calculated here, i.e., the Hamiltonian couplings $H^J_{n\Lambda n'\Lambda'}$, become

$$H^J_{n\Lambda,n'\Lambda'}(\alpha,\beta) = \int d\cos\theta\ d^J_{\Lambda\Lambda'}(\theta)<\chi_{n\Lambda}(\alpha\underset{\sim}{e}_z)|He^{-i\theta J_y}|\chi_{n'\Lambda'}(\beta\underset{\sim}{e}_z)>$$

$$= \int d\cos\theta\ d^J_{\Lambda\Lambda'}(\theta)<\chi_{n\Lambda}(\alpha\underset{\sim}{e}_z)|H|\chi_{n'\Lambda'}(\beta(\cos\theta\underset{\sim}{e}_z+\sin\theta\underset{\sim}{e}_x))> \tag{27}$$

There is no separation in angular and radial type terms and all quantities are referred to the space-fixed axes.

3. DIATOMIC SPECTRA

3.1 Symmetry Labels

In addition to the total angular momentum parity should be conserved. It is well known (see, e.g., Landau and Lifshitz, 1965) that there is a definite relationship between the total angular momentum, the total parity, and the intrinsic quantum numbers ($n\Lambda$). In order to derive this connection we have to specify some of the additional labels n. In the heteronuclear case one distinguishes electronic states which are invariant (+ states) or change sign (− states) under a reflexion through a plane containing the internuclear axis. If Λ=0 (Σ states) the + and − states are different, for $\Lambda\neq 0$ ($\Pi,\Delta,\ldots$ states) they are the $\pm$ combinations of the degenerate $\pm$ functions. In the homonuclear case the electronic parity is also conserved, one speaks of g (gerade=even) and u (ungerade=odd) states.

In the "one band" approximation the GC function corresponding to Σ and $\Lambda\neq 0$ electronic states are (from (19))

$$\begin{aligned} \psi_{JM} &= \int f(\alpha) P^{J}_{M0} \chi_0(\alpha \underset{\sim}{e}_z)\, \alpha^2 d\alpha \\ \psi_{JM} &= \int f(\alpha)\, [P^{J}_{M\Lambda}\chi_\Lambda(\alpha \underset{\sim}{e}_z) \pm P^{J}_{M-\Lambda}\chi_{-\Lambda}(\alpha \underset{\sim}{e}_z)]\alpha^2 d\alpha \end{aligned} \tag{1}$$

Indeed these functions have a definite parity $(-)^{J+s}$ where s is 0 for a + state and 1 for a − state.

In the homonuclear case good parity functions can be generated if the electronic g or u states are combined with even or odd nuclear parts

$$\Phi_{G,U} = \Phi(|\underset{\sim}{R}-\underset{\sim}{\alpha}|) \pm \Phi(|\underset{\sim}{R}+\underset{\sim}{\alpha}|) \tag{2}$$

The GCM wavefunction will have parity $(-)^{p+P}$, where p and P are 0 or 1 for gerade or ungerade functions, respectively. Working out the integrals making use of the projection technique one finds that the terms arising from $\Phi(|\underset{\sim}{R}+\underset{\sim}{\alpha}|)$ are identical up to a sign factor. More specifically the wavefunctions are proportional to

$$[1+(-)^{J+s+p+P}] \tag{3}$$

This factor determines whether a given electronic state (i.e., a given s and p) and a total angular momentum and parity are compatible.

At the same time it tells us which nuclear intrinsic states to use.
Particle statistics implies that the spatial wavefunction must be
symmetric (s) or antisymmetric (a) such that after combination with
a nuclear spin function the state changes sign or not under interchange of the nuclei depending on whether the latter are fermions
or bosons. From (3) now follows the relationship between the total
parity, the sign ($\pm$) of the electronic term and the nuclear permutation symmetry. If the level is even (odd) and positive (negative)
then it is symmetric, but when it is even (odd) and negative
(positive) it is antisymmetric under interchange of the nuclei.
The results are summarised in the figure. These diagrams can be
constructed requiring (3) to be non-vanishing. An interesting
situation occurs for spin zero nuclei. In this case there are no
antisymmetric spin states such that G basis states are obligatory.
It then follows that (3) is zero for Σ_g^+, Σ_u^- (Σ_u^+, Σ_g^-) and even (odd)
J values. Such levels are therefore non-existing, a well-known
result. In order to derive the above within the AA requires a
transformation of symmetry operators (defined in the SF frame) to
the BF frame. This is quite tedious and rather confusing (Landau-
Lifshitz, 1965).

J	J	J	J	J	J
6 —— +	6 —— −	6 $\overset{-}{\underset{+}{=}}$	6 —— +s(a)	6 —— −a(s)	6 $\overset{-a(s)}{\underset{+s(a)}{=}}$
5 —— −	5 —— +	5 $\overset{+}{\underset{-}{=}}$	5 —— −a(s)	5 —— +s(a)	5 $\overset{+s(a)}{\underset{-a(s)}{=}}$
4 —— +	4 —— −	4 $\overset{-}{\underset{+}{=}}$	4 —— +s(a)	4 —— −a(s)	4 $\overset{-a(s)}{\underset{+s(a)}{=}}$
3 —— −	3 —— +	3 $\overset{+}{\underset{-}{=}}$	3 —— −a(s)	3 —— +s(a)	3 $\overset{+s(a)}{\underset{-a(s)}{=}}$
2 —— +	2 —— −	2 $\overset{-}{\underset{+}{=}}$	2 —— +s(a)	2 —— −a(s)	2 $\overset{-a(s)}{\underset{+s(a)}{=}}$
1 —— −	1 —— +	1 $\overset{+}{\underset{-}{=}}$	1 —— −a(s)	1 —— +s(a)	1 $\overset{+s(a)}{\underset{-a(s)}{=}}$
0 —— +	0 —— −	0 $\overset{\pm}{=}$	0 —— +s(a)	0 —— −a(s)	0 $\overset{+s(a)}{\underset{-a(s)}{=}}$
Σ^+	Σ^-	Π	Σ_g^+(u)	Σ_g^-(u)	Π_g(u)

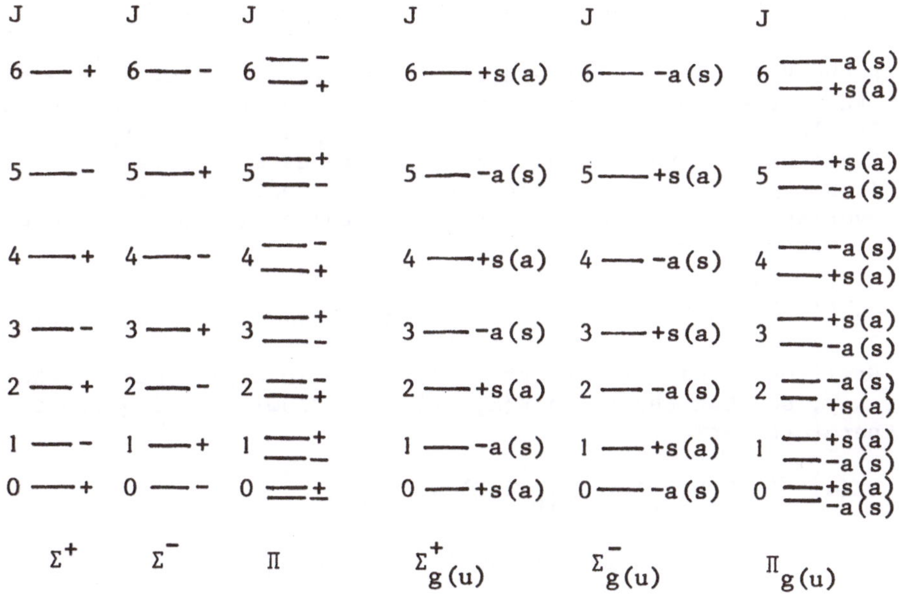

3.2 Separation of the Molecular Energies

One of the great successes of the AA is that the energy
pattern

$$E \cong E_{trans} + E_{el} + E_{vib} + E_{rot'}$$

which roughly explains the structure of the spectra, can be obtained in a simple way. Indeed, this decomposition follows easily from the assumption that the electronic, rotational and nuclear vibrational motions decouple. This is reflected in the product type of the corresponding wavefunction. The question now arises whether it is possible to approximately solve the GCM equation such that the energy can be written as in (3) without having to actually separate the dynamical motions. For this purpose we evaluate the energy of the GCM projected states. As an example we treat Σ states. The results are easily generalized to the $\Lambda \neq 0$ case. We have $(\chi_0(\alpha \underset{\sim}{e}_z) \equiv \chi_0(\alpha))$

$$E_J = \frac{\iint f(\alpha) \langle \chi_0(\alpha) | (P^J_{M0})^+ H P^J_{M0} | \chi_0(\beta) \rangle f(\beta) \alpha^2 \beta^2 d\alpha d\beta}{\iint f(\alpha) \langle \chi_0(\alpha) | (P^J_{M0})^+ P^J_{M0} | \chi_0(\beta) \rangle f(\beta) \alpha^2 \beta^2 d\alpha d\beta} \tag{4}$$

Using the properties of the projection operators we can rewrite this as

$$E_J = \frac{\iint f(\alpha) (\int \langle \chi_0(\alpha) | H R(\theta) | \chi_0(\beta) \rangle P_J(\theta) d\theta) f(\beta) \alpha^2 \beta^2 d\alpha d\beta}{\iint f(\alpha) (\int \langle \chi_0(\alpha) | R(\theta) | \chi_0(\beta) \rangle P_J(\theta) d\theta) f(\beta) \alpha \beta d\alpha d\beta} \tag{5}$$

where the $P_J(\theta)$ are the Legendre polynomials. In view of the fact that the nuclear basis states are strongly localized in space the matrix elements occuring in the above integrals are sharply peaked functions of $\alpha-\beta$ and will decrease rapidly as θ differs from 0 or π. It is therefore reasonable, in a first approximation, to replace the overlap matrix element in the dominator of (5) by a product of two Gaussians

$$e^{-r/2\ \theta^2}\ e^{-s/2\ (\alpha-\beta)^2} = \Delta_{rot}(\theta) \Delta_{vib}(\alpha-\beta) \tag{6}$$

The Hamiltonian matrix element will behave in approximately the same way, so that the ratio $K(\alpha,\beta,\theta)$ is a slowly varying function. We therefore set

$$\begin{aligned} K(\alpha,\beta,\theta) &\cong E(\alpha_0) + K_{vib}(\alpha,\beta) + K_{rot}(\theta) \\ &= E(\alpha_0) + \frac{s}{2}[B(\alpha-\alpha_0)^2 + 2A(\alpha-\alpha_0)(\beta-\alpha_0) + B(\beta-\alpha_0)^2] \\ &\quad - \frac{r}{2} C\theta^2 \end{aligned} \tag{7}$$

where $\cong$ means α and β are around α_0 and θ is in the neighbourhood of 0 or π. The constants s, r, A, ... are easily determined by expanding the intrinsic states in Taylor series around $\theta=0$ and $\alpha=\beta=\alpha_0$. One obtains

$$s = \langle \frac{\partial\chi_0}{\partial\alpha} | \frac{\partial\chi_0}{\partial\alpha} \rangle_0 \qquad r = \langle\chi_0|J_y^2|\chi_0\rangle_0$$

$$sB = \langle\chi_0|H-E(\alpha_0)|\frac{\partial^2\chi_0}{\partial\alpha^2}\rangle_0 \quad sA = \langle\frac{\partial\chi_0}{\partial\alpha}|H-E(\alpha_0)|\frac{\partial\chi_0}{\partial\alpha}\rangle_0 \tag{8}$$

$$rC = \langle\chi_0|HJ_y^2-E(\alpha_0)J_y^2|\chi_0\rangle_0$$

where $\langle\ \rangle_0$ means that the matrix elements have to be taken at $\theta=0$, $\alpha=\beta=\alpha_0$. The energy expectation value is then a sum of 3 terms

$$E_J = E(\alpha_0) + \frac{\int P_J(\theta)K_{rot}(\theta)\Delta_{rot}(\theta)d\theta}{\int P_J(\theta)\Delta_{rot}(\theta)d\theta}$$

$$+ \frac{\iint f(\alpha)\ K_{vib}(\alpha,\beta)\Delta_{vib}(\alpha,\beta)f(\beta)\alpha^2\beta^2 d\alpha d\beta}{\iint f(\alpha)\Delta_{vib}(\alpha-\beta)f(\beta)\alpha^2\beta^2 d\alpha d\beta} \tag{9}$$

Since the integrands in θ have non-zero values only around 0 and π one can insert the asymptotic expansion of the Legendre polynomials.

$$P_J(\theta) \underset{\theta\to 0}{\sim} 1 - \frac{\theta^2}{4} J(J+1) \tag{10}$$

This allows us to evaluate the second term in (9) in closed form. The function $f(\alpha)$ has so far been left arbitrary. If it is varied to minimize the last term in (9) one obtains the integral equation in the Gaussian overlap approximation and the quadratic approximation (see part I). The resulting energy spectrum under these conditions is

$$E_{vJ} = E(\alpha_0) - \Delta E_{vib} - \Delta E_{rot} + (v+\tfrac{1}{2})\omega + J(J+1)/2I$$

$$\Delta E_{vib} = A/2 \qquad \omega = (A^2-B^2)^{1/2} \tag{11}$$

$$\Delta E_{rot} = C \qquad I = r/C$$

which is of the required form (4). The terms ΔE_{vib}, ΔE_{rot} represent the energy gain due to the variational treatment of vibrations and rotations. The vibration frequency ω and the moment of inertia I depend upon the local structure of the energy surface around the equilibria $\alpha=\alpha_0$ and $\theta=0,\pi$. For the $\Lambda\neq 0$ case one has to use the expansion

$$d^J_{\Lambda\Lambda}(\theta) \underset{\theta\to 0}{\sim} 1 - \frac{\theta^2}{4}\,(J(J+1)-\Lambda^2) \qquad (12)$$

For a detailed calculation of the moments of inertia we refer to the nuclear physics literature (Verhaar, 1964).

The energy formula (11) has here been derived as a result of mathematical approximations rather than through decoupling dynamical motions. It is important to notice that although the weight functions factorize as a product of a rotational and vibrational part the corresponding wavefunctions do not. The harmonic GCM approximation (Gaussian overlap plus quadratic approximation) is therefore a non-adiabatic zeroth order problem.

It is possible to derive other familiar results in Gaussian overlap type approximations. For example, the vibrational selection ru $\Delta v=\pm 1$ follows from an expansion of the dipole kernel.

3.3 Λ Doubling

In the one band AA terms with $\Lambda\neq 0$ are doubly degenerate. In reality, however, one finds two levels of opposite parity close together. This effect is known as Λ doubling. Van Vleck (1951) has shown that the AA fails to explain this phenomenon because of the coupling terms dropped in performing the transformation to the rotating reference frame. These operators connect electronic states Λ with $\Lambda\pm 1$. Hence in a perturbation theory treatment of Λ doubling, the effect will be of order 2Λ. The energy splitting is therefore estimated to be of the order $(m/M)^{2\Lambda}=\kappa^{8\Lambda}$, where κ is the usual Born-Oppenheimer parameter, i.e., it is most important for Π terms. In the GCM approach Λ doubling is taken into account automatically via the angular momentum projection. Indeed the functions $P^J_{M\Lambda}\chi_\Lambda$ and $P^J_{M-\Lambda}\chi_{-\Lambda}$ are non-orthogonal and interacting with respect to the Hamiltonian. The eigenvalues of the resulting 2 × 2 secular equation are therefore non-degenerate. The dependence of the level splitting upon the quantum numbers J and Λ can be calculated using techniques similar to those of the previous section. One finds

$$\Delta E_J = \text{const}\,\binom{J+\Lambda}{J-\Lambda} \qquad (13)$$

where the bracket denotes a binomial coefficient. This result is identical to the one obtained by perturbation theory following the AA(J(J+1) for Π states, (J-1)J(J+1)(J+2) for Δ states, etc.). Λ doubling, although small, is experimentally observed. It is a non-adiabatic effect which is incorporated in the GCM picture due to the interaction between the JM projections of χ_Λ and $\chi_{-\Lambda}$.

3.4 The Spectroscopic Term Formula

The spectroscopic term values for diatomic molecules can be fitted with a high degree of accuracy by the following expression,

$$E^{J}_{v\Lambda} = \sum_{k,l} A^{v}_{kl}(v+\tfrac{1}{2})^{k}[J(J+1) - \Lambda^{2}]^{\ell} \tag{14}$$

The connection between the spectroscopic constants $\{A^{v}_{kl}\}$ and the coefficients in the Taylor series expansion of the adiabatic potential energy curve about its minimum was first derived by Dunham (see e.g., Goodisman, 1973). The above formula (14) is therefore commonly known as the Dunham series for the term values.

A routine application of (14) to a diatomic system (e.g., the ground electronic state of Li_2) shows that one can fit of order 10^3 spectral lines with about 100 spectroscopic constants to an accuracy of $\approx 0.01\ cm^{-1}$. The Dunham series is therefore a very powerful tool for parameterizing spectral data. From the theoretical point of view however there is a fundamental difficulty associated with (14) in that it is conventionally justified using a WKB argument applied to the Schrödinger equation for nuclear motion under the potential energy function obtained with the adiabatic approximation: in the case of Li_2 the adiabatic approximation may be in error by $\pm 10\ cm^{-1}$ $(\sim\kappa^6)$. One could therefore argue that the workability of (14) is due to the fitting procedure rather than to its actual form. Nevertheless the success of the Dunham series (14) is so striking that one is encouraged to search for a more profound justification for it. One can in fact arrive at (14) within the framework of the GCM. Starting with the harmonic approximation (6) and (7) as an unperturbed problem and applying perturbation theory to the higher order terms in the power series for the kernels, one obtains (14) with suitably defined spectroscopic constants. Moreover since the error in the GCM is typically $O(\kappa^8)$ the above mentioned objection to (14) disappears (for Li_2 this error is of the order of 0.01 - 0.001 cm^{-1}). The GCM therefore offers the possibility of comparing ab initio theory and experiment on a consistent basis through the spectroscopic constants $\{A^{v}_{kl}\}$: although such comparisons have been widely made using the conventional argument this may be an inconsistent procedure because the intrinsic errors in the theoretical expressions obtained from calculated potential energy curves may well be much greater than the accuracy of the fit to the experimental data with the Dunham series.

4. ATOM-ATOM SCATTERING

4.1 Asymptotic Conditions and Scattering Amplitudes

The basic problem in the theoretical description of collisions between many-body systems is the expansion of the asymptotic form of the wavefunction in terms of the eigenstates of the total hamiltonian. In the case of atom-atom collisions this means that one has to decompose the state

$$\Psi(r,\underset{\sim}{R}) = e^{ik_{\bar{\ell}}Z}\phi_{\overline{\ell m}}(r) + \sum_{\ell m} f(\overline{\ell m},\ell m|\omega_R)\phi_{\ell m}(r)\frac{e^{ik_\ell R}}{R} \qquad (1)$$

as a superposition of eigenstates of (2.1). Here the $\phi_{\ell m}$ are atomic states of good total electronic angular momentum. They can be formed from the eigenstates $\phi_{\ell_A m_A}(r_A)\phi_{\ell_B m_B}(r_B)$ of the Hamiltonian for the non interacting atoms A and B by coupling

$$\phi_{\ell m}(r) = \sum_{m_A,m_B} \langle\ell_A m_A \ell_B m_B|\ell m\rangle\, \phi_{\ell_A m_A}(r_A)\phi_{\ell_B m_B}(r_B) \qquad (2)$$

A standard procedure to obtain a set complete in r, ω_R space is to further couple the $\phi_{\ell m}(r)$ with spherical harmonics in ω_R. One thus obtains the angular momentum eigenstates

$$Z^{\ell L}_{JM}(r,\omega_R) = \sum_{m,M_L} \langle\ell m L M_L|JM\rangle\phi_{\ell m}(r)Y_{LM_L}(\omega_R) \qquad (3)$$

which can be used to expand the Ψ_{JM} as

$$\Psi_{JM}(r,\underset{\sim}{R}) = \sum_{\ell L} g^J_{\ell L}(R)\; Z^{\ell L}_{JM}(r,\omega_R) \qquad (4)$$

Since these functions are built up from atomic states, quantized along the space-fixed z-axis, chosen to coincide with the initial direction of relative motion, it is straightforward to match a superposition of the Ψ_{JM} as obtained in (4) to the asymptotic form (1). This gives an expression for the radial function $g^J_{\ell L}(R)$ at large R

$$g^J_{\overline{\ell L},\ell L}(R) \underset{R\to\infty}{\sim} (k_\ell)^{1/2}\,[j_L(k_\ell R)\delta_{\overline{\ell L},\ell L} - T^J_{\overline{\ell L},\ell L}\, h_L(k_\ell R)] \qquad (5)$$

where $T^J_{\overline{\ell L},\ell L}$ are the elements of the transition matrix. In terms of these constants an explicit form of the scattering amplitude can be given. This expression is rather cumbersome (in the conventions used here it can be found in Pack, 1973) and leads to a complicated form of the cross section

$$\sigma(\overline{\ell m},\ell m) = (k_\ell/k_{\overline{\ell}}) \int |f(\overline{\ell m},\ell m|\omega_R)|^2 d\omega_R \qquad (6)$$

Simpler results can be obtained by transforming the outgoing waves in (1) to the rotating frame. This leads to the so-called helicity formalism in which the quantisation axis always coincides with the internuclear vector (Child, 1974). From the transformation law (9) we obtain (initially $\overline{m}=\overline{\Lambda}$)

$$\sum_m f(\overline{\ell m},\ell m|\omega_R)\phi_{\ell m}(r) = \sum_\Lambda f(\overline{\ell\Lambda},\ell\Lambda|\omega_R)\phi_{\ell\Lambda}(r') \qquad (7)$$

where $f(\overline{\ell\Lambda},\ell\Lambda|\omega_R)$ is a new scattering amplitude giving the probability for a transition $\overline{\ell\Lambda}\to\ell\Lambda$. This form is obviously more suitable if Born-Huang type series are used to expand the asymptotic form since

$$\lim_{R\to\infty} \phi_{n\Lambda}(r'|R\underset{\sim}{e}_z) = \phi_{\ell\Lambda}(r') \qquad (8)$$

Asymptotically H_e will commute with ℓ^2 such that in the limit one of the quantum numbers n is simply the total electronic angular momentum. It is well known (Choi et al., 1977) that there exists a unitary transformation between the asymptotic radial functions $g^J_{\overline{\ell L},\ell L}(R)$ and $g^J_{\overline{\ell\Lambda},\ell\Lambda}(R)$, i.e. the $\{g^J_{\overline{\ell\Lambda},\ell\Lambda}\}$ follow from (5). The same transformation exists of course between the corresponding transition matrices $T^J_{\overline{\ell L},\ell L}$ and $T^J_{\overline{\ell\Lambda},\ell\Lambda}$.

The important simplification which arises in the helicity representation is that the scattering amplitude takes a particularly simple form

$$f(\overline{\ell\Lambda},\ell\Lambda|\omega_R) = i/(2\sqrt{k_\ell k_{\overline{\ell}}})\ (-)^{\overline{\ell}-\ell+\overline{\Lambda}-\Lambda} \sum_J (2J+1)T^J_{\overline{\ell\Lambda},\ell\Lambda}D^{J^*}_{\overline{\Lambda}\Lambda}(\omega_R) \qquad (9)$$

This result is a consequence of the fact that in the helicity picture the outgoing direction of relative motion coincides with the quantisation axis (Choi et al., 1977). The corresponding cross section is

$$\sigma(\overline{\ell\Lambda},\ell\Lambda) = (\pi/k^2_{\overline{\ell}}) \sum_J (2J+1)|T^J_{\overline{\ell\Lambda},\ell\Lambda}|^2 \qquad (10)$$

from which all physical quantities related to the collision process can be calculated.

4.2 Asymptotic GC Wavefunctions

In the previous section we outlined how the asymptotic form of the radial functions $g^{J,as}_{\ell\Lambda,\ell\Lambda}(R)$ in the Born-Huang series can be constructed. In order to apply the GCM to scattering problems we must investigate the GC states (19) as $R\to\infty$. If the wave packet type function $\Phi(|\underset{\sim}{R}-\underset{\sim}{\alpha}|)$ is sharply peaked it vanishes unless $|\underset{\sim}{R}-\underset{\sim}{\alpha}|\cong 0$. This means that the asymptotic region in R space implies a similar region in $\underset{\sim}{\alpha}$ space. More precisely it is easy to prove that for any Φ

$$\Phi(|\underset{\sim}{R}-\underset{\sim}{\alpha}|) \underset{R\to\infty}{\sim} G(R|\alpha)\delta(\omega_R-\omega_\alpha) \tag{11}$$

where $G(R|\alpha)$ is symmetric in R and α and essentially zero unless $R-\alpha\cong 0$. If the nuclear intrinsic state is a Gaussian $e^{-(R-\alpha)^2/W^2}$ (see part I) one can calculate that

$$e^{-(\underset{\sim}{R}-\underset{\sim}{\alpha})^2/W^2} \underset{R\to\infty}{\sim} (\pi W^2/R\alpha)e^{-(R^2+\alpha^2)/W^2}\ \delta(\omega_R-\omega_\alpha) \tag{12}$$

Inserting (11) into the GC wavefunctions one sees that angular and radial GC integrations decouple and that with the identification

$$\int G(R|\alpha)f^{J,as}_{\ell\Lambda,\ell\Lambda}(\alpha)\alpha^2 d\alpha = g^{J,as}_{\ell\Lambda,\ell\Lambda}(R) \tag{13}$$

the GC series becomes a Born-Huang expansion. This equation should be regarded as an integral equation for the asymptotic radial weight functions such that a superposition of GC states has the asymptotic form (1).

4.3 GC Scattering Equations

Solution of the scattering problem within the GC framework amounts to solving the coupled integral equations (2.25) under the restriction (13). The integration range of the radial GC runs from 0 to ∞. However, beyond a certain radius, say α_c, of several times the typical equilibrium distances of the bound states, $n\Lambda$, the radial weight functions can safely be identified with the solutions of (13). The coupled equations, taking into account the asymptotic conditions can therefore be written as

$$\begin{aligned}&\sum_{\Lambda} \int_0^{\alpha_c} K^J_{n\Lambda,n'\Lambda'}(\alpha,\beta|E)\ f^J_{n\Lambda,n'\Lambda'}(\beta)\beta^2 d\beta \\ &= -\sum_{\ell'\Lambda'} \int_{\alpha_c}^{\infty} K^J_{\ell\Lambda,\ell'\Lambda'}(\alpha,\beta|E)\ f^{J,as}_{\ell\Lambda,\ell'\Lambda'}(\beta)\beta^2 d\beta\end{aligned} \tag{14}$$

This is a set of inhomogeneous equations for the interior (0 to α_c) part of the radial weight functions. These unknowns must be joined with the asymptotic functions known from (13) such that (14) has to be supplemented with the matching condition

$$f^{J}_{\overline{n\Lambda},n\Lambda}(\alpha_c) = f^{J,as}_{\overline{\ell\Lambda},\ell\Lambda}(\alpha_c) \tag{15}$$

These equations are necessary since, in addition to the $f^{J}_{\overline{n\Lambda},n\Lambda}$, the transition matrix elements $T^{J}_{\overline{\ell\Lambda},\ell\Lambda}$ enter the equations as supplementary unknowns through the $f^{J,as}_{\overline{\ell\Lambda},\ell\Lambda}$.

REFERENCES

Born, M., 1951, Gött. Nachr. math. phys. 1.
Child, M.S., 1974, Molecular Collisions Theory, Academic Press, New York.
Choi, B.H., Poe, R.T. and Tang, K.T., 1977, J. Chem. Phys. 69, 411.
Goodisman, J., 1973, Diatomic Interaction Potential Theory, Academic Press, New York.
Hirschfelder, J.O. and Wigner, E.P., 1935, Proc. Nat. Acad. Sci. 21, 113.
Lathouwers, L., Van Leuven, P. and Bouten, M., 1977, Chem. Phys. Lett. 52, 439.
Lathouwers, L., 1978, Phys. Rev. A18, 2150.
Lathouwers, L., Van Leuven, P., 1978, IJQC symp. 12, 371.
Lathouwers, L., 1979, J. Phys. B12, L99.
Landau, Lifshitz, 1965, Quantum Mechanics, Pergamon Press, London.
Pack, P., 1973, J. Chem. Phys. 60, 633.
Van Vleck, J., 1929, Phys. Rev. 33, 467.
Verhaar, B., 1964, Nucl. Phys. 54, 641.
Wigner, E.P., 1959, Group Theory, Academic Press, New York.

IRREVERSIBLE AND NONLINEAR DYNAMICS OF OPEN SYSTEMS

E. B. Davies

Mathematical Institute
University of Oxford
24-29 St. Giles, Oxford, OX1 3LB, England

1. INTRODUCTION

In the generally accepted mathematical description of the dynamics of a fixed number of non-relativistic particles, the state of the system at a particular time is described by a unit vector $\psi(t)$ in a suitable complex Hilbert space $\mathcal{H}$, say $L^2(\mathbb{R}^{3n})$ for n distinguishable spinless particles, and the time evolution is determined by the Schrödinger equation

$$\frac{d\psi}{dt} = - i \, H \, \psi(t) \tag{1}$$

where the self-adjoint operator H is called the Hamiltonian of the system and is obtained by a well-known prescription. There are circumstances, arising particularly in statistical mechanics, where one has to consider more complicated states of the system, called mixed states. These are density matrices, or more precisely trace class linear operators $\rho:\mathcal{H}\to\mathcal{H}$ such that ρ is self-adjoint, non-negative and of trace equal to one. Here trace is defined by

$$\mathrm{tr}[\rho] = \sum_n \langle\rho\psi_n, \psi_n\rangle$$

where $\{\psi_n\}$ is any orthonormal basis of $\mathcal{H}$. The evolution equation for mixed states corresponding to (1) is

$$\frac{d\rho}{dt} = - i \, [H,\rho] \tag{2}$$

with solution

$$\rho(t) = e^{-iHt}\rho(0)e^{iHt} .$$

In spite of the satisfactory theoretical nature of the above description, it is usually the case for multi-particle systems that (1) and (2) are too complicated to be directly soluble either theoretically or numerically. There have therefore arisen a large number of approximations to the basic equations, some of which have become so popular as to have acquired a theoretical status of their own - I mention the Hartree-Fock and Born-Oppenheimer theories, and the mean-field model of nuclear structure. Only recently have mathematicians taken much interest in these phenomenological theories. It is helpful to separate their analysis into three problems.

(i) Do the approximate evolution equations actually possess global solutions?

(ii) If so what are the qualitative and quantitive properties of those solutions?

(iii) Under what conditions are the solutions of the approximate equations close to the solutions of the true equations, and what does "close" mean?

In the first two lectures I shall try to answer these questions for a quantum-mechanical theory of irreversible evolution of open systems. In my last lecture I shall talk about a phenomenological non-linear Schrödinger equation related to the Hartree problem and to the problem of molecular structures.

2. DEFINITION OF QUANTUM DYNAMICAL SEMIGROUPS

Quantum dynamical semigroups are relevant to the description of the time evolution of one or several quantum particles interacting with an infinite external medium called the reservoir. While a full description would involve the quantization of the external medium as a quantum field theory with an infinite number of degrees of freedom, one hopes to obtain an approximate evolution equation without carrying out this complicated procedure. In typical cases the external medium might be the electromagnetic field, the phonon field of a lattice, or a low density gas, while the system of interest might be one or more elementary particles, a radiating atom, or a molecule.

The first observation is that if the medium is in equilibrium at a positive temperature the system will often relax from an arbitrary initial state, even a pure state, to its Gibbs state, which is mixed. Clearly the evolution is irreversible and cannot be determined by a Hamiltonian, which always takes pure states to pure states. The dynamics of such a system must be specified ab initio for mixed states.

We assume that the evolution equation is of the form

$$\frac{d\rho}{dt} = Z\rho(t)$$

so that there are no memory effects. The solution

$$\rho(t) = e^{Zt}\rho(0) = T_t\{\rho(0)\}$$

is determined by linear operators T_t on the space $\mathcal{J}(\mathcal{H})$ of trace class operators on $\mathcal{H}$. Such operators T_t are sometimes called super-operators. Considerations of "conservation of probability" lead one to expect the following properties:

(i) $\rho \geq 0$ implies $T_t\,\rho \geq 0$ for all $t \geq 0$.

(ii) $\mathrm{tr}\,[T_t\rho] = \mathrm{tr}\,[\rho]$ for all $t \geq 0$.

(iii) $T_t\,T_s = T_{t+s}$ for all $s, t \geq 0$.

(iv) $t \to T_t\rho$ is continuous (in trace norm) for all $\rho \in \mathcal{J}(\mathcal{H})$.

It turns out that if one adds in further (technical?) conditions

(v) T_t is completely positive for all $t \geq 0$,

(vi) $t \to T_t$ is norm continuous,

then the generators Z can be completely determined (Lindblad, 1976; Gorini et al., 1976).

THEOREM 1 Under the above conditions

$$Z(\rho) = -i\,[H,\rho] - \frac{1}{2}[R,\rho]_+ + \sum_{n=1}^{\infty} A_n\,\rho\,A_n^* \tag{3}$$

where H is self-adjoint, R, A_n are bounded and

$$R = \sum_{n=1}^{\infty} A_n^*\,A_n.$$

For further technical developments see Davies (1976b), Davies (1977) and Evans and Lewis (1977).

In order to apply this theorem one has to develop a method of specifying the operators H, R, A_n in particular situations. Clearly H should be the Hamiltonian of the system in the absence of interactions with the external environment. The operator R is called the decay or interaction rate for the following reason. If one neglects the last term in (3) the evolution equation becomes

$$\frac{d\rho}{dt} = -i[H,\rho] - \frac{1}{2}[R,\rho]_+$$

whose solution

$$\rho(t) = e^{(-iH - \frac{1}{2}R)t}\rho(0)e^{(iH - \frac{1}{2}R)t}$$

satisfies

$$\frac{d}{dt}\,\mathrm{tr}\,[\rho(t)] = -\mathrm{tr}\,[R\,\rho(t)].$$

So R measures the decay of probability in the absence of the last term. The operators A_n describe instantaneous transitions of the system due to its interaction with the environment.

As a simple example consider the Bloch equations. On the Hilbert space $H = C^2$ we put

$$E_{11} = \begin{matrix}1 & 0\\ 0 & 0\end{matrix}, \quad E_{22} = \begin{matrix}0 & 0\\ 0 & 1\end{matrix}, \quad E_{12} = \begin{matrix}0 & 1\\ 0 & 0\end{matrix}, \quad E_{21} = \begin{matrix}0 & 0\\ 1 & 0\end{matrix}$$

The Bloch equations then take the form

$$\frac{d\rho}{dt} = -i[\omega_1 E_{11} + \omega_2 E_{22},\rho] - \frac{1}{2}[\lambda_1 E_{11} + \lambda_2 E_{22},\rho]_+$$
$$+ \lambda_1 E_{21}\rho E_{12} + \lambda_2 E_{12}\rho E_{21}$$

The constants ω_1 and ω_2 equal the energies of the excited state and ground state respectively. The constant $\lambda_1 \geq 0$ is the rate of transitions into the ground state while $\lambda_2 \geq 0$ is the rate of transitions into the excited state. Direct calculations lead to the following result.

THEOREM 2 If $\rho(0) \geq 0$ and $\mathrm{tr}\,[\rho(0)] = 1$ then $\rho(t) \geq 0$ and $\mathrm{tr}\,[\rho(t)] = 1$ for all $t \geq 0$. Moreover

$$\lim_{t\to\infty} \rho(t) = \frac{\lambda_1}{\lambda_1 + \lambda_2} E_{11} + \frac{\lambda_2}{\lambda_1 + \lambda_2} E_{22}$$

for all initial states $\rho(0)$.

We next consider the relaxation of a single mode of a boson quantum field, or a harmonic oscillator. If $[a, a^*] = 1$, the appropriate evolution equation is

$$\frac{d\rho}{dt} = -i[\omega a^* a,\rho] - \frac{\lambda}{2}[a^* a,\rho]_+ + \lambda\, a\,\rho\, a^* \qquad (4)$$

where ω is the frequency of the oscillator and $\lambda \geq 0$ is the decay rate. If Ω is the ground state then for each $z \in C$ the coherent state $\psi(z)$ is

$$\psi(z) = e^{-\frac{1}{2}|z|^2} \sum_{n=0}^{\infty} z^n a^{*n} \Omega/n!$$

The time evolution determined by (4) has a particularly simple description in terms of coherent states.

THEOREM 3 If

$$\rho(0) = \iint |\psi(x + iy)><\psi(x + iy)| \sigma(x, y) \, dx \, dy$$

then

$$\rho(t) = \iint |\psi(x_t + iy_t)><\psi(x_t + iy_t)| \sigma(x, y) \, dx \, dy$$

where

$$x_t + iy_t = e^{(-i\omega - \frac{1}{2}\lambda)t} (x + iy)$$

Hence

$$\lim_{t\to\infty} \rho(t) = |\Omega><\Omega|$$

Further developments of this example are of considerable theoretical significance in quantum optics, and may be found in Haake (1973), Davies (1976b) and references cited there.

We next turn to the second law of thermodynamics. The relative entropy of two mixed states ρ, σ on H is defined by

$$S(\rho|\sigma) = \mathrm{tr}[\rho \log \rho - \rho \log \sigma].$$

If σ is a Gibbs state

$$\sigma = e^{-\beta H} / \mathrm{tr}[e^{-\beta H}]$$

then $S(\rho|\sigma)$ is closely related to the free energy of ρ. The following theorem (Lindblad, 1975; Spohn, 1978; Spohn and Lebowitz, 1978) is further evidence that quantum dynamical semigroups are of physical interest.

THEOREM 4. Let T_t satisfy (i) - (vi) and let σ be an equilibrium state for T_t, that is $T_t\sigma = \sigma$ for all $t \geq 0$. Then $S(T_t \rho|\sigma)$ is a monotonically decreasing function of time. Moreover $S(\rho|\sigma) \geq 0$ with equality if and only if $\rho = \sigma$.

There is a substantial amount of work concerning conditions which imply that T_t has a unique equilibrium state σ, such that

$$\lim_{t\to\infty} T_t\rho = \sigma$$

for all initial states ρ. See Davies (1976b), Evans (1977) and Spohn (1976).

We conclude this section by mentioning some recent work on quantum-mechanical scattering theory (Davies, 1978c; 1979d) where the time evolution is specified by a quantum dynamical semigroup instead of the usual Hamiltonian. We claim that such a development is necessary if one wishes to allow the possibility of electron capture by a positively ionised molecule, or neutron capture by a nucleus, unless one passes to the much more sophisticated theory in which the quantized electromagnetic field is incorporated into the dynamics. In both the above problems the two bodies may combine temporarily to form an unstable composite body, which is only rendered stable upon the emission of a photon. The use of quantum dynamical semigroups allows a simple model of this emission process to be constructed, and it is shown in Davies (1979d) that capture may indeed occur in this model.

3. DERIVATION OF QUANTUM DYNAMICAL SEMIGROUPS

It is possible to take the attitude that the second law of thermodynamics is inconsistent with Hamiltonian dynamics, so an evolution equation such as (4) should be regarded as the starting point for a "new quantum statistical mechanics". I am not convinced that this is necessary, and wish to outline a possible derivation of (4) from a Hamiltonian evolution law. While the derivation is only completely rigorous for somewhat idealized problems, it does appear to show that there are no fundamental conceptual difficulties involved in the approach.

We assume that the system of particles is associated with the Hilbert space $\mathcal{H}$ and that the external medium is represented by a quantum field with an infinite number of degrees of freedom and ground state (vacuum) Ω in the Hilbert space $\mathcal{K}$. The composite Hilbert space is $\mathcal{H} \otimes \mathcal{K}$ and the Hamiltonian on $\mathcal{H} \otimes \mathcal{K}$ is

$$H = H_S + H_F + H_I$$

where H_S is the system Hamiltonian, H_F is the field Hamiltonian and H_I the interaction Hamiltonian. Standard forms for all these terms are well-known. If the system is initially in the mixed state ρ, the field is initially in its ground state, and the two are uncorrelated, then the state on $H \otimes K$ at time $t \geq 0$ is

$$e^{-iHt}(\rho \otimes |\Omega\rangle\langle\Omega|)\, e^{iHt}$$

and the state of the system is

$$\rho(t) = \mathrm{tr}_K\, [e^{-iHt}(\rho \otimes |\Omega\rangle\langle\Omega| e^{iHt}] \tag{5}$$

where the partial trace tr_K corresponds to averaging over the reservoir degrees of freedom alone. It is clear that if $\rho(0) \in J(H)$ is non-negative with trace one, then $\rho(t)$ has the same properties. The linear operator T_t on $J(H)$ defined by $T_t\rho(0) = \rho(t)$ has all the properties of a quantum dynamical semigroup except that in general

$$T_t T_s \neq T_{t+s}$$

for $s, t \geq 0$. It may be shown that $\rho(t)$ always satisfies a generalized master equation of the type

$$\frac{d\rho}{dt} = -i\,[H, \rho(t)] + \int_0^t K(s)\, \rho(t-s)\, ds \tag{6}$$

If $K(s)$ is mostly concentrated near $s = 0$, then (6) may be replaced by the approximate equation

$$\frac{d\rho}{dt} = -i\,[H, \rho(t)] + \int_0^\infty K(s)\, ds\, \rho(t) = Z\, \rho(t) \tag{7}$$

called the master equation. Physical justifications of the approximation depend upon the time scale for the relaxation or memory effects being very short compared with the time scale for the evolution of the system. On the mathematical side one tries to identify some small parameter λ (the ratio of two masses, a coupling constant, the density of a gas, the inverse of the number of nucleons in a nucleus, Planck's constant, etc.), to estimate, rigorously if possible, the difference between the solutions of (6) and (7) in terms of λ, and hence to show that the difference converges to zero as $\lambda \to 0$. Those who do not believe in the physical meaningfulness of the limit, particularly in the case $h \to 0$, may stop after having made the error estimate, and evaluate the error numerically. While I agree with their philosophical position, I am afraid I do not share their enthusiasm for numerical calculations.

There are a variety of asymptotic results of the type we want, surveyed from a physical point of view by Davies (1977) and from a pure mathematical point of view by Davies (1980). I shall describe here one of the simplest, the weak coupling limit theorem, which is not however necessarily the one relevant to the greatest number of applications. It turns out that there are two time scales, t and $\tau = \lambda^2 t$, as in van Hove's theory, and that the irreversible effects happen over the slower time scale τ. We take the Hamiltonian to be

$$H = H_S + H_F + \lambda\, Q \otimes \phi$$

where the field operator ϕ on K has zero expectation in the vacuum state Ω. We also must assume for technical reasons that the field particles do not interact with each other, so that H_F is a free Hamiltonian. In the following theorem of Davies (1974) there are also other technical hypotheses of a less important nature which we do not mention.

THEOREM 5. Let $\rho(0)$ be a mixed state on H and let $\rho_\lambda(t)$ be defined by (5). Then for all $0 \le \tau < \infty$

$$\lim_{\lambda\to 0} \|\rho_\lambda(\lambda^{-2}\tau) - \exp\{(\lambda^{-2} Z_0 + K^\#)\tau\}\rho(0)\|_1 = 0$$

where $\|\cdot\|_1$ is the trace norm, and Z_0, $K^\#$ are certain operators on $J(H)$.

In the above formula Z_0 is defined by

$$Z_0(\rho) = -i[H_S, \rho]$$

while

$$K^\#(\rho) = \lim_{a\to\infty} \frac{1}{a}\int_{s=0}^{a} \int_{t=0}^{\infty} dt\, ds.$$

$$\{-Q_{t+s} Q_s \rho\, h(t) + Q_{t+s} \rho\, Q_s \overline{h(t)} + Q_s \rho\, Q_{t+s} h(t) - \rho\, Q_s Q_{t+s} \overline{h(t)}\}$$

and the two-point function of the reservoir is

$$h(t) = \langle \phi\, e^{-iH_F t} \phi\, \Omega, \Omega\rangle$$

For the convergence of the t-integral it is necessary that

$$\int_0^\infty |h(t)|\, dt < \infty$$

and for typical situations involving infinite volume quantum fields in n space dimensions it appears to be the case that

$$h(t) = t^{-n/2}$$

as $|t| \to \infty$. For the existence of the average $^{\#}$ as $a \to \infty$ it is necessary to assume that H_S has discrete spectrum, but analogous results exist when the spectrum is continuous (Davies, 1976a) or only partly discrete as for typical atomic Hamiltonians (Davies, 1978a).

It is straightforward to show that Z_o and K commute so that the free evolution of the system, which is very rapid on the τ-time scale, and its relaxation are independent processes. If the reservoir is initially in equilibrium at the inverse temperature β, this being achieved by choosing the two-point function h(t) appropriately, then it may be shown that

$$\lim_{\tau\to\infty} e^{K^{\#}\tau}(\rho) = e^{-\beta H_S} / \operatorname{tr}[e^{-\beta H_S}]$$

for all initial states $\rho \in \mathcal{J}(\mathcal{H})$.

A number of variations on the basic theorem above are possible. One may replace the weak coupling limit by a singular coupling limit (Hepp and Lieb, 1973; Gorini and Kossakowski, 1976; Frigerio and Gorini, 1976) which is physically quite different even if mathematically closely related to the weak coupling limit (Palmer, 1977). Some progress has been made towards allowing interactions between the reservoir particles (Davies and Eckmann, 1975). Palmer (1976) has also applied the same techniques to analyse a particle interacting with a reservoir of particles in the low density limit of the reservoir.

Once one has such a basic theorem it is possible to give rigorous derivations from first principles in a suitable limit of phenomenological laser equations (Hepp and Lieb, 1973), the Bloch equations (Pulé, 1974), a model of a Josephson oscillator (Hepp, 1975a), the Onsager relations (Hepp, 1975b), a quantum Boltzmann equation (Palmer, 1976), a model of atomic radiation (Davies, 1978a), a model of heat conduction (Davies, 1977b), quantum detailed balance (Kossakowski et al., 1977), the B.C.S. model (Buffet and Martin, 1978), Kubo's linear response theory (Davies and Spohn, 1978).

It occasionally happens that the operator $K^{\#}$ in Theorem 5 is of the form

$$K^{\#}(\rho) = -i\,[V,\rho] \tag{8}$$

so that the limiting evolution is of Hamiltonian type. This usually happens when the quantum field can only experience virtual excitations, so that there can be no long-term exchange of energy.

LEMMA 6 Let H_λ be Hamiltonians on $\mathcal{H} \otimes \mathcal{K}$, and H a Hamiltonian on such that

$$\lim_{\lambda \to 0} e^{-iH_\lambda t} (\psi \otimes \Omega) = (e^{-iHt} \psi) \otimes \Omega$$

for all $\psi \in \mathcal{H}$. Then

$$\lim_{\lambda \to 0} \mathrm{tr}_{\mathcal{K}} \; e^{-iH_\lambda t} (\rho \otimes | \Omega > < \Omega |) \, e^{iH_\lambda t}$$

$$= e^{-iHt} \rho \, e^{iHt}$$

for all mixed states ρ on $\mathcal{H}$.

In an example of this type given by Davies (1979b), a fixed number of particles interact with a quantum field of massive bosons, the parameter λ being the mass ratio of the two types of particle. It is shown that in the limit as $\lambda \to 0$, the asymptotic evolution of the particles if of the type described in the above lemma, the operator V of (8) being a sum of pair potentials of Yukawa or Coulomb type between the particles, due to their exchange of virtual bosons. For a rigorous derivation an ultraviolet cutoff is needed in the model.

I wish to point out before closing this section that I have not mentioned any of the similar derivations by rigorous asymptotic analysis of classical dynamic equations. For a comprehensive account of such results, including the Lorentz gas, the Rayleigh gas, the Landau equation, the Boltzmann equation and the Vlasov equation, I can do no better than refer to Spohn (1978/1979).

4. INTRODUCTION TO NON-LINEAR SCHRÖDINGER EQUATIONS

In the last decade there has been a steadily increasing interest in non-linear equations, among which are a number of non-linear Schrödinger equations. This is a very large subject, out of which I can only select a few papers which I have come across, but it is important to realize that some applications of the NLSE are purely classical, some are semi-classical and semi-quantum-mechanical, while others are intended as somewhat speculative generalizations of quantum mechanics.

The equation

$$i \frac{\partial \psi}{\partial t} + \frac{\partial^2 \psi}{\partial x^2} + 2 \, |\psi(x)|^2 \psi(x) = 0$$

in one space dimension, is often called the NLSE and is of great interest because it possesses soliton solutions (Scott et al., 1973). The global solubility and other properties of a generalization of this, namely

$$i \frac{\partial \psi}{\partial t} = - \Delta \psi + \lambda |\psi(x)|^k \psi(x) = 0$$

in n space dimensions, has been demonstrated for various λ and k by Ginibre and Velo (1979a, 1979b, 1978). Hepp (1974) and Ginibre and Velo (1979c) also derive these equations as classical field equations in the limit $h \to 0$ of a suitable quantum field theory.

The use of NLSE's has been advocated by Mielnik (1974), Bialynicki-Birula and Mycielski (1976), Kibble (1978) and others for a possible generalization of quantum theory. On the other hand Haag and Bannier (1978) have shown that the measurement theory of such models is necessarily more classical than quantum-mechanical in some respects.

There is a large amount of mostly non-rigorous literature on the use of NLSE's to provide models of quantum-mechanical friction. While one motivation for this work seems to come from the need to find a model for deep inelastic nuclear scattering, the particular NLSE's studied are usually selected on grounds of simplicity alone. The NLSE of Kostin (1972)

$$i \frac{\partial \psi}{\partial t} = - \frac{1}{2m} \Delta \psi + \frac{\gamma}{2i} \log (\psi/\overline{\psi}) \psi$$

which exhibits frictional effects, and the closely related NLSE of Bialynicki-Birula and Mycielski (1976)

$$i \frac{\partial \psi}{\partial t} = - \frac{1}{2m} \Delta \psi - \gamma \log (|\psi|^2) \psi$$

which possesses solitary waves, have some features which render them particularly interesting.

We next mention a simple class of NLSE's, which are not local, but may nevertheless be of interest because they are exactly soluble. These equations are basically linear but with ψ-dependent coupling constants. Examples are

$$i \frac{\partial \psi}{\partial t} = \frac{1}{2m} P^2 \psi + \langle P \psi, \psi \rangle Q \psi$$

where the particle is subjected to a force proportional to its momentum, but in the opposite direction, and

$$i \frac{\partial \psi}{\partial t} = \frac{1}{2m} P^2 \psi + Q^2 \psi - 2 \langle Q, \psi \psi \rangle Q \psi + \langle Q^2 \psi, \psi \rangle \psi$$

This equation is space translation invariant, conserves momentum, possesses solitary wave solutions and may be written in the form of a Hartree equation

$$i \frac{\partial\psi}{\partial t} = -\frac{1}{2m} \Delta \psi (x) + \psi(x) \int (x-y)^2 |\psi(y)|^2 dy$$

at least for unit vectors ψ. It is closely related to the subject matter of the next section.

For further references to NLSE's of all the above types, and others, see Hasse (1975), Immele et al. (1975), Messer (1979) and Messer and Baumgartner (1978).

We finally come to the best known and motivated of the NLSE's, namely the Hartree equation, a simple example of which is

$$i \frac{\partial\psi}{\partial t} = -\frac{1}{2m} \Delta \psi + \psi(x) \int_{\mathbb{R}^3} V(x-y) |\psi(y)|^2 dy \tag{9}$$

on $L^2(\mathbb{R}^3)$. This and the related Hartree-Fock equation are used in quantum chemistry as approximations to the linear multibody Schrödinger equation, which is too complicated to be treated directly. In fact it may be rigorously shown that the solutions of (9) are asymptotically equal to certain solutions of

$$i \frac{\partial\psi}{\partial t} = \sum_{j=1}^{N} \frac{1}{2m} \Delta_j \psi + \frac{1}{2N} \sum_{i\neq j=1}^{N} V(x_i - x_j) \psi$$

on $L^2(\mathbb{R}^{3N})$, in the mean field limit $N\to\infty$. Many people have obtained such results rigorously but the treatment of Spohn (1978/1979) is the only one I know which is formulated in the above terms.

The existence of global solutions of (9) for suitable initial data is not trivial, but has been demonstrated for various classes of potentials V by Bove et al. (1976), Chadam and Glassey (1975) and Davies (1979e); the first two papers also solve the Hartree-Fock problem. The existence of eigenvectors and a ground state for (9) has also been shown (Lieb, 1977; Lieb and Simon, 1977; and other papers cited there).

5. STABILITY AND MOLECULAR STRUCTURE

Although the foundations of quantum chemistry are secure in the sense that the relevant quantum-mechanical Hamiltonians are certainly known, there remain interesting problems in the subject, which are hidden by the conventional manner of computation using the Born-Oppenheimer, Hartree-Fock and other "approximations" (Woolley, 1976, 1978; Essen, 1977). One such problem is the very definition of a molecule, which cannot be just an eigenstate of the relevant Hamiltonian. For it appears that the eigenstates of optically active molecules are superpositions of the "stable states" of the optical isomers, and that these "stable states" may be non-stationary over geological time-scales. While the standard computational techniques may identify these states correctly, the reason for their success is unclear since neither stability nor the state preparation procedure are ever mentioned. In other words one constructs certain states using the Hamiltonian and chemical insight, and declares these to be molecular states, but does not derive these states from first principles.

As a possible substitute for "chemical insight" I wish to propose a definition of molecular states based on a concept of stability. Following Davies (1979c) I define stable states of a Hamiltonian H on $L^2(\mathbb{R}^{3n})$ to be the unit vectors ψ in $L^2(\mathbb{R}^{3n})$ for which the functional

$$\langle H\psi,\psi\rangle + \alpha W(\psi) \tag{10}$$

with $\alpha > 0$ and

$$W(\psi) = \sum_{r=1}^{3n} \{ \langle Q_r^2\psi,\psi\rangle - \langle Q_r\psi,\psi\rangle^2 \}$$

has a variational local minimum. The weight $W(\psi)$ measures the variance of the distribution of ψ in position space, and the idea is that one discriminates against states, such as a superposition of two isomeric states, for which this variance is large. The undertermined parameter $\alpha > 0$ depends upon the environment of the molecule and presumably increases with its density.

Let me point out before continuing that the minimization of (10) is closely related in spirit to one of the derivations of the Gibbs state

$$\rho_\beta = e^{-\beta H} / \mathrm{tr}\,[e^{-\beta H}]$$

namely ρ_β is known to be the mixed state which minimizes

$$\mathrm{tr}\,[H\rho] + \beta^{-1}\,\mathrm{tr}\,[\rho\log\rho] \tag{11}$$

the second term again being non-linear and representing an influence of the surrounding medium.

In order to simplify the following theorems we shall consider only one simple example. We take two equal mass spin zero distinguishable particles interacting by a pair potential V, with Hamiltonian

$$H = -\frac{1}{2m}\Delta_1 - \frac{1}{2m}\Delta_2 + V(x_1 - x_2)$$

on $L^2(\mathbb{R}^6)$. We suppose that V is continuous, central, that $V(x) \to 0$ as $|x| \to \infty$ and that $V(x) < 0$ for some $x \in \mathbb{R}^3$.

THEOREM 7 If $m > 0$ is large enough then (10) possesses a global minimum ψ (called a stable ground state of the Hamiltonian) which satisfies

$$H\psi + \alpha \sum_{r=1}^{6} \{Q_r^2 \psi - 2 \langle Q_r \psi, \psi\rangle Q_r \phi\} = E\psi$$

Note that if $\alpha = 0$, then H possesses a unique rotationally invariant negative energy ground state for large enough m, but only after removal of the centre of mass kinetic energy term.

THEOREM 8 If $m > 0$ is large enough then the stable ground state is not unique and breaks rotational symmetry.

Each stable ground state is associated with a particular orientation of the molecule in the laboratory coordinate system. It is also possible for parity symmetry to be broken. It is shown by Davies (1979c) that in simple cases there is uniqueness, and hence no symmetry breaking, for small $m > 0$. Hence the system undergoes a phase transition as m increases. Similar results are obtained by varying α instead of m.

A derivation of (10) or of (11) from first principles must presumably involve regarding the molecule as an open system coupled to an infinite reservoir by a linear multibody Hamiltonian. Davies (1979c) gives a partial justification along these lines by computing the ground state of a Hamiltonian coupling the molecule to an infinite phonon field, in the Hartree approximation, in much the same manner as done for the polaron (Lieb and Yamazaki, 1958).

REFERENCES

Bialynicki-Birula, I. and Mycielski, J., 1976, Non-linear wave mechanics, Ann. Phys. 100: 62-93.

Bove, A., Da Prato, G. and Fano, G., 1976, On the Hartree-Fock time-dependent potential, Commun. Math. Phys. 49: 25-33.

Buffett, E. and Martin, P.A., 1978, Dynamics of the open BCS model, J. Stat.Phys. 18: 585-632.

Chadam, I.M. and Glassey, R.T., 1975, Global existence of solutions to the Cauchy problem for time-dependent Hartree equations, J. Math. Phys. 16: 1122-1130.

Davies, E.B., 1974, Markovian master equations, Commun. Math. Phys. 39: 91-110.

Davies, E.B., 1976a, Markovian master equations 2, Math. Ann. 219: 147-158.

Davies, E.B., 1976b, Quantum theory of open systems, Academic Press, London.

Davies, E.B., 1977, Master equations: a survey of rigorous results, Rend. Sem. Mat. Fis. Milano 47: 165-173.

Davies, E.B., 1978a, A model of atomic radiation, Ann. Inst. H. Poincaré 28A: 91-110.

Davies, E.B., 1978b, A model of heat conduction, J. Stat. Phys. 18: 161-170.

Davies, E.B., 1978c, Two-channel Hamiltonians and the optical model of nuclear scattering, Ann. Inst. H. Poincaré 29A: 395-413.

Davies, E.B., 1979a, Generators of dynamical semigroups, J. Functional Anal., in press.

Davies, E.B., 1979b, Particle-boson interactions and the weak coupling limit, J. Math. Phys. 20: 345-351.

Davies, E.B., 1979c, Symmetry breaking for a nonlinear Schrödinger equation, Commun. Math. Phys. 64: 191-210.

Davies, E.B., 1979d, Non-unitary scattering and capture, preprint.

Davies, E.B., 1979e, Some time-dependent Hartree equations, preprint.

Davies, E.B., 1980, One-parameter semigroups, Academic Press, London.

Davies, E.B. and Eckmann, J.-P., 1975, Time decay for fermion systems with persistent vacuum, Helv. Phys. Acta 48: 731-742.

Davies, E.B. and Spohn, H., 1978, Open quantum systems with time-dependent Hamiltonians and their linear response, J. Stat. Phys. 19: 511-523.

Essen, H., 1977, The physics of the Born-Oppenheimer approximation, Int. J. Quantum Chem. 12: 721-735.

Evans, D.E., 1977, Irreducible quantum dynamical semigroups, Commun. Math. Phys. 54: 293-297.

Evans, D.E. and Lewis, J.T., 1977, "Dilations of irreversible evolutions in algebraic quantum theory", Commun. Dublin Inst. Adv. Studies Ser. A, no.24.

Frigerio, A. and Gorini, V., 1976, N-level systems in contact with a singular reservoir II, J. Math. Phys. 17: 2123-2127.

Ginibre, J. and Velo, G., 1979a, On a class of nonlinear Schrödinger equations, I, the Cauchy problem, general case, J. Functional Anal., in press.

Ginibre, J. and Velo, G., 1979b, On a class of nonlinear Schrödinger equations, II, Scattering theory, general case, J. Functional Anal., in press.

Ginibre, J. and Velo, G., 1978, On a class of nonlinear Schrödinger equations, III, Special theories in dimensions 1, 2 and 3, Ann. Inst. H. Poincaré 28A: 287-316.

Ginibre, J. and Velo, G., 1979c, The classical field limit of scattering theory for non-relativistic many-boson systems, I, Commun. Math. Phys. 66: 37-76.

Gorini, V. and Kossakowski, A., 1976, N-level system in contact with a singular reservoir, J. Math. Phys. 17: 1298-1305.
Gorini, V., Kossakowski, A., and Sudarshan, E.C.G., 1976, Completely positive dynamical semigroups on N-level systems, J. Math. Phys. 17: 821-825.
Haag, R. and Bannier, U., 1978, Comments on Mielnik's generalized (nonlinear) quantum mechanics, Commun. Math. Phys. 60: 1-6.
Haake, F., 1973, "Statistical treatment of open systems by generalized master equations", Springer tracts in modern physics, Vol. 66.
Hasse, R.W., 1975, On the quantum-mechanical treatment of dissipative systems, J. Math. Phys. 16: 2005-2011.
Hepp, K., 1974, The classical limit for quantum-mechanical correlation functions, Commun. Math. Phys. 35: 265-277.
Hepp, K., 1975a, Two models for Josephson oscillators, Ann. Phys. 90: 258-294.
Hepp, K., 1975b, The Onsager relations for quantum systems weakly coupled to KMS reservoirs, Z. Phys. B20: 53-54.
Hepp, K. and Lieb, E.H., 1973, Phase transitions in reservoir-driven open systems, with applications to lasers and superconductors, Helv. Phys. Acta 46: 573-603.
Immele, J.D., Kou, K.K., and Griffin, J.J., 1975, Special examples of quantized friction, Nucl. Phys. A241: 47-60.
Kibble, T.W.B., 1978, Relativistic models of nonlinear quantum mechanics, preprint.
Kossakowski, A., Frigerio, A., Gorini, V. and Verri, M., 1977, Quantum detailed balance and KMS condition, Commun. Math. Phys. 57: 97-110.
Kostin, M.D., 1972, On the Schrödinger-Langevin equation, J. Chem. Phys. 57: 3589-3591.
Lieb, E.H., 1977, Existence and uniqueness of the minimising solution of Choquard's nonlinear equation, Stud. Applied Math. 57: 93-106.
Lieb, E.H. and Simon, B., 1977, The Hartree-Fock theory for Coulomb systems, Commun. Math. Phys. 53: 185-194.
Lieb, E.H. and Yamazaki, K., 1958, Ground state energy and effective mass of the polaron, Phys. Rev. 111: 728-733.
Lindblad, G., 1975, Completely positive maps and entropy inequalities, Commun. Math. Phys. 40: 147-151.
Lindblad, G., 1976, On the generators of quantum dynamical semigroups, Commun. Math. Phys. 48: 119-130.
Messer, J., 1979, Friction in quantum mechanics, Acta Phys. Austriaca 50: 75-91.
Messer, J. and Baumgartner, B., 1978, Nonlinear von Neumann equations for quantum dissipative systems, Z Physik 32: 103-105.
Mielnik, B., 1974, Generalized quantum mechanics, Commun. Math. Phys. 37: 221-256.
Palmer, P.F., 1976, D. Phil. thesis, Oxford.
Palmer, P.F., 1977, The singular coupling and weak coupling limits, J. Math. Phys. 18: 527-529.

Pulé, J., 1974, The Bloch equations, Commun. Math. Phys. 38: 241-256.

Scott, A.C., Chu, F.Y.F., and McLaughlin, D.W., 1973, The Soliton, a new concept in applied science, Proc. I.E.E.E. 61: 1443-1483.

Spohn, H., 1976, Relaxation of finite closed systems, Reports Math. Phys. 10: 283-296.

Spohn, H., 1978, Entropy production for quantum dynamical semi-groups, J. Math. Phys. 19: 1227-1230.

Spohn, H., 1978/1979, "Kinetic equations from Hamiltonian dynamics: the Markovian limit", Univ. of Leuven Lecture Notes.

Spohn, H. and Lebowitz, J.L., 1978, Irreversible thermodynamics for quantum systems weakly coupled to thermal reservoirs, Adv. Chem. Phys. 38: 109-142.

Woolley, R.G., 1976, Quantum theory and molecular structure, Adv. Phys. 25: 27-52.

Woolley, R.G., 1978, Further remarks on molecular structure in quantum theory, Chem. Phys. Lett. 55: 443-446.

QUASIPERIODIC AND STOCHASTIC INTRAMOLECULAR DYNAMICS:

THE NATURE OF INTRAMOLECULAR ENERGY TRANSFER

Stuart A. Rice

The Department of Chemistry and The James Franck Institute
The University of Chicago
Chicago, Illinois 60637, U.S.A.

1. INTRODUCTION

These lectures are intended to provide an elementary account of the nature of intramolecular energy transfer in isolated molecules. This subject has attracted considerable attention recently, in part because of advances in classical mechanics and in part in response to the need for analysis of the consequences of selective photo-excitation of molecules. The relevant literature is now quite extensive, and it contains contributions representing an enormous range of approaches and goals. Moreover, the techniques of analysis used range from abstract topological description of the general behaviour of a class of systems to numerical integration of the equations of motion of a particular system. In the limited time (and space) available for these lectures I can do no more than sketch some of the principle ideas advanced. Insofar as it is possible I will focus attention on the dynamics of intramolecular energy exchange, that is on the time evolution of nonstationary states of an isolated molecule.

In an earlier plenary lecture [1] I summarised the major questions associated with understanding intramolecular energy exchange as follows:

1. Under what conditions, if any, is intramolecular energy exchange slow/rapid relative to other processes, for example photon emission, or isomerization, or fragmentation?

2. How does the intramolecular energy exchange depend on the energy of the molecule and the nature of the initial excitation?

3. If there are situations for which intramolecular energy exchange is slow relative to chemical reaction, why does this behaviour occur? Does it derive from special characteristics of the molecular force field? Are there dynamical or symmetry restrictions on the spectrum of states in these cases? Are these special situations commonly or rarely found?

4. Given the answers to (3), can we devise excitation methods and reaction conditions that permit enhancement of the selectivity of the chemistry that follows?

It is now widely accepted that the vast majority of systems of coupled nonlinear oscillators display quasiperiodic motion at low energy and stochastic motion at high energy [2]. Nevertheless, it is also known that there are exceptions to this generalization [3]. Moreover, both the uniformity of behaviour in the quasiperiodic and stochastic domains of the molecular dynamics and the nature of the transition between them, are incompletely understood. Prompted by these observations I will augment the preceding list of questions with the following:

5. Are there isolated quasiperiodic trajectories embedded in the stochastic domain of the molecular dynamics?

6. If the answer to (5) is yes, is the existence of such trajectories peculiar to a small set of initial conditions? How does the intramolecular force field influence the existence or nonexistence of these trajectories?

7. If the answer to (5) is no, are there dynamical transients that are quasiperiodic on a time scale that permits interception by competition with other dynamical processes, e.g. collisions?

The relevance of these questions will become increasingly apparent as these lectures proceed. At present it is not possible to give full answers to any of these questions, but partial answers are suggested by the several analyses and points of view discussed.

Although the questions listed above have been phrased in the language of classical mechanics, the application of interest, namely the study of intramolecular energy transfer, requires a quantum mechanical description. An important component of these notes is the examination of the similarities and differences in the classical and quantum mechanical descriptions of the same nonlinear coupled oscillator model.

The notes that follow this introduction draw heavily on the work of others and are not original. In some places I have paraphrased the explanations given in other articles, and I have freely used figures from a variety of sources. Although I have referred to many papers, those by Ford [4] and Berry [5] have been primary sources. The material discussed is not balanced and has many omissions, the most important omission being the set of subjects related to Kolmogorov entropy; the reader is referred elsewhere for this material [6].

2. THE CLASSICAL MECHANICAL DESCRIPTION OF COUPLED NONLINEAR OSCILLATOR SYSTEMS

A. Some Properties of Trajectories

Given a dynamical system with N degrees of freedom described by the Hamiltonian H(p,q) and the equations of motion

$$\dot{q}_s = \frac{\partial H}{\partial p_s}, \quad \dot{p}_s = \frac{\partial H}{\partial q_s} \tag{2.1}$$

and given the initial values of the coordinates and momenta, q_s^o, p_s^o, the values of q, p at any other time t

$$q_s = q_s(t,q^o,p^o) \quad , \quad p_s = p_s(t,q^o,p^o) \tag{2.2}$$

are unique under very weak conditions. Equations (2.2) can, in principle, be solved for q_s^o, p_s^o,

$$q_s^o = q_s^o(t,q,p) \quad , \quad p_s^o = p_s^o(t,q,p), \tag{2.3}$$

which gives 2N functions of the phase space variables and the time which are constant along any trajectory of the system. Elimination of t between the equations (2.3) leaves 2N-1 functions of only the phase space variables; these functions also have the property of being constant along any trajectory. This argument establishes the existence of 2N-1 functions $C_j(q,p)$ which are integrals of the motion; attributing a set of numerical values to the C_j is equivalent to completely determining the system trajectory in phase space.

Saying that the set of 2N-1 values of C_j determine the trajectory is one thing; finding the values is quite another! Of course, every constant of the motion must satisfy the Poisson-Bracket relation

$$\{H,C_j\} = 0, \quad j = 1,2,\ldots,2N-1 \tag{2.4}$$

but the only obvious solution is $C_1 = H(q,p)$. Although it is in principle possible to stepwise find 2N-2 other functions which with C_1 form a complete set of functionally independent integrals of the motion, in practice this is impossible to execute even for very simple mechanical systems.

Note that $C_1 = H$ requires that the trajectory of a conservative system lie on the energy surface $H(q,p) = E$. In general, each of the equations

$$C_j = k_j, \qquad j = 1,2,\ldots,2N-1 \tag{2.5}$$

for given k_j defines a 2N-1 dimensional hypersurface in the 2N dimensional phase space. The trajectory of the system must lie entirely on each of these surfaces, hence is determined entirely by their hyperdimensional intersection. Put in slightly different words, fixing the value of any C_j restricts the region of phase space in which the trajectory can lie. Specification of all 2N-1 C_j reduces the allowable dimensionability from 2N to 1, which is the trajectory of the system. However, the integrals of the motion C_j are of two types [7]. Some are isolating, in the sense that the domain of phase space to which they restrict the trajectory is compact and readily partitioned from the full phase space - the language used here is loose but the geometric visualization intended should be clear. The integral $C_1 = H(q,p)$ is of this type. Others, apparently the vast majority, are nonisolating. The regions of phase space to which they restrict the trajectory pass tortuously through the full domain accessible under the isolating integrals of motion. The distinction between these two classes of integrals of the motion is evident even for the simple system of two independent harmonic oscillators whose Hamiltonian is ($m = 1$),

$$H = \frac{1}{2}(p_1^2 + \omega_1^2 q_1^2) + \frac{1}{2}(p_2^2 + \omega_2^2 q_2^2), \tag{2.6}$$

which leads to the equations of motion

$$\begin{aligned} p_i \cos \omega_i t + \omega_i q_i \sin \omega_i t &= p_i^o \\ \omega_i q_i \cos \omega_i t - p_i \sin \omega_i t &= \omega_i q_i^o \end{aligned} \qquad i = 1,2 \tag{2.7}$$

Elimination of t, for each value of i, gives

$$\begin{aligned} p_i^2 + \omega_i^2 q_i^2 &= p_i^{o2} + \omega_i^2 q_i^{o2} = \text{constant} \\ C_i &= \frac{1}{2}(p_i^2 + \omega_i^2 q_i^2) \end{aligned} \qquad i = 1,2 \tag{2.8}$$

Finally, elimination of t between the equations of motion for different $i = 1,2$ leads to a third integral of the motion, C_3. The nature of C_3 depends on the ratio ω_2/ω_1, in particular on whether this ratio is rational or irrational. In the case ω_2/ω_1 is rational the projection of the system trajectory on the $q_1 q_2$ plane is a closed curve, in fact a Lissajous figure (Fig. 2.1). In this case $C_3(q,p)$ is a multivalued function with a finite number

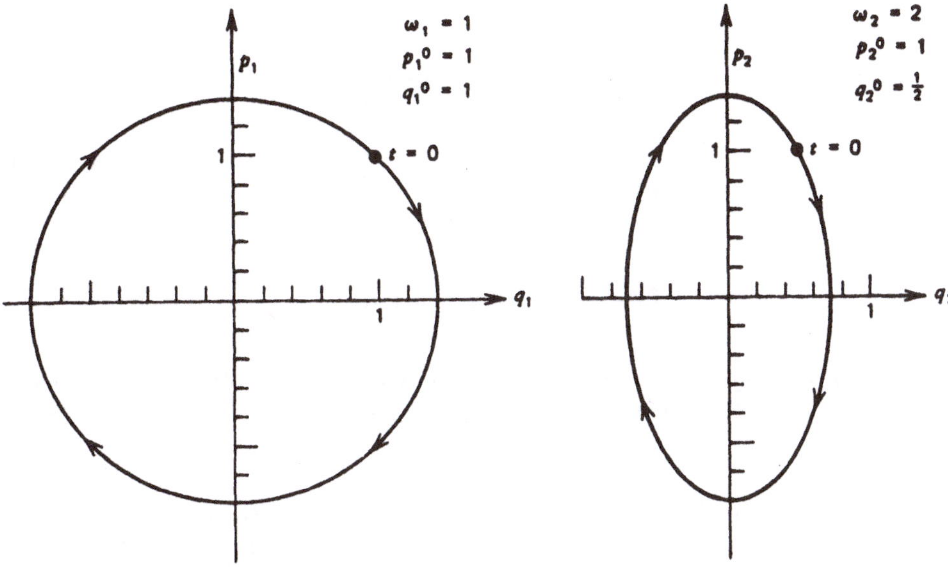

Fig. 2.1a. Trajectories of a separable two oscillator system with rational frequency ratio projected on the p_1q_1 and p_2q_2 planes.

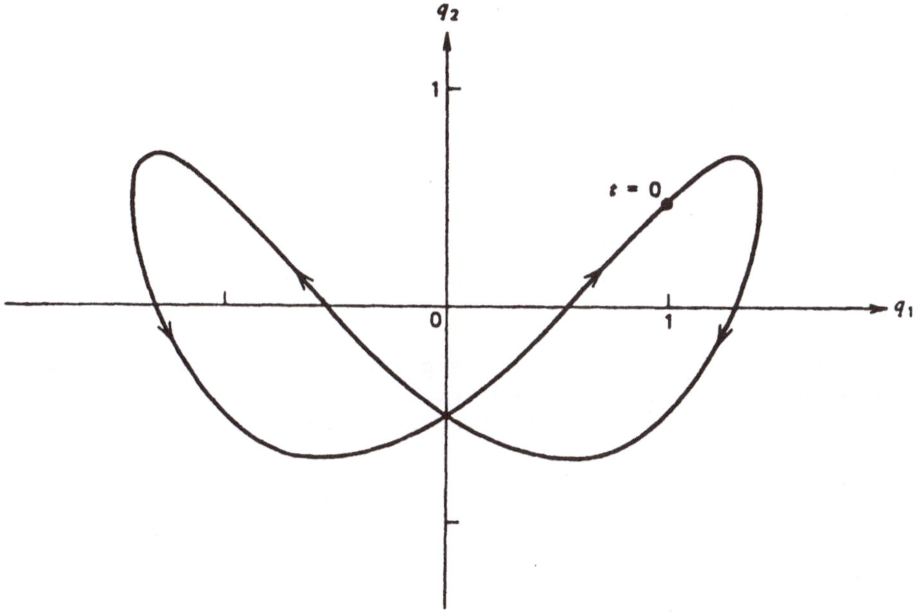

Fig. 2.1b. The trajectory of a separable two oscillator system with rational frequency ratio $\omega_2/\omega_1 = 2$ projected on the q_1q_2 plane.

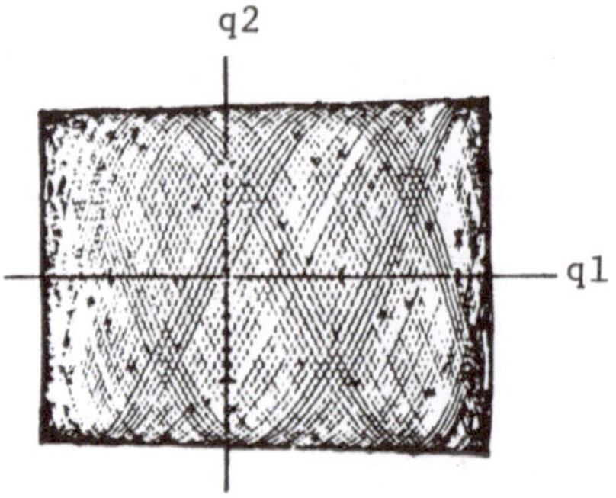

Fig.2.1c. The trajectory of a separable two oscillator system with irrational frequency ratio, projected on the q_1q_2 plane.

of branches. On the other hand, if ω_2/ω_1 is irrational, the projection of the system trajectory on the q_1q_2 plane does not generate a closed curve because there can be no rational integer set which leads to matching of the two oscillators. As a result, the motion of the system is not periodic, and the projected trajectory in the q_1q_2 plane passes arbitrarily close to each point lying within the rectangle defined by the maximum amplitudes of the two oscillators (Fig. 2.1). The trajectory thereby densely fills the accessible q_1q_2 space. Although the integral of the motion on C_3 exists, it is a pathological function, namely a multivalued function with an infinite number of branches. Note that when ω_2/ω_1 is rational the projected trajectory is a Lissajous figure that restricts the motion of the representative point to a small portion of the q_1q_2 plane. In this case C_3 is an isolating integral of the motion. But when ω_2/ω_1 is irrational the existence of C_3 does not prevent the projected trajectory from filling the energetically accessible region of the q_1q_2 plane. In this latter case C_3 is a nonisolating integral of the motion.

A different view of dynamics is given by the Hamilton-Jacobi form of mechanics. This representation of the dynamics is based on finding a canonical transformation such that

$$K = H(P), \qquad \dot{Q}_s = \frac{\partial K}{\partial P_s} = \omega_s(P), \quad \dot{P}_s = \frac{\partial K}{\partial Q_s} = 0 \tag{2.9}$$

The new momenta P_s are integrals of the motion, while the new coordinates Q_s are linear functions of time. Note that only N constants of the motion are determined by the canonical transformation, so the system trajectory is restricted to an N-1 dimensional subspace of the full phase space, but not to a smaller space. For a bounded system linear combinations of the P_s define actions J_s and their conjugate angle variables ϕ_s; the J_s and ϕ_s define the action-angle représentation of mechanics. It is only for the case of a system of independent harmonic oscillators that the ω_s are independent of P_s and constant. In more general cases we can, in principle, find $H(J,\phi)$ but the corresponding frequencies will not be independent of J.

Consider the situation in which the system Hamiltonian can be separated into

$$H(J,\phi) = H_0(J) + \varepsilon H_1(J,\phi) \tag{2.10}$$

$H_0(J)$ describes an integrable system, the trajectories of which densely cover regions of phase space. As in the example of two harmonic oscillators with ω_2/ω_1 irrational, we expect that when regions of phase space are densely covered the frequencies $\omega_i \equiv (\partial H_0/\partial J_i)$ are not related by a set of rational integers. The term $\varepsilon H_1(J,\phi)$ is a "small" perturbation. The traditional view of the influence of anharmonicity on the motion of coupled oscillators suggests that εH_1 destroys the topological structure of the trajectories corresponding to $H_0(J)$ no matter how small εH_1, if only enough time elapses. The idea is that εH_1 causes the trajectory to wander out of densely filled regions corresponding to $H_0(J)$ = constant, thereby filling all of accessible phase space.

A remarkable theorem, due to Kolmogorov, Arnold and Moser (KAM) [8] implies that the intuitive description of the trajectory just given is incorrect. The theorem says that provided ε is "sufficiently small" and $H_1(J,\phi)$ is analytic in J and ϕ in a given domain, the phase space can be separated into two regions of non-vanishing volume. One of these is small, and it shrinks to zero volume as $\varepsilon \to 0$. The larger of the two regions has the structure characteristic of $H_0(J)$. Thus, the KAM theorem asserts that for the majority of initial conditions the trajectories of the system have the same character as in the uncoupled oscillator case (Lissajous figures restricted to N-1 dimensions). There is a small region (of instability) in which the trajectories are wildly erratic and can depart drastically from the nearby confined trajectories.

To apply the KAM theorem we need to know what is "small enough" with respect to ε or, equivalently, for fixed ε how the topological behaviour of the trajectory changes as the energy of the system increases. At present all of our knowledge concerning this crucial point is derived from numerical solutions of the equations of motion of model systems [4,9]. Some hypotheses, based on analytical considerations, have been advanced to explain the results of the numerical studies, but these have followed and cannot yet replace the trajectory calculations [10].

Before examining the results of the numerical solution of the equations of motion of a typical system of coupled nonlinear oscillators, and before describing the analytical considerations advanced to interpret those results, it is desirable to give a qualitative description of the expected behaviour of trajectories in such a system. For illustrative purposes consider again a system with two degrees of freedom so that, in angle-action variables,

$$H = H_0(J_1,J_2) + V(J_1,J_2,\phi_1,\phi_2) \tag{2.11}$$

When $V = 0$, H_0 generates a motion for which J_1,J_2 = constants and $\phi_i = \omega_i(J_1,J_2)t + \phi_i^0$, $\omega_i \equiv \partial H_0/\partial J_i$. The motion of the unperturbed system is conveniently represented on a two dimensional torus where ϕ_1,ϕ_2 are the angle coordinates and J_1,J_2 the radii (Fig.2.2). If V is small enough and the Jacobians

$$\frac{\partial(\omega_1,\omega_2)}{\partial(J_1,J_2)} \neq 0 \tag{2.12}$$

KAM show that most of the unperturbed tori bearing conditionally periodic motion with incommensurate frequencies continue to exist, being only slightly perturbed by V. On the other hand, tori bearing periodic motion or very nearly periodic motion, with commensurate frequencies, or with incommensurate frequencies whose ratio is well approximated by r/s, r,s small integers, are grossly deformed by V and no longer remain close to unperturbed tori. Furthermore, although the unperturbed tori with commensurate frequencies which are destroyed by $V \neq 0$ are everywhere dense, KAM show that the majority (in the sense of measure theory) of initial conditions lead to motion on preserved tori bearing conditionally periodic motion when V is sufficiently small. Thus, KAM theory shows that for small V most initial conditions lead to nonergodic motion.

What is the character of the motion not on preserved tori? Imagine H expanded in a Fourier series [11]:

$$H = H_0(J_1,J_2) + f_{mn}(J_1,J_2)\cos(m\phi_1 + n\phi_2) + \ldots \tag{2.13}$$

In KAM theory the angle-dependent terms are eliminated by successive canonical transformations, each of which is close to the identity transformation. The final Hamiltonian is a function of transformed variables only and is "close" to the original Hamiltonian. If this can be accomplished in some general sense, one finds that the unperturbed motion, for the most part, lies on tori close to unperturbed tori.

To illustrate these ideas suppose the only important coupling term in (2.13) is f_{mn} (as displayed). Then to eliminate the term $\cos(m\phi_1 + n\phi_2)$ introduce the canonical transformation

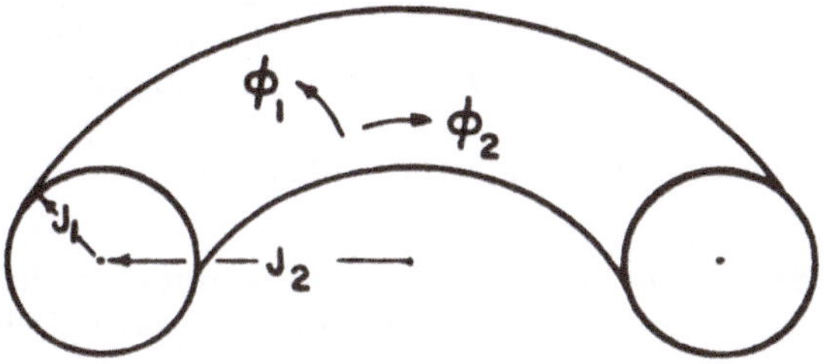

Fig.2.2a. Angle-action variables and the invariant torus for a two oscillator system.

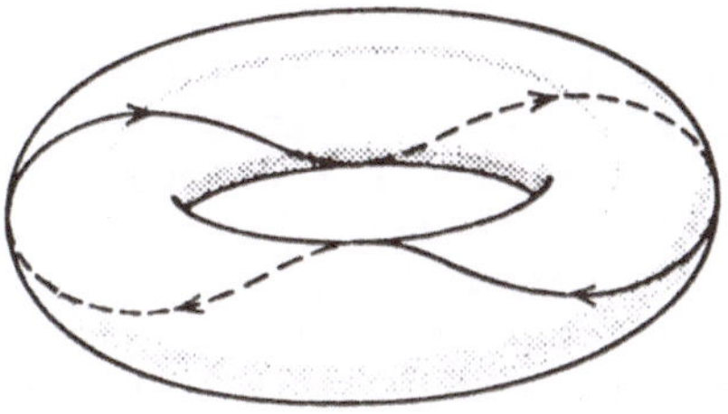

Fig.2.2b. A periodic trajectory on the torus.

$$F = S_1\theta_1 + S_2\theta_2 + B_{mn}(S_1,S_2) \sin (m\theta_1 + n\theta_2) \tag{2.14}$$

with S,θ the transformed action-angle variables, and B_{mn} to be determined. If $B_{mn} = 0$, then $S_i = J_i$, $\theta_i = \phi_i$. Applying F to H,

$$H = H_0(S_1,S_2) + \{[m\omega_1(S_1,S_2) + n\omega_2(S_1,S_2)]\, B_{mn} + f_{mn}(S_1,S_2)\} \cos (m\theta_1 + n\theta_2) + \ldots, \tag{2.15}$$

$$\omega_i(S_1,S_2) = \frac{\partial H_0}{\partial S_i} \tag{2.16}$$

To lowest order, the angle-dependent term is eliminated if

$$B_{mn}(S_1,S_2) = -\frac{f_{mn}(S_1,S_2)}{m\omega_1(S_1,S_2) + n\omega_2(S_1,S_2)} \tag{2.17}$$

which requires that $m\omega_1 + n\omega_2 \neq 0$ or be very small compared to f_{mn}. If the denominator is small compared to f_{mn}, B_{mn} is large, and the transformation is not close to the identity transformation, hence the transformed motion is not close to the unperturbed motion. Consequently, if there exists a band of frequencies ω_i for which

$$|m\omega_1(J_1,J_2) + n\omega_2(J_1,J_2)| << |f_{mn}(J_1,J_2)| \tag{2.18}$$

then the angle dependent term grossly distorts an associated zone of unperturbed tori bearing the frequencies satisfying the inequality. In general, if one Fourier component satisfies the inequality, there will be additional terms $\cos(m'\phi_1 + n'\phi_2)$ in H with ratios m'/n' sufficiently close to m/n that the inequality is also satisfied for them - hence the zone of unperturbed tori distorted by the displayed term will simultaneously be affected by many other angle dependent terms.

Note that the inequality cited is a kind of resonance relationship which, if satisfied, asserts that $\cos(m\phi_1 + n\phi_2)$ resonantly couples the oscillators when their frequencies lie in the designated bands. When V is small, hence all f_{mn} small, such resonance zones are narrow and the KAM theorem shows that the totality of all resonant zones is small relative to the measure of the allowed phase space. We expect that as V and f_{mn} increase, or as E increases the measure of the resonant zones will also increase until most of phase space is filled by them. KAM theory thereby predicts an amplitude instability for conservative nonlinear oscillator systems permitting a transition between predominantly quasi-periodic and ergodic motion.

Walker and Ford have provided a very elegant illustration of the general ideas just sketched [11]. Consider a system described by the Hamiltonian (2.11) with

$$H_0 = J_1 + J_2 - J_1^2 - 3J_1J_2 + J_2^2 \tag{2.19}$$

Set

$$q_i = (2J_i)^{1/2} \cos \phi_i, \tag{2.20}$$

$$p_i = -(2J_i)^{1/2} \sin \phi_i, \tag{2.21}$$

so that

$$J_i = \tfrac{1}{2}(p_i^2 + q_i^2) \tag{2.22}$$

and

$$\omega_1 = 1 - 2J_1 - 3J_2 = \frac{\partial H_0}{\partial J_1}, \tag{2.23}$$

$$\omega_2 = 1 - 3J_1 + 2J_2 = \frac{\partial H_0}{\partial J_2} . \qquad (2.24)$$

In order that $\omega_1 > 0$, $\omega_2 > 0$ we require E to be in the range $0 < E < 3/13$ and that the J_i lie on the branch that goes to zero with E.

We represent the motion in this system by level curves in the q_2p_2 plane. Note that the unperturbed level curves in the q_2p_2 plane are circles centred on the origin. Points on the level curves in the q_1p_1 plane, defined by $q_2 = 0$, $p_2 > 0$ (or $\phi_2 = 3\pi/2$) also lie on concentric circles. The circular level curves in either plane are enclosed by a bounding level curve representing the intersection of the energy surface with each plane.

Consider the case of a 2 - 2 resonance described by

$$H = H_0(J_1,J_2) + \alpha J_1J_2 \cos (2\phi_1 - 2\phi_2), \qquad (2.25)$$

which system has the additional constant of the motion

$$C = J_1 + J_2 \qquad (2.26)$$

Now eliminate J_2 from H using C, and set $\phi_2 = 3\pi/2)$; we obtain for the level curves in the J_1 plane

$$(3 + \alpha \cos 2\phi_1)J_1^2 - (5C + C \cos 2\phi_1)J_1 + C + C^2 - E = 0 \qquad (2.27)$$

(Recall that $H_0 = J_1 + J_2 - J_1^2 - 3J_1J_2 + J_2^2$.)

As shown in Fig. 2.3 the unperturbed circular level curves are only slightly distorted except in the 2-2 resonance zone enclosed by the self intersecting separatrix level curve. The two self-intersection points represent distinct unstable periodic solutions, while the two invariant points at the centre of each crescent region represent distinct stable periodic solutions. Since the central point of each crescent represents a distinct periodic orbit the two crescents are not a chain of two islands; the central points of an island chain represent a single periodic orbit. For all four of the mentioned periodic orbits we have

$$\dot{J}_1 = 0, \ \dot{J}_2 = 0,$$

$$\dot{\phi}_1 - \dot{\phi}_2 = 0. \qquad (2.28)$$

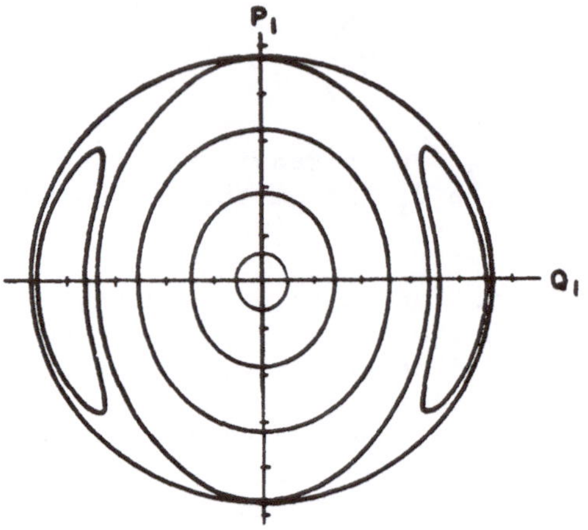

Fig. 2.3. The surface of section for the system with Hamiltonian

$$H = H_0 + J_1 J_2 \cos(2\phi_1 - 2\phi_2)$$

$$H_0 = J_1 + J_2 - J_1^2 - 3J_1J_2 + J_2^2$$

at an energy corresponding to an isolated 2-2 nonlinear resonance. (From G.H. Walker and J. Ford, Phys. Rev. 188, 416 (1969).)

For the stable periodic orbits

$$J_1 = \frac{5+\alpha}{1+\alpha} J_2, \tag{2.29}$$

$$\phi_1 - \phi_2 = \frac{\pi}{2} \text{ or } \frac{3\pi}{2}, \tag{2.30}$$

while for the unstable periodic orbits

$$J_1 = \frac{5-\alpha}{1-\alpha} J_2, \tag{2.31}$$

$$\phi_1 - \phi_2 = 0 \text{ or } \pi \tag{2.32}$$

Note that $2\omega_1 = 2\omega_2$ implies that

$$J_1 = 5J_2 \tag{2.33}$$

Thus, just as expected from the condition stated in Eq. (2.18), the 2-2 resonance zone of highly distorted tori occurs in a neighbourhood of the unperturbed torus bearing the frequencies $2\omega_1 = 2\omega_2$. Now substituting $J_1 = 5J_2$ into $H_0 = J_1 + J_2 - J_1^2 - 3J_1J_2 + J_2^2$ we find on the unperturbed 2-2 torus that

$$J_1 = \frac{5}{13}[1 - (1 - \frac{13}{3}E)^{1/2}], \tag{2.34}$$

$$J_2 = \frac{1}{13}[1 - (1 - \frac{13}{3}E)^{1/2}] \tag{2.35}$$

Consequently the unperturbed 2-2 torus and the perturbed 2-2 resonance zone exist for all allowed energies $0 < E < \frac{3}{13}$. As the energy increases from zero the 2-2 resonance moves out from the origin and increases in width.

The next higher resonance after the 2-2 is the 2-3 or 3-2 resonance. Consider the Hamiltonian

$$H = H_o(J_1,J_2) + \beta J_1^{3/2} J_2 \cos(3\phi_1 - 2\phi_2), \tag{2.36}$$

which has the additional constant of the motion

$$C = 2J_1 + 3J_2 \tag{2.37}$$

The level curves in the J_1 plane are

$$E = \frac{1}{3}C + \frac{1}{9}C^2 + (\frac{1}{3} - \frac{13}{9}C)\, J_1 + \frac{13}{9}J_1^2 - \frac{1}{3}\beta(CJ_1^{3/2} - 2J_1^{5/2})\cos 3\phi_1. \tag{2.38}$$

As shown in Fig. 2.4 the 3-2 resonance zone consists of a chain of three islands (points at the centres of each crescent represent a single stable solution). Similarly the three self intersecting points on the separatrix represent a single unstable periodic solution. Again setting

$$\left.\begin{aligned} \dot{J}_1 &= 0, \quad \dot{J}_2 = 0, \\ 3\dot{\phi}_1 - 2\dot{\phi}_2 &= 0, \end{aligned}\right. \tag{2.39}$$

we find

$$\left.\begin{aligned} J_2 &= \frac{1 + 2J_1^{3/2}}{13 + \frac{9}{12}J_1^{1/2}}, \\ 3\phi_1 - 2\phi_2 &= \pi,\ 3\pi,\ 5\pi \end{aligned}\right\} \text{ stable orbit} \tag{2.40}$$

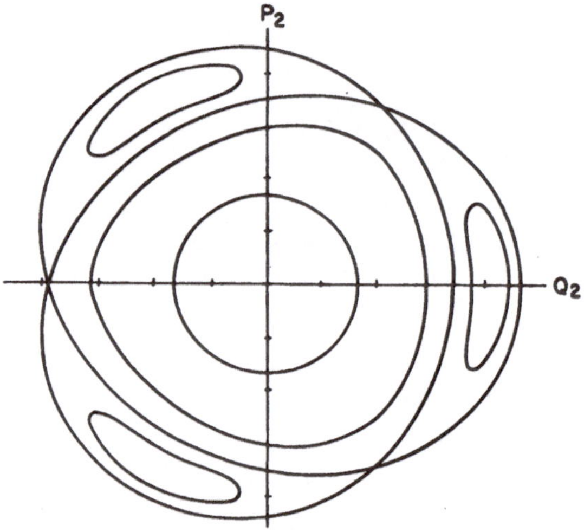

Fig. 2.4. The surface of section for the system with Hamiltonian

$$H = H_o + \beta J_1^{3/2} J_2 \cos(3\phi_1 - 2\phi_2)$$

$$H_o = J_1 + J_2 - J_1^2 - 3J_1 J_2 + J_2^2$$

at an energy corresponding to an isolated 3-2 nonlinear resonance. (From G.H. Walker and J. Ford, Phys. Rev. 188, 416 (1969).)

$$\left.\begin{aligned} J_2 &= \frac{1 - 2J_1^{3/2}}{13 + \frac{9}{2}J_1^{1/2}}, \\ 3\phi_1 - 2\phi_2 &= 0,\ 2\pi,\ 4\pi \end{aligned}\right\} \quad \text{unstable orbit} \tag{2.41}$$

As expected, the 3-2 resonance zone lies near to the unperturbed 3-2 zones. Setting $3\omega_1 = \omega_2$, H_0 yields

$$J_1 = \frac{5}{13} - \left(\frac{3}{13} - E\right)^{1/2}, \tag{2.42}$$

$$J_2 = \frac{1}{13}, \tag{2.43}$$

as the values of J_1, J_2 on the unperturbed torus, to be compared with the perturbed values. Since $J_1 = \frac{1}{2}(p_1^2 + q_1^2) > 0$,

$$E = \frac{3}{13} - \left(J_1 - \frac{5}{13}\right)^2 \geq \frac{14}{169} \approx 0.08 \tag{2.44}$$

At $E = \frac{14}{169}$, $J_1 = 0$, hence the unperturbed 3-2 torus and the 3-2 resonance zone appear abruptly at the origin of the J_1 plane. They appear abruptly in the J_2 plane when the bounding level curve moves out to $J_2 = 1/13$, which is when $E = 14/169$. The corresponding 2-3 resonance is generated from

$$H = H_0(J_1,J_2) + \beta J_1 J_2^{3/2} \cos(2\phi_1 - 3\phi_2), \tag{2.45}$$

which has the additional constant of the motion

$$C = 3J_1 + 2J_2. \tag{2.46}$$

Level curves in the J_2 plane are found from

$$E = \frac{1}{3} - \frac{1}{9}C^2 + (\frac{1}{3} - \frac{5}{9}C)J_2 + \frac{23}{9}J_2^2 + \beta[\frac{2}{3}J_2^{5/2} - \frac{1}{3}CJ_2^{3/2}] \cos 3\phi_2. \tag{2.47}$$

Just as for the 3-2 resonance case, the 2-3 resonance zone appears in the J_2 plane about the unperturbed 2-3 torus, which appears abruptly at $E = 0.16$. The level curves are similar to those of the 3-2 case, except that the chain of three islands appears in the J_2 plane.

In general, it is found that:

(i) A Hamiltonian of the form

$$H = H_0(J_1,J_2) + f_{mn}(J_1,J_2) \cos (m\phi_1 + n\phi_2)$$

has an "extra" well defined constant of the motion:

$$C = nJ_1 - mJ_2.$$

(ii) An m-n resonance for $m \neq n$ introduces a chain of m islands in the J_1 plane and a chain of n islands in the J_2 plane. (Islands are ovals surrounding points representing stable periodic orbits.)

(iii) Isolated resonances distort the unperturbed tori by introducing, in pairs, new stable and unstable periodic orbits.

(iv) An m-n resonance zone appears abruptly, in general at some $E > 0$, and is bounded by a separatrix which passes through the unstable periodic solutions.

(v) The m-n resonance zones decrease in size rapidly as m and n increase.

Thus far we have commented only on the consequences of the existence of isolated nonlinear resonances. In general we must expect that a system of coupled nonlinear oscillators will support many different nonlinear resonances of the types described, and that at some energy the different resonance zones will begin to overlap. Walker and Ford [11] and Ford and Lunsford [12] have studied simple models which fully confirm this inference. When the resonant zones corresponding to the several nonlinear resonances overlap there is gross distortion of the level curves on the q_2p_2 plane and higher order nonlinear resonances also begin to strongly influence these curves (see next section). The net result is a macroscopic distortion of the system tori, which then implies a macroscopic instability of the trajectory. Clearly, the energy at which there is an onset of overlap of nonlinear resonances is a plausible boundary between quasi-periodic and grossly irregular motion [13]. We shall return to this point in Section 4.

B. Numerical Studies of the Dynamics of Nonlinear Oscillator Systems

In a sense, it can be said that numerical solution of the equations of motion of some system is intended to reveal the consequences of the breakdown of integrability and the lack of isolating integrals of the motion other than the energy. Poincaré introduced a representation of the results of trajectory analysis which permits visualization of these consequences [14]. This representation, which is most useful for two dimensional systems, portrays the motion on a so-called (Poincaré) surface of section. We have already used this representation, without naming it, in the discussion of the level curves in the q_2p_2 plane for a system with nonlinear resonances. Now we generalize our description. Consider for simplicity, a Hamiltonian of the form (m = 1)

$$H = \frac{1}{2}(p_1^2 + p_2^2) + f(q_1,q_2). \tag{2.48}$$

For fixed energy, H = E, (2.48) has only three independent variables. One surface of section is defined by the intersection of H = E with $q_1 = 0$; in that plane the coordinates are p_2 and q_2. To each point in the surface of section there corresponds a unique value of p_2 and q_2, and of E and $q_1 = 0$. Then p_1 is determined except for sign since, from (2.48),

$$p_1 = \pm [2E - p_2^2 - f(q_2,0)]^{1/2}. \tag{2.49}$$

A given trajectory of a bound system will repeatedly cross the surface of section, since that trajectory must repeatedly pass through $q_1 = 0$, half the passages with $p_1 > 0$ and half with $p_1 < 0$. We now recognize two possibilities. If there exist isolating integrals of the motion other than the energy, such as C_3 with

ω_2/ω_1 rational for the Hamiltonian (2.6), the system trajectory lies on a hypersurface of smaller dimensionality than the energy surface. This hypersurface intersects the surface of section in a smooth closed curve - closed because the motion is periodic. In contrast, if there are no isolating integrals of the motion other than the energy, the intersections of the trajectory with the surface of section will cover that surface. The pattern of intersections will appear random, but is in fact not, since the trajectory satisfies the deterministic equation of motion. The two cases described are schematically sketched in Fig. 2.5.

The surface of section representation has another characteristic that will prove useful in our considerations. Since any point initially on the intersection of a trajectory and the surface of section will repeatedly cross the surface of section at, in general, different points, at each crossing the entire surface of section is mapped into itself. Of course the time taken for the recrossing will be different for each point of the surface of section, but that does not alter the character of the mapping. Given that the dynamics is described by the Hamiltonian equations of motion, this mapping is area preserving (this property is related to Liouville's theorem).

Let T be the operator that maps the point $Z^{(0)}$ on the surface of section into the point $Z^{(1)}$:

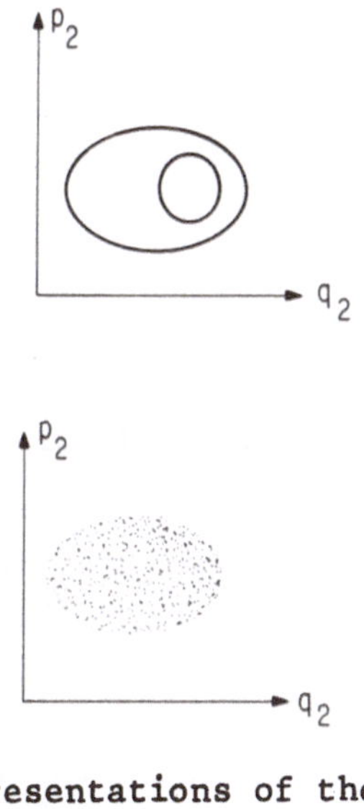

Fig. 2.5. Schematic representations of the surface of section in the quasi-periodic region and the stochastic region.

$$Z^{(1)} = T\,Z^{(0)} \tag{2.50}$$

A point $Z^{(0)}$ is called a fixed point when $Z^{(1)} = Z^{(0)}$. It can be shown that when T maps a compact simply connected subset S of the plane onto itself, then S contains a fixed point of T. In general the mappings of surfaces of section for the dynamical systems of interest to us will have many fixed points. The behaviour of the trajectories in the vicinities of these fixed points will be shown to determine the overall behaviour of the dynamical system.

To begin a description [15] of the nature of the fixed points of the mapping T we note that a set of points U is said to be invariant if TU = U, and for an area preserving mapping it is sufficient that TZ belongs to U if Z belongs to U. A fixed point of a mapping is called stable if every neighbourhood of this fixed point contains an invariant neighbourhood. Suppose, now, that a fixed point of the mapping T is located at $Z \equiv (p,q) = (0,0)$, which can always be achieved by a translation of the coordinate system. Then in the neighbourhood of this fixed point the mapping T can be represented by the linear equations

$$\begin{pmatrix} p^{(1)} \\ q^{(1)} \end{pmatrix} = \begin{pmatrix} a & b \\ c & d \end{pmatrix} \begin{pmatrix} p^{(0)} \\ q^{(0)} \end{pmatrix} \tag{2.51}$$

with

$$\det(T) = \begin{Vmatrix} a & b \\ c & d \end{Vmatrix} = 1, \tag{2.52}$$

since T is an area preserving mapping. Thus the eigenvalues of T, denoted as λ_1 and λ_2, satisfy the condition

$$\lambda_1\lambda_2 = 1 \tag{2.53}$$

If both λ_1 and λ_2 are real, and $\lambda_2 = \lambda_1^{-1} \neq 1$, the fixed point is called hyperbolic; if λ_1 and λ_2 are imaginary conjugates the fixed point is called elliptic; and if $\lambda_1 = \lambda_2 = \pm 1$ the fixed point is called parabolic. The reason for these designations, and the characteristics of each of these kinds of fixed points, will be made apparent below.

Recall that the trajectory representing the motion of a coupled oscillator system lies on a torus, and that in the case when the equations of motion are integrable the intersection of the trajectory with the surface of section generates simple closed curves. We have defined the surface of section in p_2q_2 space. In angle-action space the transverse radius of the torus is, say, J_2, and the corresponding angle is ϕ_2 (see Fig. 2.2); then the invariant curves on the surface of section are circles. A mapping of the system trajectory conserves J_2 but changes ϕ_2. For the case that the time interval between crossings of the surface of section is $\tau = 2\pi/\omega_1$ and the initial angular coordinate is $\phi_1^{(0)}$, the angular coordinate on the next crossing is

$$\phi_1^{(1)} = \phi_1^{(0)} + \omega_2 t = \phi_1^{(0)} + 2\pi\frac{\omega_2}{\omega_1} \tag{2.54}$$

where the ratio of frequencies

$$\frac{\omega_2}{\omega_1} = \alpha(J_2) \tag{2.55}$$

is, at constant energy, a function of J_2 only. The topology of successive intersections of the trajectory of an integrable system with the surface of section is, therefore, described by the twist mapping

$$J_2^{(1)} = TJ_2^{(0)} = J_2^{(0)}$$

$$\phi_2^{(1)} = T\phi_2^{(0)} = \phi_2^{(0)} + 2\pi\alpha(J_2^{(0)}) \tag{2.56}$$

The function $\alpha(J_2)$ is called the rotation number. Under the twist mapping, which is area preserving, circles map into circles, and radius vectors into curved arcs passing through the origin. Suppose the mapping of the trajectory is changed to

$$J_2^{(1)} = T_\varepsilon J_2^{(0)} = J_2^{(0)} + \varepsilon f(J_2^{(0)}, \phi_2^{(0)})$$

$$\phi_2^{(1)} = T_\varepsilon \phi_2^{(0)} = \phi_2^{(0)} + 2\pi\alpha(J_2^{(0)}) + \varepsilon g(J_2^{(0)},\phi_2^{(0)}) \tag{2.57}$$

where the functions f and g have period 2π in $\phi_2^{(0)}$, are zero when $J_2^{(0)} = 0$ and are related such that $\det(T_\varepsilon) = 1$. The condition $f = 0$, $g = 0$ when $J_2^{(0)} = 0$ guarantees that the origin of the coordinates remains a fixed point despite the change in the form of the mapping, and the condition on the determinant of T_ε guarantees that it is area preserving. The KAM theorem, applied to the relationship between invariant curves generated by the two mappings, T and T_ε, states that most of the points in the surface of section lie on smooth invariant curves which are distortions of the invariant circles generated by the twist mapping T. These smooth invariant curves are sections of tori which are likewise distorted from the original tori. The only possible exceptions to this behaviour correspond to points that are near tori on which $\alpha(J_2^{(0)} = \omega_2/\omega_1$ is rational ("commensurable" tori). On commensurable tori the trajectories are closed orbits. It is easy to see that every point on the circle with rotation number

$$\alpha(J_2^{(0)}) = \frac{r}{s} \qquad \text{(r,s integers)} \tag{2.58}$$

is a fixed point of the twist mapping

$$\begin{aligned} T^s J_2^{(0)} &= J_2^{(0)} \\ T^s \phi_2^{(0)} &= \phi_2^{(0)} + 2\pi s\alpha(J_2^{(0)}) \\ &= \phi_2^{(0)} + 2\pi r\phi_2^{(0)} \\ &= \phi_2^{(0)} \pmod{2\pi} \end{aligned} \tag{2.59}$$

The Poincaré-Birkhoff theorem [16] states that, in general, an even multiple of s (say 2ks, k = 1,2,...) fixed points remains when the mapping is changed from T to T_ε. This theorem gives supplementary information to that gained from the KAM theorem, which says nothing about the case of transformations of commensurable tori. Note that not all fixed points are preserved under the change $T \to T_\varepsilon$; all but a finite number are destroyed.

The nature of a fixed point is determined by the heaviour of nearby invariant curves. Consider, first, the case that the eigenvalues of the mapping T are imaginary:

$$\begin{aligned} \lambda_1 &= e^{i\alpha} \\ \lambda_2 &= \lambda_1^* = e^{-i\alpha}. \end{aligned} \tag{2.60}$$

In this case T can be written in the form

$$T = \begin{pmatrix} \cos\alpha & \sin\alpha \\ -\sin\alpha & \cos\alpha \end{pmatrix} \tag{2.61}$$

which corresponds to a rotation through the angle α. Thus in this case T is a twist mapping, and the invariant curves are circles. In the more general case that λ_1 and λ_2 are complex, rather than pure imaginary, the invariant curves are ellipses. It is for this reason that a fixed point of this type is designated elliptic. Trajectories in the vicinity of elliptic fixed points are stable, since any trajectory that crosses the surface of section near such a point will still be near it after arbitrarily many iterations of the transformation (Fig. 2.6).

Consider, next, the case that the eigenvalues of T are real numbers, where say $|\lambda_1| > 1$. Then T can be written in the form

$$T = \begin{pmatrix} 1/\lambda & 0 \\ 0 & \lambda \end{pmatrix} \qquad (2.62)$$

In the neighbourhood of such a point the invariant curves satisfy the generic equation pq = constant, hence are hyperbolic. When $\lambda_1 > 0$ the fixed point is said to be of the ordinary hyperbolic type. In this case the successive crossings of the surface of section remain on one invariant branch. When $\lambda_1 < 0$ the fixed point is said to be a hyperbolic point with reflection, since successive crossings of the surface of section jump back and forth between opposite branches. In either case trajectories in the vicinity of a hyperbolic fixed point are unstable in the following sense: any point on a trajectory that passes near to, but not coincident with a hyperbolic fixed point, will map far away from the fixed point (Fig. 2.7).

Finally, in the special case when $\lambda_1 = \lambda_2 = \pm 1$, the fixed point is said to be parabolic. When $\lambda_1 = \lambda_2 = +1$, T can be reduced to the form

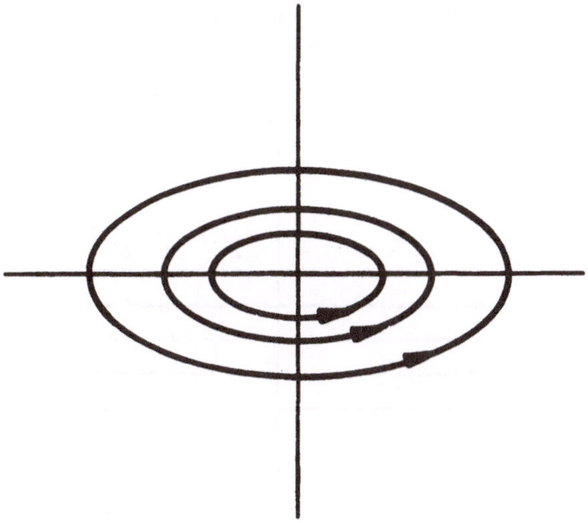

Fig. 2.6. Schematic representation of trajectories near an elliptic fixed point.

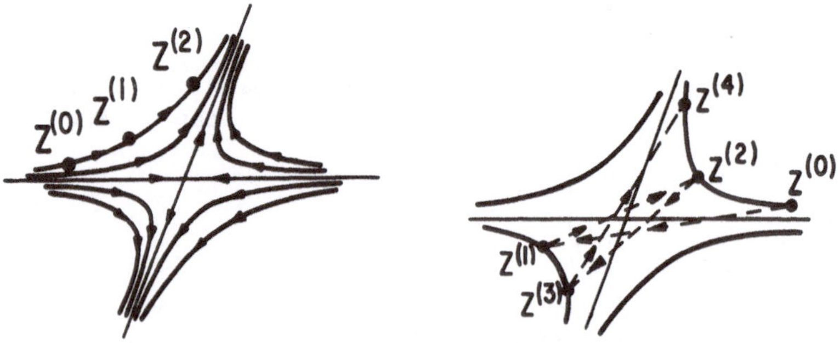

Fig. 2.7. a. Schematic representation of trajectories near an ordinary hyperbolic fixed point.
b. Schematic representation of trajectories near a hyperbolic fixed point with reflection.

$$\left(T = \begin{matrix} 1 & 0 \\ C & 1 \end{matrix} \right) \tag{2.63}$$

with C a constant. The invariant curves are, therefore, straight lines (Fig. 2.8). It is possible for there to be a line of parabolic fixed points. For example, the twist mapping (2.56) generates such a line whenever the curve to which it is applied has a rational rotation number. When the mapping is then changed to that shown in (2.57), what were parabolic fixed points of the mapping (2.56) become a finite set of alternating hyperbolic and elliptic fixed points, since the eigenvalues of the mapping (2.57) are not equal to unity under the addition of the terms in εf and εg. When ε is

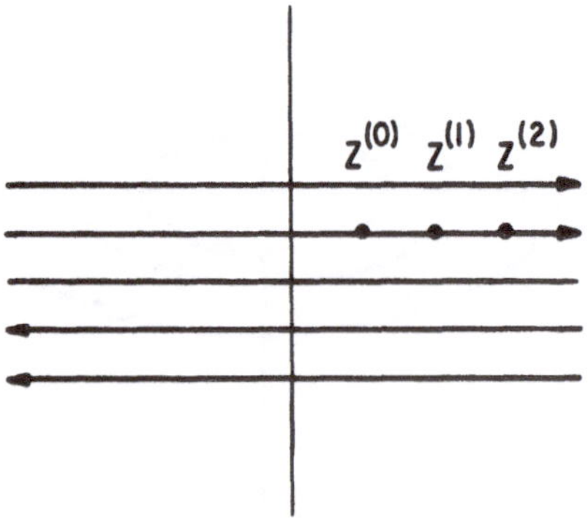

Fig. 2.8. Schematic representation of trajectories near a parabolic fixed point.

small these eigenvalues are close to unity, so the hyperbolic fixed points are of the ordinary (not the reflecting) type. Therefore, of the 2 ks fixed points of T^s that remain after the breakup of the curve with rational rotation number r/s, ks are elliptic and ks are hyperbolic, the two types forming an alternating sequence.

Application of both the Poincaré-Birkhoff and the KAM theorems to the behaviour of trajectories in the vicinity of an elliptic fixed point establishes the following: There are closed invariant curves corresponding to motion on tori with irrational frequency ratio and, where there used to be invariant curves of the pure twist mapping (2.56) corresponding to rational rotation number, the use of the twist mapping (2.57) with $\varepsilon \neq 0$ generates a new structure of fixed points, half of which are elliptic. In the neighbourhoods of these new elliptic fixed points there are invariant curves corresponding to irrational rotation number and, where there used to be invariant curves of the pure twist mapping (2.56) corresponding to rational rotation number, ... and so on ad infinitum.

The structure near hyperbolic fixed points is different. At any hyperbolic fixed point P_H four curves meet, two ingoing, (H_+) and two outgoing (H_-) (Fig. 2.9). If

$$\lim_{s\to\infty} T^s Z \to P_H, \tag{2.64}$$

then Z is on H_+, and if

$$\lim_{s\to\infty} T^{-s} Z \to P_H, \tag{2.65}$$

then Z is on H_-. Clearly, points on H_+ approach P_H indefinitely slowly, and points on H_- recede from P_H indefinitely slowly, as $s \to \pm\infty$ respectively. Consider following H_+ and H_- away from P_H. For an integrable system the arcs join smoothly, since any point on this invariant curve can be thought of as having started at P_H, mapped along H_- and then along H_+ back to P_H after a double infinity of iterations of the mapping. When P_H is a fixed point of T^s, P_H

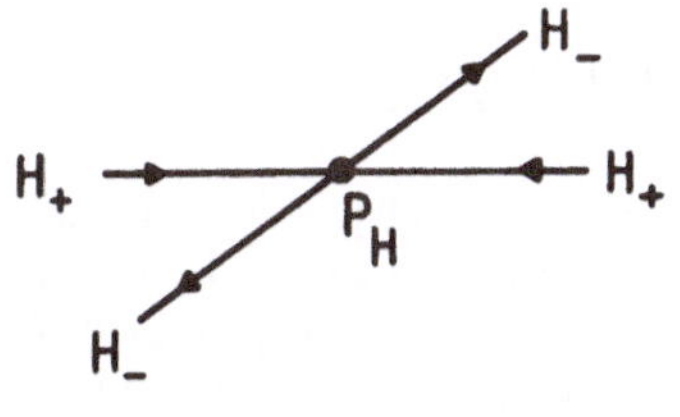

Fig. 2.9. Schematic representation of invariant curves meeting at a hyperbolic fixed point.

is one member of a set of s hyperbolic fixed points corresponding to an unstable closed orbit and the outgoing curve H_- from P_H joins smoothly onto the ingoing curve H_+ belonging to a neighbouring hyperbolic fixed point. However, this kind of smooth joining is the exceptional behaviour. In general, instead, the arcs H_+ and H_- intersect at what is called a homoclinic point (if the arcs belong to the same hyperbolic fixed point or to different fixed points of the same unstable closed orbit) or a heteroclinic fixed point (if the arcs belong to two hyperbolic fixed points not associated with the same closed orbit).

The behaviour of the curves H_+ and H_- beyond a homoclinic point is remarkably complicated. Consider one such point Z, and the effect of T^s on trajectories in the neighbourhood of Z. Because T is a continuous mapping all the successively generated neighbourhoods must be similar near to Z. Therefore the existence of one homoclinic point implies the existence of an infinity ($s \to \infty$) of others. The consequence is that H_+ forms a series of loops intersecting H_-, and vice versa. Indeed, since T is area preserving, every point on H_- between two intersections with H_+ is itself a further intersection. Thus, each point on an "early loop" of, say, H_- maps onto "late loops" that are more and more complicated and wander over more and more of the surface of section, so the point Z will eventually map arbitrarily close to any other point in the surface of section. In this region of phase space neither smooth invariant curves nor tori exist. To sum up, near each rational invariant curve there are hyperbolic fixed points with associated apparently chaotically wandering curves, and elliptic fixed points surrounded with invariant curves which repeat the entire structure ad infinitum (Fig. 2.10). There is, then, an intimate interlacing of regions of integrable and nonintegrable motions. The latter are so chaotic that the motion can be considered stochastic.

We are, finally, ready to examine the results of numerical solution of the equations of motion of a system of coupled nonlinear oscillators. Several such systems have been studied; we choose two to illustrate the range of dynamical behaviour possible.

Consider, first, a system described by the Hamiltonian

$$H(q,p) = \frac{1}{2}(p_1^2 + q_1^2) + \frac{1}{2}(p_2^2 + q_2^2) + q_1^2 q_2 - \frac{1}{3}q_2^3, \qquad (2.66)$$

introduced by Henon and Heiles [9]. This Hamiltonian represents two oscillators coupled by an interaction that is cubic in the amplitude; it describes bounded motion for $E < 1/6$, and unbounded motion for $E > 1/6$. The trajectories corresponding to (2.66) have been determined by Ford [4], with the results shown in Figs. 2.11-2.13. In these figures, which represent the surface of section

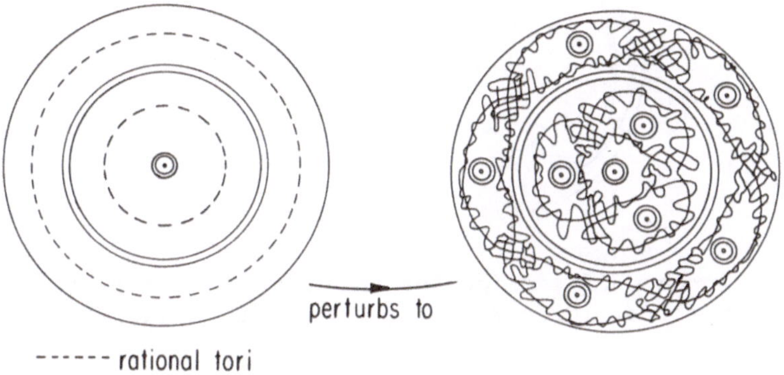

Fig. 2.10. Schematic representation of the change under perturbation of trajectories associated with rational invariant tori. (After M. Berry, in Topics in Nonlinear Dynamics, A.I.P. Conference Proc. 46, N.Y. (1978), p.16.)

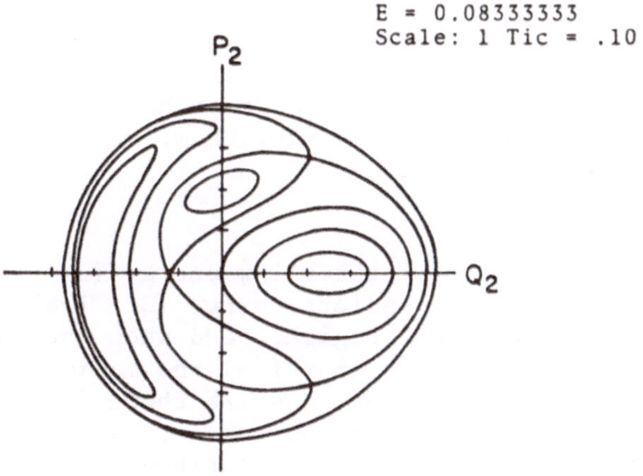

Fig. 2.11. Surface of section for the Henon-Heiles system with E = 1/12. (From J. Ford, Adv. Chem. Phys. 24, 155 (1973).)

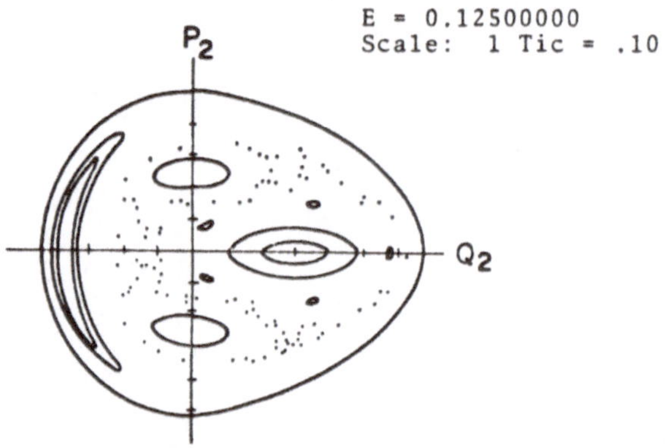

Fig. 2.12. Surface of section for the Henon-Heiles system with E = 1/8. (From J. Ford, Adv. Chem. Phys. 24, 155 (1973).)

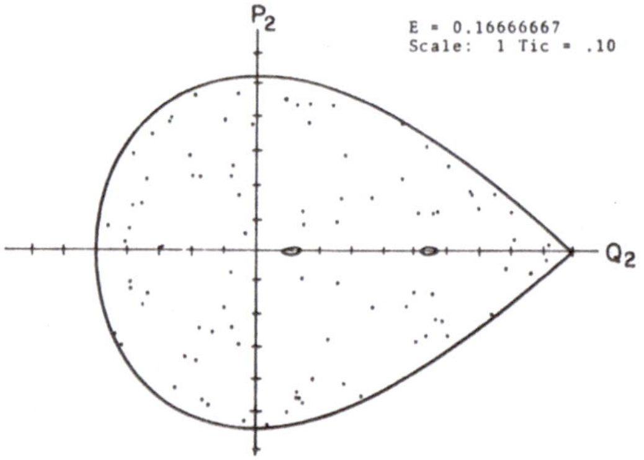

Fig. 2.13. Surface of section for the Henon-Heiles system with E = 1/6. (From J. Ford, Adv. Chem. Phys. 24, 155 (1973).)

for the system, it is clear that for low energy (E = 1/12) the motion is periodic, for intermediate energy (E = 1/8) it is mostly periodic with some nonperiodic regions, and when the energy is close to the dissociation limit (E = 1/6) the motion is apparently nonperiodic. Therefore, for, say, E < 1/8 there is an isolating integral of the motion other than the energy, but not for all E < 1/6, which is just the behaviour described by the KAM theorem.

It is worthwhile examining the calculations portrayed in Figs. 2.11-2.13 in further detail. Consider again the surface of section representation shown in Fig. 2.11. In this case E = 1/12.

Every trajectory yields a smooth locus of intersections with the surface of section, indicating that the system is deep within the region of KAM stability. At the centre of each region of ovals on the q_2 or p_2 axes is an elliptic fixed point of the mapping generated by a stable periodic orbit, and each oval is an invariant curve generated by those quasiperiodic orbits of H lying on smooth KAM surfaces. The self-intersection points of the self-interacting separatrix curve are hyperbolic fixed points of the mapping generated by unstable periodic orbits. Orbits started elsewhere on this separatrix curve generate mapping iterates which asymptotically approach one or another of the unstable fixed points. Finally, all the allowed trajectories for H at E = 1/12 intersect the q_2 p_2 plane at or within the outermost oval shown in the figure; points outside this bounding curve yield unphysical negative values for p_1^2.

When the system energy is increased to E = 1/8 large areas of the allowed q_2 p_2 plane are still covered with invariant curves generated by trajectories lying on smooth KAM surfaces, but the remaining area is almost uniformly covered by the intersection points of each (and almost every) orbit initiated in this area (Fig. 2.12). Note that the random looking splatter points is generated by a single trajectory; almost any other orbit initiated in the interior of this region would generate a similar pattern of points. It is found that in this region of disintegrated invariant curves initially close points in the plane map apart exponentially (Fig. 2.14). The computer evidence indicates that a dense set of unstable fixed points exists throughout such regions.

Finally, for the surface of section representation shown in Fig. 2.13 the energy is E = 1/6. The mapping now reveals a more or less uniform distribution of intersections over the allowed region of the q_2 p_2 plane, i.e., the entire allowed region consists of disintegrated invariant curves.

We conclude that even in a mechanical system as simple as that described by the Henon-Heiles Hamiltonian (2.66) there is a smooth transition for nonstatistical to stochastic behaviour of the system's trajectories (in this case as E is increased). To connect this behaviour with our discussion of the properties of area preserving mappings, consider again Fig. 2.12, especially the region containing ovals surrounding the central fixed point on the positive q_2 axis. Sequential mapping iterates of an initial point (q_2,p_2) lying on an oval rotate about the central fixed point and the average angle of rotation (in radians) is the rotation number α of the invariant oval curve. The value of α associated with each invariant curve varies smoothly as one progresses out from the central fixed point. The KAM theorem leads to the conclusion that invariant curves surround the central fixed point provided, among other things, the associated rotation numbers are irrationals satisfying the inequality

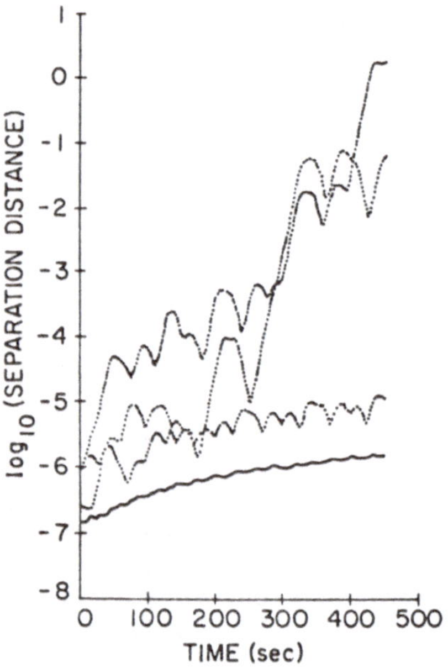

Fig. 2.14. The separation distances versus time for two initially close trajectories of the Henon-Heiles system. E = 1/8. The lower curves are for trajectory pairs in the quasi-periodic region, the upper curves for trajectory pairs in the stochastic region. (From J. Ford, Adv. Chem. Phys. 24, 155 (1973).)

$$|\omega - \frac{\ell}{k}| > \frac{\varepsilon}{k^{5/2}} \tag{2.67}$$

for all integers ℓ, k, where ε is a constant independent of k. Between these nondense invariant curves, where one might expect to find a dense set of invariant curves with rational rotation numbers α, the Poincaré-Birkhoff theorem states that in general only remnants of such α rational invariant curves remain, in the form of interleaved fixed points, half being elliptic and half being hyperbolic Thus, an extremely accurate numerical calculation of the trajectories in a neighbourhood of the central elliptic fixed point on the positive q_2 axis of Fig. 2.12 should reveal a pattern like that in Fig. 2.10. An indication of this structure for the Henon-Heiles system at E = 1/8 appears in Fig. 2.15, which shows only a few of the hundreds of calculated fixed points found at this energy.

In summary, it is the ovals surrounding the central fixed point that correspond to the invariant curves generated by quasi-periodic orbits on KAM surfaces. Between these invariant curves

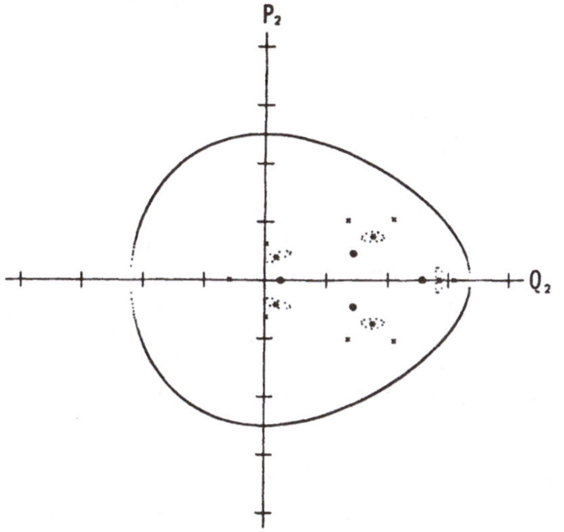

Fig. 2.15. Selected fixed points of the mapping generated by the Henon-Heiles Hamiltonian at energy E = 1/8. The innermost set of four fixed points located at the dots inside small circles are elliptic members of the T^4 family. The fixed points located at the asterisks are elliptic members of the T^5 family. Surrounding each member of this T^5 family is a set of eleven hyperbolic fixed points belonging to T^{55}. Finally, the eight fixed points located at the x-symbols are hyperbolic fixed points belonging to T^8. With the exception of the T^8 family each family of fixed points shown represents half of an alternating elliptic-hyperbolic set. However, all 16 members of the two T^8 families, which lie in the stochastic region are hyperbolic. (From J. Ford, Adv. Chem. Phys. 24, 155 (1973).)

on the equivalent KAM surfaces there exists the structure shown in Fig. 2.10, and more! The fixed point families lying between the invariant curves form a dense set; some of these families are alternating hyperbolic-elliptic, but others are alternating hyperbolic-hyperbolic. The separatrices emanating from each hyperbolic point no longer smoothly connect adjacent hyperbolic fixed points as the separatrix in Fig. 2.11 appears to do. Not only do separatrices from the same family intersect each other but they also intersect the separatrices belonging to nearby hyperbolic points. In regions containing these many intersecting separatrices, two close initial points locally map exponentially apart. This exponential behaviour is especially marked in those very narrow annular regions containing only hyperbolic fixed points. Thus the phase space for the Henon-Heiles system is always pathologically divided into sets of stochastic and nonstochastic trajectories. At low energies the motion is almost all nonstatistical; as the energy increases the originally microscopic regions containing stochastic trajectories also increase in size, eventually covering most of the allowed phase space. The transition is analytically smooth, although rather abrupt (at about $E = 0.11$) on the scale of the dissociation energy for the Henon-Heiles Hamiltonian.

We can also make a connection with the argument that multiple overlap of nonlinear resonances is the source of such behaviour in oscillator systems. For the Henon-Heiles system the motion for a specified trajectory involves two frequencies, ω_1 and ω_2, which depend on initial conditions (the system is anharmonic). Since the unperturbed frequencies are equal we should expect resonant energy exchange between the oscillators in those initial condition regions for which $\omega_1 \approx \omega_2$. Thus, at $\omega_1 = \omega_2$ one should find a periodic orbit surrounded by a region of resonant energy exchange. This is what Fig. 2.11 shows about the central "$\omega_1 = \omega_2$" fixed points. Further, in this nonlinear system there are also resonances of the type $n\,\omega_1 = m\,\omega_2$ for suitable initial conditions. Again we expect to find a periodic orbit for initial conditions such that $n\,\omega_1 = m\,\omega_2$ surrounded by (perhaps) small regions wherein there is energy exchange. This is, essentially, the picture partially verified in Fig. 2.15. As usual, for small E it is expected that all resonances except $\omega_1 = \omega_2$ have (almost) undetectable widths; as E increases the width of each resonance also should increase, thereby destroying the intervening smooth KAM surfaces and, eventually, "overlapping" each other. In such overlap regions, containing many resonances, the final system state is extremely sensitive to the initial state because of the numerous intervening trajectory "collisions" with unstable (and perhaps stable) periodic orbits.

As already mentioned, several other systems of coupled nonlinear oscillators display behaviour similar to that of the Henon-Heiles system. It is very important to note, however, that not all such systems behave like the Henon-Heiles system. Consider the Toda Hamiltonian [17]

$$H(q,p) = \tfrac{1}{2}(p_1^2 + p_2^2) + \frac{1}{24}\left[\exp(2q_2 + 2\sqrt{3}q_1) + \exp(2q_2 - 2\sqrt{3}q_1) + \exp(-4q_2)\right] - \frac{1}{8}. \tag{2.68}$$

It is easily shown that expansion of (2.68) to third order in q_1 and q_2 generates the Henon-Heiles Hamiltonian (2.66). The trajectories corresponding to (2.68) were studied by Ford [18]. The results, presented as surfaces of section, are shown in Fig. 2.16. Given the exponential nonlinearity of H, the most plausible guess as to the motion under H is that the only isolating integral is

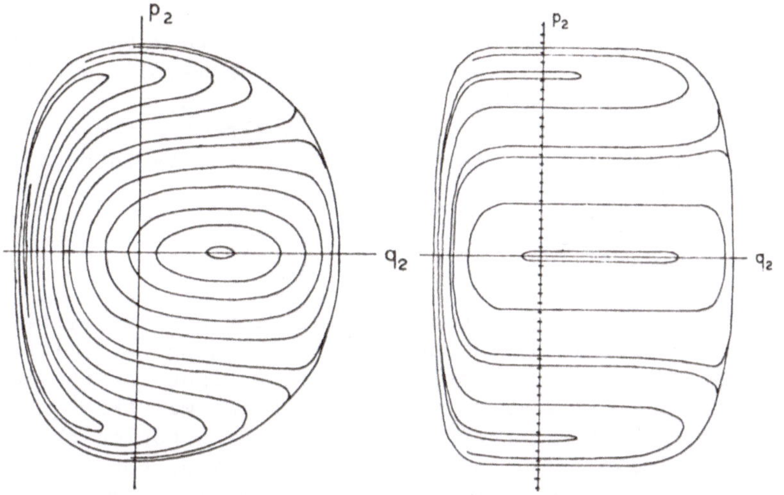

Fig. 2.16. Surface of section for the Toda system. On the left E = 1, on the right E = 256. (From J. Ford, Fundamental Problems in Statistical Mechanics, Ed. E.D.G. Cohen, North Holland, Amsterdam, Vol. 3 (1975), p.215.)

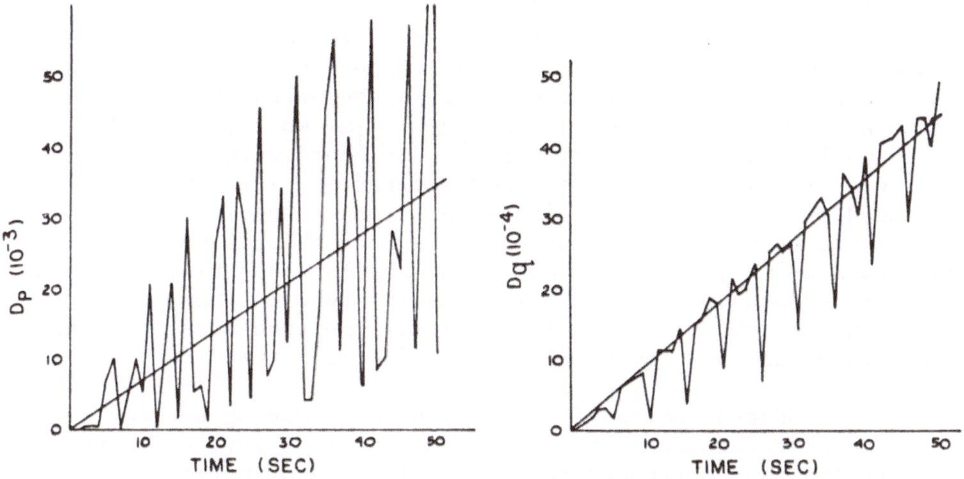

Fig. 2.17. The separation versus time in momentum and position space for two initially close trajectories in the Toda system. (From J. Ford, Fundamental Problems in Statistical Mechanics, Ed. E.D.G. Cohen, North Holland, Amsterdam, Vol. 3 (1975), p.215.)

the energy. That expectation is wrong! The surfaces of section for all energies clearly show evidence of periodic behaviour and two trajectories started near one another separate only linearly in time (Fig. 2.17). Although this behaviour was discovered from numerical integration of the equations of motion, it was later shown by Henon that other isolating integrals of the motion do exist, and that (2.68) corresponds to a completely integrable case, despite the nonlinearity.

C. Perturbation Theory

Given that we wish to study the dynamics of some system of nonlinear oscillators, described by a non-integrable Hamiltonian, can perturbation theory be used to find invariants of the motion? A vast literature deals with the attempt to answer this question [19]. The straightforward application of perturbation theory to the solution of the classical equation of motion, that is expansion of the solution as a power series in some small parameter about a reference solution for a separable system, followed by successive solution of the equations for the first, second, ..., nth order corrections, runs into a serious difficulty. In the nonlinear oscillator system the frequencies depend upon the amplitude of motion or, put another way, the nonlinearities alter the frequencies of the system. The existence of this frequency shift is not accounted for in the individual terms in the simplest form of perturbation theory, with the consequence that secular terms arise in the perturbation expansion. A secular term is one that grows linearly with time. Clearly, if any such term appears in the perturbation solution that solution can be valid only for a very limited time. The perturbation theory approach also suffers from the difficulty that successive terms in the perturbation series contain denominators of the form $m_i\omega_i - m_j\omega_j$ which, for ω_i/ω_j irrational, become arbitrarily small for those values of $m_j/m_i = r/s$ that closely approximate the frequency ratio. The nature of this second difficulty has been discussed in the preceding sections. Systematic methods of dealing with the secular problem have been developed by many investigators; an excellent description of the various techniques, and the relationships between them, can be found in the text by Nayfeh [19]. I will mention here only the ideas involved in a few of these techniques.

The Lindstedt-Poincaré method [20] of removing secular terms from the perturbation expansion of the solution involves development of both the system frequencies and amplitudes of motion in powers of the strength of the perturbation which measures the difference between the equation of motion of interest and that of an integrable system. For example, in the case of a single nonlinear oscillator the perturbation parameter is the coefficient of the nonlinear term in the equation of motion. Given these expansions of the frequency and amplitude, to each order in the perturbation parameter the coefficients are chosen so as to remove the secular terms in the perturbation expansion of the amplitudes.

The method of multiple time scales, developed by Storock [21], Frieman [22], Nayfeh [23], and Sandri [24], is based on the observation that the perturbation expansion of the solution to the equation of motion implies that the system can be thought of as having many different dynamical time scales, which are of the order εt, $\varepsilon^2 t$, $\varepsilon^3 t$, ... This follows from the way in which the time domain for which a solution is valid is systematically extended in the order-by-order removal of secular terms via the Lindstedt-Poincaré method. Given that the solution of the equation of motion depends explicitly on t, εt, $\varepsilon^2 t$, ..., as well as ε itself, the time derivative in the equation of motion is transformed according to

$$\frac{d}{dt} = \frac{\partial}{\partial t} + \varepsilon\frac{\partial}{\partial(\varepsilon t)} + \varepsilon^2 \frac{\partial}{\partial(\varepsilon^2 t)} + \ldots , \tag{2.69}$$

each term representing a successively slower variation of the system. Substitution of this form of the time derivative, and the expansion of the solution

$$x(t;\varepsilon) = \sum_{m=0}^{M-1} \varepsilon^m x_m(t, \varepsilon t, \ldots, \varepsilon^m t) + \theta(\varepsilon^M t), \tag{2.70}$$

into the equation of motion, followed by equating coefficients of like powers of ε, leads to a set of equations for the x_m. The solutions to the equations for the x_m contain arbitrary functions of the m time scales $\varepsilon^m t$; the condition $(x_m/x_{m-1}) < \infty$ for all t, εt, ..., $\varepsilon^m t$ provides the closure needed to determine the x_m, and it is equivalent to requiring the elimination of the secular terms in the perturbation expansion of the solution to the equation of motion. A discussion of the relationship between these two methods and the Liouville equation description of N-body mechanics can be found in a paper by Wilson and Rice [24a].

Leaving aside the difficulties associated with the appearance of secular terms and small denominators, there are two principle categories into which the many proposed perturbation techniques fall. In the so-called method of averaging, developed by Van der Pol [25], Krylov, Bogolubov and Mitropolski [26], and Kruskal [27], the time variation of the frequencies of the oscillator system is taken to be small relative to the frequencies themselves. When this is the case the drift of the frequencies with time can be separated from the oscillatory motion by transformation to a new set of variables. This transformation consists of a systematic expansion of the variables in powers of a parameter ε that describes the deviation of the oscillator system frequencies from the constant values they have in the limit $\varepsilon = 0$. To lowest order in ε the required equations involve an average over the periods of the system frequencies. If the perturbation procedure is carried out as far as terms of order ε^n the solution to the original equation which is generated is correct to order ε^{n+1}, and that solution is valid for a time of the order of ε^{-1}. Thus, this method yields a solution that is asymptotically correct as $\varepsilon \to 0$.

In another method, associated with the contributions of Birkhoff [16], von Zeipel [28], Hori [29], Deprit [30], and Kamel [31], the Hamiltonian of the system is written in the form

$$H = H_2 + \varepsilon H_3 + \varepsilon^2 H_4 + \dots , \tag{2.71}$$

where H_2 is purely quadratic in the phase space variables, H_3 is cubic, etc. The procedure used to analyse the motion of the system involves the construction of near identity canonical transformations which reduce (2.71) to a simpler form, for which an invariant can be found. This canonical transformation is carried out step-by-step, as in the other systematic perturbation theory methods mentioned above.

My purpose in introducing these ideas from perturbation theory is derived from the following questions:

(i) Given that an N-oscillator system has quasiperiodic behaviour for $E < E_c$ and stochastic behaviour for $E > E_c$, are there isolated quasi-periodic solutions embedded in the stochastic domain?

(ii) If there are not solutions which are asymptotically quasi-periodic, are there solutions which are approximately quasi-periodic for some period of time?

(iii) If either of the above situations prevails can the corresponding solutions be obtained by the methods of perturbation theory?

These questions are of more than academic interest when one attempts to understand intramolecular vibrational relaxation in isolated molecules.

The results of numerical integration of model equations of motion suggest, but do not establish, that even at the dissociation limit there remain very small regions of quasi-periodic motion embedded in the stochastic domain (see Fig. 2.3). A more satisfying, but not quite complete answer to question (i) is contained in the work of Birkhoff [16]; he proved the existence of an infinite number of periodic orbits near every homoclinic point. Danby showed [32], by numerical calculation for a model system, that all these orbits originate from stable regular families. That is, a stable periodic orbit (represented by invariant points around a central invariant point) becomes unstable as the perturbation to the integrable system grows, yet at the same time there is generated another stable family, and so on. It is possible that the range of the perturbation over which the new family is stable is smaller than the original, and so on, so that the total range of the perturbation over which all the nested descendants of the original stable family is stable, is finite.

The trajectories just described are regular periodic orbits in the sense that they can be generated from the orbits of the unperturbed system by varying continuously their shape as a function of the perturbation. There is, in addition, another class of trajectories, called irregular orbits; these are independent of the regular orbits and they appear only for a value of the perturbation greater than some threshold. The irregular periodic orbits cannot be continuously transformed into the orbits of the unperturbed system. A pair of irregular families starts at the threshold value of the perturbation, ε_{min}, one of them stable and the other unstable. At some larger value of ε the stable family generates, in turn a new set of stable and unstable families, and so on. It is possible, as for the regular families, that above some value of ε all nested families of orbits are unstable. Finally, the irregular periodic orbits seem to play a role in promoting ergodicity in that they can join regions of phase space that are associated with different regular periodic orbits.

The following two examples illustrate the character of the quasiperiodic and the periodic solutions embedded in the stochastic domain.

Contoupolos [33] has made an extended study of stable orbits for the system with the Hamiltonian (m = 1)

$$H = \frac{1}{2}(p_1^2 + p_2^2) + \frac{1}{2}(Aq_1^2 + Bq_2^2) - \varepsilon q_1 q_2^2. \tag{2.72}$$

This system has also been studied by Barbanis [34] and, like the Henon-Heiles system, has a low energy quasiperiodic domain and a higher energy stochastic domain. The Hamiltonian permits dissociation at the energy

$$E_D = \frac{A^3}{8\varepsilon^2} \tag{2.73}$$

for the case A = B. The onset of stochastic behaviour is a relatively insensitive function of ε. It is found that $E_c/E_D \approx 0.76$ for $0.05 < \varepsilon < 0.20$ when A = 0.1. Contoupolos studied the stable orbits of this system as a function of ε. He found that stable orbits existed deep inside the region where stochastic motion is the dominant behaviour. In somewhat more detail, he concluded that:

"(a) The main families of periodic orbits for very small perturbations ε are only: (1) the $\bar{x} \equiv A^{\frac{1}{2}}$ x axis and (2) the "central" periodic orbit near the $\bar{y} \equiv B^{\frac{1}{2}}$ y axis, intersecting the $\bar{x}$ axis perpendicularly at one point, if the ratio $A^{\frac{1}{2}}/B^{\frac{1}{2}}$ = irrational. If $A^{\frac{1}{2}}/B^{\frac{1}{2}}$ is a rational n/m then we also have two resonant periodic orbits (one stable and one unstable) making n and m oscillations along the $\bar{x}$ and $\bar{y}$ axes, respectively. If $A^{\frac{1}{2}}/B^{\frac{1}{2}}$ is "near" a rational n/m then resonant periodic orbits appear as the perturbation increases beyond a minimum value $\varepsilon > 0$.

(b) As ε increases more and more, families intersecting the $\bar{x}$ axis less than a given number of times appear. If ε is not very large most of these families are regular, i.e. they originate at the stable main families above. Families originating at the "central" family or the stable resonant family of periodic orbits have their minimum (in other problems their maximum) ε at the point of intersection; i.e., the tangent of the characterisitic of the new family is, in most cases, perpendicular at the point of intersection.

(c) Every regular periodic orbit makes n and m oscillations along the $\bar{x}$ and $\bar{y}$ axes, respectively, and has a rotation number n'/m=2-n/m [more generally n'/m=-n/n (modulo 1)].

(d) From every stable family intersecting the $\bar{x}$ axis m times, new families with the same rotation number are formed, intersecting the $\bar{x}$ axis 2m or 3m, etc., times. These are the families of second "genre" according to the terminology of Poincaré.

(e) Besides the regular families of periodic orbits we have many irregular families not connected with the main families above. Such families are in general, but not always, unstable. In general such irregular families are composed of two branches in quite different parts of the diagram $(\varepsilon, \bar{x})$. In many cases, one cannot correspond unambiguously a rotation number to them, but even if we can assign them a rotation number, it is not adjacent to the rotation number of nearby regular families.

(f) Irregular families of orbits intersecting the $\bar{x}$ axis less than a given number m of times appear only for relatively large ε, larger than the perturbation at which the corresponding regular families (with the same m) appear. The appearance of irregular families seems to be connected with the "dissolution" of the invariant curves of nonperiodic orbits.

(g) Beyond the escape perturbation, i.e., for $\varepsilon > \varepsilon_{esc}$ there are many escape regions, containing escaping orbits. The range of values of $\bar{x}$ for which we have escaping orbits increases as ε increases. However, there are left some nonescape regions containing periodic orbits. We may even have stable periodic orbits (and nearby tube orbits of positive measure) for ε much larger than the escape perturbation. Further, every orbit escaping after a very large time may be considered nonescaping for practical reasons. We conclude that if, in a given potential field, a moving point has energy greater than the escape energy (but not extremely large), it may not escape for very long times or for all times.

(h) Near the escape regions, a large number of new families of periodic orbits appear. Therefore the transition between orbits escaping too fast (after, say, 1 intersection with the $\bar{x}$ axis, except the initial point) and nonescaping periodic orbits is too abrupt. There are not many intermediate cases of orbits escaping after 2, 3, ... intersections, until we reach nonescaping orbits. However, these nonescaping orbits are highly unstable. Therefore small perturbations may make them escaping.

(i) No family of periodic orbits seems to end at an escape orbit or at the family $\bar{y} = 0$, or another family. All families seem to extend up to $\varepsilon = \infty$. This is the case whenever $A^{1/2}/B^{1/2} > 2/3^{1/2}$. If $A^{1/2}/B^{1/2} > 2/3^{1/2}$, families appearing for small ε originate at the family $\bar{y} = 0$, reaching their maximum ε at or near the "central" family. Even in this case, however, these are families extending to large values of ε, and presumably up to $\varepsilon = \infty$."

In the system of units used by Contoupolos, the escape perturbation is $\varepsilon_{esc} = 4.602$. Figures 2.18-2.21 show in increasing detail the nature of the orbits, and the intermingling of stable and unstable behaviour deep in the stochastic region. Figure 2.22 shows the behaviour of some orbits well above the escape perturbation.

Helleman and Bountis [35] have developed a rapidly convergent variational method for the construction of periodic solutions of the equation of motion, analytic in the time, and with arbitrary period. When applied to the Henon-Heiles system they obtain what appears to be a dense set of one parameter families of periodic solutions wherever the motion is bounded, including the domain where stochastic trajectories dominate the surface of section. The approach to finding periodic solutions is based on the following idea. Consider a system with two degrees of freedom. Instead of specifying an orbit in the usual manner, that is by the initial positions and momenta, one can, for closed orbits, specify the fundamental frequencies and initial phases. When a closed orbit describes the motion in a system with two degrees of freedom the fundamental frequencies are commensurable; the ratio of the frequencies, $m_1/m_2 \equiv \sigma$, is then a rational fraction and the Poincaré recurrence time of the system is just $T_r = \omega_r^{-1} = m_2\omega_1^{-1} = m_1\omega_2^{-1}$ where ω_r is the greatest common frequency of the system. Helleman and Bountis specify the periodic orbits by the value of T_r and the numbers $2m_1$ and $2m_2$ of zeroes of the velocities within one recurrence period. For given m_1 and m_2 there is an infinite one parameter

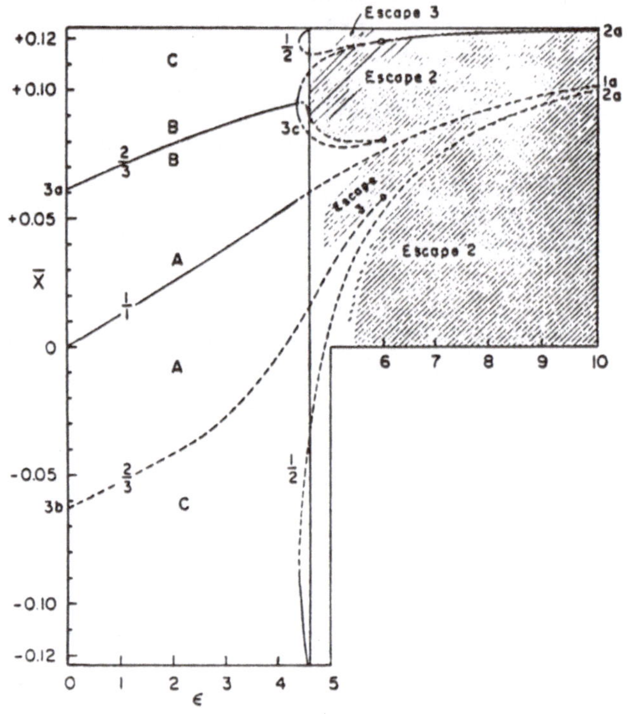

Fig. 2.18. Characteristic curves of the four main families of periodic orbits and the family 1/2: The family 1/1 gives the central periodic orbits intersecting the $\bar{x}$ axis only once perpendicularly. The boundaries $\bar{x} = \pm 0.12363$ represent the family $\bar{y} = 0$. The families 2/3 represent resonant periodic orbits with rotation number 2/3. Solid lines represent stable orbits, dashed lines unstable ones. Periodic orbits in the region A, B, C are generated from the families 1/1, the stable family 2/3, and $\bar{x} = 0$, respectively. The vertical straight line at $\varepsilon = 4.602$ represents the escape perturbation. The dashed regions represent orbits escaping before the 2nd or 3rd intersection with the $\bar{x}$ axis (besides the initial point). (From G. Contoupolos, Astron. J. 75, 96 (1970).)

family of frequency pairs; the parameter is ω_r, i.e., $\omega_1 = m_2\omega_r$ and $\omega_2 = m_1\omega_r$. The solution to the equation of motion is developed from Fourier series of the form

$$q_1(t) = \sum_n A_n e^{in\omega_r t}, \quad q_2(t) = \sum_n A_n' e^{in\omega_r t} \qquad (2.74)$$

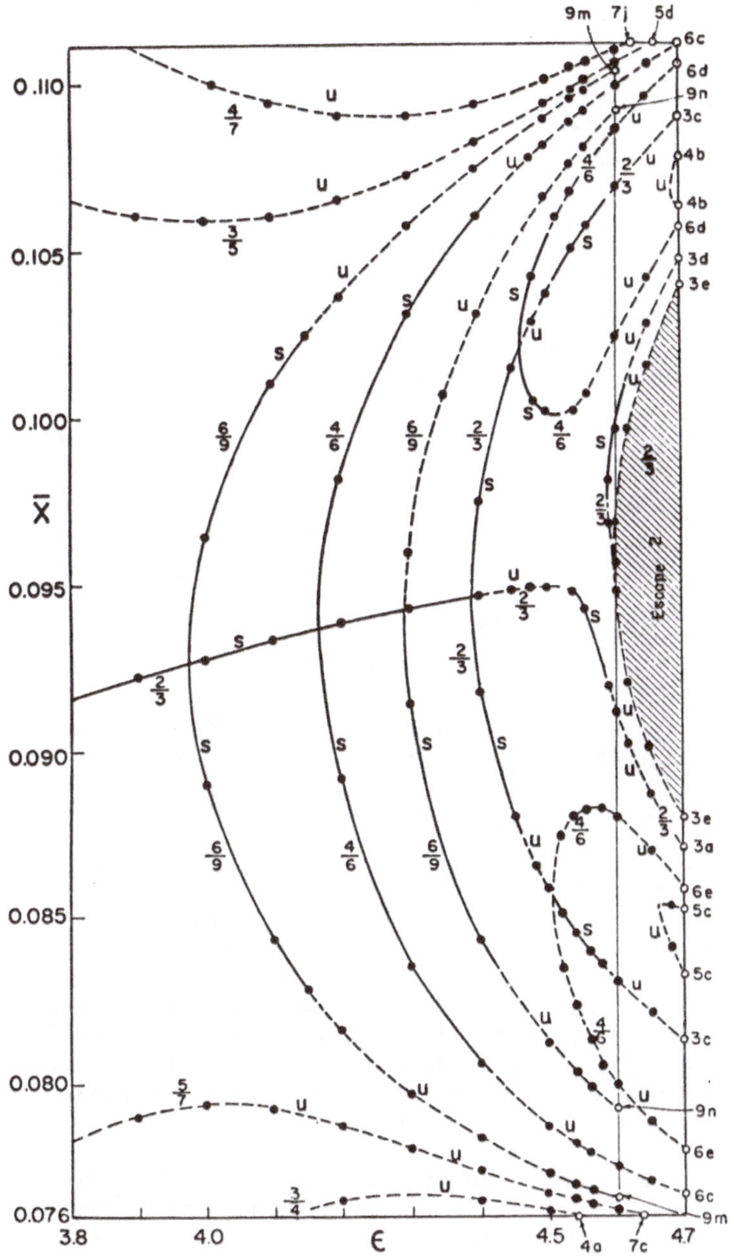

Fig. 2.19. Part of Fig. 2.18 in detail. The vertical line at ε = 4.602 represents the escape perturbation. (From G. Contoupolos, Astron. J. 75, 96 (1970).)

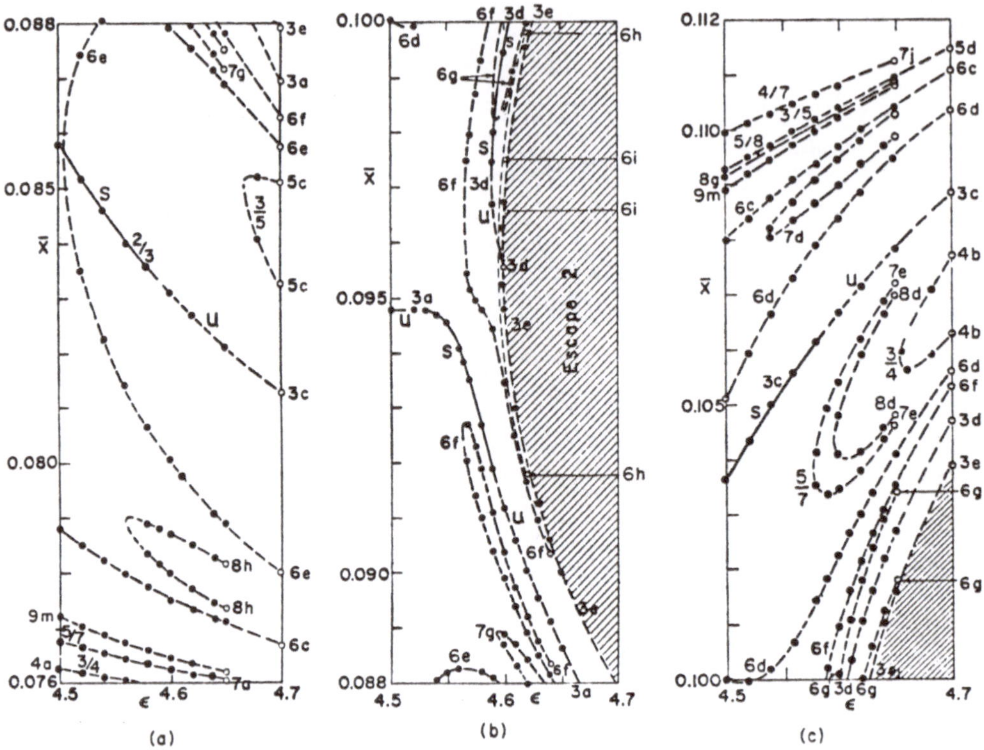

Fig. 2.20. Parts of Fig. 2.19 in detail. (From G. Contoupolos, Astron. J. 75, 96 (1970).)

Helleman and Bountis found the coefficients in (2.74) by a convergent iterative procedure. Of course, the dominant coefficients in (2.74) must be those associated with the fundamental solution $\omega_1 = m_2\omega_r$, $\omega_2 = m_1\omega_r$. From the solutions found one works backwards, for fixed m_2/m_1, to find the initial positions and momenta. Note that, by restricting attention to those families of periodic orbits parameterized by T_r with fixed $m_2/m_1 \equiv \sigma$, the Fourier series representation of the solution is free of problems associated with small divisors.

Figure 2.23 shows some of the results obtained by Helleman and Bountis for the Henon-Heiles system with the small amplitude frequencies $\omega_1 = \omega_2$. The figure shows the loci of periodic solutions for constant σ and varying T_r; along each curve T_r varies continuously, but T_r changes discontinuously as one goes from one value of σ to a neighbouring value. The sensitivity of T_r to changes in σ is extreme, the discontinuous jumps being of many orders of magnitude in T_r for miniscule changes in σ. Of course, the curves for

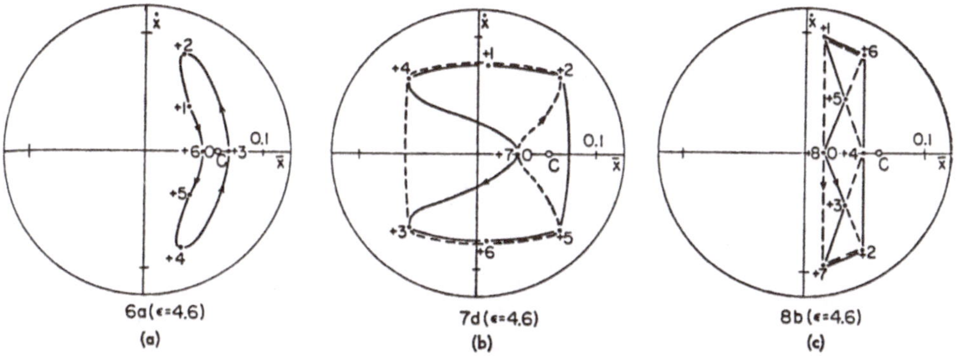

Fig. 2.21. The points of intersection of three periodic orbits of the families 6a, 7d, 8b by the surface of section $\bar{y} = 0$ (of the space $\bar{x}\dot{x}\bar{y}$) for $\bar{y} > 0$. The initial point is 0 and successive points are numbered. C represents the "central" periodic orbit. The outermost circle is $\bar{x}^2 + \dot{x}^2 = 2E$. The points of intersection may be joined by one or more smooth curves. (From G. Contoupolos, Astron. J. 75, 96 (1970).)

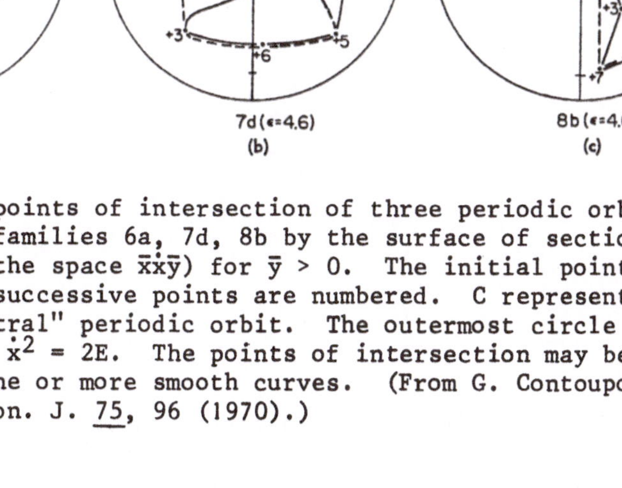

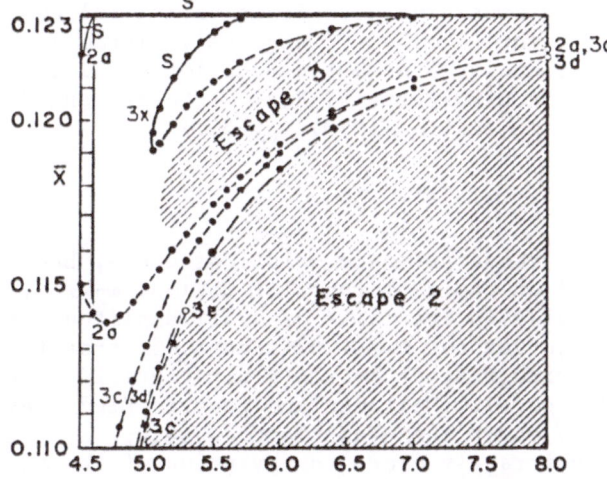

Fig. 2.22. Characteristic curves of the families 2a, 3c, 3d, 3e and 3x from ε = 4.5 up to ε = 8: The family $\bar{y} = 0$ is stable up to about ε = 7.0, while the family 3x contains a stable branch from its minimum ε (ε ≈ 5.0) up to the termination (starting) point (ε ≈ 5.8).

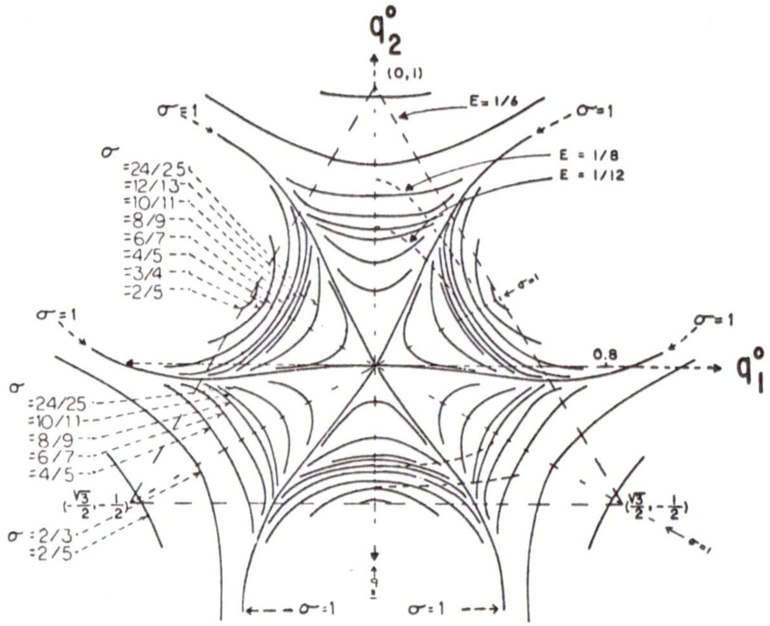

Fig. 2.23. Converged initial displacements q_1^o, q_2^o of the periodic solutions of the Henon-Heiles system (with $p_1(0) = 0 = p_2(0)$) at different values of $\sigma \equiv m_2/m_1$; only the "primary" solutions are used (i.e., m_2, m_1 relative prime). Note the confluence of σ-curves (near $\sigma = 1$) resulting in a 'sensitive dependence on initial conditions' there, yielding 'stochastic regions'. (From R. Helleman and T. Bountis, Stochastic Behavior in Classical and Quantum Hamiltonian Systems, Eds. G. Casati and J. Ford, Springer-Verlag, Berlin (1979), p.353.)

$\sigma \neq 1$ represent those periodic solutions where, because the frequencies depend on the amplitudes of motion, $\omega_2 \neq \omega_1$ even though the equality holds in the limit of small amplitude of motion. Note that the curves for $\sigma \neq 1$ bifurcate off that for $\sigma = 1$, and are symmetric with respect to the equipotentials of H (the dotted lines in the figure). Helleman and Bountis find that the curves for different σ cover the $q_1^o q_2^o$ plane densely; a σ curve can be constructed arbitrarily close to any point of the plane by σ-interpolation using only rational values of σ since the ordering of the nested σ-curves varies "continuously" with σ.

The preceding example addresses the question of the existence of quasiperiodic trajectories deep in the stochastic domain, but not whether these trajectories can be calculated by perturbation theory. Gustavson [36] has developed an algorithm for constructing formal integrals of a Hamiltonian system with N degrees of freedom, and applied this to the Henon-Heiles system. The method uses a succession of canonical transformations to reduce the Hamiltonian to normal form. For a given system, truncated power series then represent approximate integrals of the motion. Some results are shown in Figs. 2.24-2.26, which display the Poincaré surfaces of section for E = 1/12, 1/8 and 1/6, respectively. These figures should be compared with Figs. 2.11-2.13 in the same order. We note that at low energy, say E = 1/12, the perturbation method yields results in excellent agreement with direct numerical integration of the equation of motion. At higher energy, say 1/6, the perturbation method suggests the existence of quasiperiodic trajectories not found in the numerical integration of the equation of motion.

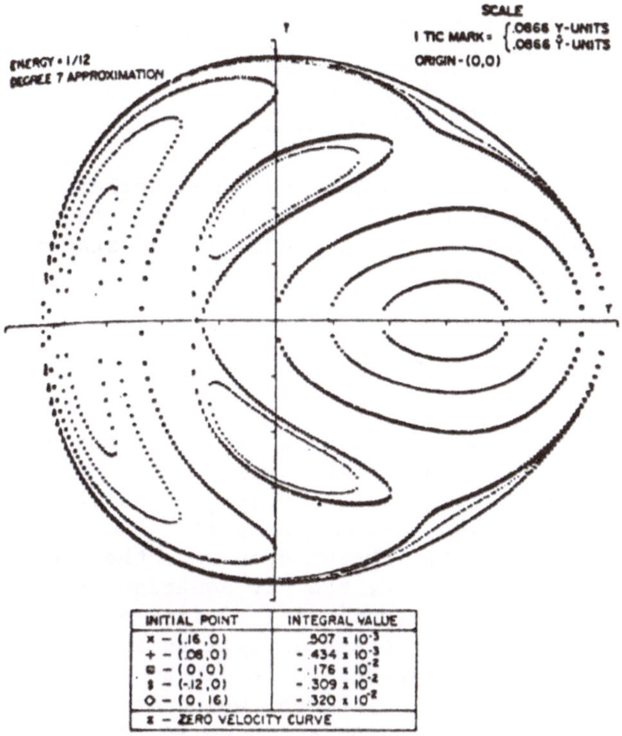

Fig. 2.24. Surface of section for the Henon-Heiles system calculated by successive canonical transformation to Birkhoff normal form. E = 1/12. (From F.G. Gustavson, Astron. J. 71, 670 (1966).)

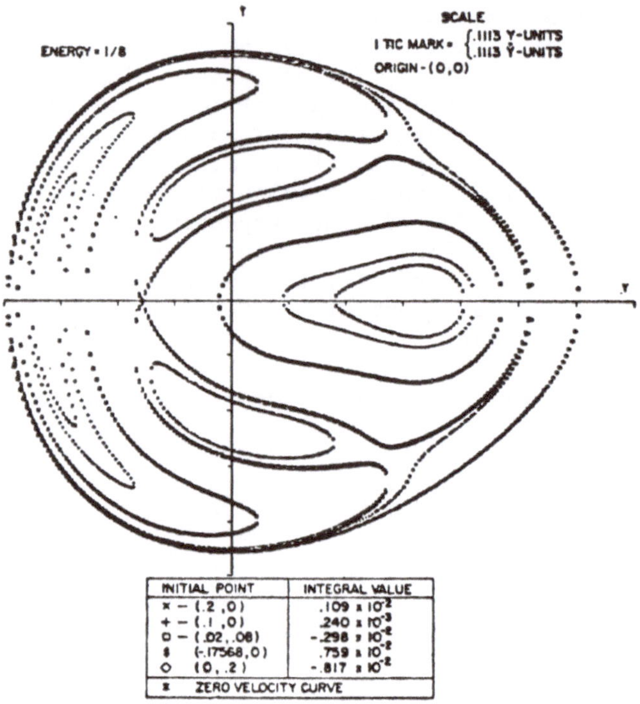

Fig. 2.25. Surface of section for the Henon-Heiles system calculated by successive canonical transformation to Birkhoff normal form. E = 1/8. (From F.G. Gustavson, Astron. J. 71, 670 (1966).)

What interpretation should be attached to the perturbation theory results in the high energy domain? The key point to make is that a truncated power series representation of a Hamiltonian, which is an approximation to some "complete" Hamiltonian, can correspond to an integrable system even when the complete Hamiltonian does not. For example, the KdV equation arises as an approximation to the hydrodynamic description of water waves in a shallow canal, and in other cases as well as an approximation to the exact dynamics of the process. It can be shown that the KdV equation has solitary wave solutions, which are invariant wave form motions in the system, and these are seen in real experiments, hence the approximation is a good representation of the phenomenon described. Similarly, it appears that the truncated Gustavson Hamiltonian, which is an approximation to the complete Henon-Heiles Hamiltonian, describes a system which has quasiperiodic motion where the Henon-Heiles system dynamics is dominantly stochastic.

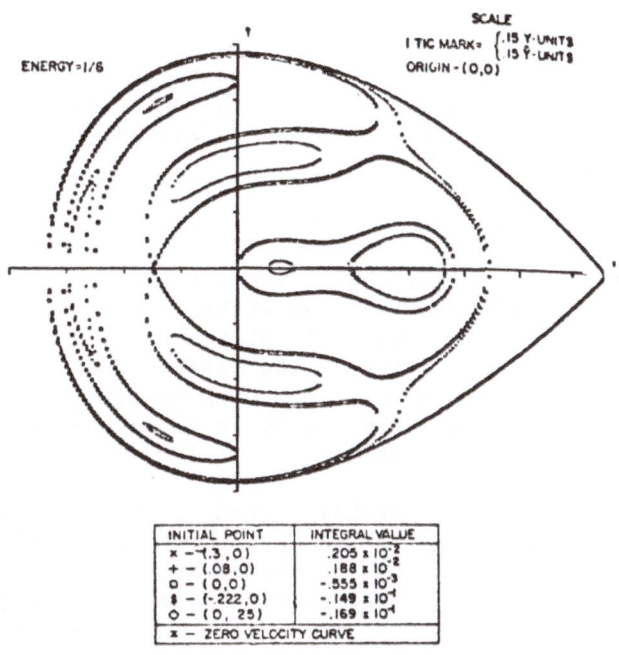

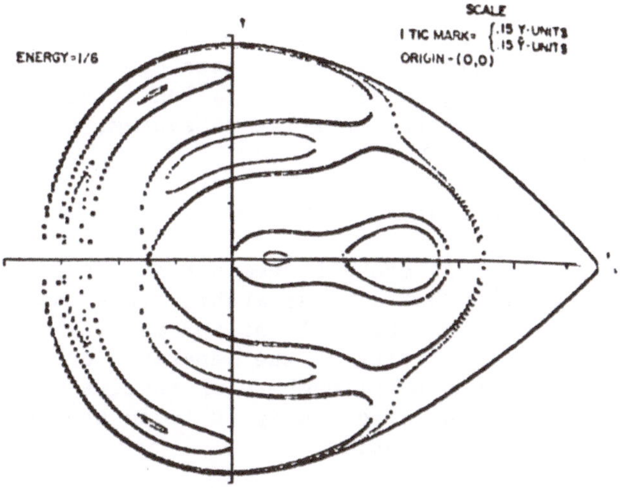

INITIAL POINT	INTEGRAL VALUE
× – (.3,0)	$.205 \times 10^{-2}$
+ – (.08,0)	$.188 \times 10^{-2}$
▫ – (0,0)	$-.555 \times 10^{-3}$
$ – (-.222,0)	$-.149 \times 10^{-1}$
○ – (0, 25)	$-.169 \times 10^{-1}$
x – ZERO VELOCITY CURVE	

Fig. 2.26. Surface of section for the Henon-Heiles system calculated by successive canonical transformation to Birkhoff normal form. E = 1/6. (From F.G. Gustavson, Astron. J. 71, 670 (1966).)

It is tempting to speculate that these quasiperiodic motions are long lived, and possibly, though not necessarily, asymptotically stable.

It is also worth noting that the view of the system dynamics given by the Poincaré surface of section is likely to overemphasize the stochastic nature of the dynamics, since very small perturbations of the trajectory on an invariant torus can lead to wildly erratic intersections with the surface of section even though the motion in a large part of the phase space is smooth and quasiperiodic when viewed overall as a trajectory on the torus. If so, the use of perturbation theory to calculate invariants of the motion is akin to analytic continuation of the invariant solutions. This point of view reappears when the quantum mechanics of motion in the stochastic region is analysed.

D. Summary

In this section I have discussed the classical mechanics of coupled nonlinear oscillator systems. Although some aspects of the analysis were quite general, the detailed illustrations given focussed attention on the two oscillators system, a point to which I will return below.

It has been shown that:

(i) A separable N-oscillators system has at least N isolating integrals of the motion, but a nonseparable system will in general have less than N isolating integrals of the motion. Of course, given that the dimensionalities of the separable and nonseparable systems are the same, the total numbers of integrals of the motion of each are the same; the discrepancy is accounted for by non-isolating integrals of the motion.

(ii) The motion of the separable system is quasiperiodic, the trajectory being restricted to lie on a torus in the phase space of the system. When the system is not separable some of the trajectories are quasiperiodic, but not all. These other trajectories are not restricted to lie on a torus, although they, of course, lie on the energy surface.

(iii) In the case that the system is separable the motion can be described simply in terms of action-angle variables and the corresponding Hamilton-Jacobi equation possesses a complete integral that generates the transformation of representation into angle-action coordinates. In contrast, the Hamilton-Jacobi equation for the nonseparable system does not have a complete integral which can generate a transformation to angle-action variables; in this case angle-action variables do not exist.

(iv) The KAM theorem establishes that most of the quasiperiodic trajectories of some unperturbed system survive under sufficiently small perturbation, but that there is also generated a set of stochastic trajectories. The relative weight of the stochastic trajectories grows as the energy increases, and they eventually fill all of the accessible phase space (the energy surface).

(v) The onset of stochastic behaviour appears to be relatively sharp, at an energy E_c, but in fact there are some stochastic trajectories even at low energy, and there are periodic and quasiperiodic trajectories embedded in the stochastic domain.

(vi) Even when periodic trajectories exist in the stochastic domain they do not reflect global invariants of the system because their properties, e.g. periods of the orbits, are extremely sensitive to very small changes in the initial conditions.

There is a difference between systems with two and systems with more than two degrees of freedom. In the former case stochastic trajectories can be isolated from one another in the system phase space by the intervention of a torus supporting a quasiperiodic motion. In the latter case, assuming the energy to be the only isolating integral of the motion, the quasiperiodic trajectories do not isolate the stochastic trajectories from one another. All of the regions of the phase space in which stochastic motion occurs are connected into a network which permeates the entire phase space (called the Arnold web) [8]. Elements of the Arnold web are arbitrarily close to every point of the phase space, hence if the initial conditions start the phase point on the Arnold web the motion of the point will be stochastic and will eventually reach every region of the energy surface. The process by which this takes place, namely along a stochastic trajectory inside the Arnold web, is called Arnold diffusion. It is generally believed that the Arnold web covers the energy surface whenever $N > 2$, but Contoupolos, Galgani and Giorgilli [37] have shown that in a three oscillator version of the Henon-Heiles system there are noncommunicating stochastic regions of the phase space. One of these stochastic regions is large, the others small. In addition there are regions of the phase space in which the motion is quasiperiodic. Using a perturbation theory construction of integrals of the motion, Contoupolos et al. conclude that there are two isolating integrals of the motion other than the energy in the regions of quasiperiodic trajectories, one extra isolating integral of the motion in the small regions with stochastic trajectories, and no extra isolating integrals of the motion in the large region with stochastic trajectories.

Clearly, there remains an enormous amount to be learned about the classical mechanics of N-oscillator systems.

3. QUANTUM MECHANICAL DESCRIPTION OF COUPLED NONLINEAR OSCILLATOR SYSTEMS

The results of studies of the classical mechanics of coupled oscillator systems leads, obviously, to three questions about the quantum mechanical description of the same systems. These are:

(i) Is there a change in the description of the stationary states of a system of coupled oscillators at some energy E_c analogous to the change in classical trajectory from quasiperiodic to stochastic?

(ii) Assuming the answer to (i) is yes, what is the nature of the stationary states for $E > E_c$, and how are these related, if at all, to properties of classical trajectories in the stochastic domain?

(iii) If intramolecular redistribution of energy is regarded as one of several competing processes, how is its rate influenced by the nature of the states when $E > E_c$?

Several approaches to answering these questions are discussed in the following text.

A. The Nordholm-Rice Analysis [38]

The first step in searching for an analogue of the KAM transition from quasiperiodic to stochastic behaviour in a quantum mechanical description of a coupled oscillator system is the selection of a criterion to signal that transition. It turns out that the choice of criterion is not trivial, since it quickly drags into the analysis rather esoteric, yet important, properties of quantum states. The existing formulations of quantum ergodic theory provide no assistance for the situations of interest to us, so at present the necessary criterion is postulated on the basis of analogies with the behaviour of systems described by classical mechanics. Nordholm and Rice [38] suggest that the important question is how an initially localized excitation spreads over the energy surface of the system, and they propose to classify a state of the system as ergodic if an excitation initially not uniformly distributed over the energy surface becomes uniformly distributed as $t \to \infty$. This suggestion is deliberately constructed to be in direct correspondence with the usual definition of ergodicity on a classical energy surface.

Implementation of the Nordholm-Rice criterion is based on the principle that the eigenvalues of an arbitrarily complicated Hamiltonian can be computed by use of an expansion in a complete set of basis functions and evaluation of matrix elements, though an actual calculation can be technically difficult and very tedious. Given this methodology, the Nordholm-Rice criterion for the onset of stochastic behaviour has the advantage of being easily tested for any given set of basis states. It has the disadvantage that the conclusion of a test for this kind of ergodicity is, in general, basis dependent. Of course, a basis dependence can be connected to reality if there exists an excitation mechanism that prepares the system in one or more of the basis states.

The details of the Nordholm-Rice analysis can be found in Ref. [38], which also contains an expanded discussion of the difficulty of selecting a criterion for the transition to stochastic behaviour in a quantum mechanical description of a coupled oscillator system.

It should be noted that the Henon-Heiles system, and several others of those studied, do not have bound states in the quantum mechanical description. These systems do have resonances in the region of the potential energy surface where there would be classical bound states. In the Nordholm-Rice calculations, and those of Stratt, Handy and Miller [39], Pomphrey [40], Jaffe and Reinhardt [41], and all others that I know of, the distinction between resonances and bound states has been neglected. The calculations which employ a basis set expansion use functions that span only that part of the space where classical bound states exist, and the several semiclassical quantization schemes use a generalized Bohr-Sommerfeld-Wilson rule on action integrals defined with respect to periodic or quasiperiodic orbits generated from classically bound orbits of a reference system. Thus, all of the calculations reported to date approximate resonances with stationary state functions in a truncated Hilbert space. I shall return to this point later.

Nordholm and Rice studied several model systems of coupled nonlinear oscillators, including the Henon-Heiles system so thoroughly described in Section 2. Some of the systems have only algebraic nonlinearities; others have exponential nonlinearities such as is characteristic of a Morse potential function. Some of the examples have degenerate small amplitude modes; others have nondegenerate small amplitude modes. The basis states of the representation were, for each system studied, chosen to be the states of the corresponding decoupled set of harmonic oscillators. The calculations lead to the following conclusions:

(i) There is nonergodic behaviour, in general, below a critical energy (possibly a critical energy region). The nonergodicity is more marked and persistent for nondegenerate systems than for equivalent degenerate systems.

(ii) For degenerate systems there are occasional global states interspersed in the local states. (A global state is one with amplitude uniformly distributed over the energy surface.)

(iii) For low energy, typically less than half the dissociation energy E_D, the asymptotic distribution of amplitude over the equi-energetic basis states is not very sensitive to the coupling.

(iv) Initial states with comparable excitation in all oscillators tend to evolve to global states, whereas extremal initial states with excitation mostly localized in one oscillator tend to evolve to local states.

(v) For high energy, typically above half the dissociation energy, the asymptotic distribution of amplitude over the basis states is very sensitive to the coupling. In this case the final states achieved tend to have mixed character (e.g., a wide but uneven spread of overlaps).

(vi) From a very crude analysis of the time evolution of the model systems it is estimated that in these cases it takes of the order of one to ten vibrational periods for an initially localized non-stationary state to achieve its asymptotic form.

Clearly, the quantum dynamics of coupled oscillator systems is analogous to the classical dynamics of these systems. In particular, a KAM-like transition exists. Note also how conclusion (iv) is very like the prediction that if energy is initially localized in one oscillator it requires a higher total energy to reach the region of overlap of nonlinear resonances than if the same initial energy is spread over several oscillators (see Section 4B).

What is the character of the stationary states in the stochastic region? One answer to this question is given by the coefficients in the basis set expansion. A more graphic description is obtained from the nodal patterns of the wave functions. For one dimensional systems the nodes of the wave function and the quantum number of the state have a one-to-one correspondence. What are nodal points in the one dimensional case become nodal surfaces in higher dimensional cases. Consider, for simplicity, a two dimensional system, in which case there is some specific pattern of nodal lines which corresponds to a given eigenstate. For example, if the two dimensional system is separable these nodal lines form a regular rectangular grid (Fig. 3.1a). In contrast, Pechukas [42] showed that if the two dimensional system is not separable the nodal lines need not, in general, intersect to form a simple grid; instead there can be avoided intersections such as shown in Fig. 3.1b, leading to an overall pattern like that shown in Fig. 3.1c. It is an obvious suggestion that the nodal lines will intersect in the domain where the classical trajectories are quasiperiodic, and will have some complicated, broken-up pattern in the domain where the classical trajectories are stochastic. Stratt, Handy and Miller [39] have examined the nodal structure of the eigenfunctions of the Barbanis Hamiltonian

$$H = \frac{1}{2}(p_1^2 + p_2^2) + \frac{1}{2}(1.6q_1^2 + 0.9q_2^2) - 0.8q_1q_2^2 \qquad (3.1)$$

for which the small amplitude frequencies are obviously $(1.6)^{\frac{1}{2}}$ and $(0.9)^{\frac{1}{2}}$. The method of computation was that of Nordholm and Rice, namely expansion in the harmonic oscillator functions of the uncoupled counterpart of (3.1), generated by the first two terms of (3.1). Some of the results are shown in Figs. 3.2 and 3.3. In these diagrams the regions in which the wave function is positive are black, and the regions in which the wave function is negative are blank. The remaining dotted regions are those in which the wave function is very small, hence noise in the numerical calculation prevents assignment of a definite sign. For the pattern shown in Fig. 3.2, which is in the domain where the corresponding classical trajectory is quasiperiodic, it is easy to visualize two curved

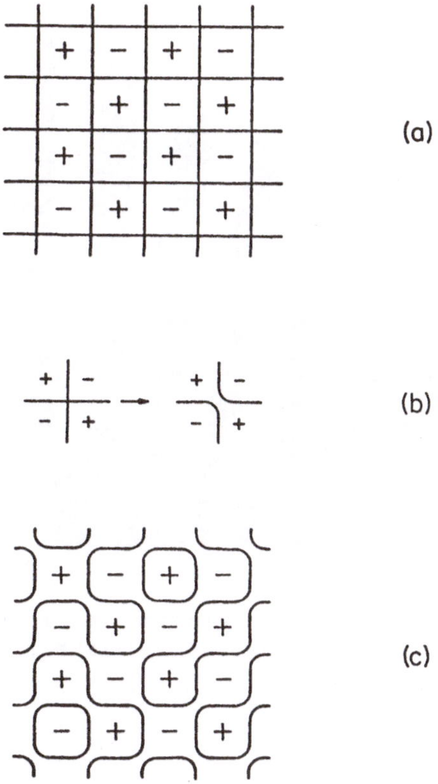

Fig. 3.1. Schematic representation of the nodal lines for a two oscillator system. Case (a) corresponds to separable oscillators, case (b) shows how introduction of inter-oscillator coupling distorts a nodal crossing, and case (c) is a possible nodal pattern for coupled oscillators (after ref. 39).

orthogonal axes along which black and blank spaces alternate. When this is possible the numbers of nodes corresponding to the separable coordinates (by construction the curved axes are orthogonal) give the quantum numbers of the independent components of the eigenstates. On the other hand, the distribution of black and blank spaces in Fig. 3.3, which is at an energy in the domain where the corresponding classical trajectory is stochastic, seems to involve several nodal patterns superimposed. In this case it is not easy, and possibly impossible, to find a set of curved orthogonal axes that will neatly separate it into regularly alternating sequences of black and blank spaces. But this failure does not mean one cannot find some coordinate system in which quantum numbers can be assigned (see later), only that whatever the character of that coordinate

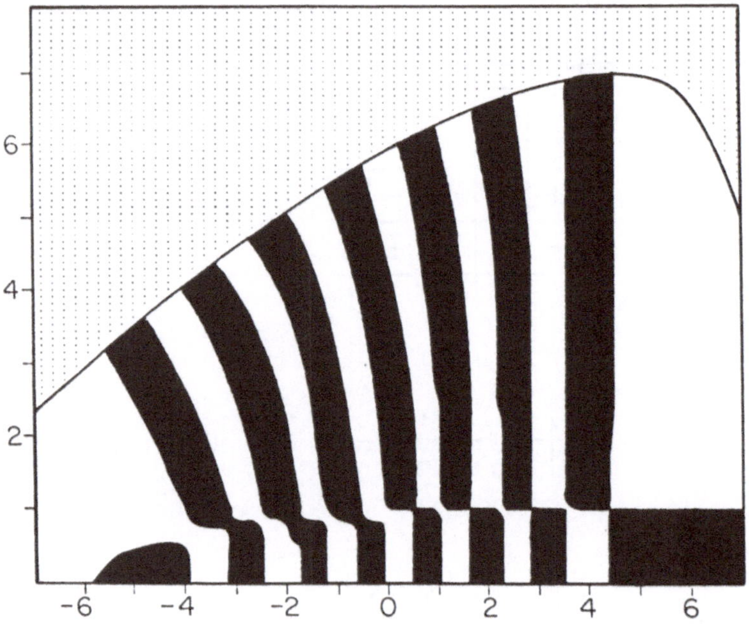

Fig. 3.2. The nodal pattern for an eigenstate of the Hamiltonian (3.1) at an energy below the KAM-like transition (after ref. 39).

system it has a complicated representation in terms of the coordinates of the harmonic oscillators of the uncoupled reference system.

Just as in the "local-global" categorization used by Nordholm and Rice, Stratt, Handy and Miller find that some low lying states have greatly distorted nodal patterns, some high lying states have almost regular patterns, and that there are borderline cases which are hard to assign to either category. The overall correlation with the nature of the corresponding classical trajectory is greatly strengthened by the observation that the expectation values of q_1^2 and q_2^2 are very large in the stochastic region, and have smaller values (comparable to those expected for harmonic oscillators with quantum numbers n_1 and n_2, respectively) in the quasiperiodic region.

Stratt, Handy and Miller suggest that some of the difficulties associated with the basis set dependence of the Nordholm-Rice criterion for stochasticity can be overcome by transforming to "natural orbitals". In a system with two degrees of freedom the wave function corresponding to the Nordholm-Rice choice of basis

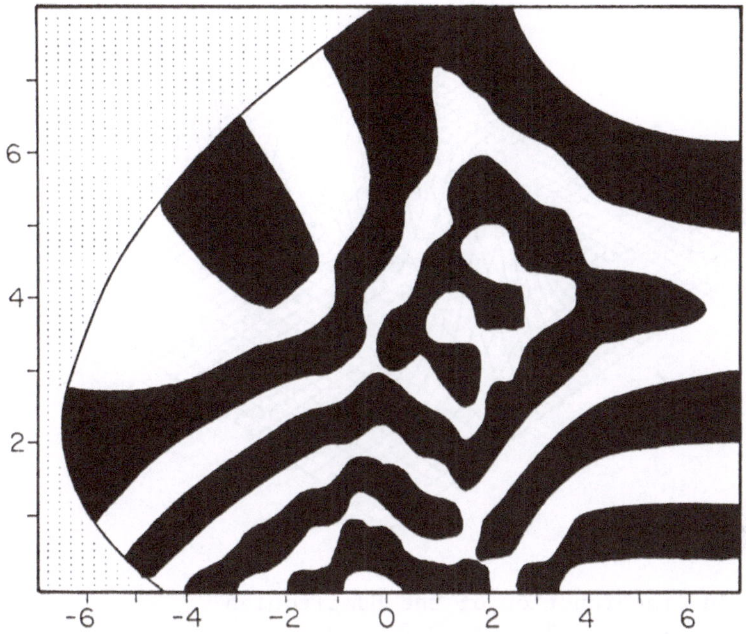

Fig. 3.3. The nodal pattern for an eigenstate of the Hamiltonian (3.1) at an energy above the KAM-like transition (after ref. 39).

functions is a bilinear combination of harmonic oscillator functions. The natural orbitals for this system are those for which the wave function has a diagonal representation; these can be shown to be just the linear combinations of harmonic oscillator functions that give the most rapid convergence of the expansion of the wave function. Stratt, Handy and Miller find that in the quasiperiodic domain one of the natural orbitals dominates the wave function, whereas in the stochastic domain many contribute significantly to the wave function. These results confirm completely the Nordholm-Rice analysis, which was based on the bilinear expansion in the harmonic oscillator basis functions, and suggest that basis function sensitivity of the onset of stochasticity may be a less serious problem than originally thought.

As to the wave functions themselves, Figs. 3.4 and 3.5 shows eigenfunctions of the Henon-Heiles Hamiltonian in the quasiperiodic region and the stochastic region, respectively. These wave functions computed by Marcus, Noid and Koszykowski [43] , are for the case that the small amplitude frequencies are in the ratio $\omega_1/\omega_2 = 2$.

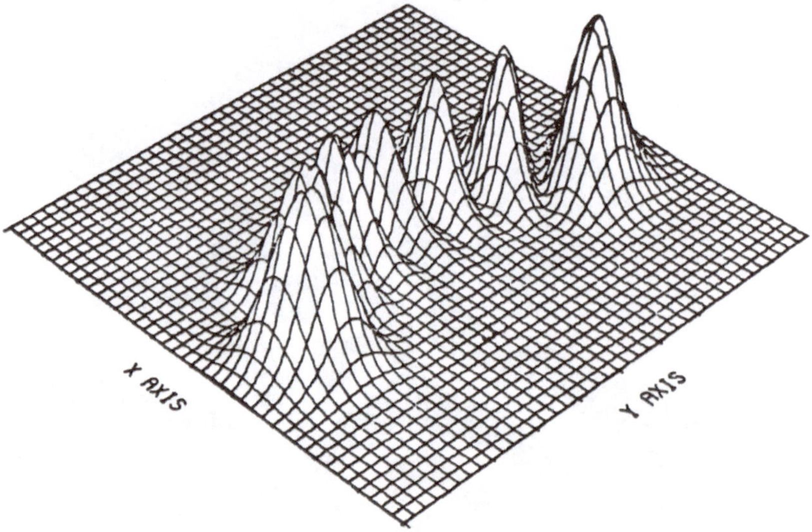

Fig. 3.4. An eigenfunction of the Hamiltonian

$$H = \frac{1}{2}(p_x^2 + p_y^2) + \frac{1}{2}(\omega_x^2 x^2 + \omega_y^2 y^2) + \lambda x(y^2 + \eta x^2)$$

at an energy where the classical motion is quasiperiodic. In this case $\omega_x = 2\omega_y$. (From R.A. Marcus, D.W. Noid and M.L. Koszykowski in Stochastic Behavior in Classical and Quantum Hamiltonian Systems, eds. G. Casati and J. Ford, Springer-Verlag, Berlin (1979).)

B. Semiclassical Quantization of Coupled Oscillator Systems

Although it is, in principle, always possible to calculate the eigenvalues of a Hamiltonian operator by expansion in a suitable complete set of basis functions and evaluation of the matrix elements thereby generated, for many purposes, this is not an intellectually satisfying procedure. For example, this method provides no information on the relationship between the stationary states of a system and the corresponding classical trajectories. Given the qualitative indication that a KAM type transition occurs in the quantum mechanical description of a coupled oscillator system, there is reason to examine other methods for finding the eigenvalues of the Hamiltonian operator. The most widely used class of "other methods" is semiclassical quantization. Although the purpose of much of the work on the semiclassical methodology is to calculate eigenvalues as easily as possible, I shall ignore that aspect of the theory. Rather, I shall focus attention on a necessary ingredient of the semiclassical methodology, namely a scheme for generating, classifying and selecting periodic and quasiperiodic trajectories of a

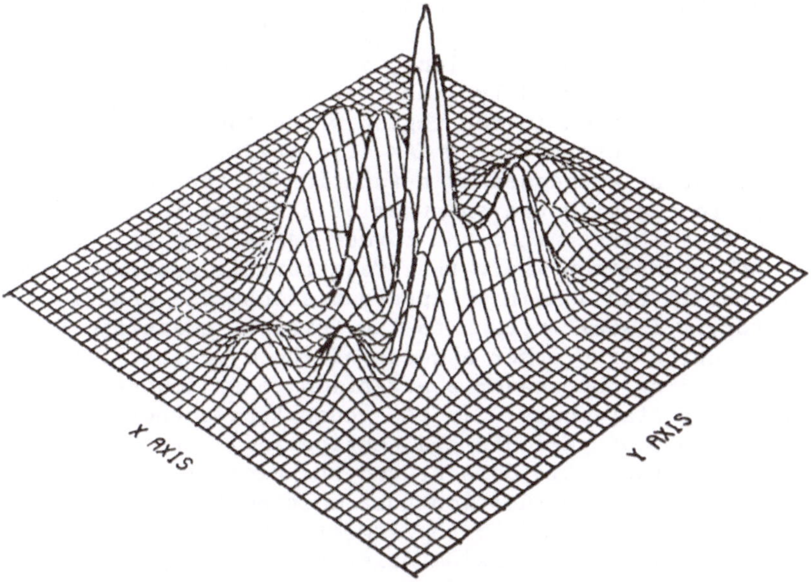

Fig. 3.5. An eigenfunction of the Hamiltonian described in the caption to Fig. 3.4 where the classical motion is stochastic. (From R.A. Marcus, D.W. Noid and M.L. Koszykowski in Stochastic Behavior in Classical and Quantum Hamiltonian Systems, eds. G. Casati and J. Ford, Springer-Verlag, Berlin (1979).)

system. It is just this intermediate part of the semiclassical analysis that can be used to give a vivid connection between the eigenstates of a Hamiltonian operator and the trajectories of the corresponding classical system.

Percival [44], Berry [45], Miller [46], Marcus [47], Delos [48] and Reinhardt [41] have advanced the semiclassical theory of bound states. Very important early work by Gutzwiller dealt with periodic orbits [49]. These investigators have adopted the Einstein generalization of the Bohr-Sommerfeld-Wilson quantization condition, namely [50]

$$J = \frac{1}{2\pi} \oint \sum_{k=1}^{N} p_k \, dq_k = (n + \delta) \hbar , \tag{3.2}$$

where the integral over the invariant differential sum $\Sigma\, p_k dq_k$ is along closed curves in coordinate space that need not be classical trajectories. The usual Bohr-Sommerfeld-Wilson condition is defined for each separable coordinate q_k, and the corresponding action integral is taken around one cycle of the motion of the coordinate

q_k. The Bohr-Sommerfeld-Wilson condition depends on the choice of coordinates whereas Einstein's condition, which defines an integration over all N action functions, does not depend on the choice of coordinates. It is readily seen that the Einstein condition can be applied to motion in the region where the trajectory is quasiperiodic, but not above the KAM transition, since in that domain p_k cannot be expressed as a function of the q_k. This has led Percival to classify the spectrum of a system into the categories regular and irregular, the former pertaining to the domain below the KAM transition where the trajectory is quasiperiodic, and the latter to the domain of apparent stochastic behaviour of the trajectory. I shall return to the Percival classification, and its connection with other aspects of the dynamics, later. For the present, I draw the reader's attention to an important exception to the preceding statements. In the general case there are a set (of measure zero?) of periodic and quasiperiodic trajectories embedded in the stochastic domain, and the Einstein condition can be applied to them. The question that must be answered is: Does the set of states generated by the semiclassical quantization of the embedded quasiperiodic trajectories exhaust the set of stationary states of the Hamiltonian of the system?

Percival's development [44] of the semiclassical theory of coupled oscillator systems starts from the time independent formulation of Hamilton's equations, namely

$$\underset{\sim}{\omega}\cdot\frac{\partial \underset{\sim}{q}}{\partial \underset{\sim}{\phi}} = \frac{\partial H(\underset{\sim}{p},\underset{\sim}{q})}{\partial \underset{\sim}{p}} \tag{3.3}$$

$$\underset{\sim}{\omega}\cdot\frac{\partial \underset{\sim}{p}}{\partial \underset{\sim}{\phi}} = -\frac{\partial H(\underset{\sim}{p},\underset{\sim}{q})}{\partial \underset{\sim}{q}} \tag{3.4}$$

with, in this section, the shorthand notation $\underset{\sim}{\omega}\cdot\frac{\partial}{\partial \underset{\sim}{\phi}} = \sum_i \omega_i \frac{\partial}{\partial \phi_i}$ and similarly $\frac{\partial}{\partial \underset{\sim}{p}} = (\frac{\partial}{\partial p_1}, \ldots, \frac{\partial}{\partial p_N})$.

If the system Hamiltonian is written in the form

$$H = \sum_{n=1}^{N} \frac{1}{2}(p_n^2 + \lambda_n^2 q_n^2) + \varepsilon V(\underset{\sim}{q}) , \tag{3.5}$$

and the coordinates and momenta are expanded in Fourier series,

$$\underset{\sim}{q}(\underset{\sim}{J},\underset{\sim}{\phi}) = \sum_{\underset{\sim}{\ell}} \tilde{\underset{\sim}{q}}_{\underset{\sim}{\ell}} (\underset{\sim}{J})\, e^{i\underset{\sim}{\ell}\cdot\underset{\sim}{\phi}} , \tag{3.6}$$

$$\underset{\sim}{p}(\underset{\sim}{J},\underset{\sim}{\phi}) = \sum_{\underset{\sim}{\ell}} \tilde{\underset{\sim}{p}}_{\underset{\sim}{\ell}} (\underset{\sim}{J})\, e^{i\underset{\sim}{\ell}\cdot\underset{\sim}{\phi}} . \tag{3.7}$$

Then Eqs. (3.3) and (3.4) become

$$i(\underset{\sim}{\ell}\cdot\underset{\sim}{\omega})\tilde{\underset{\sim}{q}}_{\ell} = \tilde{\underset{\sim}{p}}_{\ell} , \tag{3.8}$$

$$i(\underset{\sim}{\ell}\cdot\underset{\sim}{\omega})\tilde{\underset{\sim}{p}}_{\ell} \cdot = \tilde{\underset{\sim}{f}}^{HO}_{\ell} + \varepsilon\tilde{\underset{\sim}{f}}^{V}_{\ell} . \tag{3.9}$$

where the Fourier components of the force have been separated into a harmonic oscillator term

$$\tilde{\underset{\sim}{f}}^{HO}_{\ell} = (-\lambda_1^2\tilde{q}_{1\ell}, \ldots, -\lambda_N^2\tilde{q}_{N\ell}) \tag{3.10}$$

and a coupling term

$$\tilde{\underset{\sim}{f}}^{V}_{\ell} = \left(\frac{1}{2\pi}\right)^N \int_0^{2\pi} d\underset{\sim}{\phi}\, \left(-\frac{\partial V}{\partial \underset{\sim}{q}}\right) e^{i\underset{\sim}{\ell}\cdot\underset{\sim}{\phi}} \tag{3.11}$$

Equations (3.8) and (3.9) are to be regarded as a set of simultaneous equations for the $\tilde{\underset{\sim}{q}}_{\ell}$ and $\tilde{\underset{\sim}{p}}_{\ell}$. Elimination of $\tilde{\underset{\sim}{p}}_{\ell}$ between them gives

$$(\lambda_k^2 - (\underset{\sim}{\ell}\cdot\underset{\sim}{\omega})^2)\tilde{q}_{k\underset{\sim}{\ell}} = \varepsilon\tilde{f}^{V}_{k\underset{\sim}{\ell}} , \quad (k = 1, \ldots, N) \tag{3.12}$$

The solution to Eq. (3.12) is obtained by iteration, using as the zeroth order approximation the solution for $\varepsilon = 0$, i.e., the solution for uncoupled harmonic oscillators. This zeroth order solution is used first to calculate the zeroth order Fourier coefficient of the coupling force, via Eq. (3.11), and that coefficient is used in the $|\ell| = 1$ term of Eq. (3.12) to calculate the first order frequencies. These are

$$\omega_k^{(1)} = \lambda_k^2 - \varepsilon \frac{\tilde{f}^{V}_{k\underset{\sim}{\ell}}}{\tilde{q}^{(0)}_{k\underset{\sim}{\ell}}} , \quad (|\ell| > 1,\ k = 1, \ldots, N) \tag{3.13}$$

The remaining member of Eq. (3.12), for $|\ell| > 1$, are solved for the first order Fourier coefficients of the coordinates:

$$\tilde{q}^{(1)}_{k\underset{\sim}{\ell}} = \frac{\varepsilon\tilde{f}^{V(0)}_{k\underset{\sim}{\ell}}}{\lambda_k^2 - (\underset{\sim}{\ell}\cdot\underset{\sim}{\omega}^{(1)})^2} \tag{3.14}$$

The calculation is closed, at this level, by using specific values of the action to fix $\tilde{q}^{(1)}_{k1}$. This step imposes the initial conditions on the oscillator system. The entire process is then repeated until the desired accuracy of convergence is achieved.

Chapman, Garrett and Miller [46], and Schatz [51], use the Born [52] periodic generating function approach to define suitable canonical transformation of a coupled oscillator Hamiltonian to angle-action variables; Jaffe and Reinhardt [41] have presented an improved rapidly converging version of this method. Let the Hamiltonian for the oscillator system be

$$H = H_o(p,q) + \varepsilon V(q) \tag{3.15}$$

where $H_o(p,q)$ describes an integrable system. Let $\underset{\sim}{J}^o$, $\underset{\sim}{\phi}^o$ be the action-angle variables of H_o, and rewrite (3.15) in terms of these variables:

$$H = H_o(\underset{\sim}{J}^o) + \varepsilon V(\underset{\sim}{J}^o,\underset{\sim}{\phi}^o) \ . \tag{3.16}$$

The equations of motion are then

$$\mu^{(0)}(\underset{\sim}{J}^o,\underset{\sim}{\phi}^o) = \dot{\underset{\sim}{J}} = -\varepsilon \frac{\partial V(\underset{\sim}{J}^o,\underset{\sim}{\phi}^o)}{\partial \underset{\sim}{\phi}^o} \tag{3.17}$$

$$\omega^{(0)}(\underset{\sim}{J}^o,\underset{\sim}{\phi}^o) = \omega^o(\underset{\sim}{J}^o) + \varepsilon \frac{\partial V(\underset{\sim}{J}^o,\underset{\sim}{\phi}^o)}{\partial \underset{\sim}{J}^o} , \tag{3.18}$$

where $\omega^o(J^o) = (\partial H_o(\underset{\sim}{J}^o)/\partial \underset{\sim}{J}^o)$. Although $\underset{\sim}{J}^o$ and $\underset{\sim}{\phi}^o$ are the proper actions and angles with which to describe the uncoupled oscillators, they are not the action-angle variables of the full Hamiltonian. However, Eqs. (3.17) and (3.18) show that $\underset{\sim}{J}^o$ and ϕ^o are "almost" the correct action-angle variables, the deviation being of order ε. Jaffe and Reinhardt therefore introduce another transformation, this one designed to generate the action-angle variables of H to order ε^2, and then another transformation, and so on. After n such transformations

$$H = H^{(n)}(\underset{\sim}{J}^{(n)}) + \varepsilon^{2^n} V^{(n)}(\underset{\sim}{J}^{(n)},\underset{\sim}{\phi}^{(n)}) , \tag{3.19}$$

$$\mu^{(n)}(\underset{\sim}{J}^{(n)},\underset{\sim}{\phi}^{(n)}) = \dot{J}^{(n)} = -\varepsilon^{2^n} \frac{\partial V^{(n)}}{\partial \underset{\sim}{\phi}^{(n)}} , \tag{3.20}$$

$$\omega^{(n)}(\underset{\sim}{J}^{(n)},\underset{\sim}{\phi}^{(n)}) = \dot{\phi}^{(n)} = \omega_o^{(n)}(\underset{\sim}{J}^{(n)}) + \varepsilon^{2^n} \frac{\partial V^{(n)}}{\partial \underset{\sim}{J}^{(n)}} , \tag{3.21}$$

so the variables generated are the action-angle variables of H correct to order ε^{2^n}.

The key step in this analysis is the first transformation from $\underset{\sim}{J}^o$, $\underset{\sim}{\phi}^o$ to $\underset{\sim}{J}^{(1)}$, $\underset{\sim}{\phi}^{(1)}$. The generating function is

$$F_1(\underset{\sim}{\phi}^o,\underset{\sim}{J}^{(1)}) = \underset{\sim}{\phi}^o\cdot\underset{\sim}{J}^{(1)} + \varepsilon W^{(1)}(\underset{\sim}{\phi}^o,\underset{\sim}{J}^{(1)}) , \tag{3.22}$$

$$W^{(1)}(\underset{\sim}{\phi}^o,\underset{\sim}{J}^{(1)}) = \sum_{\underset{\sim}{\ell}\neq 0} W^{(1)}_{\underset{\sim}{\ell}}(\underset{\sim}{J}^{(1)})\, e^{i\underset{\sim}{\ell}\cdot\underset{\sim}{\phi}^o} \tag{3.23}$$

which yields

$$\underset{\sim}{J}^o = \underset{\sim}{J}^{(1)} + \varepsilon \sum_{\underset{\sim}{\ell}\neq 0} i\underset{\sim}{\ell}\, \tilde{W}^{(1)}_{\underset{\sim}{\ell}}(\underset{\sim}{J}^{(1)})\, e^{i\underset{\sim}{\ell}\cdot\underset{\sim}{\phi}^o} , \tag{3.24}$$

$$\underset{\sim}{\phi}^{(1)} = \underset{\sim}{\phi}^o + \varepsilon \sum_{\underset{\sim}{\ell}\neq 0} \left(\frac{\partial \tilde{W}^{(1)}_{\underset{\sim}{\ell}}(\underset{\sim}{J}^{(1)})}{\partial \underset{\sim}{J}^{(1)}}\right) e^{i\underset{\sim}{\ell}\cdot\underset{\sim}{\phi}^o} , \tag{3.25}$$

and it can be shown that

$$\tilde{W}^{(1)}_{\underset{\sim}{\ell}}(\underset{\sim}{J}^{(1)}) = -\left(\frac{V^{(0)}_{\underset{\sim}{\ell}}(\underset{\sim}{J}^o)}{i\underset{\sim}{\ell}\cdot\underset{\sim}{\omega}^{(0)}(\underset{\sim}{J}^o)}\right)_{\underset{\sim}{J}^o=\underset{\sim}{J}^{(1)}} , \tag{3.26}$$

$$\frac{\partial \tilde{W}^{(1)}_{\underset{\sim}{\ell}}}{\partial \underset{\sim}{J}^{(1)}} = -\left(\frac{\partial \tilde{V}^{(0)}_{\underset{\sim}{\ell}}(\underset{\sim}{J}^o)/\partial \underset{\sim}{J}^o}{i\underset{\sim}{\ell}\cdot\underset{\sim}{\omega}^{(0)}(\underset{\sim}{J}^o)}\right)_{\underset{\sim}{J}^o=\underset{\sim}{J}^{(1)}} , \tag{3.27}$$

with

$$V(\underset{\sim}{J}^o,\underset{\sim}{\phi}^o) = \sum_{\underset{\sim}{\ell}\neq 0} \tilde{V}^{(0)}_{\underset{\sim}{\ell}}(\underset{\sim}{J}^o)\, e^{i\underset{\sim}{\ell}\cdot\underset{\sim}{\phi}^o} . \tag{3.28}$$

The new Hamiltonian is of the same form as the old Hamiltonian, except that the coefficient of the coupling term is now ε^2. Therefore the same transformation can be applied again. The generating function of the nth transformation is, in terms of the variables of the $n-1^{th}$ Hamiltonian

$$\tilde{W}^{(n)}_{\underset{\sim}{\ell}}(\underset{\sim}{J}^{(n)}) = -\left(\frac{\tilde{V}^{(n-1)}_{\underset{\sim}{\ell}}(\underset{\sim}{J}^{(n-1)})}{i\underset{\sim}{\ell}\cdot\omega_o^{(n-1)}(\underset{\sim}{J}^{(n-1)})}\right)_{\underset{\sim}{J}^{(n-1)}=\underset{\sim}{J}^{(n)}} \tag{3.29}$$

$$\frac{\partial \tilde{W}_{\ell}^{(n)}}{\partial J^{(n)}} = - \frac{\partial \tilde{V}_{\ell}^{(n-1)}/\partial J^{(n-1)}}{i\ell\cdot\omega_o^{(n-1)}(J^{(n-1)})} \quad J^{(n-1)}=J^{(n)} \tag{3.30}$$

Jaffe and Reinhardt call their procedure a classical Van Vleck transformation.

For the present purposes the most interesting semiclassical quantization method is that of Swimm and Delos [48]. They have shown that the eigenvalues of the Hamiltonian operator of a coupled oscillator system can be very accurately calculated by transforming the classical Hamiltonian to the Birkhoff normal form, and then quantizing via the Bohr-Sommerfeld-Wilson rule. It was mentioned in Section 2C that the Hamiltonian of a coupled oscillator system could be brought into so called normal form by a succession of canonical transformations. When in normal form the Hamiltonian is a function of the variables combined as $\frac{1}{2}(P_k^2 + Q_k^2)$, i.e. a function of one dimensional harmonic oscillator Hamiltonians. Clearly, if a Hamiltonian can be brought into normal form, the motion is quasiperiodic.

The succession of transformations required to generate the normal form of the Hamiltonian in general diverges because of small divisors in the coefficients. This divergence arises from the existence of stochastic trajectories which cannot be described by the normal form representation of the Hamiltonian, and which are dense in at least part of the accessible range. Nevertheless, as shown by Gustavson [36], a truncated series of transformations generates a Hamiltonian which is in normal form up to some well defined order in ε, and which is an approximation to the original Hamiltonian. This truncated Hamiltonian describes quasiperiodic trajectories which are good approximations to the quasiperiodic trajectories of the original Hamiltonian.

In brief, the Gustavson procedure involves the following steps, illustrated here for the Henon-Heiles Hamiltonian written in the form

$$\begin{aligned} H &= \frac{1}{2}(p_1^2 + p_2^2 + \omega_1^2 q_1^2 + \omega_2^2 q_2^2) + \varepsilon q_2(q_1^2 + \eta q_2^2) \\ &= H^{(2)} + H^{(3)} \end{aligned} \tag{3.31}$$

It is assumed that the ratio ω_2/ω_1 is irrational. The substitutions

$$p_k \to \omega_k^{1/2} p_k' \tag{3.32}$$

$$q_k \rightarrow \omega_k^{-1/2} q_k' \tag{3.33}$$

give

$$H^{(2)} = \frac{1}{2}(p_1'^2 + q_1'^2)\omega_1 + \frac{1}{2}(p_2'^2 + q_2'^2)\omega_2 , \tag{3.34}$$

$$H^{(3)} = \frac{\varepsilon}{\omega_1 \omega_2^{1/2}} q_1'^2 q_2' + \frac{\varepsilon\eta}{\omega_2^{3/2}} q_2'^3 . \tag{3.35}$$

The new Hamiltonian is further reduced by the sequential application of canonical transformations defined by a generating function $W^{(s)}$ which has the property

$$Q = q' + \frac{\partial W^{(s)}}{\partial P} \tag{3.36}$$

$$P = p' - \frac{\partial W^{(s)}}{\partial q'} \tag{3.37}$$

$W^{(s)}$ is a homogeneous polynomial of degree s, where $s > 3$. The transformation of variables of degree s leaves unchanged those terms of degree less than s, but changes all other terms. We have, then,

$$H(p',q') = \Gamma(P,Q) \tag{3.38}$$

The new Hamiltonian $\Gamma(P,Q)$ is now separated into terms of different degrees and $P + \partial W^{(s)}/\partial q'$ and $q' + \partial W^{(s)}/\partial P$ expanded about P and q', respectively. Collection of terms of equal degree then gives

$$H^{(i)}(P,q') = \Gamma^{(i)}(P,q') , \qquad i < s \tag{3.39}$$

$$\hat{D}\, W^{(s)}(P,q') = \Gamma^{(s)}(P,q') - H^{(s)}(P,q') , \quad i = s \tag{3.40}$$

$$\Gamma^{(i)}(P,q') = H^{(i)}(P,q') + \sum_j \frac{1}{j!} \left[\left(\frac{\partial W^{(s)}}{\partial q'}\right)^j \left(\frac{\partial^{|j|} H^{(\ell)}}{\partial P^j}\right) - \left(\frac{\partial W^{(s)}}{\partial P}\right)^j \left(\frac{\partial^{|j|} \Gamma^{(\ell)}}{\partial q'^j}\right) \right] \tag{3.41}$$

with the constraints on the indices $\ell - |j| + |j|(s - 1) = i$, $1 < |j| < \ell < i$, $\ell > 2$, $s > 3$ and $j! = j_1! j_2!$. Finally, $\hat{D}$ is the operator

$$\hat{D} = -\omega_1\left(q_1'\frac{\partial}{\partial P_1} - P_1\frac{\partial}{\partial q_1'}\right) - \omega_2\left(q_2'\frac{\partial}{\partial P_2} - P_2\frac{\partial}{\partial q_2'}\right) \tag{3.42}$$

and

$$H^{(i)}(P,q') \equiv [H^{(i)}(p,q')]_{p=P}\ , \tag{3.43}$$

$$\Gamma^{(i)}(P,q') \equiv [\Gamma^{(i)}(P,Q)]_{Q=q'} \tag{3.44}$$

solve (3.40) q' and P are transformed so as to diagonalize $\hat{D}$:

$$P_k = \frac{1}{\sqrt{2}}(\eta_k + i\xi_k)\ , \tag{3.45}$$

$$q_k' = \frac{1}{\sqrt{2}}(\eta_k - i\xi_k)\ . \tag{3.46}$$

It is found that

$$\Phi_{\ell_1\ell_2m_1m_2} = \eta_1^{\ell_1}\eta_1^{\ell_2}\xi_1^{m_1}\xi_2^{m_2}$$

are the eigenfunctions of D corresponding to the eigenvalues $i\sum_k \omega_k(m_k - \ell_k)$, and that

$$W^{(s)} = \hat{D}^{-1}[\Gamma^{(s)} - H^{(s)}]. \tag{3.47}$$

$W^{(s)}$ is kept finite by choosing $\Gamma^{(s)}$ to cancel those terms in $H^{(s)}$ which would have vanishing denominator in (3.47) when $\Gamma^{(s)}$ and $H^{(s)}$ are expanded in the basis of the eigenfunctions of $\hat{D}$. There is no difficulty in doing this for ω_2/ω_1 irrational; for ω_2/ω_1 rational a slightly more elaborate procedure is needed. In the case that ω_2/ω_1 is irrational the end result is that the normal form can be written as a sum of products of one dimensional harmonic oscillator Hamiltonians, and the quantization can be effected to yield analytic formulae. In the case that ω_2/ω_1 is rational additional terms which cannot be written as products of one dimensional Hamiltonians appear in the normal form, and while one degree of freedom can be quantized analytically the other must be quantized by a one dimensional (generally numerical) phase integration.

All of the semiclassical quantization schemes described give eigenvalues in excellent agreement with these calculated by direct diagonalization of the Hamiltonian operator. That fact I believe to be an important clue to the interpretation of the relationship between quantum mechanical states and classical trajectories of a system of coupled oscillators. I will return to this point below.

C. Conjectures on the Relationship Between Quantum Mechanical States and Classical Trajectories

The studies of the quantum states of a system of coupled oscillators described in Section 3B lead to two inferences:

(i) There appears to be an analogue of the KAM transition in the quantum mechanical description of a nonintegrable system.

(ii) The states generated by semiclassical quantization are derived from quasiperiodic trajectories embedded in the stochastic domain of the classical dynamics.

Note that inferences (i) and (ii) are not specifically contradictory, but they are not reinforcing. If inferences (i) and (ii) are both correct the quantum mechanical description of the coupled oscillator system must differ from the classical mechanical description more than previously thought to be the case.

The results which lead to inference (i) derive from the definition of quantum ergodicity posed by Nòrdholm and Rice [38], or the equivalent posed by Handy, Stratt and Miller [39]. The inference is also supported by calculations of Pomphrey to be described below. Berry [45] has picked up a suggestion made by Nordholm and Rice and developed an analysis of the Wigner function representation of a system of coupled oscillators. For states in the quasiperiodic domain of phase space the classical limit of the Wigner function is a delta function on the corresponding invariant torus. In this case there is a simple connection between quasiperiodic trajectories and stationary states, as might have been expected from the Bohr-Sommerfeld-Wilson quantization condition for separable systems. Berry is then led to the natural conjecture that the Wigner function corresponding to a stochastic trajectory spreads over the stochastic region of the phase space, so that a surface of section for the quantum state ought (?) to show randomly distributed maxima and minima. There is, at present, not even a clue as to what form of randomness (e.g., a Gaussian random distribution) to expect in this surface of section. Nevertheless, if the Wigner function on the surface of section is very irregular, it is reasonable to connect stochastic trajectories and quantum states in a fashion analogous to that used to connect quasiperiodic trajectories and quantum states.

Berry's study of the semiclassical limit of the Wigner function has many interesting features; The reader is referred to the original paper for details [45]. Here we shall consider only the conjecture just described concerning the Wigner function on the surface of section. Hutchinson and Wyatt [53] have constructed the Wigner function for the Henon-Heiles system with Hamiltonian

$$H = \frac{1}{2}(p_1^2 + p_2^2) + \frac{1}{2}(0.49q_1^2 + 1.69q_2^2) - 0.10(q_1q_2^2 - q_1^3) . \tag{3.48}$$

The dissociation energy is, in this case, $E_D = 11.46$, and the onset of stochastic behaviour is estimated to be $E_c = 5.73$ using the Brumer-Duff-Toda criterion (see Section 4). The method of calculation is the same as that used by Nordholm and Rice. The surface of section Wigner function distribution is defined by

$$\Psi_W(q_1,p_1;q_2=0) = \int_{-\infty}^{\infty} \Psi_W(q_1,q_2=0,p_1,p_2)\, dp_2 \tag{3.49}$$

with the usual definition

$$\Psi_W \equiv (\pi\hbar)^{-N} \int d\underset{\sim}{X} e^{-2i\underset{\sim}{p}\cdot\underset{\sim}{X}/h}\, \psi(\underset{\sim}{q} + \underset{\sim}{X})\, \psi^*(\underset{\sim}{q} - \underset{\sim}{X}) . \tag{3.50}$$

Figure 3.6 shows the ground state surface of section ($E = 0.00$), Fig. 3.7 the surface of section for $E = 4.78$, which is close to but below E_c, and Figs. 3.8 and 3.9 the surfaces of section for $E = 6.28$ and $E = 9.96$, close to but above E_c and near the dissociation limit, respectively. In contrast to the conjecture made by Berry, these distributions are very regular!

The results of the Wigner function calculation are not the only ones to show regularity in the wavefunction not immediately apparent in the basis set expansion used to infer the analogue of the KAM transition. Figure 3.8 shows that the wavefunction of a state in the energy domain where the classical dynamics is dominantly stochastic is rather regular in appearance. Certainly, it is not evident to the eye that the wavefunction in coordinate space has a random component in its amplitude or phase.

Although I originally believed that it was correct, I now think that Berry's conjecture focusses attention on the wrong trajectories. It is correct that the quasiperiodic orbits embedded in the stochastic domain have small measure, but it is arguable that it is only these that are pertinent in the comparison with the eigenstates of the Hamiltonian operator. The argument, which is based on the accuracy of the several semiclassical quantization schemes, is as follows. All of these schemes start from a quasiperiodic trajectory and generate new quasiperiodic trajectories. The calculations of Helleman and Bountis [35] suggest that solutions of this type are dense in the stochastic region, although of very small measure relative to the stochastic solutions. For the calculations thus far reported, the complete agreement, both in number and value, between the eigenvalues determined by semiclassical quantization and by diagonalization of the Hamiltonian operator then suggests

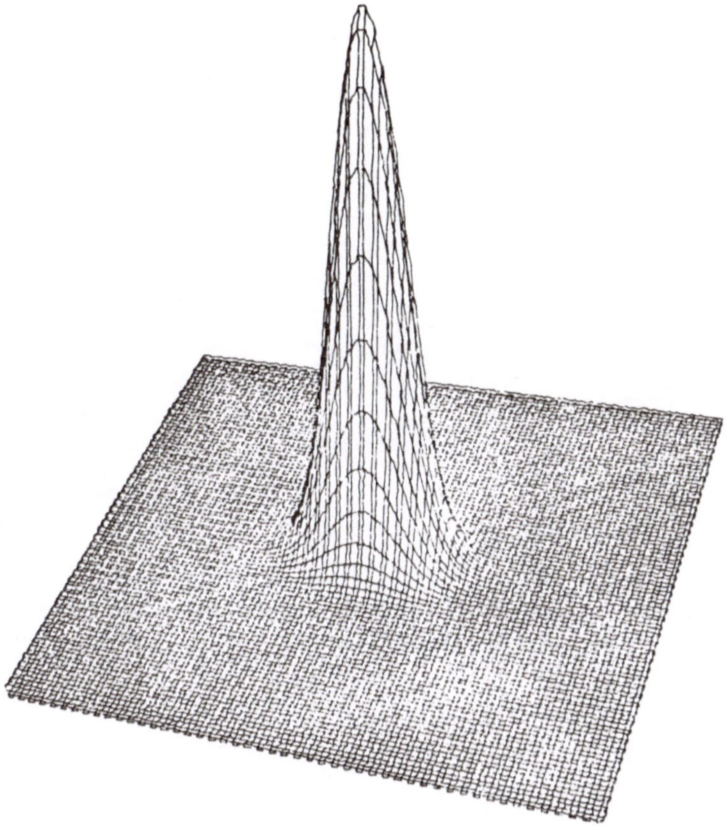

Fig. 3.6. Surface of section for the Wigner function (3.49) at $E = 0.00$. (From J.S. Hutchison and R.E. Wyatt, private communication.)

that there is a one-to-one correspondence between quasiperiodic trajectories and stationary states of the system. Or, put another way, the intricate substructure of the classical trajectories that leads to stochastic behaviour plays no role in determining the stationary states of the system.

The just described point of view is rather close to that stated by Einstein and the recent conjectures of Percival [44]. It remains the case that the Einstein quantization condition cannot be used for the overwhelming majority of classical trajectories above the KAM transition since in that domain p_k cannot be expressed as a function of the q_k. What has not been adequately emphasized, however, is that there are periodic and quasiperiodic solutions embedded in the stochastic domain, and the Einstein quantization

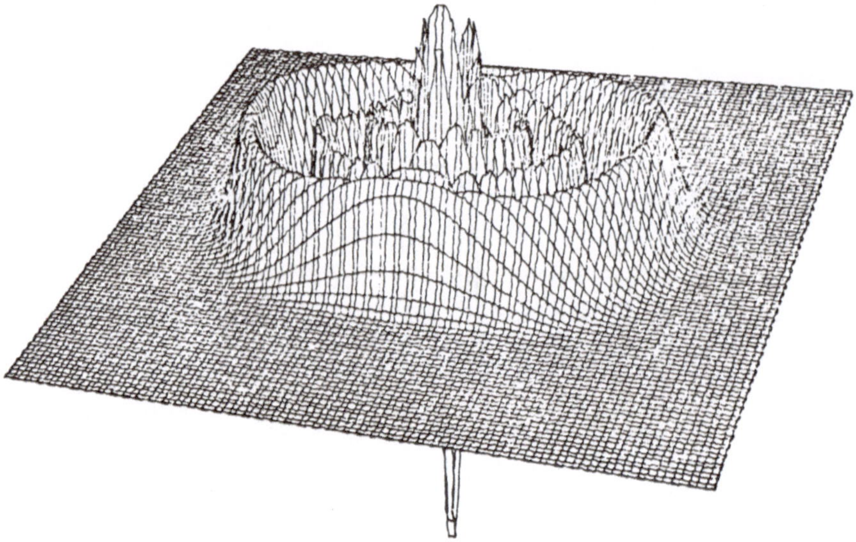

Fig. 3.7. Surface of section for the Wigner function (3.49) at E = 4.78. (From J.S. Hutchison and R.E. Wyatt, private communication.)

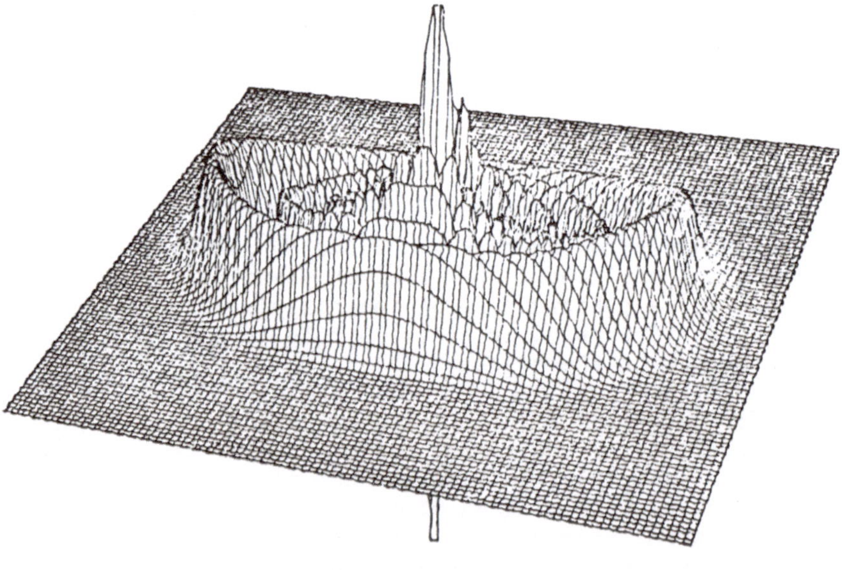

Fig. 3.8. Surface of section for the Wigner function (3.49) at E=6.28. (From J.S. Hutchison and R.E. Wyatt, private communication.)

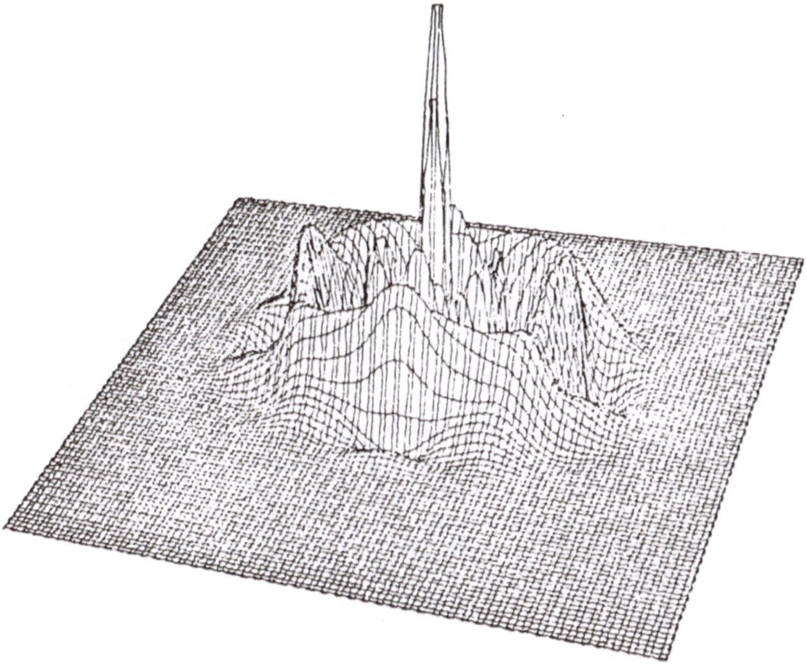

Fig. 3.9. Surface of section for the Wigner function (3.49) at $E = 9.96$. (From J.S. Hutchison and R.E. Wyatt, private communication.)

condition does apply to these. As already stated, the numerical evidence is that these states exhaust the bound state spectrum of the Hamiltonian operator. As to Percival's perceptive remarks, the Helleman and Bountis results illustrate that the nature of the quasiperiodic solutions is inordinately sensitive to the initial conditions, just as inferred by Percival. A further connection between conclusion (iii) and (v) of Nordholm and Rice and Percival's notion of regular and irregular spectra has been made in a calculation by Pomphrey [40] He studied the sensitivity of the parameterized Henon-Heiles Hamiltonian ($m = 1$)

$$H = \frac{1}{2}(p_1^2 + p_2^2) + \frac{1}{2}(q_1^2 + q_2^2) + \alpha(q_1^2 q_2 - \frac{1}{3}q_2^3) \tag{3.51}$$

to the value of α. In this case the dissociation energy is $1/6\alpha^2$. Pomphrey computed the eigenvalues of (3.51) for the range $0.090 <$ $\alpha < 0.086$ and examined the sensitivity of the spectrum as a function of the energy. This sensitivity is measured by the second difference

$$\Delta_i \equiv |\{E_i(\alpha + \Delta\alpha) - E_i(\alpha)\} - \{E_i(\alpha) - E_i(\alpha - \Delta\alpha)\}| \quad (3.52)$$

Perturbation theory yields the result

$$\Delta_i \sim \theta(\Delta\alpha^3) . \quad (3.53)$$

The calculations show that for $E < 16 = 0.74\ E_D$ all second differences are very small. This is the regular region of the spectrum, corresponding to localized asymptotic distribution over the basis states, and to quasiperiodic motion in the classical limit. For $E < 16$ eigenvalues are found with corresponding Δ_i orders of magnitude larger, i.e., the spectrum is very sensitive to small changes in α. This is the irregular region of the spectrum, corresponding to global asymptotic distribution over the basis states, and to apparently stochastic motion in the classical limit (see Fig. 3.10). It is also illuminating to compare the coverage of the surface of section by the apparently stochastic trajectory with the region of Hilbert space wherein the spectrum is very sensitive to the coupling. For the classical Henon-Heiles Hamiltonian the total area covered by unstable trajectories up to energy E is

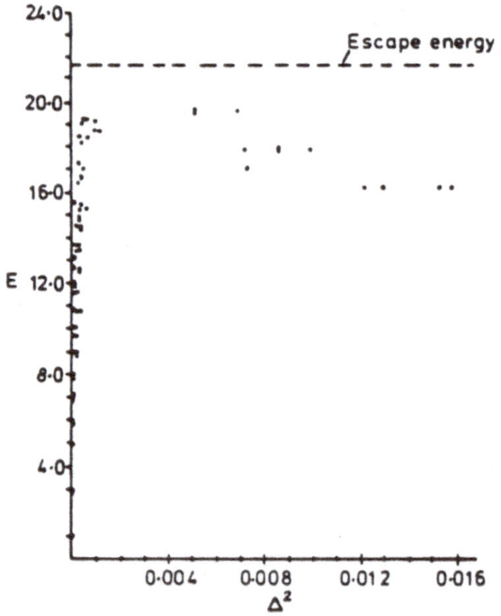

Fig. 3.10. Sensitivity of the eigenvalues of the Henon-Heiles Hamiltonian to small changes in coupling. (From N. Pomphrey, J. Phys. B. 7, 1909 (1974).)

$$I(E) = \int^{E} \alpha_I(E)dE \tag{3.54}$$

$$\alpha_I(E) = 0,\ E < 0.68E_D; \quad \alpha_I(E) = 3.125(\frac{E}{E_D}) - 2.125,\ E > 0.68E_D$$

where "total area" means the relative area of the surface of section. The quantity is to be compared with

$$S(E) = \frac{1}{E_D} \sum_i^{E} n_I(E_i) \langle \Delta E_i \rangle , \tag{3.55}$$

corresponding to the part of Hilbert space where the spectrum is very sensitive to a change in α. Here $n_I(E) = 1$ if E_i is very sensitive to the value of α, $n_I(E_i) = 0$ otherwise. Also,

$$\langle \Delta E_i \rangle = \frac{1}{2}(E_{i+1} - E_{i-1}) \tag{3.56}$$

As shown in Fig. 3.11 the quantum mechanical results follow, qualitatively, the shape of the classical curve. (I(E) and S(E) have different dimensionality, hence cannot agree quantitatively.)

Thus far attention has been focussed on the similarities between the classical and quantum dynamics of the several systems discussed. Heller [40a] in an important contribution, has pointed out the existence of interference effects that lead to a fundamental difference between the quantum mechanical and classical mechanical behaviour of systems of coupled nonlinear oscillators. He showed, using only group theoretical arguments, that nonlinear oscillator systems which have symmetries leading to degenerate quantum states do not transfer energy equivalently to rigorously equivalent phase space locations. In particular, if the initial state is a wave packet, Heller shows that the time averaged probability of finding the system in the initial state is larger than that of finding the system in states which are symmetrically equivalent. This behaviour, which is independent of energy, contradicts the behaviour expected when the corresponding classical mechanical trajectory is stochastic, since in the latter case the time averaged probabilities for finding the system in symmetrically equivalent states are equal.

The interference effects, destructive in some regions of phase space and constructive in other regions, represent only one of the fundamental differences between the quantum mechanical and classical mechanical descriptions of a system. An equally important difference arises from the nature of the preparation and observation processes. Kosloff and Rice have demonstrated that the means that must be used to verify the prediction that the time averaged probabilities of finding the system in symmetrically equivalent states are not the

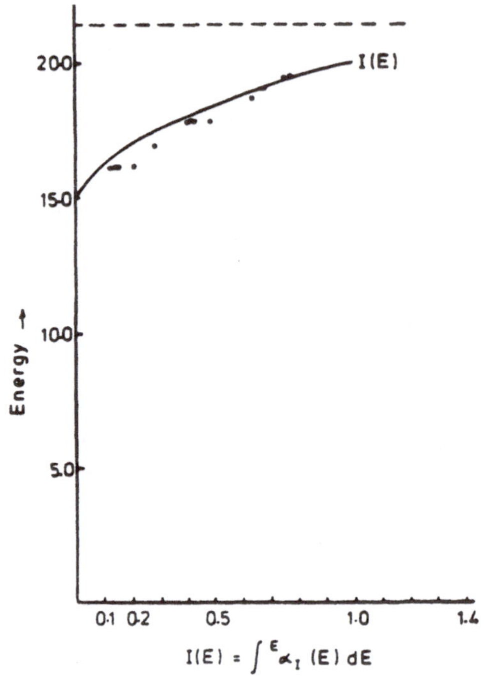

Fig. 3.11. Comparison of the area of surface of section covered by nonperiodic trajectories and the domain of Hilbert space where the eigenvalues are sensitive to small changes in coupling. (From N. Pomphrey, J. Phys. B. 7, 1909 (1974).)

same, destroy the basis for the lack of equality of time averaged amplitudes [40b]. Thus the complete quantum mechanical description, including the nature of the preparation and observation processes, does not predict asymptotic behaviour which is necessarily different from that of the classical mechanical description.

Heller's argument can be summarized as follows. Let ψ_a be the wavefunction of the initial state. The time averaged probability of finding the system in the state ψ_b is

$$P(a|b) = \lim_{T\to\infty} \frac{1}{T} \int_0^T \mathrm{Tr}(\hat{\rho}_a(t)\hat{\rho}_b)dt, \tag{3.57}$$

where $\hat{\rho}_a = |a\rangle\langle a|$, and $\hat{\rho}_b = |b\rangle\langle b|$ are the density operators corresponding to ψ_a and ψ_b. Suppose the eigenfunctions of the system Hamiltonian are ϕ_{ni}, where n labels the energy and i the degeneracy. Then

$$\psi_a = \sum_{n,i} a_{ni}\,\phi_{ni}\,,$$
$$\psi_b = \sum_{m,i} b_{mi}\,\phi_{mi}\,, \tag{3.58}$$

so that (3.57) can be rewritten in the form

$$\begin{aligned} P(a|b) &= \sum_n \sum_m \left| \sum_i a_{ni}\, b_{mi} \right|^2 \\ &= \sum_n \left| \underset{\sim}{a}_n \cdot \underset{\sim}{b}_n \right|^2 \end{aligned} \tag{3.59}$$

Heller now compares P(a|a) with P(a|Ra), where R is a symmetry operation. Because R is represented by a unitary matrix which preserves length, it is found that

$$|\underset{\sim}{a}_n \cdot \underset{\sim}{b}_n|^2 = |\underset{\sim}{a}_n \cdot \Gamma(R)\underset{\sim}{a}_n|^2 \le |\underset{\sim}{a}_n \cdot \underset{\sim}{a}_n|^2\,, \tag{3.60}$$

which implies

$$P(a|a) \ge P(a|Ra)\,. \tag{3.61}$$

The prediction implied by (3.61) is not meaningful unless a means of preparation and observation is specified. We now note that in order to differentiate between symmetry equivalent states of an isolated system it is necessary to break the isolation and the symmetry, i.e., the preparation and the measurement processes necessarily introduces into the total Hamiltonian for the system and the preparation or measuring apparatus a term which lifts the degeneracy of the states of the isolated system Hamiltonian. When that degeneracy is lifted Heller's argument ceases to be valid.

Consider, as an example, the Henon-Heiles Hamiltonian. The potential energy surface in this case has three fold symmetry and belongs to the group C_{3v}:

$$v(r,\theta) = \frac{1}{2}r^2 + \frac{1}{3}r^3 \sin 3\theta\,. \tag{3.62}$$

For simplicity only the influence of the measurement process will be discussed; a similar argument to the one that follows holds for the preparation process. We imagine using a two level system as a measuring apparatus to differentiate the three equivalent states [40c, 40d]; the spin up state of the measuring apparatus will be correlated with ψ_a and the spin-down position with ψ_b. The initial state of the measuring apparatus is, then, spin up.

The interaction between the system and the measuring apparatus is taken to be

$$\hat{H}_I = g(t) \sin (\frac{\theta}{2}) (\hat{\sigma}_x - \frac{1}{2}) , \tag{3.63}$$

where g(t) specifies the time dependence of the probing interaction and $\hat{\sigma}_x$ is the appropriate angular momentum operator of the two level measurement apparatus. A measurement is made as follows: the system is probed, subject to the interaction Hamiltonian (3.63), for a period such that

$$\int_0^t g(t)dt = \frac{2\pi}{\sqrt{3}} . \tag{3.64}$$

As a result of this probing, the final state of the measuring apparatus records the time averaged probability P(a b) in a system in which the threefold degeneracy has been lifted. Although the form chosen for $\hat{H}_I$ in (3.63) is specific, the principle implied is generally valid; we conclude that Eq. (3.61) describes a situation which is necessarily disturbed by the observation process.

The result of this analysis of the influence of observations on the asymptotic distribution of amplitude in a quantum mechanical system only shows that the interference effects in a system with degenerate states that would prohibit quasi-ergodic behaviour are perturbed by observation; the result does not show that quasi-ergodic behaviour must be observed, nor does it show that it is impossible to observe the consequences of (3.61). Imagine an ensemble of systems with interaction (3.63) prepared in a coherent state. Suppose the measurement process is carried out, at different times $t_1 < t_2 < \ldots$ on different replicas of the ensemble. Although each measurement on one replica alters the amplitude distribution in that replica, rendering it useless for further measurements, we imagine that in the absence of measurements (3.61) is valid so that measurements on different replicas will yield behaviour different from that predicted if (3.61) is not valid. The validity of this interpretation will depend on the nature of the preparation process and of the observation made, and each case must be examined for special characteristics.

The last topic to be considered in this subsection is the nature of the absorption spectrum below and above the KAM-like transition in the harmonic oscillator basis state representation of the eigenstates of a coupled oscillator system. Stratt, Handy and Miller [39] have computed the expected infrared spectrum for a system with the Hamiltonian (3.1); typical results are shown in Fig. 3.12. Cases (a) and (b) are for excitation to quantum states below the transition, (c) to a borderline state, and (d) to a state above the

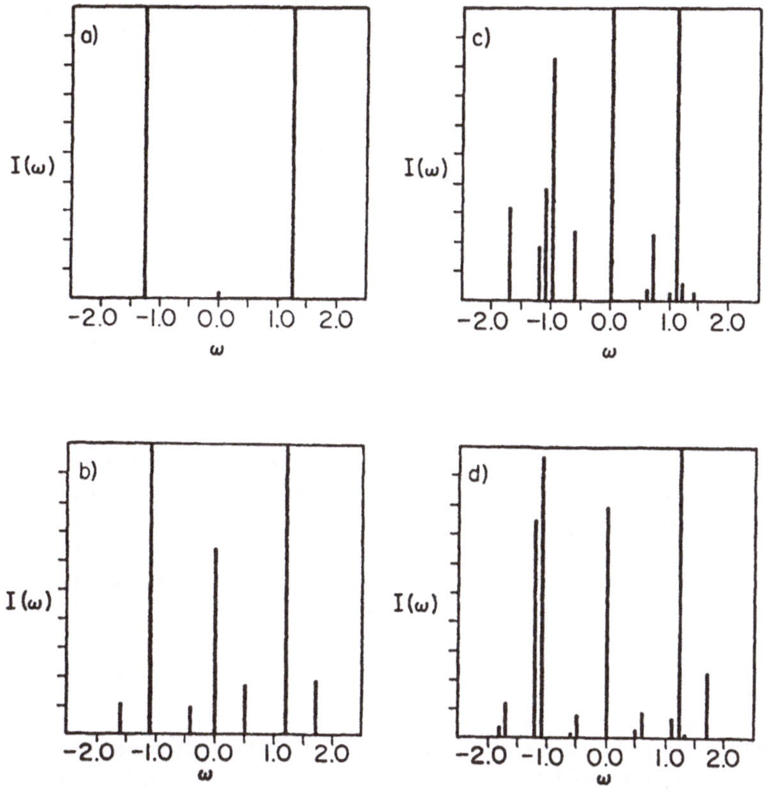

Fig. 3.12. The power spectra of selected eigenstates of the Hamiltonian (3.1). Cases (a) and (b) are for energies below the KAM-like transition, case (c) is borderline, and case (d) is for an energy above the transition (from ref. 39).

transition. There is an obvious trend from "regular" to "irregular" behaviour just as suggested by Percival.

Given the spectrum it is possible to devise a thought-experiment that creates a pure excited eigenstate of the coupled oscillator system (neglecting radiative damping). What is required is coherent excitation at the transition frequencies shown in, say, Fig. 3.12d. This can, in principle, be achieved by splitting up a single laser beam to drive several tunable sources with phase coherence. The difficulty is, of course, that in a real molecule we do not know which transitions to excite coherently. It is also necessary to enter the caveat that the coupling of vibrational and rotational motions in a molecule could alter the nature of the conclusions drawn from the studies of non-rotating coupled oscillator systems. However, if one could excite high energy molecular eigenstates (neglecting radiative dampling) it is conceivable that the ensuing chemistry is different from that of incoherently excited molecules.

One further comment: the nature of the intramolecular energy exchange can influence the rate of a unimolecular reaction. At first sight it appears that, because the classical KAM transition typically appears for $E < E_D$, the rate of fragmentation should be accurately accounted for by a statistical model. A deeper examination reveals that the matter is not so simple. First, the very nature of the irregular spectrum suggests that a decomposition rate might not be a monotone function of the energy. Second, resonances in the localized states of the bond that breaks could conceivably be derived from non-ergodic states of the molecule interspersed sparsely in the ergodic region of states. Third, the matrix elements coupling different vibrations of the molecule might vary over such a large range that only a subset of all vibrations is effectively coupled on the time scale of the reaction. All of these problems have been addressed, though none can be said to be "solved". The reader is referred to Refs. [55-63], which is a sampling of the approaches and ideas employed in this problem.

E. Summary

We have seen that if a coupled oscillator system has classical mechanical quasiperiodic trajectories for $E < E_c$ and stochastic trajectories for $E > E_c$, and if the eigenstates of that system are represented as a superposition of harmonic oscillator basis states, then there is a "KAM-like" change in the amplitudes of the contributing basis functions at about E_c. Yet the wavefunction itself, and the corresponding Wigner function do not display any sign of "randomness", and the evidence from semiclassical quantum theory suggests that only a subset of the quasiperiodic trajectories embedded in the stochastic trajectories correlate with the eigenstates of the system. Thus, despite the fact that the measure of these quasiperiodic trajectories is very small compared to the measure of all trajectories, they are the pertinent ones.

The idea that only the quasiperiodic trajectories correlate with the eigenstates of the system has been proposed many times. The argument (not a proof) I like best is due to Freed [64] who, from a careful consideration of the semiclassical approximation to the Green's function of the system, shows that interference effects destroy the contributions to the action of all but the quasiperiodic trajectories.

What, then, is the meaning of the KAM-like behaviour of the coefficients in the harmonic oscillator basis function expansion of the eigenstate? I do not know, and I think it is important to find out, especially if we wish to learn how to prepare a molecule in a state which will evolve in a way that produces interesting chemistry. I suggest it is plausible that the KAM-like behaviour of the representation of the wavefunction in some basis set is characteristic of any projection onto a set of localized states, and that as the "size" of the wavefunction increases, more and more localized basis functions are required for an adequate representation. If the "size" of the wavefunction increases more rapidly above some energy E_c, the behaviour of the projections onto the localized basis states will show an apparent change from "localized" to "global" character. Quasiperiodic trajectories with nearly the same energy can have rather different spatial extension, so one might expect rapid variation in the basis state coefficients as in the Nordholm-Rice calculation. It seems likely that the quasiperiodic trajectories embedded in the stochastic domain of the classical dynamics have just these properties.

4. CRITERIA FOR THE ONSET OF STOCHASTIC BEHAVIOUR

The calculations described in the preceding sections demonstrate the existence of a transition from quasiperiodic to stochastic behaviour in a Hamiltonian system. In this section we examine the various theories that have been advanced to account for and predict the onset of stochastic behaviour as a function of the energy of the system.

A. The Brumer-Duff-Toda Local Instability Criterion [65 , 66]

It has already been mentioned that the change from linear to exponential divergence of adjacent trajectories as time increases is characteristic of transition from quasiperiodic to stochastic behaviour of the system. Brumer and Duff [65], and independently Toda [66], assume that the existence of a local instability in a trajectory generates a global long time instability. Their arguments proceed as follows.

Consider a Hamiltonian system with N degrees of freedom. Imagine two sets of initial conditions, very nearly the same, corresponding to starting two trajectories from adjacent points. The distance betweeen these trajectories, one of which we call the reference trajectory, is measured by

$$\begin{aligned} d_j(t) &= q_j'(t) - q_j(t) , \\ d_{j+N}(t) &= p_j'(t) - p_j(t) . \end{aligned} \tag{4.1}$$

The differences $d_\ell(t)$ have time dependences determined by

$$\frac{d}{dt}[d_j(t)] = m_j^{-1}[p_j'(t) - p_j(t)] ,$$
$$\frac{d}{dt}[d_{j+N}(t)] = -\left[\frac{\partial V}{\partial q_j'} - \frac{\partial V}{\partial q_j}\right] . \tag{4.2}$$

To simplify Eq. (4.2) Brumer and Duff expand p_j' about p_j and $(\partial V/\partial q_j')$ about $(\partial V/\partial q_j)$ and linearize the resultant equations. It is found that

$$\dot{\underset{\sim}{d}}(t) = S(t)\, \underset{\sim}{d}(t) , \tag{4.3}$$

where $\underset{\sim}{d}(t)$ is the column vector of the time derivatives of the $d_\ell(t)$ and $S(t)$ is the $2N \times 2N$ matrix defined by

$$S(t) = \left[\begin{array}{c|c} O & \begin{matrix} m_1^{-1} & & & \\ & m_2^{-1} & & \\ & & \ddots & \\ & & & m_N^{-1} \end{matrix} \\ \hline A(t) & O \end{array}\right] , \tag{4.4}$$

where O is an $N \times N$ null matrix and $A(t)$ is the $N \times N$ matrix of the second derivatives $-(\partial^2 V/\partial q_i \partial q_j)$. Let the vector $\underset{\sim}{d}(t)$ be transformed to a new vector $\underset{\sim}{h}(t)$ by the definition

$$\underset{\sim}{d}(t) \equiv T(t)\, \underset{\sim}{h}(t) , \tag{4.5}$$

so that $\dot{\underset{\sim}{h}}(t)$ satisfies the equation

$$\dot{\underset{\sim}{h}}(t) = C(t)\, \underset{\sim}{h}(t) \tag{4.6}$$

with

$$C(t) = T^{-1}(t)\,[S(t)\, T(t) - \dot{T}(t)]. \tag{4.7}$$

The structure of Eq. (4.7) suggests clearly the motivation for these substitutions. The transformation $T(t)$ is chosen to diagonalize $S(t)$. If $T^{-1}(t)T(t)$ varies only slowly with t, then the qualitative behaviour of $\underset{\sim}{h}(t)$, hence $\underset{\sim}{d}(t)$, is determined by the eigenvalues of $S(t)$, which are the elements of $C(t)$. If the elements of $C(t)$ are all imaginary $\underset{\sim}{h}(t)$ is an oscillatory function of t, whereas if at least one is real and positive $\underset{\sim}{h}(t)$ grows exponentially with t. Brumer and Duff present qualitative arguments to support the notion that $T^{-1}(t)T(t)$ is slowly varying in time, and Toda simply drops the $T(t)$ term in (4.7) to achieve the same simplification. For Hamiltonians of the type we consider, in which the potential energy is a function of the coordinates only, the eigenvalues of $S(t)$ are also functions of the coordinates only, and are determined by the roots of the equation

$$\det\,[B - \lambda_k^2 1] = 0 , \tag{4.8}$$

where B is the N × N symmetric matrix of the second derivatives of the potential energy, $-(m_i m_j)^{-\frac{1}{2}}\,(\partial^2 V/\partial q_i \partial q_j)$, and 1 is the N × N unit matrix. Since B is symmetric the eigenvalues λ_k will occur in pairs $(-\lambda_k, \lambda_k)$ and be either real or pure imaginary. From the elements of B it is clear that the λ_k depend explicitly on the system masses and the nature of the potential energy surface. They also depend implicitly on the system energy and the several momenta since these define the coordinate space region sampled by a trajectory. It is this implicit energy dependence that is used to define the critical energy for stochastic behaviour, E_c. In particular, E_c is that energy required for a trajectory to sample some region of coordinate space within which $\mathcal{Re}\ \lambda_k > 0$. For very simple systems E_c can be calculated analytically, but for most systems of interest it must be calculated numerically. Details of the methodology used in the numerical calculations can be found in the original papers.

In Table 4.1 are displayed values of E_c for several model systems with N = 2. Comparison of the entries shows that in many cases the Brumer-Duff-Toda criterion yields an estimate of E_c that agrees with the results of numerical integration of the equations of motion. There are, however, cases in which it fails completely. The two glaring exceptions in the Table 4.1 are, first, the class of systems with integrable equations of motion, and, second, the unequal mass Toda system.

The accuracy of the Brumer-Duff-Toda method for estimating E_c depends on three assumptions: (1) First, and most important, it is assumed that a trajectory will become stochastic if it passes through a region of local instability. There is, at present, no obvious justification for this assumption. Indeed, examples are known in which the mapping that generates the trajectory creates balancing dilations and contractions in different portions of phase

Table 4.1. Critical Energies for Model Problems as Determined from Exact Dynamics*

Type	V(x,y)	m_1	m_2	E_c	E_c (Variational Equations)
Henon-Heiles	$\frac{1}{2}(x^2+y^2) + x^2y - \frac{1}{3}y^3$	1	1	1/12 - 1/8	1/12
Barbanis	$\frac{0.1}{2}(x^2+y^2) - 0.1\, x^2y$	1	1	0.075	0.00625
Tredgold	$\frac{1}{2}[\frac{3}{2}a(x^2+y^2) - \frac{3}{4}c(x^2+y^2)^2]$	$\frac{3}{2}$	$\frac{3}{2}$	$6a^2/24c$	$5a^2/24c$
Thiele-Wilson	$[1-\exp -(x+y)/2]^2 +$ $[1-\exp -(y-x)/2]^2$	$\frac{4}{11}$	$\frac{4}{3}$	≥0.6 for symmetric	≥0.5 for symmetric; 0.25 for unrestricted
Modified Henon-Heiles	$\frac{1}{2}(x^2+y^2) + x^2y + \frac{1}{3}y^2$	1	1	∞	1/24
Toda	See Eq. (2.68)	equal mass		∞	∞
		unequal mass		varies with mass	∞
(2-2) + (2-3) Resonance	Combination of terms from Eqs. (2.25) and (2.45)			∿ 0.2095	--
Henon-Heiles 12	$\frac{1}{2}(x^2+y^2) - \frac{1}{2}x^2y^2$	1	1	>0.5	>0.5

*From P. Brumer, Advances in Chemical Physics, in press.

space. The assumption that exponential divergence of adjacent trajectories occurs everywhere in those regions of phase space where some one $Re\ \lambda_k > 0$, and not just in the immediate vicinity of the point about which the expansion is made, has been tested by trajectory studies in a model of AB + C collisions [67]. Some typical results are shown in Figs. 4.1 and 4.2. Fig. 4.1 shows that two adjacent trajectories diverge exponentially as t increases, and Fig. 4.2 shows the corresponding magnitude of one of the eigenvalues of S in the region of space sampled by the reference trajectory. In this figure real eigenvalues are shown by the solid line, imaginary eigenvalues by the dashed line. A comparison of Figs. 4.1 and 4.2 shows that $|\underset{\sim}{d}(t)|/|\underset{\sim}{d}(0)|$ increases exponentially when $Re\ \lambda_1(t) > 0$, $Im\ \lambda_1(t) = 0$, but not when $Re\ \lambda_1(t) = 0$, $Im\ \lambda_1(t) > 0$. It is also found that the slope of log $[|d(t)|/d(0)]$ vs t in regions where $|d(t)|$ increases exponentially is well approximated by the average value

$$< \lambda_1 > = \frac{1}{t_f - t_i} \int_{t_i}^{t_f} \lambda_1(t)dt , \qquad (4.9)$$

where t_f, t_i, define the length of the interval over which $Re\ \lambda_1(t) > 0$. This numerical equivalence establishes a correlation between the rate of adjacent trajectory divergence in the neighbourhood of a local instability, which is determined by the eigenvalues of $S(t)$, and the global rate of adjacent trajectory divergence. Although similar results have been found from trajectory analyses of several systems, there are also found to be cases, in bound systems, in which trajectories expected to be locally unstable are not because of the influence of nearby periodic orbits. Despite these exceptions, the most commonly encountered situation is one in which the onset of stochastic behaviour is closely correlated with the penetration of the trajectory into a region of configuration space for which $Re\ \lambda_k > 0$.

The remaining assumptions of the Brumer-Duff-Toda analysis are associated with its execution:

(2) The regions of local instability are found by linearizing Eq. (4.2).

(3) It is assumed that both $T(t)$ and $T^{-1}(t)T(t)$ are sufficiently slowly varying with t to not qualitatively change the time dependence of $\underset{\sim}{d}(t)$.

Cerjan and Reinhardt [68] have pointed out that, since the reference trajectory and its adjacent partner both evolve in time, $A(t)$ in Eq. (4.4) is an implicit function of time through its explicit dependence on an origin on the reference trajectory. In the Brumer-Duff-Toda analysis this implicit time dependence is

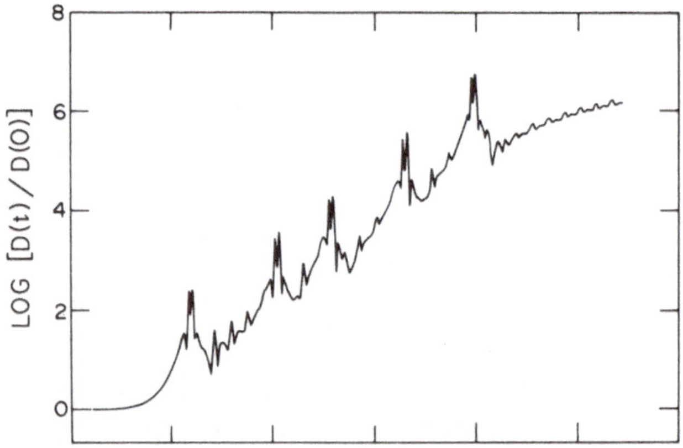

Fig. 4.1. $\ell n\ D(t)/D(0)$ vs t for a typical trajectory in an A + BC collinear collision. The system Hamiltonian is

$$H = \frac{11}{8}p_1^2 + \frac{3}{8}p_2^2 + [1 - \exp\{-(q_1 + q_2)/2\}]^2 + [1 - \exp(-q_1 - q_2)]^2$$

with q_1 the motion along the asymmetric stretch and q_2 the motion along the symmetric stretch of the linear ABC. The initial conditions correspond to the BC diatom being in the quasiclassical ground state, and the initial relative kinetic energy being 0.1 eV. The masses correspond to $H + H_2$. (From J.W. Duff and P. Brumer, J. Chem. Phys. 67, 4898 (1977).)

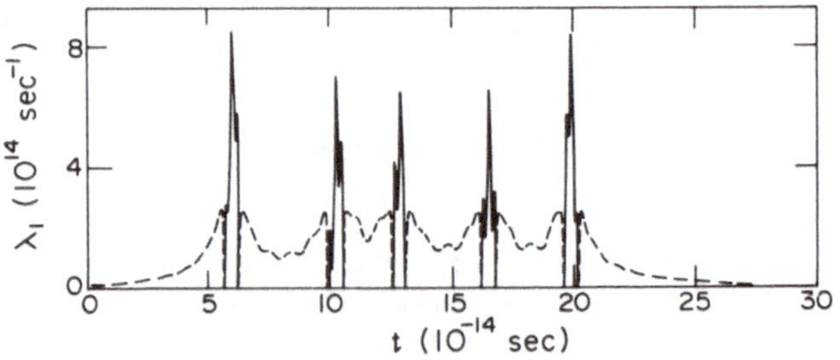

Fig. 4.2. λ_1, an eigenvalue of S(t), encountered by the unperturbed trajectory of the pair which leads to Fig. 4.1. The dashed curves denote real values of $\lambda_1(t)$ and the solid curves imaginary values of $\lambda_1(t)$. (From J.W. Duff and P. Brumer, J. Chem. Phys. 67, 4898 (1977).)

neglected, which is equivalent to searching for exponential divergence of the adjacent trajectory from a fixed point in phase space, rather than from a point on the simultaneously evolving reference trajectory. The net result is that the stability properties of neighbouring trajectories are tacitly assumed to follow the reference trajectory adiabatically. Cerjan and Reinhardt eliminate this approximation by modifying the procedure embodied in assumptions (2) and (3) above. They determine the exact critical points of the full nonlinear difference dynamics equation, Eq. (4.2), and then perform a linear stability analysis at each critical point. At a stable (elliptic) critical point the eigenvalues of the mapping are imaginary, and at an unstable (hyperbolic) critical point they are real and positive. Cerjan and Reinhardt find a richer spectrum of possible behaviour than is obtained from the Brumer-Duff-Toda analysis. In some cases the lowest energy critical point changes from elliptic to hyperbolic at E_c; then the trajectory becomes stochastic for $E > E_c$. In other cases higher order critical points, whose positions vary with the energy, become degenerate with the lowest critical point and have opposite stability characteristics; then the trajectory is stable or, at worst, weakly unstable for some energy far above E_c. When applied to the same systems as studied by Brumer and Duff, Cerjan and Reinhardt find that for each system the eigenvalue corresponding to the critical point at $\underset{\sim}{d}(t) = 0$ is the same as that deduced from Eq. (4.8), but that higher order critical points play an important role in the dynamics of systems for which the eigenvalues of (4.8) give an erroneous prediction of instability. In particular, they find that in integrable systems there are compensating critical points that alternate elliptic and hyperbolic character as the energy increases through the value of E_c determined from Eq. (4.8). For example, although the critical point corresponding to $\underset{\sim}{d}(t) = 0$ for the anti-Henon-Heiles system changes from elliptic to hyperbolic at $E = 1/24$, a higher critical point becomes degenerate with this one at $E = 1/24$, and it has the opposite stability characteristics. Thus there is only a second order instability, corresponding to the existence of the higher order critical point, which in general influences the dynamics for $E > 1/24$. We conclude, in agreement with the observation that the anti-Henon-Heiles system is integrable, there will not be a macroscopic trajectory instability in this system.

In summary, the local instability criterion for global instability, as developed by Brumer, Duff and Toda, and modified by Cerjan and Reinhardt, predicts that the existence of a first order instability in the difference trajectory $\underset{\sim}{d}(t)$ implies stochastic behaviour of the trajectory for $E > E_c$, and that second order instabilities do not generate a stochastic trajectory. To date this scheme satisfactorily accounts for the behaviour of systems with $N = 2$; it has not yet been applied to enough cases to assess its range of validity.

Despite its successes, the method of locating instabilities in trajectories described in this section is incomplete in several respects. For example, it is known that there are isolated hyperbolic fixed points below E_c in a number of systems. Although the measure associated with these points is very small, their existence reflects a feature of the system dynamics that is completely ignored by the analyses described in this section. Also, the methodology described gives no clue as to why stable closed orbits stabilize trajectories in neighbouring regions of phase space, as is found to occur in some cases. Cerjan and Reinhardt conjecture that assumption of an adiabatically developing instability relative to a reference trajectory does not allow for "restabilization" of the trajectory as it reenters a stable domain of the phase space. If the reentry occurs with proper phase relative to the reference trajectory, the initially destabilized trajectory can be propagated without growth of the instability. Behaviour of just this type has been found in a system studied by Benettin, Brambilla and Calgani.

Hänsel [68a,68b] has used a Lyupanov-type criterion to discuss the instability of a trajectory in the stochastic domain. The method is related to the subject of Kolmogorov entropy, so will not be discussed here.

B. The Chirikov Overlapping Nonlinear Resonances Instability Criterion [69]

We have already seen, via the study of a model Hamiltonian, that an isolated nonlinear resonance in a dynamical system leads to a distortion of the invariant curves in the Poincaré surface of section. Each resonance centre is located by the condition $m_1\omega_1(J_1) = m_2\omega_2(J_2)$ (m_1, m_2 integers), and it influences the system dynamics in a region of phase space with nonzero volume; that region, which is associated with the domain of distortion of the toroidal surfaces characteristic of the dynamics in the absence of the nonlinear resonance, is called a resonance zone. The size of a resonance zone is, generally, a decreasing function of the order of the nonlinear resonance, hence it is reasonable to assume that the low order resonances have the largest influence on the qualitative character of the system dynamics. Following this line of reasoning, Zaslavskii and Chirikov conjectured that the system dynamics becomes stochastic when the energy is such that the resonance zones arising from independent low-order nonlinear resonances overlap [70]. Indeed, the model system described by Ford [11], which was discussed in Section 2, shows this behaviour when $\alpha \neq 0$ and $\beta \neq 0$ (see Fig. 4.3). For $E < 0.2095$ the 2-2 and 2-3 resonance zones do not overlap, whereas for $E > 0.2095$ they do, and the invariant curves in the Poincaré surface of section show a corresponding change from simple arcs to erratic, disjoint points of intersection.

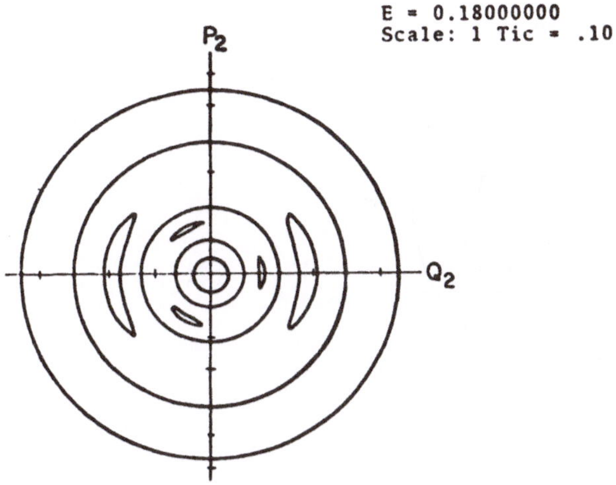

(a)

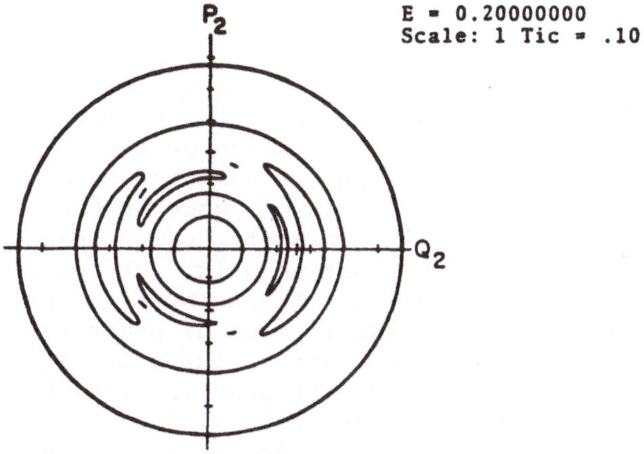

(b)

Fig. 4.3. Surfaces of section for the system with

$$H = H_o + J_1J_2 \cos(2\phi_1 - 2\phi_2) + J_1J_2^{3/2} \cos(2\phi_1 - 3\phi_2)$$

$$H_o = J_1 + J_2 - J_1^2 - 3J_1J_2 + J_2^2.$$

(a) E = 0.18000; (b) E = 0.20000; (c) E = 0.20950.
(From G.H. Walker and J. Ford, Phys. Rev. <u>188</u>, 416 (1969).)

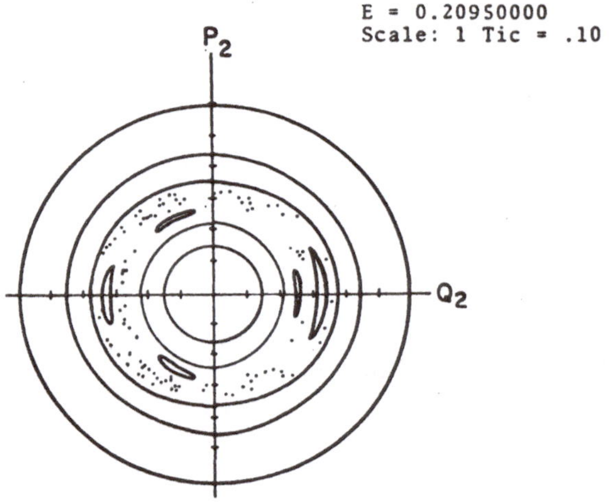

Fig. 4.3(c)

To quantify this criterion of instability for a general N-oscillator system Chirikov writes the system Hamiltonian in the form

$$H(J,\phi) = H_o(J) + \sum_{\underset{\sim}{m}} V_{\underset{\sim}{m}} e^{i\underset{\sim}{m}\cdot\underset{\sim}{\phi}} \tag{4.10}$$

where $\underset{\sim}{m} = (m_1, m_2, \ldots, m_N)$ and $\underset{\sim}{\phi} = (\phi_1, \phi_2, \ldots, \phi_N)$. He then asserts that close to the m_1-m_2 nonlinear resonance all other terms in the sum of (4.10) can be neglected. The residual Hamiltonian is, next, transformed by expansion about the position of the m_1-m_2 resonance centre into a form isomorphous with the Hamiltonian of a pendulum. The size of the m_1-m_2 nonlinear resonance zone is estimated to be equal to the size of the cell in action-frequency space enclosed by the separatrix of the isomorphous pendulum motion. This choice is motivated by the observation that the separatrix of the pendulum motion is the boundary between two regions of phase space, in one of which the motion is oscillatory and in the other of which it is rotation (Fig. 4.4). Moreover, the pendulum motion is a harmonic oscillation only when the amplitude is small, or what is the same thing the maximum momentum is small. As the amplitude is increased the frequency of oscillation decreases, and it becomes zero on the separatrix. Similarly, for any given value of the momentum in the domain where it is large there is some characteristic rotation frequency, and as the energy is decreased that frequency decreases and becomes zero on the separatrix. The separatrix has two branches which intersect at homoclinic points which define the position of unstable equilibrium. These intersections can be thought of as a unique "trajectory", since in the absence of any perturbation a pendulum would remain in the corresponding configuration forever.

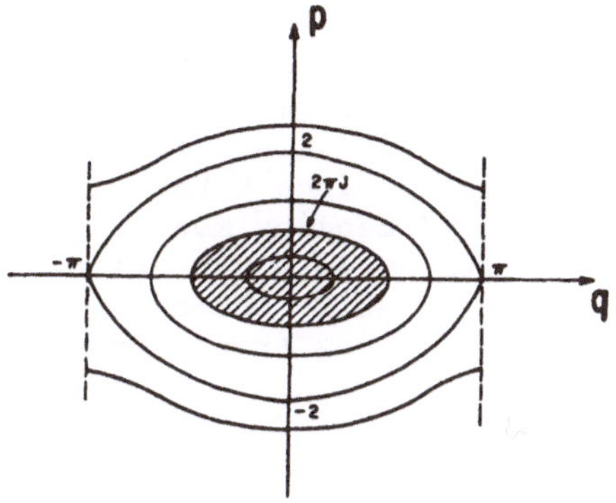

Fig. 4.4. Trajectories of the motion of a pendulum in phase space.

Clearly, if the motion of the pendulum can be described by a trajectory very close to the separatrix, on either side, a small perturbation can grossly alter the dynamics by converting an oscillation into a rotation, or vice versa.

Rather than outline the analysis given by Chirikov, which has been mostly used to describe motion under Hamiltonians which include an externally applied oscillatory perturbation, we shall examine a model introduced by Oxtoby and Rice [71], which is intended to elucidate the relationships between nonlinear resonances and statistical behaviour in intramolecular energy exchange.

The molecular model considered consists of a "critical" nonlinear oscillator driven by coupling to "other" vibrational modes; rotation-vibration interaction is neglected. Oxtoby and Rice choose bond-angle coordinates to describe the motions in the system.

When energy transfer due to interaction between non-bonded atoms is neglected the potential energy is separable in bond-angle coordinates, and the kinetic energy is of the form

$$KE = \frac{1}{2}\sum_i \sum_j p_i \, G_{ij} \, p_j \; . \tag{4.11}$$

G_{ij} depends on the bond lengths, angles and constituent masses. Suppose G_{ij} is approximated by its value for the equilibrium configuration of the molecule, denoted G^o_{ij}. Then the molecular Hamiltonian has the simple form

$$\begin{aligned} H &= H_o + V \\ &= \sum_i [U_i(q_i) + \tfrac{1}{2}G^o_{ii}p_i^2] + \sum_i \sum_{>j} G^o_{ij}p_ip_j \; . \end{aligned} \tag{4.12}$$

When G^o_{ij} vanishes, as in the coupling of bond bending and bond stretching in a linear molecule, the next higher term in G^o_{ij} is retained and V is only slightly more complicated. Note that V is pairwise additive between modes. Also, for a large molecule many of the coupling terms are zero since G has matrix elements only between nearby stretching and bending modes. Furthermore the perturbation averages to zero over a vibrational period. Then the dominant effect of V will arise through near resonant coupling of two modes. Oxtoby and Rice now transform to action-angle variables. Then V becomes

$$V = \sum_i \sum_{<j} V^{ij}(J_i,\phi_i,J_j,\phi_j)$$

$$2V^{ij} = \sum_{m,n} V^{ij}_{mn}(J_i,J_j)\, e^{i(m\phi_i + n\phi_j)} + \text{C.C.} \tag{4.13}$$

Recall that a resonance occurs when

$$\frac{d}{dt}(m\phi_i + n\phi_j) = 0 , \tag{4.14}$$

or

$$m\omega_i(J_i) + n\omega_j(J_j) = 0. \tag{4.15}$$

For given total energy in the two interacting modes i,j, the resonance condition defines a resonance centre - denoted J^r_i, J^r_j - in phase space for each m,n. Following the general description given above, close to the resonance centre the nonresonant terms in V^{ij} can be neglected since they are small relative to resonant term. Then, near a resonance the total Hamiltonian simplifies to

$$H_r \approx \omega_i' \frac{(\Delta J_i)^2}{2} + \omega_j' \frac{(\Delta J_j)^2}{2} + V^{ij}_{mn}(J^r_i,J^r_j)\cos(m\phi_i + n\phi_j) \tag{4.16}$$

$$\Delta J_i \equiv J_i - J^r_i ,$$

$$\omega_i' \equiv \left(\frac{d\ i}{dJ_i}\right)_{J_i=J^r_i} \tag{4.17}$$

By virtue of conservation of energy in the two intersecting oscillators

$$\Delta J_j = \frac{\partial J_j}{\partial J_i}\Delta J_i , \tag{4.18}$$

hence

$$H_r = [\omega_i' + \omega_j'(\frac{\partial J_j}{\partial J_i})^2] \frac{(\Delta J_i)^2}{2} + V_{mn}^{ij}(J_i^r, J_j^r) \cos(m\phi_i + n\phi_j) \tag{4.19}$$

Note that $(\partial J_j/\partial J_i)$ is to be calculated subject to conservation of energy in the pair of oscillators i,j.

The reduced form for H_r (4.19) is, as mentioned above, the Hamiltonian for a simple pendulum. When ΔJ_i is small enough the restoring force in V_{mn}^{ij} will pull the resonance back toward the resonance centre. The width of the resonance is determined by the range of J_i over which the "pendulum" is stable. This is

$$(\Delta J_i)_{mn} = \left(4 \frac{|V_{mn}^{ij}|}{|\omega_i' + (\frac{\partial J_j}{\partial J_i})^2 \omega_j'|} \right)^{1/2}, \tag{4.20}$$

and the width in energy space is

$$(\Delta E_i)_{mn} \approx (\frac{\partial E_i}{\partial J_i})_{J_i = J_i^r} (\Delta J_i)_{mn} = \left(\frac{|V_{mn}^{ij}|}{|\omega_i' + (\frac{\partial J_j}{\partial J_i})^2 \omega_j'|} \right)^{1/2} 4\omega_i(J_i^r) \tag{4.21}$$

When the energy lies within $(\Delta E_i)_{mn}/2$ of E_i^r the system will be stabilized and will oscillate around the resonance centre. Since for every choice of m,n there is a resonance, the set of resonance centres is dense in action space. But, the resonance widths decrease rapidly as m,n increase so that resonance overlap considerations can be restricted to only the first few resonances.

Oxtoby and Rice propose that the molecular dynamics can be qualitatively classified according to the locations and widths of the nonlinear resonances (4.15). Phase space is then divided into three parts:

(i) The representative point lies outside all resonances. Then the energies of the different vibrational modes change slowly.

(ii) The representative point moves under the influence of a single resonance. In this case the energies in the resonantly coupled modes can, under certain conditions, change rapidly, but the motion is periodic, and the resonance is stabilized.

(iii) The representative point is simultaneously influenced by several resonances: numerical studies indicate that then the trajectories appear to behave stochastically, erratically filling the phase space.

The behaviour described under (iii) leads to the contention that the stochasticity arising from overlap of nonlinear resonances leads to rapid energy exchange between vibrational modes. Oxtoby and Rice have studied this contention by examining simple models. For example, suppose that resonance between a pair of oscillators dominates the approach to stochasticity. Take

$$H_o = \sum_i [U_i(q_i) + \frac{1}{2} G^o_{ii} p_i^2] , \tag{4.22}$$

$$U_i(q_i) = D_i[1 - \exp(-q_i/a_i)]^2 \tag{4.23}$$

where D_i is the bond dissociation energy and a_i defines the length scale for the potential. The frequency of small amplitude motion is

$$\Omega_i = (\frac{2D_i G^o_{ii}}{a_i^2})^{1/2} \tag{4.24}$$

In action-angle variables H_o is

$$H_o = \sum_i [D_i - D_i(1 - \Omega_i J_i/2D_i)^2] , \tag{4.25}$$

$$\omega_i \equiv (\frac{\partial H}{\partial J_i}) = \Omega_i(1 - \Omega_i J_i/2D_i) . \tag{4.26}$$

Since

$$E_i = D_i - D_i\ (1 - \frac{\Omega_i J_i}{2D_i})^2 \tag{4.27}$$

one can write

$$\omega_i(E_i) = \Omega_i(1 - \frac{E_i}{D_i})^{1/2} . \tag{4.28}$$

The corresponding momentum is

$$p_i = (\frac{2D_i}{G^o_{ii}})^{1/2} \frac{\omega_i}{\Omega_i} (1 - \frac{\omega_i^2}{\Omega_i^2}) \frac{\cos\phi_i}{1 + (1 - \omega_i^2/\Omega_i^2)^{1/2} \sin\phi_i} \tag{4.29}$$

where $\phi_i \equiv \omega_i t + \Theta_i$ is the angle variable of the canonical transformation. Continuing, (4.21) becomes,

$$(\Delta E_i)_{mn} = 8\sqrt{2}\left(\frac{D_i D_j (G^o_{ij})^2}{\Omega_i^2 \Omega_j^2 G^o_{ii} G^o_{jj}}\right)^{1/4}\left(\frac{D_i \omega_i^3 \omega_j^3}{\Omega_i^2 \omega_j^2 + \Omega_j^2 \omega_i^2}\right)$$

$$\times\left(\frac{1 - \omega_i/\Omega_i}{1 + \omega_i/\Omega_i}\right)^{m/4}\left(\frac{1 - \omega_j/\Omega_j}{1 + \omega_j/\Omega_j}\right)^{n/4} \qquad (4.30)$$

and

$$m\omega_i(E_i) = n\omega_j(E_j) \qquad (4.31)$$

The results of numerical calculations for this model reveal the following:

(i) In typical two oscillator resonance dominated dynamics there will be significant nonrandom behaviour over the entire energy range up to D_i. Slow energy redistribution occurs because the system can be "trapped" for many vibrational periods near the centres of isolated nonlinear resonances (see Figs. 4.5 and 4.6).

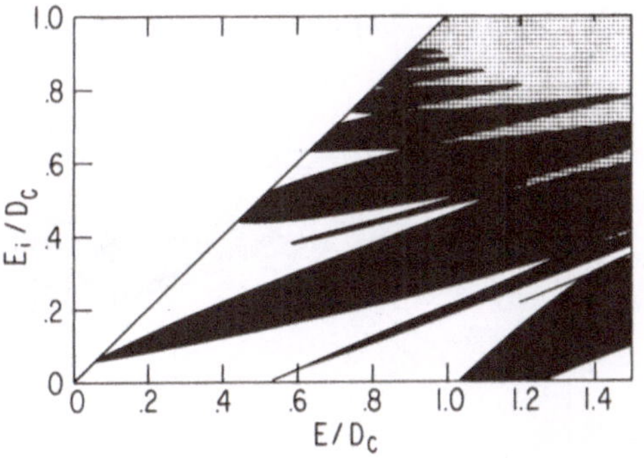

Fig. 4.5. Resonance structure for CC stretch (1000 cm^{-1}) coupled to CH stretch (2900 cm^{-1}); CH dissociation energy is 2D, where D is the CC dissociation energy. Black areas indicate single resonances; cross-hatched areas indicate resonance overlap. E is the total energy in the two oscillators, E_i the energy in the critical CC stretching mode. (From D.W. Oxtoby and S.A. Rice, J. Chem. Phys. 65, 1676 (1976).)

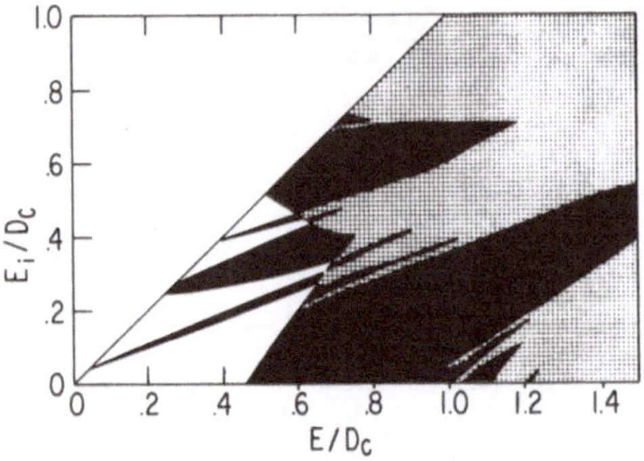

Fig. 4.6. Resonance structure for CC stretch (1000 cm^{-1}) coupled to another CC stretch (1300 cm^{-1}); the second CC dissociation energy is 1.5 D. (From D.W. Oxtoby and S.A. Rice, J. Chem. Phys. 65, 1676 (1976).)

(ii) As the total energy increases the relative volume of phase space occupied by overlapping resonances increases, so that energy redistribution becomes the behavioural norm (see Figs. 4.5, 4.6 and 4.7).

(iii) If one bond dissociation energy becomes large, the corresponding oscillator becomes more harmonic, leading to a decrease in the width and (especially) number of nonlinear resonances.

(iv) As the two frequencies Ω_i, Ω_j move apart, the number of nonlinear resonances decreases, especially so at low energy.

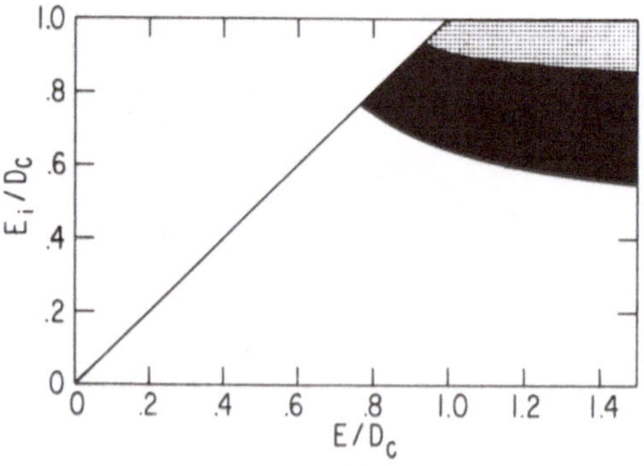

Fig. 4.7. Resonance structure for CC stretch (1000 cm^{-1}) coupled to CCH bend (400 cm^{-1}); the bending mode is taken to be harmonic. (From D.W. Oxtoby and S.A. Rice, J. Chem. Phys. 65, 1676 (1976).)

(v) Resonance widths are proportional to

$$[(G_{ij}^{o})^2/G_{ii}^{o}G_{jj}^{o}]^{1/4}$$

so that large changes in masses and bond angles are necessary to affect the resonances and their overlap.

(vi) Harmonic bending modes (especially low frequency modes) are strongly coupled to the critical bond stretch only when the critical mode is very close to dissociation. Then, a likely pathway for energy transfer to a breaking bond involves, first, transfer from other bond stretching modes and, second, only when the critical bond is close to dissociation, transfer from the bending modes (see Fig. 4.7).

Values of the critical energy for the transition to stochastic behaviour are $E_c \approx 0.85D_c$ for the model shown in Fig. 4.5 and $E_c \approx 0.65D_c$ for the model shown in Fig. 4.6.

What happens when there are many nonlinear oscillators? For two interacting oscillators the zeroth order energy surface is one dimensional. In this case there is only one path from a given point to any other, hence narrow resonances do not overlap. For $N > 2$ oscillators, the energy surface is of dimension $N - 1$, and centres of resonances are not points, but rather $N - 2$ dimensional surfaces.

(vii) Many oscillator case: Suppose one oscillator is singled out for attention. If a large amount of energy is put into this oscillator relaxation will proceed in two stages. First there will occur transfer of energy to a few modes directly coupled to the critical oscillator and, second, transfer from these modes to the rest of the molecule. We expect the second process to be, basically, "statistical" in character. The same argument applies to the reverse process, namely energizing some vibrational mode. The analysis of the first process leads to consideration of an ensemble of one-dimensional resonance distributions, each member of the ensemble corresponding to a different distribution of non-critical energy amongst the $N - 1$ oscillators directly coupled to the critical oscillator. The general conclusions concerning stochasticity in the two oscillator case are found to remain valid in the many oscillator case (see Fig. 4.8). It is found that as the number of oscillators increases, the energy of each one (on average) decreases. Consequently, the noncritical oscillators are closer to harmonic and each one has fewer and narrower resonances with the critical oscillator. Thus resonance overlap need not increase with the number of oscillators directly coupled to the critical oscilllator (see Fig. 4.8).

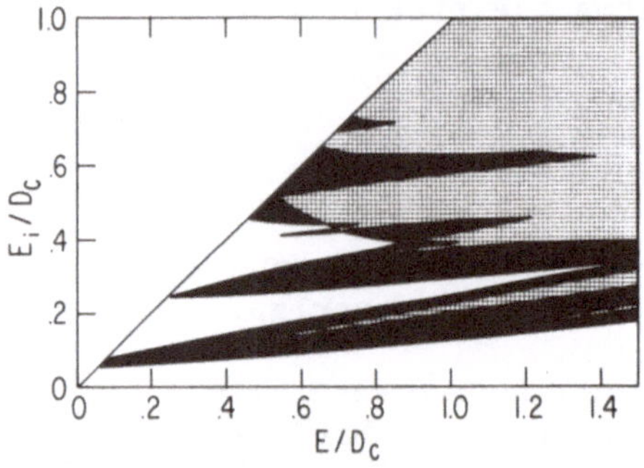

Fig. 4.8. Resonance structure for the CC stretch (1000 cm^{-1}) coupled equally to the three modes of Figs. 4.4 - 4.7. E is the total energy in all four modes, E_i the energy in the critical stretching mode. (From D.W. Oxtoby and S.A. Rice, J. Chem. Phys. 65, 1676 (1976).)

A direct confirmation of the Oxtoby-Rice analysis can be obtained by comparison with studies of dissociating trajectories on an energy surface. Identifying the onset of stochasticity with overlap of nonlinear resonances is in agreement with Bunker's [72] molecular dynamics studies of linear triatomic molecules and the existence of a random distribution of lifetimes against fragmentation (see Fig. 4.9).

C. The Autocorrelation Function Diagnostic for the Onset of Stochastic Behaviour

Mo has proposed to use the change in character of the time dependence of the autocorrelation function of a suitable dynamical variable as a diagnostic for the onset of stochastic behaviour in a Hamiltonian system [73]. The genesis of this idea is the following: If two trajectories, corresponding to slightly different initial conditions, lie in the conditionally periodic region of the phase space, then starting close to one another at t = 0 they move apart linearly as t increases. If the two trajectories lie in the stochastic region of phase space, under the same conditions on the initial state they move apart exponentially as t increases.

The autocorrelation function of the displacement along one trajectory behaves differently. In the quasiperiodic domain of the phase space it is an oscillatory function, but in the stochastic region of the phase space it decays, the latter behaviour representing the loss of correlation between the initial value of the position and the value at time t as t increases. The autocorrelation function of the velocity behaves analogously in these two regions of the phase space. Mo's calculation is designed to locate the

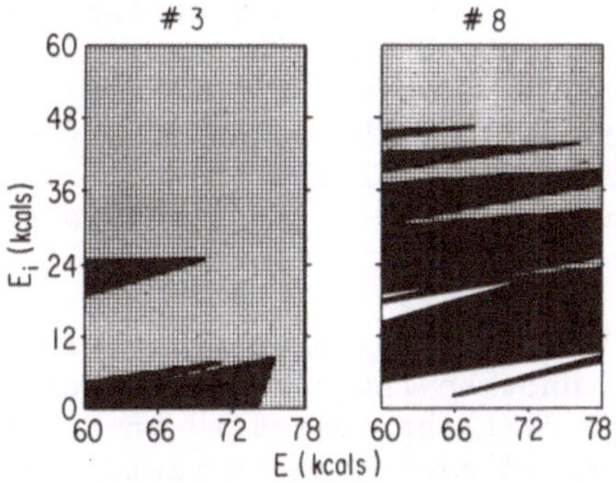

Fig. 4.9. Resonance structures for Bunker's model #3 (a random lifetime distribution case) and #8 (a random case). Critical oscillator dissociation energies in both cases are 60 kcal/mole. Bunker investigated the energy range from 62.5 - 75 kcal. (From D.W. Oxtoby and S.A. Rice, J. Chem. Phys. 65, 1676 (1976).)

onset of the change in character in the generalized velocity autocorrelation function defined by

$$D(t) = \langle \dot{d}_o(t)\dot{d}_o(0)\rangle , \tag{4.32}$$

where

$$d_o = \sum_{i=1}^{N} (p_i^2 + q_i^2) \tag{4.33}$$

is the distance from (p,q) to the origin of the phase space. Note that (4.32) is a mixed correlation function, since it contains the autocorrelation functions of velocity and force and the cross correlation function between the velocity and force, all along one trajectory.

Mo starts with the Liouville equation, defines an inner product on the microcanonical energy shell, and then develops an equation of motion for D(t) using the projection operator and continued fraction representation of Mori [74]. This representation is truncated at low order to yield practical equations for D(t). If the motion of the system is stable for most of the initial conditions there exists a series of canonical transformations which reduce the Hamiltonian to integrable form. Then D(t) has the structure

$$D(t) = \sum_i a_i \cos \theta_i t , \tag{4.34}$$

with the a's and θ's real valued functions of E_o and ε (where ε is defined by $H = H_o(p,q) + \varepsilon V(p,q)$). Only cosine terms appear because of time reversal invariance. When ε increases for fixed E_o, or E_o for fixed ε, some θ_i become complex or pure imaginary and $D(t)$ decays or grows exponentially. The value of ε for which $\theta_i \to$ complex is taken as the critical point for the change from quasiperiodic to stochastic behaviour. Note this condition is global, hence stronger than the local exponential separation of trajectories.

Mo asserts that $D(t)$ grows exponentially with t in the stochastic domain, but this cannot be the case. The definition of $D(t)$ requires that $d_1(0)$ and $d_1(t)$ be evaluated on the same trajectory, then their product averaged over trajectories. Consequently, if the trajectory becomes stochastic $d_1(t)$ becomes uncorrelated with $d_1(0)$, and $D(t)$ must decay. Tabor [75] has made calculations of $\langle d_o(0)d_o(t)\rangle$, $\langle q(0)q(t)\rangle$ and $\langle p(0)p(t)\rangle$ for the Henon-Heiles system and shown that the expected behaviour is found. The important point is that the time dependence of $D(t)$ changes at E_c, hence the appearance of complex θ_i signals the onset of stochastic behaviour.

Although Mo's calculations seem to give the correct value of E_c for the Henon-Heiles model, the exact result for the pure harmonic case, and for the case $H = \frac{1}{2}(p_1^2 + q_1^2 + p_2^2 + q_2^2) - \frac{1}{2}q_1^2q_2^2$ no transition as also found in Henon and Heiles' integration of the corresponding equations of motion, I suspect there is a flaw in her numerical work. This suspicion arises from the apparent verification that $D(t)$ grows exponentially above E_c, which cannot be correct. Yet it remains a fact that the predictions published agree with those obtained by other means. Clearly, more work is required to clarify this situation and the potential value of Mo's algorithm.

D. The Greene Orbital Instability Criterion and the Helleman-Bountis Periodic Orbital Construction [76, 35]

The KAM theorem states that tori supporting orbits with incommensurate frequencies are deformed by a nonlinear perturbation but, at least for small enough perturbation or low enough energy, not destroyed. In contrast, tori supporting closed orbits, that is those for which the orbital frequencies are commensurate, are, in general, destroyed under a nonlinear perturbation. An outline of the pattern of dissolution of the corresponding invariant curves was given in Section 2B. Greene's criterion [76] for the onset of stochastic behaviour is based on the idea that the breakup of an invariant curve representing quasiperiodic motion is associated with a sudden change from stability to instability of nearby periodic orbits. In particular, it is contended that when the ks elliptic fixed points of the first stage breakup of the invariant curve with rational frequency ratio themselves become unstable, the invariant curves corresponding to quasiperiodic orbits which are close to them also become unstable. Since a periodic orbit is closed it can be characterized by the ratio of fundamental frequencies, the rotation number. The "closeness" of a quasiperiodic orbit to some one periodic orbit can then be measured by representing the irrational frequency ratio of the quasiperiodic orbit as a continued fraction. Successive truncations of that continued fraction give the rotation numbers of closed orbits that come successively "closer" to the particular quasiperiodic orbit. By following the stability of the closed orbit as its rotation number approaches the irrational value for the quasiperiodic orbit Greene can predict the breakup of the latter, and also predict the onset of global stochastic behaviour of the trajectory.

Execution of the analysis suggested by Greene requires consideration of the nature of the mapping of the Poincaré surface of section onto itself; the reader is referred to the original references for the details [76]. When applied to the "kicked pendulum" model, the predicted and (numerically) observed onsets of stochastic behaviour are found to be in good agreement. When applied to the Henon-Heiles system, the method generates the known features of periodic solutions, their mode of dissolution, etc.; it has not yet been used to calculate E_c for this case.

In Section 2C we sketched the ideas used by Helleman and Bountis [35] to compute the periodic orbits of a system. For the Henon-Heiles system they found the periodic solutions were dense, but inordinately sensitive to initial conditions. Each point along a curve with $\sigma = 1$ (see Section 2C) represents an initial condition for an orbit that closes in one revolution in phase space, with corresponding interpretation for the curves with other values of σ, all of which are rational. As the energy of the system increases the σ-curves tend to bunch more and more sharply as shown in Fig. 4.10. This is called σ-confluence by Helleman and Bountis. It has already been remarked that a given σ-curve is a smooth function of the recurrence time T_r, but that the recurrence time changes discontinuously as one moves from one to another σ-curve. This implies that a small change in initial conditions, which generates a small change in σ, leads to a dramatic change in the orbital topology. Inevitably, then, the denser the clustering of σ-curves, the greater the change in trajectory for a small change in initial conditions. Since the recurrence times of these orbits are very different, the distance between two trajectories initially very close will grow rapidly (exponentially?) for the case that there is σ-confluence of the periodic orbits. Hence, σ-confluence can be used as a criterion for the onset of stochastic behaviour of the system trajectories. The analysis reveals the regions of phase space where stochastic behaviour appears; for the Henon-Heiles system it correctly accounts for the fact that stochastic behaviour first occurs in the vicinity of those fixed points with $\sigma = 1$.

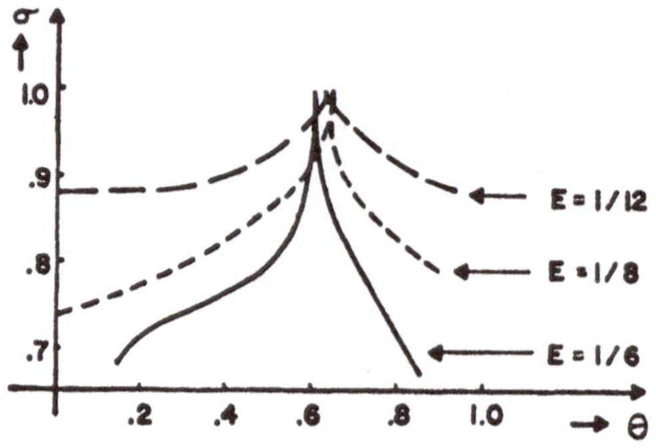

Fig. 4.10. Variation of $\sigma = m_2/m_1$ along an equipotential of H at energy E as a function of the polar angle θ about the origin in the $p_1^o q_1^o$ plane. Note how very small changes in θ generate large changes in σ when confluence of trajectories occurs. (From H.G. Helleman and T. Bountis, Stochastic Behaviour in Classical and Quantum Hamiltonian Systems, eds. G. Casati and J. Ford, Springer-Verlag (1979), p.353.)

In its current stage of development the Helleman-Bountis method does not predict the value of E_c.

E. Summary

Each of the methods discussed in this section gives some insight into the nature of the transition from quasiperiodic to stochastic motion of the trajectory of an isolated Hamiltonian system; each also suffers from some deficiency. It must also be remembered that for the purpose of understanding intramolecular energy relaxation in particular cases it is equally important to have information about the location of the transition and about the periodic and quasiperiodic trajectories embedded in the stochastic domain. Leaving aside questions of technical difficulty, the orbital stability analysis of Greene, if it can be combined with the method of generating periodic solutions put forward by Helleman and Bountis, seems to hold the most promise for providing information both on E_c and on the nature of the embedded trajectories. In contrast, the generalized Brumer-Duff-Toda-Cerjan-Reinhardt local instability method involves the least labour for the calculation of E_c, but is incapable of yielding any information about the trajectories for $E > Ec$. The Zaslavskii-Chirikov overlapping nonlinear resonances criterion for locating E_c is, in some ways, the easiest to grasp, but it too fails to give information about possible periodic or quasiperiodic orbits for $E > E_c$. Moreover, the algorithm used to analyse a dynamical system, namely reduction of the system

Hamiltonian to that of a generalized pendulum, is likely much less accurate for the location of E_c than is possible within the concept that E_c occurs when nonlinear resonance zones overlap.

From the point of view of molecular mechanics an important lack in all the methods discussed arises from omission of the nature of the excitation process. Given that most coupled oscillator systems will undergo the quasiperiodic to stochastic transition in their dynamics, and given that isolated (and even sometimes dense) quasiperiodic and periodic trajectories permeate the stochastic domain, it is important to learn if those trajectories can be picked out by some preparation process. Without that knowledge it is not possible to account for the variety of intramolecular relaxation processes, nor to turn them to useful purposes.

REFERENCES

1. S.A. Rice in Advances in Laser Chemistry, ed. A. Zewail, Springer-Verlag, Berlin (1978), p.2.
2. See, for example, Topics in Nonlinear Dynamics, ed. S. Jorna, A.I.P. Conf. Proc. 46, New York (1978).
3. The Toda system, described in Section 2, is one such exception.
4. J. Ford, Adv.Chem.Phys. 24, 155 (1973). J. Ford, in Fundamental Problems in Statistical Mechanics, Vol. 3, ed. E.D.G. Cohen, North Holland, Amsterdam (1975), p.215.
5. M.V. Berry in Ref. 2, p.16.
6. See G. Benettin, C. Froeschle and J.P. Scheidecker, Phys. Rev. A19, 2454 (1979), and references cited therein.
7. A good description of isolating and nonisolating integrals of the motion is given in G. Contoupolos, Astrophys. J. 138, 1297 (1963).
8. See, for example, V.I. Arnold and A. Avez, Ergodic Problems of Classical Mechanics, Benjamin, N.Y. (1968), Y.M. Treve in Ref. 2, p.147.
9. M. Henon and C. Heiles, Astron. J. 69, 73 (1964).
10. See Section 4.
11. G.H. Walker and J. Ford, Phys. Rev. 188, 416 (1969).
12. J. Ford and G.H. Lunsford, Phys. Rev. A 1, 59 (1970).
13. G.M. Zaslavskii and B.V. Chirikov, Sov. Phys. Usp. 14, 549 (1972). B.V. Chirikov, Research Concerning the Theory of Nonlinear Resonances and Stochasticity, Novosibirsk, U.S.S.R. (1969), unpublished, translated as CERN 71-40 Geneva (1971).
14. H. Poincaré, Les Methodes Nouvelles de la Mecanique Celeste, Dover, N.Y. (1957).
15. I have made extensive use of the very clear presentation of M.V. Berry, Ref. 5, in the following paragraphs.

16. G.D. Birkhoff, Dynamical Systems, Am. Math. Soc., Providence, R.I., revised edn. (1966).
17. M. Toda, Prog. Theor. Phys. Suppl. 45, 174 (1970); 59, 1 (1976). M. Toda, Phys. Repts. Phys. Lett. C. (Netherlands) 18, 1 (1975).
18. J. Ford, S.D. Stoddard and J.S. Turner, Prog. Theor. Phys. 50, 1547 (1973).
19. Two books that give excellent treatments are: A.H. Nayfeh, Perturbation Methods, Wiley, N.Y. (1973); G. Giacaglia, Perturbation Methods in Nonlinear Systems, Springer-Verlag, Berlin (1972).
20. A. Lindstedt, Astron. Nach. 103, Col. 211 (1882); H. Poincaré, Ref. 14.
21. P.A. Sturrock, Plasma Hydromagnetics, Stanford Univ. Press, Stanford, Calif. (1962).
22. E.A. Frieman, J. Math. Phys. 4, 410 (1963).
23. A.H. Nayfeh, J. Math. and Phys. 44, 368 (1965).
24. G. Sandri, Nuovo Cimento B36, 67 (1967).
24a. H.R. Wilson and S.A. Rice, J. Chem. Phys. 49, 1697 (1968).
25. B. Van der Pol, Phil. Mag. 43, 700 (1926).
26. N. Krylov and N.N. Bogoliubov, Introduction to Nonlinear Mechanics, Princeton Univ. Press, Princeton, N.J. (1947). N.N. Bogoliubov and Y.A. Mitropolski, Asymptotic Methods in the Theory of Nonlinear Oscillations, Gordon and Breach, N.Y. (1961).
27. M. Kruskal, J. Math. Phys. 3, 806 (1962).
28. H. Von Zeipel, Ark. Mat. Astron. Fysik (Stockholm 11, No.1, 1; No. 7, 1 (1916).
29. G.I. Hori, Publ. Astron. Soc. Japan 18, 287 (1966); 22, 191 (1970).
30. A. Deprit, Celest. Mech. 1, 12 (1969).
31. A.A. Kamel, Celest. Mech. 1, 190 (1969); 3, 90 (1970); 4, 397 (1971).
32. J.M.A. Danby, Celest.Mech. 8, 273 (1973).
33. G. Contoupolos, Astron. J. 75, 96 (1970).
34. B. Barbanis, Astron. J. 71, 415 (1966).
35. R. Helleman and T. Bountis in Stochastic Behaviour in Classical and Quantum Hamiltonian Systems, Springer-Verlag, Berlin (1979), p.353.
36. F.G. Gustavson, Astron. J. 71, 670 (1966).
37. G. Contoupolos, L. Galgani and A. Giorgilli; Phys. Rev. A18, 1183 (1978).
38. K.S.J. Nordholm and S.A. Rice, J. Chem. Phys. 61, 203 (1974); 61, 768 (1974).
39. R.M. Stratt, N.C. Handy and W.H. Miller, submitted to J. Chem. Phys.
40. N. Pomphrey, J. Phys. B7, 1909 (1974).
40a. E.J. Heller, Chem. Phys. Lett. 60, 338 (1979).
40b. R. Kosloff and S.A. Rice, submitted to Chem. Phys. Lett.
40c. A. Peress, Am. J. Phys. 42, 886 (1974).

40d. R. Kosloff, Adv. Chem. Phys., to be published.
41. C. Jaffe and W. Reinhardt, submitted to J. Chem. Phys.
42. P. Pechukas, J. Chem. Phys. 57, 5577 (1972).
43. R.A. Marcus, D.W. Noid and M.L. Koszykowski, in Stochastic Behaviour in Classical and Quantum Hamiltonian Systems, ed. G. Casati and J. Ford, Springer-Verlag, Berlin (1979), p.283.
44. For a review see I.C. Percival, Adv. Chem. Phys. 36, 1 (1977); I.C. Percival, J. Phys. A7, 794 (1974); I.C. Percival and N. Pomphrey, J. Phys. B31, 97 (1976).
45. M.V. Berry, Phil. Trans. Roy. Soc. (London) A287, 237 (1977); J. Phys. A10, 2083 (1977); M.V. Berry and M. Tabor, Proc. Roy. Soc. (London) A349, 101 (1976).
46. S. Chapman, B.C. Garrett and W.H. Miller, J. Chem. Phys. 64, 502 (1976); W.H. Miller, J. Chem. Phys. 63, 936 (1975).
47. D.W. Noid and R.A. Marcus, J. Chem. Phys. 62, 2119 (1975); W. Eastes and R.A. Marcus, J. Chem. Phys. 61, 4301 (1974).
48. R.T. Swimm and J.B. Delos, J. Chem. Phys. in press; also in Stochastic Behaviour in Classical and Quantum Hamiltonian Systems, eds. G. Casati and J. Ford, Springer-Verlag, Berlin (1979), p.306.
49. M.C. Gutzwiller, J. Math. Phys. 8, 1979 (1967); 10, 1004 (1969); 11, 1971 (1970); 12, 343 (1971).
50. The quantity δ is usually $\frac{1}{2}$, but can be different. See, for example, Ref. 5.
51. G.C. Schatz and T. Mulloney, J. Phys. Chem. 83, 989 (1979).
52. M. Born, The Mechanics of the Atom, G. Bell & Sons, London. (1927).
53. J.S. Hutchinson and R.E. Wyatt, private communication.
54. These investigations are discussed in Section 4.
55. K.S.J. Nordholm and S.A. Rice, J. Chem. Phys. 62, 157 (1975).
56. C.A. Parr and A. Kuperman, from C.A. Parr, Ph.D. thesis, California Institute of Technology (1968).
57. See, for example, S.A. Rice in Excited States, Vol. II, ed. E.C. Lim, Academic Press, N.Y. (1975), p.112.
58. K.G. Kay and S.A. Rice, J. Chem. Phys. 57, 3041 (1972).
59. E.J. Heller and S.A. Rice, J. Chem. Phys. 61, 936 (1974).
60. M. Muthukumar and S.A. Rice, J. Chem. Phys. 69, 1619 (1978).
61. W.M. Gelbart, S.A. Rice and K.F. Freed, J. Chem. Phys. 57, 4699 (1972).
62. K.G. Kay, J. Chem. Phys. 61, 5205 (1974).
63. L. Van Hove, Physica 21, 517 (1955); 23, 441 (1957); 25, 268 (1959).
64. K.F. Freed, Disc. Faraday Soc. 55, 68 (1973).
65. P. Brumer and J.W. Duff, J. Chem. Phys. 65, 3566 (1976). J.W. Duff and P. Brumer, J. Chem. Phys. 67, 4898 (1977).
66. M. Toda, Phys. Lett. A48, 335 (1974).
67. J.W. Duff and P. Brumer, J. Chem. Phys. 67, 4898 (1977).
68. C. Cerjan and W. Reinhardt, submitted to J. Chem. Phys.

68a. K.D. Hänsel, Chem. Phys. 33, 35 (1979).
68b. K.D. Hänsel, preprint.
69. B.V. Chirikov, A Universal Instability of Many-Dimensional Oscillator Systems, preprint; Research Concerning the Theory of Nonlinear Resonances and Stochasticity, CERN 71-40 Geneva (1971).
70. G.M. Zaslavskii and B.V. Chirikov, Sov. Phys. Usp. 14, 549 (1972).
71. D.W. Oxtoby and S.A. Rice, J. Chem. Phys. 65, 1676 (1976).
72. D. Bunker, J. Chem. Phys. 37, 393 (1962); 59, 4621 (1973).
73. K.C. Mo, Physica 57, 445 (1972).
74. H. Mori, Prog. Theor. Phys. 34, 399 (1965).
75. M. Tabor, private communication.
76. J.M. Greene, J. Math. Phys. 9, 760 (1968); 20, 1183 (1979); KAM Surfaces Computed from the Henon-Heiles Hamiltonian, preprint.

MANIFESTATIONS OF PARITY VIOLATIONS IN ATOMIC AND MOLECULAR SYSTEMS

Robert A. Harris

Department of Chemistry
University of California
Berkeley, California, 94720, USA

A. GENERAL ASPECTS OF PARITY

The conservation of parity is just a fancy way of saying that there is no fundamental way of distinguishing left from right. Another way of putting it is that the mirror image of any experiment may be performed. To paraphrase Feynman and L. Frank Baum (Feynman, 1963, 1965; Baum, 1900): In the Wizard of Oz, the Wizard opened up the left side of the tin woodman's breast and put in a heart. Suppose we communicated with an alien Wizard who was made of matter, obeyed the same laws as we, and was in a closed palace. Suppose we asked it to construct a tin woodman. We could give it all of the instructions up to which side to put the heart.

The above view was considered correct until the discoveries of Lee and Yang (1956) changed this vision forever. We could now tell the alien to make an exact copy of the tin woodman.

In these lectures we shall discuss new developments which give rise to alternative ways of informing the creature how to construct exact replicas of the woodman.

The operation of parity or inversion is discussed in every quantum mechanics book in the world (see for example, Landau and Lifshitz, 1958; Sakurai, 1964), so we shall only briefly review the subject. Classically the operation of inversion turns every point $\underset{\sim}{r}$ into $-\underset{\sim}{r}$. Thus, the operation may be constructed by a mirror reflection in a plane followed by a rotation by π perpendicular to the plane. As one knows one may either change the coordinate system, or invert the object. Both are consistent views. We shall usually invert the object. It is clear from the definition that parity is not a meaningful operation in two dimensions, only in 1 or 3 dimensions.

In quantum mechanics, the operation of inversion is described by the parity operator, P, a unitary and hermitian operator. Thus, the expression of inversion and its subsequent return of a point to its original place by a second inversion is given by,

$$P^2 = 1 \tag{1}$$

and

$$P^{-1} = P = P^{+} . \tag{2}$$

The eigenstates and eigenvalues of parity are readily found through the use of (1). That is, if $|\lambda>$ is an eigenstate of parity, then,

$$P|\lambda> = \lambda|\lambda> , \tag{3}$$

and

$$P^2|\lambda> = \lambda^2|\lambda> . \tag{4}$$

However, equation (4) relates λ to λ^2,

$$\lambda^2 = 1 \tag{5}$$

or

$$\lambda = \pm 1 \tag{6}$$

Thus, the eigenstates of parity are either even or odd under inversion.

The parity operator changes the operator $\underset{\sim}{r}$ into $-\underset{\sim}{r}$, hence we may write

$$P^{+}\underset{\sim}{r}P = -\underset{\sim}{r} , \tag{7}$$

or

$$[P,\underset{\sim}{r}]_{+} = 0 . \tag{8}$$

Other vectors such as momentum, $\underset{\sim}{p}$, also anticommute under the operation of parity. Vectors such as angular momentum, $\underset{\sim}{L}$, and spin, $\underset{\sim}{S}$, do not change sign under inversion. These are "axial vectors" or "pseudovectors". These vectors commute with the parity operator,

$$[P,\underset{\sim}{L}] = 0 . \tag{9}$$

Now to an important point germaine to our future work. We may construct two classes of scalars: firstly the ordinary scalars, f, which commute with P,

$$[f,P] = 0 . \tag{10}$$

Two such scalars are inner products of vectors with vectors, or axial vectors with axial vectors. The second class of scalars anticommute under the operation of parity. They are pseudo or axial scalars, g. These scalars anticommute with P,

$$[P,g]_+ = 0 . \tag{11}$$

Pseudoscalars may be formed from the inner product of a vector with an axial vector. Thus, $\underset{\sim}{r}\cdot\underset{\sim}{S}$ is an axial scalar.

The conservation of parity expresses itself in the fact that P commutes with the Hamiltonian, H, where up until the 1950s H was any Hamiltonian. Thus, the assumption of parity conservation is

$$[P,H] = 0 . \tag{12}$$

It is clear that as long as there are no degeneracies, the eigenstates of H may be labelled by their parity signature. If there are degeneracies, eigenstates of H need not be eigenstates of P. The simplest example in the world of this latter situation is a free particle in 1-dimension. Now H is given by

$$H = \frac{\underset{\sim}{p}^2}{2m} . \tag{13}$$

Hence,

$$[\underset{\sim}{p},H] = 0 , \tag{14}$$

and

$$[P,H] = 0 \tag{15}$$

however

$$[P,\underset{\sim}{p}]_+ = 0 . \tag{16}$$

There is a twofold degeneracy.

Amongst the infinity of states we may either construct momentum eigenstates, $|\pm|p|\rangle$, which under P are transformed into one another,

$$P|\pm|p|\rangle = |\mp|p|\rangle, \tag{17}$$

or parity eigenstates which are linear combinations of momentum eigenstates,

$$P\frac{1}{\sqrt{2}}\{|+|p|>\pm|-|p|>\} = \frac{1}{\sqrt{2}}\{|+|p|>\pm|-|p|>\} \qquad (18)$$

This is just a complicated way of writing running vs. standing waves.

So far we have discussed the parity operation explicitly only for a single particle. It is clear that for many particles the parity operator is the product of the parity operator for each particle, that is for N particles, P is given by,

$$P = \prod_{i=1}^{N} P_i \; . \qquad (19)$$

If an N-particle state is a simple product of single particle states, each with parity eigenvalues $(-1)^{p_i}$, where $p_i = 0, 1$ for even or odd parity respectively, then the overall state has parity $(-1)^{\sum_i p_i}$.

In order to resolve certain paradoxes in elementary particle physics, Lee and Yang (1956) suggested that for the weak interactions parity is not conserved. That is, if H contains a part involving the weak forces,

$$[H,P] \neq 0 \; . \qquad (20)$$

Now an immediate consequence of this is β-decay (Commins, 1973). This process is essentially,

$$^{60}Co \rightarrow {}^{60}Ni^{+} + e^{-} + \bar{\nu} \; . \qquad (21)$$

If parity is not conserved, when the Co nuclear spins are lined up parallel to a magnetic field, then the e^- should be emitted preferentially either forward, or backward relative to the field, (see, for example, Feynman, 1961). This is readily seen in Figure 1. A mirror reverses the direction of the field, hence the direction of the spins. Thus, the mirror experiment changes the direction of the emitted electrons from, e.g., antiparallel to parallel. That is, if parity were conserved, there could be no preferential direction for the emitted electrons relative to the field. But experiment shows the electrons always to be emitted preferentially antiparallel to the field. The dotted arrow in Figure 1 represents reality.

It is clear from the analysis, that since the electrons are emitted antiparallel to the field, the handedness of the solenoid

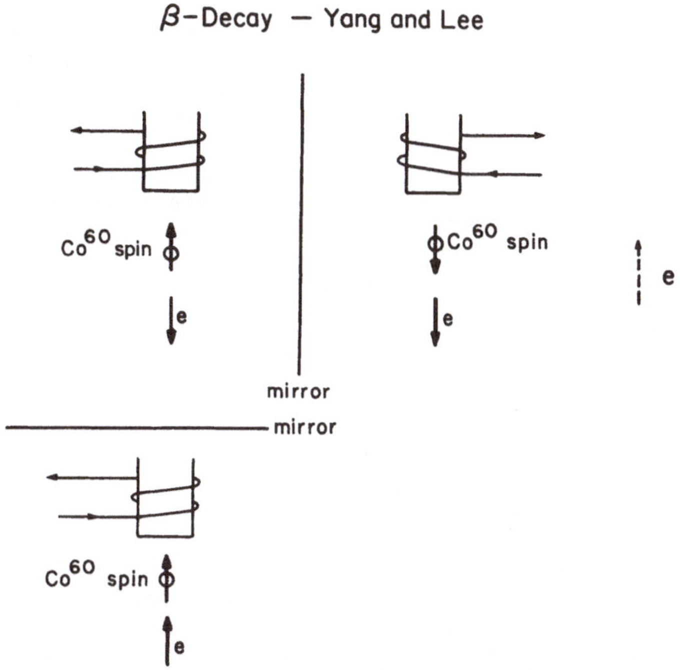

Fig. 1. Mirror experiments, neither of which represents reality.

relative to a heartless tin woodman is fixed depending on whether or not while facing forward the electrons are emitted towards the feet or head.

Although the electrons from β-decay may be "used" to fix handedness, for all practical purposes the influence of the weak interactions on atomic processes is nil. This lack of influence of the weak interactions may be traced to the fact that they are charged. Hence, one must go to higher order to obtain an effective interaction between electrons and nuclei.

B. PARITY VIOLATING NEUTRAL CURRENTS AND THE NONRELATIVISTIC PARITY VIOLATING POTENTIAL

Weinberg and Salam (Weinberg, 1967, 1971, 1972; Salam, 1968) created a "unified theory of weak and electromagnetic interactions". A consequence of this theory is that there is an interaction between electrons and protons and neutrons, and between electrons with one another which has a nonzero nonrelativistic limit. This interaction does not conserve parity.

The detailed construction of the Weinberg-Salam theory is outside the realm of my competence, so I just describe the most elementary aspects of the reasoning and write down the potentials (see for example, Abers and Lee, 1973).

Consider an interaction between fermions mediated by bose particles of rest mass m. If we use elementary perturbation theory (Henley and Thirring, 1962), then the interaction energy between two sources which absorb and emit these bose particles at a distance r away from one another is a Yukawa potential,

$$V(r) \sim \frac{e^{-r/\lambda}}{r}, \tag{22}$$

where λ is the Compton wavelength of the particle,

$$\lambda = \frac{\hbar}{mc}. \tag{23}$$

A more general theory may be constructed based upon the fermions being represented by currents. Thus the electromagnetic interaction between charged particles is due to the coupling of the currents to photons. The parity violating interaction arises in the same way because the "neutral currents" of the charged particles (electrons, photons, etc.) are coupled to a particle called the Z-particle. As these interactions give rise to potentials (22) which depend upon the wavelength λ, it is clear what happens. As $m \to 0$, (photons), $\lambda \to \infty$, and we arrive at a potential

$$V(r) \to \frac{1}{r}, \tag{24}$$

i.e., Coulomb's law. As $m \to \infty$ which is the case for the supposedly very massive Z (35 - 100 BEV), $\lambda \to 0$, and the potential becomes,

$$\lim_{\lambda \to 0} \frac{e^{-r/\lambda}}{r} \to \delta(r), \tag{25}$$

which is a contact potential. Thus, for our "neutral current" interactions, we may expect that only functions of the momentum $\underset{\sim}{p}$ and spin $\underset{\sim}{S}$, of the electron will appear in combination with $\delta(r)$ in the nonrelativistic limit.

Now according to fundamental ideas the electron current operator which appears, $\underset{\sim}{j}_e$ must be of the form

$$\underset{\sim}{j}_e \sim \underset{\sim}{V}_e + \underset{\sim}{A}_e, \tag{25}$$

where $\underset{\sim}{V}_e$ and $\underset{\sim}{A}_e$ are vector and axial vector currents respectively. Similarly, for a proton, we have

$$\underset{\sim}{j}_p \sim \underset{\sim}{V}_p + \underset{\sim}{A}_p \ . \tag{26}$$

Thus the interaction between the two symbolically has the structure

$$\underset{\sim}{j}_e \cdot \underset{\sim}{j}_p = \underset{\sim}{A}_e \cdot \underset{\sim}{A}_p + \underset{\sim}{V}_e \cdot \underset{\sim}{V}_p + \underset{\sim}{A}_e \cdot \underset{\sim}{V}_p + \underset{\sim}{V}_e \cdot \underset{\sim}{A}_p \ . \tag{27}$$

The last two terms are odd under the parity operation. Hence it is they that give rise to the violations in atoms and molecules of which we speak. Bouchiat (Bouchiat and Bouchiat, 1974a,b, 1975) carried out the calculation which determined the parity violating interaction between an electron and a nucleus in the nonrelativistic limit. The arguments are similar to the ones I presented earlier, only because of the fact that these interactions are due to currents, vectors must appear. We present only the results, as the details of Bouchiat's calculations are technical, and give no additional insight beyond what I've said.

The potential may be divided into two parts, which we shall denote as V_1 and V_2 respectively,

$$V(\underset{\sim}{r}) = V_1(\underset{\sim}{r}) + V_2(\underset{\sim}{r}) \ . \tag{28}$$

$V_1(\underset{\sim}{r})$ arises from the term $\underset{\sim}{A}_e \cdot \underset{\sim}{V}_p$ and has the form,

$$V_1(\underset{\sim}{r}) \sim 10^{-15} \text{e.v.} \ Q_w [\underset{\sim}{\sigma} \cdot \underset{\sim}{p}, \delta^3(\underset{\sim}{r})]_+ \ , \tag{29}$$

where p is the electron's momentum operator, σ is the electron spin operator and all quantities in the anticommutator are dimensionless in units of the Bohr radius. Q_w is the "Weinberg" parameter,

$$Q_w \equiv -[(4\sin^2\theta_w - 1)Z + N], \tag{30}$$

Z is the proton number, and N is the neutron number. θ_w is the "Weinberg angle" and $\sin^2\theta_w$ is assumed to have a value between 0.2 and 0.4. $V_2(\underset{\sim}{r})$ which arises from $\underset{\sim}{V}_e$ and $\underset{\sim}{A}_p$ has the non-relativistic limit,

$$V_2(\underset{\sim}{r}) \sim 10^{-15} \text{e.v.} \ [1 - 4\sin^2\theta_w](\underset{\sim}{\sigma} \cdot \underset{\sim}{\sigma}_n)[\underset{\sim}{\sigma} \cdot \underset{\sim}{p}, \delta^3(\underset{\sim}{r})]_+ \ . \tag{31}$$

In equation (31) $\underset{\sim}{\sigma}_n$ is the spin operator for the nucleus.

Let us examine the potential V. We see immediately that it is the presence of the momentum operator of the electron which creates the parity violation of the atomic system. Next we note that with exception of very light atoms, $V_1(\underset{\sim}{r})$ is the dominant term because it depends linearly upon the neutron and proton numbers. Indeed, we note that given the assumed range of $\sin^2\theta_w$,

$V_2(\underset{\sim}{r})$ will be extremely small. In fact it will vanish if $\sin^2\theta_W = 1/4$.

The potential $V(\underset{\sim}{r})$ connects one-electron states of opposite parity. Indeed, because of the structure of $V(\underset{\sim}{r})$ being of the form $[\underset{\sim}{p},\delta(\underset{\sim}{r})]_+$, only s and p states have nonvanishing matrix elements. Only s states are nonvanishing at the nucleus, and only p states have a derivative at the nucleus. This property of $V(\underset{\sim}{r})$ severely constrains the type of effects which may occur.

On the level of these lectures, the origin of quantities such as the "Weinberg angle" must be ignored. In addition, to the potentials V_1 and V_2 there is an additional interaction between electrons with one another which we shall ignore here, although perhaps some future experiment will detect it.

C. ATOMIC MANIFESTATIONS OF PARITY VIOLATIONS

What are the manifestations of the parity violating potential, $V(\underset{\sim}{r})$, in atoms? And which atoms exhibit the effects maximally?

Clearly the shifts in energy due to $V(\underset{\sim}{r})$ are uninteresting and unobservable. Thus, we must look at effects of spectroscopy other than excitation energies. The scattering of photons for example. There are a number of different things that may happen. Forbidden transitions now become weakly allowed. For example $s \rightarrow s$ single photon transitions exactly forbidden in spinless nonrelativistic quantum mechanics now have a small $p \rightarrow s$ amplitude. A second, and the most interesting manifestation of parity violation in atoms, is the appearance of chirality in atoms. Namely, one can look for the optical activity, or circular dichroism, in atoms. We shall spend most of our time discussing this class of experiments. Finally, there are two types of scattering that depend upon the atoms in an entirely passive way, namely the scattering of polarized electrons or neutrons from atoms, in particular H and D atoms.

Two experiments have been carried out which seem to indicate that $V(\underset{\sim}{r})$ exists and is of the form predicted. We shall describe these experiments in a simple nonrelativistic manner. We shall not discuss the experimental details or the relativistic aspects in any detail. Only the essential physics. We choose the completed experiments, because they have really happened. However, we shall briefly mention or reference some proposed experiments, and in particular, discuss the theoretically ideal situations H and D.

In order to describe the experiments as well as what we shall say about molecules, we must briefly describe optical activity and circular dichroism (Rosenfeld, 1928; Eyring, Walter and Kimball, 1944; Stephan, 1958; Atkins and Barron 1968). As all of the systems we shall deal with are tiny compared with the wavelength of the light being used, only the lowest multipole approximation need be considered. In this situation circular dichroism is the interference between electric and magnetic transition amplitudes which arise from the difference between the power absorbed for left and right circularly polarized light. Similarly, optical rotation arises from the interference between electric and magnetic dipole transition amplitudes which occur in the forward scattering of radiation. At any given frequency the optical rotation, Φ, is defined as the number of degrees/cm a thin infinite slab of homogeneous density of material rotates linearly polarized radiation. Quantitatively, Φ has the form (away from resonance),

$$\Phi \propto \sum_f{}' \frac{\omega^2 2\mathrm{Im}\langle i|\underset{\sim}{\mu}|f\rangle\cdot\langle f|\underset{\sim}{m}|i\rangle}{\omega_{if}^2 - \omega^2} . \tag{32}$$

i and {f} are the initial and intermediate states of the system respectively, and $\underset{\sim}{\mu}$ and $\underset{\sim}{m}$ are the electric and magnetic dipole operators respectively. For a given transition, the circular dichroism has the form,

$$\theta_{i\to f} \propto 2\mathrm{Im}\langle i|\mu|f\rangle\cdot\langle f|m|i\rangle . \tag{33}$$

Now the total power absorbed for a given transition is

$$\alpha_{i\to f} \propto |\langle i|\underset{\sim}{\mu}|f\rangle|^2 + |\langle i|\underset{\sim}{m}|f\rangle|^2 \tag{34}$$

The constants of proportionality in (33) and (34) are the same. Thus, the ratio which is of importance is

$$\delta = 2\mathrm{Im}\frac{\langle i|\underset{\sim}{\mu}|f\rangle\cdot\langle f|\underset{\sim}{m}|i\rangle}{|\langle i|\underset{\sim}{\mu}|f\rangle|^2 + |\langle i|\underset{\sim}{m}|f\rangle|^2} . \tag{35}$$

On dimensional grounds, since $|m| \sim \alpha|\mu|$, where α, the fine structure constant, is 1/137, $\delta \sim 10^{-3}$.

The fundamental property of θ or Φ is that it vanishes if $|i\rangle$ and $|f\rangle$ are eigenstates of P. This is readily seen from usage of P,

$$\begin{aligned}\langle i|\underset{\sim}{\mu}|f\rangle\cdot\langle f|\underset{\sim}{m}|i\rangle &= \langle i|P^+P\underset{\sim}{\mu}P^+P|f\rangle\cdot\langle f|P^+P\underset{\sim}{m}P^+P|i\rangle \\ &= -\langle i|P^+\underset{\sim}{\mu}P|f\rangle\cdot\langle f|P^+\underset{\sim}{m}P|i\rangle .\end{aligned} \tag{36}$$

If $P|i> = \pm|i>$, and similarly for $|f>$, then the term must be zero. Hence θ and Φ uniquely measure parity violation, or chirality.

The one other property of θ and Φ is the vanishing of the sum over all states of the circular dichroism strength, or $Im<i|\underset{\sim}{\mu}|f> \cdot <f|\underset{\sim}{m}|i>$,

$$\sum_f' Im<i|\underset{\sim}{\mu}|f> \cdot <f|\underset{\sim}{m}|i> = Im<i|\underset{\sim}{\mu} \cdot \underset{\sim}{m}|i> - Im<i|\underset{\sim}{\mu}|i> \cdot <i|\underset{\sim}{m}|i>$$

$$= 0 \tag{37}$$

Thus, the circular dichroism must be negative for some transitions in any system. Only detailed calculations can _a priori_ tell what the sign is for any given transition.

With the exception of the suggested H and D experiments, all attempted measurements of the circular dichroism or optical rotation have concentrated on the heavy atoms. The reason as mentioned earlier is that matrix elements of $V_1(\underset{\sim}{r})$ are largest for heavier atoms. This may crudely be seen as follows. $V_1(\underset{\sim}{r})$ explicitly goes as Z. Now the matrix element of $V_1(\underset{\sim}{r})$ is essentially $Z|\psi(0)|^2(1/r)$, where r is near the nucleus. Near the nucleus

$$|\psi(0)|^2 \sim Z , \tag{38}$$

and

$$\frac{1}{r} \sim Z . \tag{39}$$

Thus the matrix elements go as Z^3.

The successful atomic experiment has been carried out by Commins and his group (Conti et al., 1977) on the thallium atom. They looked at the E1 forbidden, but relativistically weakly allowed M1 transition (Chu et al., 1977) $6^2P_{1/2} \to 7^2P_{1/2}$ transition in ^{81}Tl. The experimental, and carefully determined theoretical, value of the M1 transition (Neuffer and Commins, 1977), is

$$<7^2P_{1/2}|m_z|6^2P_{1/2}> = - (2.11 \pm 0.30) \times 10^{-5} \mu_B . \tag{40}$$

Of course,

$$<7^2P_{1/2}|\mu_z|6^2P_{1/2}> = 0 . \tag{41}$$

We now calculate the effect of $V(r)$, or really $V_1(r)$, which will give rise to a weak E1 transition amplitude, that is,

$$<6^2P_{1/2} > \to |6^2P_{1/2}> + a_1|S> \equiv |\overline{6^2P_{1/2}} > , \tag{42a}$$

and

$$|7^2P_{1/2}> \rightarrow |7^2P_{1/2}> + a_2|S> \equiv |\overline{7^2P_{1/2}}> , \quad (42b)$$

where, e.g.,

$$a_1|S> \equiv -\sum_S \frac{|S><S|V_1|6^2P_{1/2}>}{\Delta E} \quad (43)$$

The measured, and calculated transition amplitudes with various Weinberg angles give

$$<\overline{6^2P_{1/2}}|\mu_z|\overline{7^2P_{1/2}}> \sim i(2.3 \pm 0.9) \times 10^{-8}\ \mu_B , \quad (44)$$

for the weakly allowed E1 transition. Thus, even though δ is approximately

$$\delta \sim \frac{2\mathrm{Im}<\overline{6^2P_{1/2}}|\mu_z|\overline{7^2P_{1/2}}>}{|<6^2P_{1/2}|m_z|7^2P_{1/2}>} , \quad (45)$$

the reverse ratio, the ratio is still 10^{-3}. That is, we have

$$\delta_{theor} \sim (+2.6 \pm 0.9) \times 10^{-3} , \quad (46)$$

and

$$\delta_{exp} \sim (+5.2 \pm 2.4) \times 10^{-3} . \quad (47)$$

We may approximately determine the optical rotation of Tl assuming that the probe frequency is one-half the resonance frequency. A short calculation gives a rotation of $\sim +10^{-8}$ degrees $cm^{-1}\rho$, where ρ is the molar density. We have assumed that the $6P_{1/2} \rightarrow 7P_{1/2}$ transition saturates the rotation, which may be somewhat in error.

We see that the alien can construct a person with the heart on the left side using Tl as a guide. All it does is send linearly polarized radiation propagating from the back to the front. The alien's "polaroid" will be at a maximum of $+10^{-8}$ degrees to the right after 1 cm of 1 molar Tl. Thus, left will be the opposite of the polaroid direction facing front.

Similar experiments have been attempted on Cs and Bi (Lewis et al., 1976) without success. A clear virtue of Tl is that it has essentially one valence electron, which is also true of Cs.

The Cs transition is in the infrared. As the Cs is prepared in an oven of a few thousand degrees, the resulting black-body radiation contributes to any noise. In addition, the transition rate is about 10% that of Tl. It has been suggested that the Bi experiment uses a theory which overestimates the transition amplitude (Carter et al., 1979). However, this could only change the result by a factor of 3.

The other successful experiment involves the scattering of longitudinally polarized electrons from H and D (C. Prescott et al., 1978, Physics Today, 1978). The experiment is at highly relativistic energies, $\sim$ 19 GeV, and also fully exhibits the precession of electron spin due to the fact that the magnetic moment of the electron is not 2.

Here we content ourselves with a nonrelativistic treatment which exhibits the fundamental idea. The scattering in spin space of an electron off an unpolarized target can only involve the vectors $\underset{\sim}{k}_1$ and $\underset{\sim}{k}_2$, the incident and final wave vectors of the electron, and $\underset{\sim}{\sigma}$, the spin. The t matrix must have the form,

$$t \sim t_o \underset{\sim}{1} + (\underset{\sim}{k}_1 \times \underset{\sim}{k}_2)\cdot\underset{\sim}{\sigma} t_1 + (\underset{\sim}{k}_1 + \underset{\sim}{k}_2)\cdot\underset{\sim}{\sigma} t_2, \tag{48}$$

if time reversal invariance holds (Gottfried, 1966). If parity is conserved, $t_2 = 0$. We therefore consider the difference between $|+\rangle$ and $|-\rangle$ longitudinally polarized incident electrons and everything out. That is,

$$\delta \equiv \delta_+ - \delta_- = \mathrm{Tr}\{t^+ \sigma_z t\} . \tag{49}$$

A short calculation shows that δ has the form

$$\delta = 2k_1 t_o t_2 (1 + \frac{k_2}{k_1} \cos\theta) , \tag{50}$$

where θ is the angle between the incident and final wave vectors. The asymmetry is observed and is totally consistent with the precession (transverse differences are negligible), and the Weinberg-Salam model and the Tl experiment.

At this stage we must say that all systems are chiral! The ideal system for light would be H and D; calculations are exact. A number of experiments have been proposed (Lewis and Williams, 1976; Hinds and Hughes, 1977, and references therein). Most of these ideas exploit the $^2S_{1/2}{}^2P_{1/2}$ crossings which occur at finite magnetic field. Thus, these effects are limited only by the Lyman α lifetime. I could never get to molecules if I discussed these ideas, which involve many lovely quantum mechanical interference properties.

D. MOLECULAR MANIFESTATIONS OF PARITY VIOLATION

Our earlier discussion has centred on parity violations on the most fundamental level. Now we shall examine chiral molecules, first in the absence of parity violating neutral currents and then in their presence. We shall see that a number of interesting concepts arise.

The forces that are involved in molecules are the same as those which govern atoms except that there is more than one nucleus. In the absence of neutral currents, the Hamiltonian of the entire system of nuclei and electrons commutes with the Parity operator for all of these particles, namely

$$[H,P] = 0 . \tag{51}$$

Thus, as in atoms, all of the eigenstates of molecules <u>may be</u> taken to be eigenstates of parity. By the definition of optical activity given previously, no molecules which are in eigenstates of H and P can be optically active, yet handed molecules abound. We present are living testimony to this fact. What is going on here?

Handed molecules are the manifestation of a symmetric double well potential, and hence a two state system which degenerates to an extremely high degree, indeed, totally in the absence of neutral currents. These two states are the lowest mean energy left and right isomers. It is also remarkable that nature has provided us with the tool of optical activity as a method of distinguishing between the two states.

We shall consider only the simplest version of handed molecules. This is a version that was understood in the earliest days of quantum mechanics and is really identical with the description of the inversion of NH_3 (Hund, 1927; Rosenfeld, 1928; Herzberg, 1945; Weston, 1954; Townes and Schalow, 1955). We shall focus on systems like NHDT, or hindered biphenyls. We may define a coordinate "x" which describes the various positions of the nuclei as they "invert" or undergo hindered rotation. Thus, x is the height of the N atoms above the H-D-T plane, or it is the angle of rotation between the two phenyl groups. It may be the rotation of the π-electrons relative to one another in a strained ethylene. The important point is that the states depend upon the coordinate x and all other degrees of freedom.

Everything we describe is exact. Surely, a system being optically active has nothing to do with the Born-Oppenheimer approximation. However, for convenience and also because the description is so adequate, let us assume the Born-Oppenheimer approximation, and assume we are in the lowest electronic, vibrational eigenstates. Hence we can consider the set of states descriptive of the coordinate x as being the solutions of the Schroedinger equation for a potential V(x), which is a "classic" double well potential.

The eigenstates of this potential are $\phi_i(x)$. Now the important thing is that $\phi_i(x)$ are eigenstates of parity,

$$P\phi_i(x) = \pm\phi_i(x) . \tag{52}$$

The overall wavefunctions for the ground electronic state are $\psi_o(\underset{\sim}{r},x)\phi_i(x)$. $\psi_o(\underset{\sim}{r},x)$ is a solution of the electronic Schroedinger equation, and only Coulomb's law is involved.

If the wells are shallow and the separation between them is not too great, then the low lying eigenstates come in pairs, and the higher states whose energies lie above the hump are just normally distributed, like higher lying oscillator states. Consider now the opposite extreme, where the wells are very deep and widely separated. In this case, there are doubly degenerate pairs of eigenstates. Because of the strict double degeneracy, the eigenstates need not be eigenstates of parity.

Let us now assume that the barrier height and width are both finite. The lowest two states are ϕ_+ and ϕ_-, and they are split by an energy. These two states are eigenstates of parity. We define the states ϕ_L and ϕ_R as follows:

$$\phi_L \equiv \frac{1}{\sqrt{2}}(\phi_+ + \phi_-) , \tag{53}$$

and

$$\phi_R \equiv \frac{1}{\sqrt{2}}(\phi_+ - \phi_-) . \tag{54}$$

We shall define the molecule as being optically active when it is in either ϕ_L or ϕ_R. These states cannot be stationary states because the degeneracy of the infinitely separated double well is broken. That is, tunnelling may now occur. Now given the definitions of ϕ_L and ϕ_R, one may show that the mean energy,

$$\langle\phi_L|H|\phi_L\rangle = \langle\phi_R|H|\phi_R\rangle = \bar{\varepsilon} , \tag{55}$$

is the minimum mean energy for any degenerate pair which can be converted into one another by the parity operation. That is, given any pair of states U_L, U_R then, e.g.

$$\langle U_L|H|U_L\rangle \geq \langle\phi_L|H|\phi_L\rangle \tag{56}$$

If there is a gap between ϕ_L, ϕ_R and the next pair which is fairly large, as there in general will be, we assume that our definition of the handed states, i.e., ϕ_L and ϕ_R is reasonable. States with greater geometric localization may be constructed, but we know that they will spread greatly with time and probably loose their handedness rapidly.

Given the system in state ϕ_L at time t = 0, it will oscillate back and forth between itself and ϕ_R as a function of time. That is,

$$\phi(t) = \cos\delta t \phi_L + i\sin\delta t \phi_R. \tag{57}$$

Such a state never racemizes. The population oscillates with time - a quantum beat. However this is not like NH_3 where the inversion beat manifests itself in spectroscopy. We may measure the optical activity as a function of time.

Because the optical activity involves virtual electronic excitation one may consider the interaction of the radiation with the molecule as instantaneously achieving its golden rule rate and adiabatically changing with time as the molecule oscillates between left and right. This separation of scales of time, as it were, is 15 orders of magnitude and well justified. Hence, if the system begins in ϕ_L at time t - 0, after a time t the optical activity will be given by $\Phi(t)$, where

$$\Phi(t) = \Phi\cos 2\delta t \tag{58}$$

If it started in ϕ_R, it would be

$$\Phi(t) = -\Phi\cos 2\delta t . \tag{59}$$

Φ is the maximum value of the optical activity. Any mixture of populations would give

$$\phi(t) = [n_L - n_R] \cos 2\delta t \Phi . \tag{60}$$

It is important to note that the optical activity of a non-interacting system of molecules will oscillate forever if the populations are not initially equal. This is a quantum beat phenomenon, and in principle observable. The optical activity is zero only when $t = (m+1)\pi/4\delta$, where m is even. There can never be racemization. It is possible to have a form of racemization only if different molecules in the ensemble are prepared in, e.g., left, at different initial times in a random manner, then the linear superposition of oscillations loses its phase coherence.

We see that our description of optical activity is really the classical two state analysis. Thus $|L\rangle$ and $|R\rangle$ may be considered eigenstates of the Pauli spin matrix σ_z,

$$\sigma_z|{}^L_R\rangle = \pm|{}^L_R\rangle \tag{61}$$

In addition, the system may be described by a density matrix

$$\rho(t) = \frac{1}{2}(\underset{\sim}{1} + \underset{\sim}{P}(t)\cdot\underset{\sim}{\sigma}) , \tag{62}$$

where $\underset{\sim}{P}(t)$ is the polarization vector, and the measurement of optical activity or circular dichroism is equivalent to measuring the operator

$$\hat{\Phi} \equiv \Phi\sigma_z .$$

The Hamiltonian we have been describing is just

$$H = \delta\sigma_x . \tag{63}$$

Thus,

$$\Phi(t) = \mathrm{Tr}\rho(0)e^{iHt}\sigma_z e^{-iHt}\Phi \tag{64}$$

$$= \mathrm{Tr}\rho(0)[\cos 2\delta t\sigma_z + \sin 2\delta t\sigma_y] \Phi$$

which for

$$\rho(0) = \frac{1}{2}(1 + \sigma_z) \tag{65}$$

gives the earlier result.

The value of δ may be obtained from theory using W.K.B., or whatever one wishes. We may note some experimental figure for inversion frequencies of nonchiral molecules

$^{14}NH_3$	∿	24,000 MHz
ND_3	∿	1,600 "
PH_3	∿	0.14 "
AsH_3	∿	$\frac{1}{2}$ cycle/year!

Thus we may expect δ's which range from MHz to the age of the universe.

The presence of the potential which is the manifestation of the parity violating neutral currents gives rise to a correction to the potential $V(x)$ which breaks the symmetry of the double well (letokhov, 1975). This point is of the utmost conceptual importance. What are some manifestations of this breaking of the fundamental symmetry between "left" and "right"?

Let us assume that $\delta = 0$, so that L and R really refer to eigenstates of H which last forever. In this case we have the following:

(a) The energy levels of L and R molecules will differ.

(b) Optical activity of an L molecule will differ slightly from that of an R molecule.

(c) The interaction between R and R will differ from that between L and L, and mixtures will differ from one another.

(d) Finally, "L life" will have metabolic processes slightly different from "R life".

By the way, it is absurd to imagine that the handed origin of life was influenced by the neutral currents! All of the above effects are absolutely miniscule, puny relative to other processes. There is one process, however, in which in at least some molecules the parity violation is commensurate, or larger, and that is what we shall now discuss (Harris and Stodolsky, 1978).

Let us incorporate the parity violation into our tunnelling problem. Consider all of the eigenstates of the symmetric double well. Let us label these states $\phi_{\pm i}$, where $\pm$ is the parity signature. We shall be interested in the effect of a parity violating potential $V_1(x)$ on the lowest two states $\phi_{\pm 0}(x)$. Clearly the potential is extremely weak, also it is odd in the inversion coordinate. Thus, we may expect that in the two state model the Hamiltonian becomes

$$H = \varepsilon\sigma_z + \delta\sigma_x , \tag{66}$$

where ε is

$$\varepsilon = \langle\phi_L|V_1(x)|\phi_L\rangle = -\langle\phi_R|V_1(x)|\phi_R\rangle . \tag{67}$$

This is an extremely good approximation. Now $V_1(x)$ connects in 1st order the states $\phi_{0\pm}(x)$ to the higher states. The additional term contributes to the time development of the optical activity. The weight of these higher states is of the order $V_1/\Delta \simeq 10^{-15}$ which will be insignificant. Hence our two state model is totally valid. Thus, for all processes we use H as above.

We now determine $\sigma_z(t)$. It is a rotation about a line to the x-z plane,

$$\Phi(t) = \mathrm{Tr}\rho_0 e^{i(\varepsilon\sigma_z+\delta\sigma_x)t}\sigma_z e^{-i(\varepsilon\sigma_z+\delta\sigma_x)t}\Phi$$

$$= \Phi\frac{\varepsilon^2 + \delta^2\cos 2\sqrt{\varepsilon^2+\delta^2}\,t}{\varepsilon^2+\delta^2} \qquad (68)$$

What happens is that the optical activity oscillates asymmetrically. The breaking of the degeneracy may change the energies by a small amount, but the wave functions are changed greatly. Notice the following: if

$$\varepsilon \geq \delta, \quad \Phi(t)/\Phi \geq 0 . \qquad (69)$$

Indeed, as $\varepsilon/\delta \to \infty$, the individual isomers L and R become infinitely stable to within an error of higher states. As ε/δ grows, the system oscillates but the depth of oscillation becomes more and more shallow.

Now assume $\varepsilon \leq \delta$.

$$\frac{\Phi(t)}{\phi_m} \geq \frac{-1 + (\varepsilon/\delta)^2}{1 + (\varepsilon/\delta)^2} . \qquad (70)$$

In the limit $\varepsilon/\delta \to \infty$, the original symmetry is restored. We see that the degree of asymmetry is $O(\varepsilon/\delta)^2$. It may be $O(1)$. Hence we have a commensurate process of elementary particles and molecular physics, which could be a unique merging of these two usually disparate fields.

The potential $V_1(x)$ may be calculated by 2nd order perturbation theory. We couple the parity violating potential $V(\underset{\sim}{r})$ to the spin-orbit coupling. It may be readily shown (Harris, Stodolsky, 1978; Zel'dovich et al., 1977) that ε is approximately

$$\varepsilon \sim Z^5 \chi\, 10^{-18} \text{ e.v.}, \qquad (71)$$

"χ" is a chirality factor which ranges between 10^{-2} and 10^{-6} (Rein et al., 1979). The Z^5 comes from the Z^3 which appears in the weak interaction matrix element, and Z^2 which is the spin-orbit contribution. Thus, even though the effect is small, it may be observable, at least in a beam.

Hund (1927) resolved what he called the "paradox of the optical isomers" - their apparent stability even though they are not in the ground state, by showing that tunnelling for complex systems is very long. Our discussion shows that at some point the weak interaction will lift the exact L-R degeneracy and hinder the tunnelling.

In the region $\varepsilon/\delta >> 1$, ϕ_L and ϕ_R will essentially become exact eigenstates. At zero temperature, in vacuum, even a simple handed isomer will stay handed essentially forever. Thus, at least for some isolated molecules, the ultimate answer to the paradox lies in the weak interaction.

E. CONCLUSIONS

It is remarkable that parity is violated. Or perhaps, it is not so amazing at all. Perhaps the violation of parity is a manifestation of a deeper symmetry - like the elliptical orbit being symptomatic of spherical symmetry. I don't know; have no idea.

Two brilliant experiments seem to confirm the existence of parity violating neutral currents. Perhaps they are the last gasp of atomic physics as a scientific region of surprises. Certainly the work of the Bouchiats breathed new life into atomic problems.

In molecules that are not prepared in handed states, parity violation reveals itself in forms entirely analogous to that which occurs in atoms: circular dichroism, degeneracies split, etc.

It is in molecules which may be prepared in a handed condition where parity violation has its most unusual and subtle effects. The fundamental symmetry between left and right is broken. This breakage, albeit tiny, leads to the asymmetric tunnelling oscillation which in its own right is interesting, and is a remarkable confluence of molecular and elementary particle processes.

In certain handed molecules, the parity violating interaction leads to an unexpected resolution of the "paradox" of the stability of isomers. Perhaps such processes will never be measured in the laboratory due to the difficulties of preparation and isolation. It may be that only in outer space may these oscillations occur without hindrance. They are, however, real.

In conclusion, it is pleasing to know that we may use radiation, atoms and molecules in order to tell our isolated alien friends which hand is the dreamer.

ACKNOWLEDGEMENTS

I would have been ill-equipped to discuss any aspects of parity violation had it not been for what I learned from Professors Leo Stodolsky and Eugene Commins. In addition, I wish to acknowledge the partial support of the NSF. Finally, I wish to thank my wife, Christine, for her imagery.

REFERENCES

Abers, E.S. and Lee, B.W., 1973, Physics Reports 96, 1.
Atkins, P.W., Barron, L.D., 1968, Proc. Roy. Soc. A304, 303.
Baum, L.F., 1900, The Wizard of Oz, p.186, Reilly and Lee.
Bouchiat, C.C. and Bouchiat, M.A., 1974a, Journal de Phys. (Paris), 35, 899.
Bouchiat, C.C. and Bouchiat, M.A., 1974b, Phys. Letters 48B, 111.
Bouchiat, C.C. and Bouchiat, M.A., 1975, Journal de Phys. (Paris) 36, 493.
Carter, S.S., and Kelly, H.P., 1979, Phys. Rev. Letters 42, 966.
Chu, S., Commins, E.D. and Conti, R., 1977, Phys. Letters 60A, 96.
Commins, E.D., 1973, Weak Interactions, McGraw Hill.
Conti, R., Buchsbaum, P., Chu, S., Commins, E.D. and Hunter, L., 1979, Phys. Rev. Letters, 42, 343.
Eyring, H., Walter, J. and Kimball, G.E., 1944, Quantum Chemistry, John Wiley & Sons.
Feynman, R.P., 1961, Theory of Fundamental Processes, W.A. Benjamin.
Feynman, R.P., 1963, Lectures on Physics, Vol. 1, Addison Wesley.
Feynman, R.P., 1965, The Character of Physical Law, MIT Press.
Gottfried, K., 1966, Quantum Mechanics, W.A. Benjamin.
Harris, R. and Stodolsky, L., 1978, Phys. Letters 78B, 313.
Henley, E.M. and Thirring, W., 1962, Elementary Quantum Field Theory, Chapter 9, McGraw-Hill.
Herzberg, G., 1945, Infrared and Raman Spectra, Van Nostrand, p.224.
Hinds, E.A. and Hughes, V.W., 1977, Phys. Letters 67B, 487.
Hund, F., 1927, Z. Phys. 43, 805.
Landau, L.D. and Lifshitz, E.M., 1958, Quantum Mechanics, Pergamon Press.
Lee, T.D., and Yang, C.N., 1956, Phys. Rev. 104, 254.
Letokhov, V., 1975, Phys. Letters 53A, 275.
Lewis, R.R. and Williams, W.L., 1976, Phys. Letters 59B, 70.
Neuffer, D.V. and Commins, E.D., 1977, Phys. Rev. A16, 844.
Physics Today, Sept. 1978, American Institute of Physics, p.17.
Prescott, C.Y. and 19 other names, 1978, Phys. Letters 77B, 347.
Rein, D.W., Hegstrom, R.A. and Sandars, P.G.H., 1979, Phys. Letters 71A, 499, and references therein.
Rosenfeld, L., 1928, Z. Physik 52, 161.
Sakurai, J.J., 1964, Invariance Principles and Elementary Particles, Princeton University Press.
Salam, A., 1968, in Elementary Particle Theory, ed. N. Svartholm, Almquist Forlag, Stockholm.
Stephan, M.J., 1958, Proc. Camb. Phil. Soc. 54, 81.
Townes, C.H. and Schalow, A.L., Microwave Spectroscopy, McGraw Hill.
Weinberg, S., 1967, Phys. Rev. Letters 19, 1264.
Weinberg, S., 1971, Phys. Rev. Letters 27, 1688.
Weinberg, S., 1972, Phys. Rev. D5, 1412.
Weston, R., 1954, J. Am. Chem. Soc. 76, 2645.
Zel'dovich, B. Yu., Sakyan, D.B., and Sobel'man, I.I., 1977, JETP Letters 25, 94.

HIGH RESOLUTION LASER SPECTROSCOPY OF MOLECULES

W. Demtröder

Fachbereich Physik
Universität Kaiserslautern
6750 Kaiserslautern, Germany

1. INTRODUCTION

The application of lasers to molecular spectroscopy has indeed revolutionized this field in many respects. Lasers have set new standards with regard to attainable sensitivity and spectral resolution. New sub-Doppler techniques have been developed which overcome the resolution limit imposed on classical spectroscopy by the Doppler-width of molecular lines. The sensitivity limit has been pushed down to the detection of molecular concentrations below the ppb (parts per billion) level. New insights into the nonlinear interaction of molecules with light can be gained by the application of powerful lasers. Multiphoton-spectroscopy, stimulated Raman-scattering or four-wave mixing techniques are only a few examples where new areas of molecular physics have been opened by laser spectroscopy.

Of particular interest is the dynamics of excited molecules. The question is, how the excitation energy, selectively pumped into the molecule by absorption of a photon, is redistributed among the different degrees of freedom. In this field lasers, which deliver short light pulses in a narrow spectral band are ideal instruments to study these problems in detail. Lifetime measurements, studies of radiationless transitions and of collisional deactivation of excited molecules are examples, which have already been successfully attacked by laser spectroscopy.

In these lectures we will restrict the discussion of laser spectroscopy to the three following subjects:

1. Detection sensitivity in molecular laser spectroscopy
2. Spectroscopic techniques with sub-Doppler-spectral resolution
3. Time resolution and lifetime measurements.

While most of the different methods can be covered only briefly, we discuss in more detail one example of a Doppler-free spectroscopic technique, known as polarization spectroscopy, because this sensitive method has some definite advantages for high resolution molecular spectroscopy.

2. TUNABLE COHERENT LIGHTSOURCES

Tunable coherent light sources can be realized in different ways. One possibility relies on such laser media which exhibit gain over a broad spectral range. Wavelength-selecting elements inside the laser resonator restrict laser oscillation to a narrow spectral interval [1] and the laser wavelength may be continuously tuned across the gain profile by varying the transmission maxima of these elements. Dye lasers and excimer lasers are examples of this type of tunable devices.

Another possibility of wavelength tuning is based on the shift of energy levels in the active medium by external perturbations, which cause a corresponding spectral shift of the gain profile and therefore of the laser wavelength. This level shift may be effected by an external magnetic field (spin-flip-Raman-lasers [2] and Zeeman-tuned gas-lasers) or by temperature - or pressure - changes (semiconductor lasers [3]).

The third possibility of generating coherent radiation with tunable wavelength uses the principle of optical frequency mixing. When the radiations from two lasers with frequencies ω_1 and ω_2 are superimposed in a medium with a sufficiently large, nonlinear part of its susceptibility, each atom is induced to forced oscillations and generates radiation with the sumfrequency $\omega_1+\omega_2$ and the difference frequency $\omega_1-\omega_2$. All these secondary waves generated at the location of each atom will have the correct phase to interfere constructively with each other, if the phase velocities of primary wave and secondary waves are identical (phase-matching condition). The resultant macroscopic wave with frequency $\omega_1+\omega_2$ or $\omega_1-\omega_2$ is emitted into a direction where the phase matching condition is fulfilled. In favourable cases the intensity of the secondary wave may reach up to 50% of the incident intensity. If the frequency ω_1 or ω_2 of the incident radiation can be tuned, the difference-or sumfrequency exhibits the same absolute tuning range, provided the phase-matching condition can be always fulfilled over this tuning range [4].

An interesting development, based on frequency mixing in crystals with nonlinear susceptibility is represented by the optical parametric oscillator [5]. A pump wave with frequency ω_p and wave-vector $\underset{\sim}{k}$ splits into two waves with frequencies ω_i and ω_s and wave-vectors $\underset{\sim}{k}_i$ and $\underset{\sim}{k}_s$ such that energy and momentum are conserved, which means that $\omega_p = \omega_i + \omega_s$ and $\underset{\sim}{k}_p = \underset{\sim}{k}_i + \underset{\sim}{k}_s$. By choosing the proper angle of incidence against the optical axis of the birefrigent crystal or by controlling the refractive index n through the crystal temperature, the idler and signal frequencies ω_i and ω_s can be varied within wide ranges.

Instead of frequency mixing in nonlinear crystals the Raman-effect in molecular gases can be used to produce Stokes or Anti-stokes frequencies $\omega_R = \omega_p + m.\omega_n$ where ω_n corresponds to a molecular eigenfrequency and ω_R to the Raman line. Tuning of the pump frequency ω_p allows a tunable ω_R which may cover a wide range in the infrared, if ω_p is in the visible region (Raman laser [6]).

The experimental realisation of these tunable coherent light sources if of course determined by the spectral range for which they are to be used. For the particular spectroscopic problem one has to decide which of the possibilities, summarized above, represents the optimum choice. The experimental expenditure depends substantially on the desired tuning range, on the achievable output power and last not least on the realized spectral bandwidth.

Many of the most widely used tunable coherent infrared sources use various semiconductor materials, either directly as active laser medium (semiconductor lasers) or as nonlinear mixing device (spin-flip-Raman-laser, frequency-difference generation). Alkali halides with different types of colour centres provide the material for the colour centre lasers [7], while birefringent crystals, such as Lithium Niobate are used in parametric oscillators and near infrared optical frequency mixing devices. Gases at high pressure where the pressure broadening exceeds the line spacing, offer another possibility of gain media for continuously tunable infrared gas lasers.

For wavelength tuning of semiconductor lasers all those parameters may be varied which determine the energy gap between the upper and lower laser level. Mostly a temperature change produced by an external cooling system or by a current change is used to generate a wavelength shift. The drawback of temperature changes induced by altering the electric current is the variation of the laser output power with current. It is therefore advantageous to use lasers with a closed refrigeration system which allows to alter the temperature over a wide range without changing the current through the semiconductor. With specially designed heterostructure semiconductor diode lasers cw-operation at room temperature has been achieved [7a]. Often an external magnetic field or a mechanical

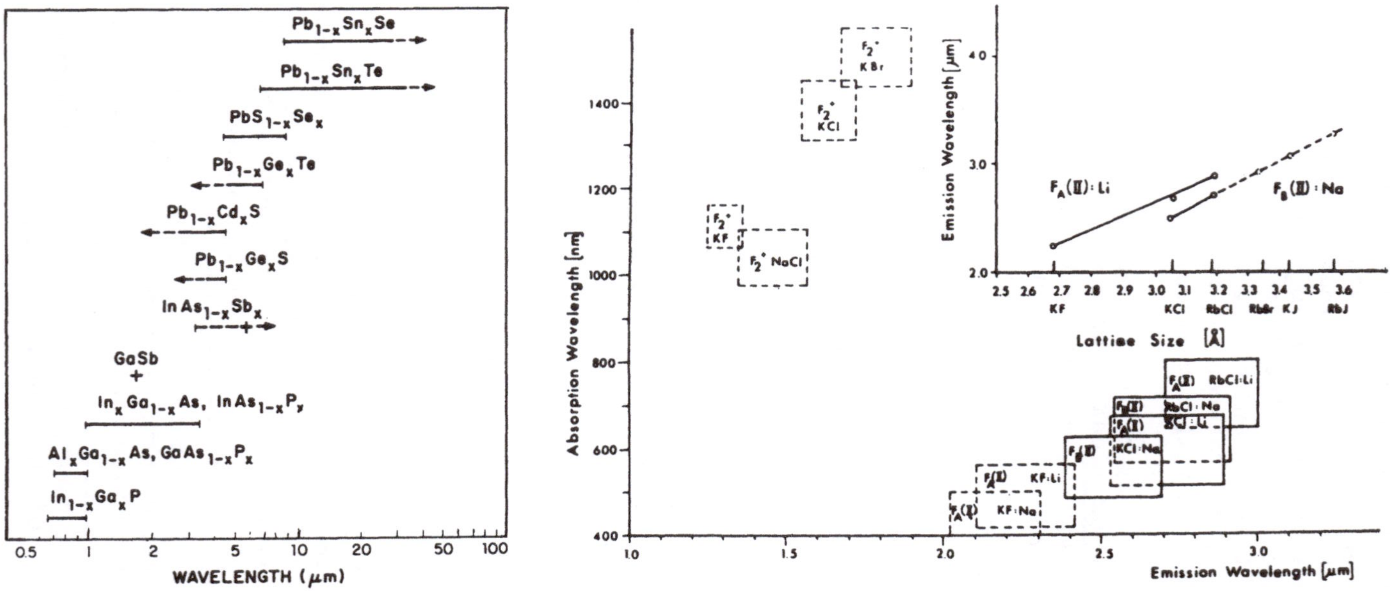

Fig. 1. a) Wavelength ranges covered by various semiconductor diode laser materials (from ref. 9a).

b) Emission and Absorption Bands of Various Colour Centre Crystals (from ref. 10).

pressure applied to the semiconductor are employed for wavelength tuning.

In general, however, no truly continuous tuning over the whole gain profile is possible. After a tuning over a range of about one wavenumber mode-hops occur if the resonator length is not altered synchroneously with the maximum of the gain profile. For the realisation of continuous tuning it is therefore necessary to use external resonator mirrors with a distance d which can be independently controlled. This implies, however, a larger distance d and therefore a smaller free spectral range. To achieve single mode oscillation an additional etalon has to be inserted into the resonator. Furthermore the endfaces of the semiconductor must be antireflection-coated because the large reflection coefficient of the uncoated surfaces (with $n = 2.5$ the reflectivity becomes already 0.2) causes large reflection losses. Such single mode semiconductor lasers have been built already [8]. By a combined variation of temperature and external magnetic field the continuous tuning range can be greatly increased.

In its advanced form the semiconductor laser represents a compact IR-laser spectrometer which combines intense light source and high resolution monochromator. The cw-output-powers of good lasers range from microwatts up to several milliwatts and the spectral resolution reaches from several MHz into the Kilohertz range, depending on the expenditure and efforts with the frequency stabilisation. Despite the expensive closed cycle refrigerator which can cool down to liquid helium temperatures such a compact laser spectrometer can easily compete with a Fourier-spectrometer regarding price, signal to noise ratio and spectral resolving power. There are numerous examples where semiconductor lasers have been successfully used in high resolution infrared spectroscopy [9].

The recently developed colour centre laser [10] is based on the excitation of colour centres in alkali-halide crystals by absorption of argon- or krypton laser radiation. The subsequent infrared fluorescence extends over a broad spectral range and tunable cw laser oscillation has been achieved with cooled crystals. The tuning range covers the near infrared region around 3 μm and cw output powers of many milliwatts have been reported (Fig. 1).

In the visible range dye-lasers in their various modification are by far the most widely used types of tunable lasers [11]. Their active media are organic dye molecules solved in liquids, which display strong broad-band fluorescence spectra under excitation by visible or uv-light. With different dyes the overall spectral range where cw- or pulsed laser operation has been achieved, extends from 300 nm to 1,2 μm.

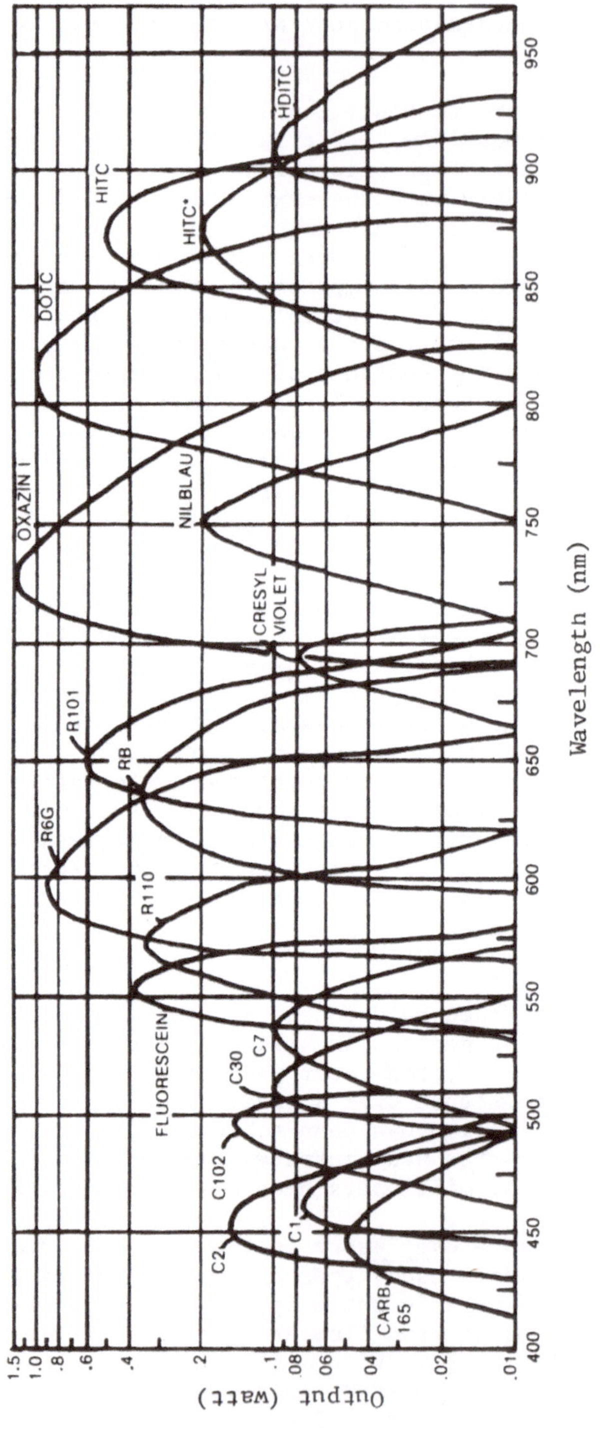

Fig. 2. Tuning range of various organic dye-solutions. (Courtesy Spectra-Physics)

The experimental realisations of dye-lasers use either flash-lamps, pulsed lasers or cw-lasers as pumping source. Recently several experiments on pumping of dye-molecules in the gas phase by high energy electrons have been reported [12]. Up to now, however, no electron pumped gas phase dye-laser could be realized. With optical pumping by a N_2-laser, however, stimulated emission from various dyes in the vapour phase has been observed.

Coarse tuning of the dye-laser wavelength can be accomplished with a birefringent Lyot-filter [13] or with a prism inside the resonator. With a three-element Lyot-filter a bandwidth $\Delta\lambda$ below 0.1 nm can be realized. If the Lyot-filter is continuously turned by a slow gear drive, the wavelength of the dye-laser is continuously tuned across the whole gain profile, which amounts for example for Rhodamin 6 G, to about 90 nm. Such a dye-laser represents a device with medium resolution which combines intense light source and monochromator. Figure 2 shows the tuning range for different laser dyes, which illustrates that the whole spectral region between 400 and 900 nm is already accessible to cw dye lasers.

Tunable single mode dye lasers with a bandwidth of a few MHz ($\hat{=}$ 10^{-4} cm^{-1}), which are continuously tunable over a spectral range of 30 GHz ($\hat{=}$ 1 cm^{-1}) are already commercially available [13a]. Multimode dye lasers with a bandwidth of about 0.1 Å can be even tuned over several hundred angströms.

Besides colour centre lasers the tunable single mode cw-dye laser is probably the most promising laser type for sub-Doppler laser spectroscopy. Great efforts have been therefore spent in many laboratories to increase output power, tuning range and frequency stability. Various resonator configurations, pump geometrics and designs of the dye flow system (jet streams) have been successfully tried to realize optimum dye-laser performance.

For single mode operation additional selecting elements have to be inserted in the resonator. In most designs two FPI-etalons with different free spectral ranges are used. Continuous tuning of the single mode laser demands synchroneous control of cavity length and transmission maxima of the selecting elements (Fig. 3). The optical pathlength of the cavity can be conveniently tuned by turning a tilted plane-parallel glass-plate, within a small interval around the Brewster angle [14].

The essential characteristics of dye-lasers is their broad homogeneous gain profile. Under ideal experimental conditions the homogeneous broadening allows that all excited dye-molecules may contribute to the gain at a single frequency. This implies that under single mode operation the output power should be not much lower than the multimode power, provided the selecting intracavity

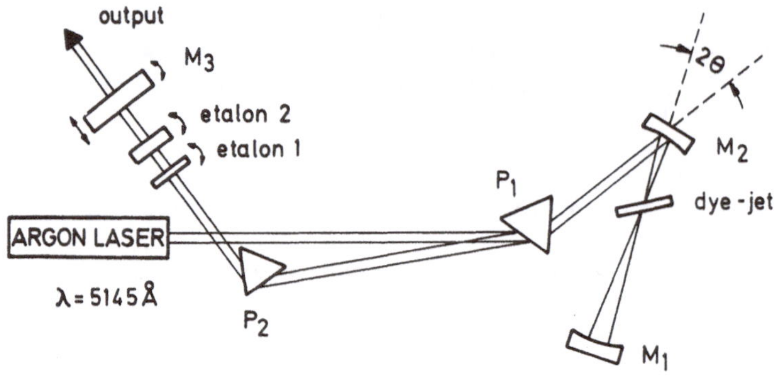

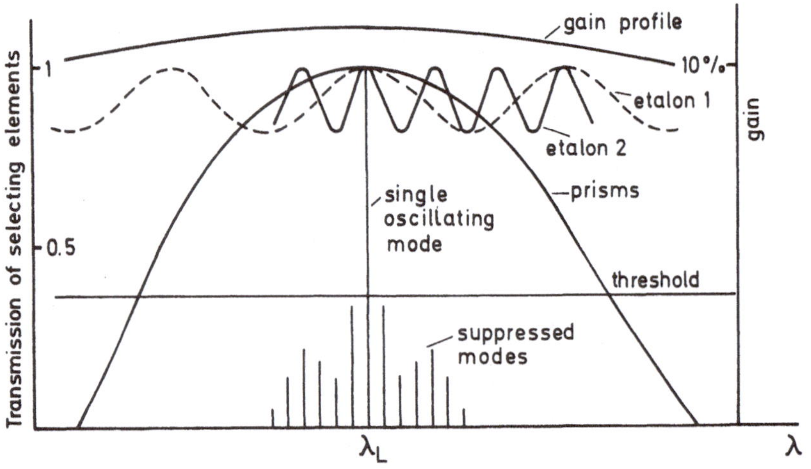

Fig. 3. Frequency dependent gain of the active medium and transmission curves of the two etalons.

elements do not introduce large additional losses. The frequency fluctuations of a single-frequency "free running" laser without active stabilization system are mainly determined by density fluctuations of the dye-jet and by mechanical jitter of the optical components and the average "linewidth" is limited to about 10 - 100 MHz, depending on the special design. In order to achieve better stability, the laser frequency has to be locked to a stable but tunable reference (e.g., an external Fabry Perot interferometer). Figure 4 illustrates such a stabilized tunable "dye-laser spectrometer" with a frequency stability of better than 1 MHz.

A severe drawback of standing-wave-resonator configurations for dye-lasers is the spatial hole burning effect [14b] which impedes single mode operation and prevents that all excited

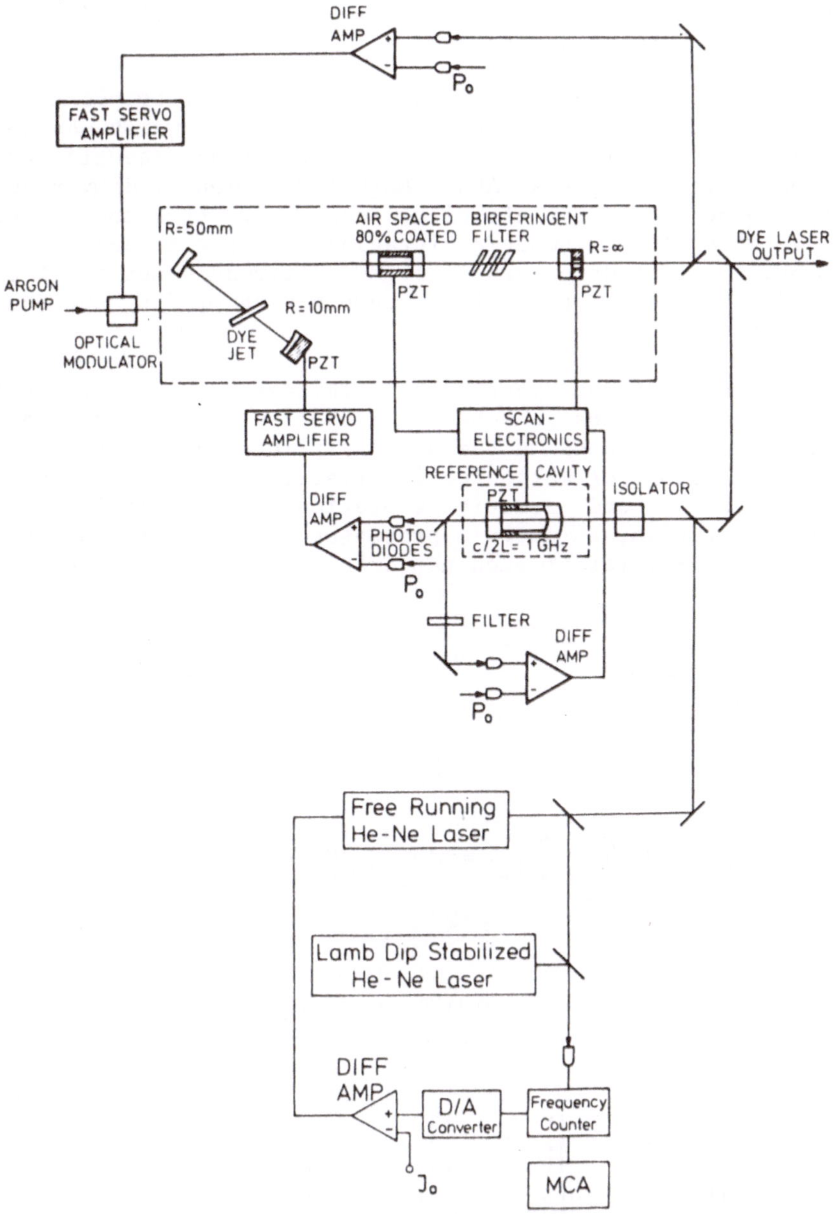

Fig. 4. Schematic arrangement of a stabilized tunable single mode cw dye-laser (from ref. 14a).

molecules within the active mode volume contribute to the gain of a single mode. Ring resonators with optical diodes allowing laser oscillation on unidirectional travelling waves, overcome this drawback. These ring-lasers allow single mode operation with less selecting elements and higher output powers. Alignment and resonator design are, however, more sophisticated than in standard resonators for standing waves [14c].

Ring lasers have the additional advantage that by proper choice of mirror radii and distances a second beam waist may be generated at a convenient location within the ring resonator where a nonlinear crystal (such as ADA = $NH_4H_2AsO_4$ ammonium dihydrogen arsenid) can be placed for efficient intracavity frequency doubling (Fig. 5). This gives a tunable ultraviolet source around λ = 250-300 nm. Output powers of 50 mW multimode power and 4 mW single mode UV-power at argon laser pump powers of 7 W have been already achieved [14d].

Besides the various types of tunable lasers, discussed in the foregoing sections, tunable coherent radiation sources have been developed which are based on the nonlinear interaction of intense radiation with atoms or molecules in crystals or in liquid and gaseous phases. Second harmonic generation, sum- or difference-frequency generation, parametric processes or stimulated Raman scattering are examples of such nonlinear optical mixing techniques,

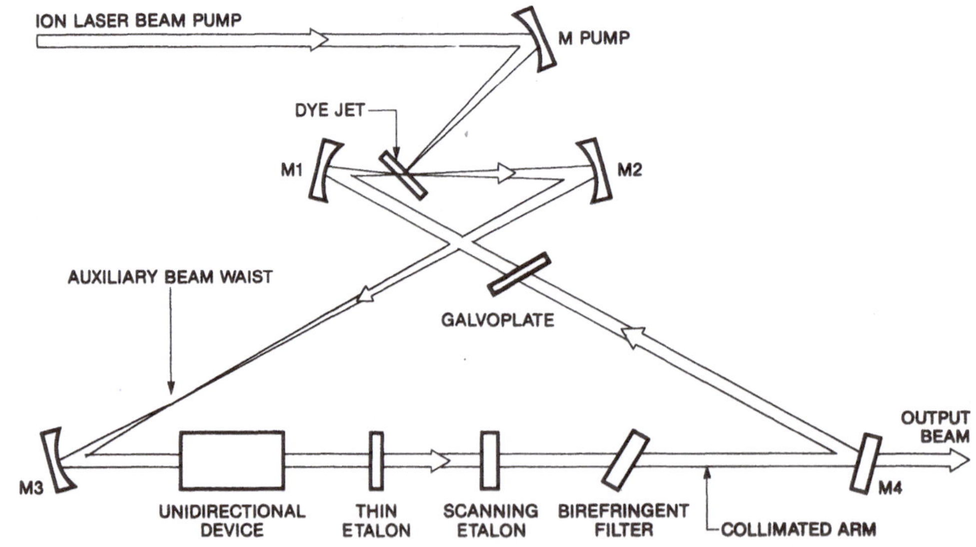

Fig. 5. Ring laser with auxiliary beam waist for intracavity frequency doubling. (Courtesy Spectra-Physics)

which have succeeded in covering the whole spectral range from the vacuum ultraviolet (VUV) up to the far infrared with sufficiently intense tunable coherent sources [15]. For a more detailed representation of these devices the reader is referred to the literature [16a,b].

3. ADVANTAGES OF LASERS IN SPECTROSCOPY

In order to illustrate the advantages of spectroscopy with tunable lasers, we will at first compare it with the conventional absorption spectroscopy which uses incoherent radiation sources. Figure 6 gives schematic diagrams of both methods.

In classical absorption spectroscopy radiation sources with a broad emission continuum are preferred (e.g., high pressure Hg-arcs, Xe-flash lamps etc.). The radiation is collimated by a lens L_1 and passes through the absorption cell. Behind a dispersing instrument for wavelength selection (spectrometer or interferometer) the intensity $I_T(\lambda)$ of the transmitted light is measured as a function of the wavelength λ (Fig. 6a). By comparison with a reference beam $I_R(\lambda)$ (which can be realized for instance, by shifting the absorption cell alternatively out of the light beam) the absorption spectrum

$$I_A(\lambda) = a\, I_R(\lambda) - I_T(\lambda)$$

can be obtained, where the constant a takes into account different wavelength independent losses of I_R and I_T (e.g., reflections of the cell walls). The spectral resolution is generally limited by the resolving power of the dispersing spectrometer. Only with large and expensive instruments (e.g., Fourier-spectrometers) the Doppler-limit may be reached.

The detection sensitivity of the experimental arrangement is defined by the minimum absorbed power which can be still detected. In most cases it is limited by the detector noise and by intensity fluctuations of the radiation source. Generally the limit of detectable absorption is reached at relative absorptions $\Delta I/I \geq 10^{-4} - 10^{-5}$. This limit can be only pushed further down in favourable cases by using special sources and lock-in detection or signal averaging techniques.

Contrary to radiation sources with broad emission continua used in conventional spectroscopy, tunable lasers offer radiation sources in the spectral range from the UV to the IR with extremely narrow bandwidths and with spectral power densities which may exceed that of incoherent light sources by many orders of magnitude.

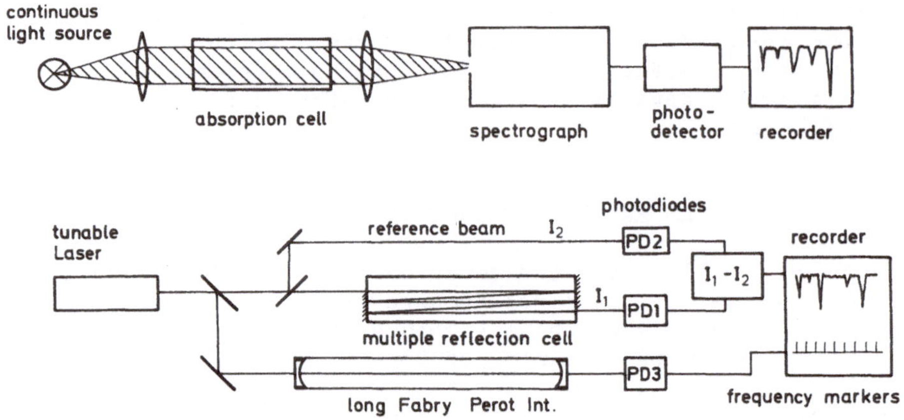

Fig. 6. Absorption-spectroscopy with lasers compared with a conventional design using incoherent radiation sources

In several regards laser absorption spectroscopy corresponds to microwave spectroscopy where klystrons or carcinotrons instead of lasers represent tunable coherent radiation sources. Laser spectroscopy transfers many of the techniques and advantages of microwave spectroscopy to the infrared, visible and ultraviolet spectral ranges.

The advantages of absorption spectroscopy with tunable lasers may be summarized as follows:

1. No monochromator is needed since the absorption coefficient $\alpha(\omega)$ and its frequency dependence can be directly measured from the difference $\Delta I(\omega) = a\, I_R(\omega) - I_T(\omega)$ between the intensities of reference beam and transmitted beam (Fig. 6b). The spectral resolution is higher than in conventional spectroscopy. With tunable single mode lasers it is only limited by the linewidths of the absorbing molecular transitions. Using Doppler-free techniques even sub-Doppler resolution can be achieved.

2. Because of the high spectral power density of many lasers the detector noise is generally negligible. Intensity fluctuations of the laser which limit the detection sensitivity, may be essentially suppressed by intensity stabilisation. This furthermore increases the signal to noise ratio and therefore enhances the sensitivity.

3. The detection sensitivity increases with increasing spectral resolution $\omega/\Delta\omega$ as far as $\Delta\omega$ is still larger than the linewidth $\delta\omega$ of the absorption line. This can be seen as follows: The relative intensity attenuation per absorption path length $x = 1$ is

$$\Delta I/I = \int_{\omega_0 - \delta\omega/2}^{\omega_0 + \delta\omega/2} \alpha(\omega).I(\omega)d\omega \Big/ \int_{\omega_0 - \Delta\omega/2}^{\omega_0 + \Delta\omega/2} I(\omega)d\omega \tag{1}$$

with the approximation:

$$\int_{\omega_0 - \Delta\omega/2}^{\omega_0 + \Delta\omega/2} I(\omega)d\omega \simeq \bar{I}.\Delta\omega \qquad \int_{\omega_0 - \delta\omega/2}^{\omega_0 + \delta\omega/2} \alpha.Id\omega = \bar{\alpha}.\bar{I}.d\omega$$

this yields:

$$\Delta I/I = \bar{\alpha}.\frac{\delta\omega}{\Delta\omega} \tag{2}$$

Decreasing the resolvable spectral interval $\Delta\omega$ from 10 $\delta\omega$ to $\delta\omega$ therefore increases the detection sensitivity approximately by a factor of 10.

4. Because of the good collimation of a laser beam, long absorption paths can be realized by multiple reflection back and forth through the absorption cell. Disturbing reflections from cell walls or windows which may influence the measurements, can be essentially avoided (for example by using Brewster-endwindows). Such long absorption paths enable the measurement of transitions even with small absorption coefficients. Furthermore pressure broadening can be reduced by using low gas pressure. This is especially important in the infrared region where the Doppler-width is small and the pressure broadening may become the limiting factor for the spectral resolution.

5. If a small fraction of the laser output is sent through a long Fabry-Perot-Interferometer with a separation d of the mirrors, the photodetector P D 3 receives intensity peaks each time the laser frequency ω_L is tuned to a transmission maximum at $\omega_L = 2\pi.m.c/2d$. These peaks serve as accurate wavelength markers which allow to calibrate the separation of adjacent absorption lines. With d = 1 m the frequency separation $\Delta\nu$ between successive transmission peaks becomes $\Delta\nu = c/2d = 150$ MHz, corresponding to a wavelength separation of 10^{-4} nm at $\lambda = 550$ nm. With a semiconfocal FP I the free spectral range is c/8d, which gives $\Delta\nu = 75$ MHz for d = 0.5 m.

6. The laser frequency may be stabilized onto the centre of an absorption line. With "wavemeters", based on long distance interferometry, it is possible to measure the wavelength λ_L of the laser with an absolute accuracy of 10^{-8} or better.

This allows to determine the molecular absorption lines with the same accuracy.

7. It is possible to tune the laser wavelength very rapidly over a spectral region where molecular absorption lines have to be detected. With electro-optical components for instance, pulsed dye lasers can be tuned over several wavenumbers within a microsecond. This opens new perspectives for spectroscopic investigations of short lived intermediate radicals in chemical reactions. The capabilities of classical flash-photolysis may be considerably extended using such rapidly tunable laser sources.

8. An important advantage of absorption spectroscopy with tunable single mode lasers stems from their capabilities to measure line profiles of absorbing molecular transitions with high accuracy. In case of pressure broadening the determination of line profiles allows to derive information about the interaction potential of the collision partners. In plasma physics this technique is widely used to determine electron and ion densities and temperatures.

9. In fluorescence spectroscopy and optical pumping experiments the high intensity of lasers allows to achieve an appreciable population in selectively excited states, which may be comparable to that of the absorbing ground states. The small laser-linewidth favours the selectivity of optical excitation and results in favourable cases in the exclusive population of single molecular levels. Because of the large line density of most molecular spectra fortuitous coincidences between laser line and molecular asborption line are generally found even with fixed frequency lasers which cannot be tuned. Therefore argon lasers, CO_2-lasers or other intense laser lines have been successfully used to selectively excited molecular levels and to produce a large population in these levels. These advantageous conditions allow to perform absorption- and fluorescence spectroscopy of excited states and to transform spectroscopic methods, such as microwave- or rf-spectroscopy, up to now restricted to electronic ground states, also to excited states.

4. HIGH SENSITIVITY DETECTION METHODS

The general method to measure absorption spectra is based on the determination of the absorption coefficient $\alpha(\omega)$ from the spectral intensity

$$I_T(\omega) = I_o \exp(-\alpha(\omega)\cdot x) \tag{3}$$

which is transmitted through an absorbing path length x. For small absorption ($\alpha \cdot x \ll 1$) we can use the approximation $\exp(-\alpha \cdot x) \approx 1-\alpha \cdot x$ and equation (3) can be reduced to

$$I_T(\omega) \approx I_o(1 - \alpha(\omega) \cdot x) \tag{4}$$

With a reference intensity $I_R = I_o$ as produced for example by a 50% beam splitter in Fig. 6 with a reflectivity R = 0.5 one can measure the absorption coefficient

$$\alpha(\omega) \cdot x = (I_R - I_T(\omega))/I_R \tag{5}$$

from the difference $I_R - I_T(\omega)$.

In case of very small values of $\alpha \cdot x$ this method cannot be very accurate since it measures a small difference of two large quantities and already small fluctuations of the splitting ratio can severely influence the measurement. Therefore several different techniques have been developed which allow to increase the sensitivity and accuracy of absorptions measurements by several orders of magnitude. In the following we will discuss some of these techniques.

The first method is based on a frequency modulation of the monochromatic incident wave. It has not been designed specifically for laser spectroscopy but was taken from microwave spectroscopy where it is a standard method. The laser frequency ω_L is modulated at a modulation frequency f, which sweeps ω_L periodically from ω_L to $\omega_L + \Delta\omega_L$. When the laser is tuned through the absorption spectrum, the difference $I_T(\omega_L) - I_T(\omega_L + \Delta\omega_L)$ is detected with a lock-in amplifier (phase-sensitive detector) tuned to the modulation frequency f. If the modulation sweep $\Delta\omega_L$ is sufficiently small, the first term of the Taylor expansion

$$I_T(\omega_L + \Delta\omega_L) - I_T(\omega_L) = \left(\frac{dI_T}{d\omega}\right) \cdot \Delta\omega_L + \frac{1}{2}\left(\frac{d^2 I}{d\omega^2}\right)\Delta\omega_L^2 + \ldots \tag{6}$$

is dominant. This term represents the <u>first derivative</u> of the absorption spectrum, as can be seen from eq. (5). When I_R is independent of ω we obtain:

$$\frac{d\alpha(\omega)}{d\omega} = -\frac{1}{x \cdot I_R} \cdot \frac{dI_T}{d\omega} \tag{7}$$

The advantage of the frequency modulation is the possibility of <u>phase sensitive detection</u>, which restricts the frequency response of the detection system to a narrow frequency interval centred at the modulation frequency f. Frequency independent background absorption from cell-windows and background noise, due to fluctuations of the laser intensity or of the density of absorbing molecules are essentially reduced. Regarding signal to noise ratio and achievable sensitivity the frequency modulation technique is superior to an intensity modulation of the incident radiation. The frequency of a single mode laser can be readily modulated when an ac-voltage is applied to the piezo on which a resonator mirror is mounted.

Another very sensitive method directly monitors the absorbed energy rather than relying on a difference measurement $(I_R - I_T)$. The energy $I_o \cdot \alpha x \cdot A$ absorbed per second in a volume $V = A \cdot x$ can be either converted into fluorescence energy and monitored with a fluorescence detection system (excitation spectroscopy) or it can be converted by collisions into thermal energy with a resultant temperature and pressure rise, which is monitored by a sensitive microphone (photo-acoustic spectroscopy). The third method is based on the sensitive dependence of the laser intensity on absorption losses inside the laser resonator. When the absorbing sample is placed inside the laser resonator, the laser intensity I_L decreases by $\Delta I_L = q^* \alpha(\omega) \cdot x \cdot I_L$ where the factor q^* which gives the enhancement of the intensity change compared with the same absorption outside the cavity, can become very large.

The most sensitive methods are capable of counting single absorbed photons. They use as monitor the ionization of atoms and molecules from highly excited states, which had been populated by absorption of laser photons. Compared to the conventional absorption measurements these sensitive detection methods represent a remarkable progress. The sensitivity limit has been pushed down from detectable relative absorptions $\Delta\alpha/\alpha \approx 10^{-5}$ to about $\Delta\alpha/\alpha \leq 10^{-17}$. We now discuss some of these methods in more detail.

1. Excitation Spectroscopy

A very high sensitivity in the visible and ultraviolet region can be achieved if the absorption of laser photons is monitored through the laser excited fluorescence. When the laser wavelength λ_L is tuned to an absorbing molecular transition $E_i \rightarrow E_k$ the number of photons absorbed per sec is

$$n_a = N_i \cdot \sigma_{ik} \cdot \Delta x \cdot n_L \tag{8}$$

where n_L is the number of incident laser photons, σ_{ik} the absorption cross-section per molecule and N_i the density of molecules in the absorbing state E_i (Fig. 7).

The number of fluorescence photons emitted per second from the excited level E_k is

$$n_{Fl} = N_k A_k = n_a \cdot \eta_k \tag{9}$$

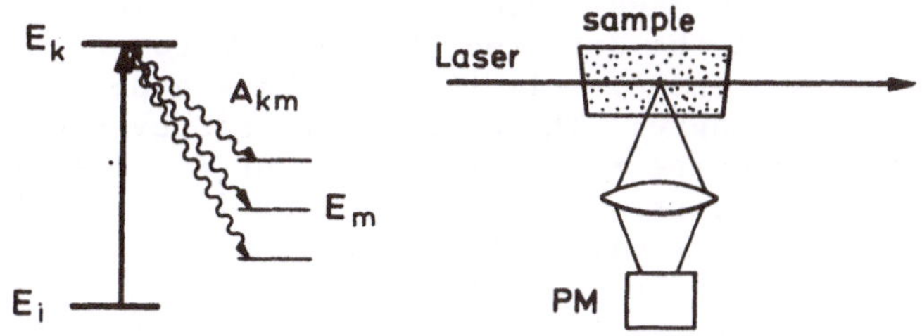

Fig. 7. Schematic diagram for explanation of excitation spectroscopy.

where $A_k = \sum_m A_{km}$ stands for the total spontaneous transition probability to all levels with $E_m < E_k$. The quantum efficiency $\eta_k = A_k/(A_k + R_k)$ gives the ratio of the spontaneous transition rate to the total deactivation rate which may also include a radiationless transition rate R_k (e.g., collision induced transitions). For $\eta_k = 1$ the number n_A of fluorescence photons emitted per sec equals the number n_a of photons absorbed per sec under stationary conditions.

Unfortunately only a fraction δ of the fluorescence photons emitted into all directions can be collected on the photomultiplier cathode where again only the fraction $\eta_{Ph} = n_{Pe}/n_{Ph}$ of these photons produce on the average n_{Pe} photo-electrons. The quantity η_{Ph} is called the quantum efficiency of the photo-cathode. The number n_{PE} of photo-electrons counted per second is then

$$n_{PE} = n_a \cdot \eta_k \cdot \eta_{Ph} \cdot \delta = N_i \; n_L \cdot \sigma_{ik} \cdot \Delta x \cdot \eta_k \cdot \eta_{Ph} \cdot \delta \tag{10}$$

Modern photomultipliers reach quantum efficiencies of $\eta_{Ph} = 0.2$. With a carefully designed optics it is possible to achieve a collection factor $\delta = 0.1$. Using photon counting techniques and cooled multipliers (dark pulse rate < 10 counts/s) counting rates of $n_{PE} = 100$ counts/s are already sufficient to obtain a signal to noise ratio $S/R \geq 8$ at integration times of 1 s.

Inserting this figure for n_{PE} into (10) illustrates that with $\eta_k = 1$ absorption rates of $n_a = 5 \cdot 10^3$/s can be already measured quantitatively. Assuming a laser power of 1 Watt at a wavelength $\lambda = 500$ nm, which corresponds to a photon flux of $n_L = 3 \cdot 10^{18}$/s, this implies, that it is possible to detect a relative absorption of $\Delta I/I \leq 10^{-14}$! When placing the absorbing probe inside the cavity where the laser power is q· times larger (q ∿ 10 to 100) this impressive sensitivity may be even further enhanced. Excitation spectroscopy has its highest sensitivity in the visible, ultraviolet and near infrared region. With increasing wavelength

λ_L the sensitivity decreases because of the following reasons: Equation 10 shows that the detected photon-electron rate n_{PE} decreases with η_k, η_{Ph} and δ. All these numbers generally decrease with increasing wavelength: The quantum efficiency η_{Ph} and the attainable signal to noise ratio is much lower for infrared than for visible photodetectors. By absorption of infrared photons vibrational-rotational levels of the electronic ground state are excited with radiative lifetimes which are generally several orders of magnitude larger than those of excited electronic states. At sufficiently low pressures the molecules diffuse out of the observation region before they radiate. This diminishes the collection efficiency δ. At higher pressures the quantum efficiency η_k of the excited level E_k is decreased because collisional deactivation competes with radiative transitions. Under these conditions opto-acoustic detection may become preferable.

2. Opto-Acoustic Spectroscopy

This sensitive technique of measuring small absorptions is mainly applied when minute concentrations of molecular species have to be detected in the presence of other components at higher pressure. An example is the detection of spurious pollutant gases in the atmosphere. Its basic principle may be summarized as follows [17]:

The beam of a tunable laser is sent through the absorption cell (Fig. 8). If the laser is tuned to an absorbing molecular transition $E_i \rightarrow E_k$, part of the molecules in the lower level E_i will be excited into the upper level E_k. By collisions with other atoms or molecules in the cell these excited molecules may transfer their excitation energy $(E_k - E_i)$ completely or partly into translational, rotational or vibrational energy of the collision partners. At thermal equilibrium this energy is randomly distributed onto all degrees of freedom, causing an increase of thermal energy and with it a rise of temperature and pressure at a constant density in the cell. When the laser beam is chopped at frequencies below 10 KHz, periodical pressure variations appear in the absorption cell which can be detected with a sensitive microphone placed at the inner side of the cell. The output signal from the microphone is proportional to the absorbed laser energy and allows therefore to determine the absorption coefficient. Because this method uses the conversion of photon energy into periodical pressure variations it is called photo-acoustic spectroscopy and the device is named spectraphone.

The idea of the spectraphone is very old and had already been demonstrated by Bell and Tyndal in 1881. However, the impressive detection sensivity obtained nowadays could be only achieved with the development of lasers, sensitive capacitance microphones, low

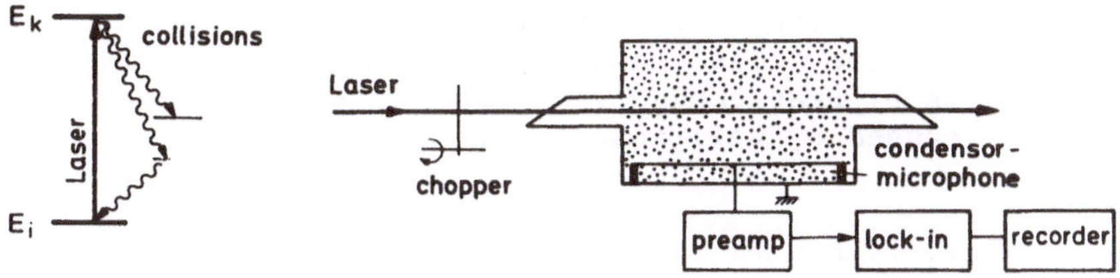

Fig. 8. Spectraphone and level diagram for opto-acoustic spectroscopy.

noise amplifiers and lock-in techniques. Concentrations down to the ppb range (parts per billion = 10^{-9}), at total pressures of 1 torr to several atmospheres, are readily detectable with a modern spectraphone.

As far as saturation effects can be neglected the acoustic signal S

$$S = C N_i \cdot \sigma_{ik}(\omega) \cdot \Delta x \cdot P_L (1 - \eta_k) \cdot S_M \tag{11}$$

is proportional to the density N_i of absorbing molecules in level E_i, the absorption cross section σ_{ik} and the pathlength Δx, the incident mean laser power P_L and the sensitivity S_M of the microphone. The signal decreases with increasing quantum efficiency η_k (which gives the ratio of emitted fluorescence photons to absorbed laser photons). The factor C depends on the spectraphone parameters. Modern condensor microphones with low noise FET preamplifier and phase sensitive detection achieve signals of larger than 1 V/torr with a background noise of 3×10^{-8} V at integration times of 1 sec. This sensitivity allows detection of pressure amplitudes below 10^{-7} torr, and is in general not limited by the electronic noise but by another disturbing effect: Laser light reflected from the cell windows or scattered by aerosols in the cell may be partly absorbed by the walls and contributes to a temperature increase. The resulting pressure rise is of course modulated at the chopping frequency and is therefore detected as background signal.

There are several ways to reduce this effect. Antireflection coatings of the cell windows, or, in case of linearly polarized laser light, the use of Brewster-windows minimize the reflections. An elegant solution chooses the chopping frequency to coincide with an acoustic resonance of the cell. This results in a resonant amplification of the pressure amplitude which may be as large as 100 fold. This experimental trick has the additional advantage, that those acoustic resonances can be selected, which correspond to standing pressure waves, with their maximum at the location of the laser beam in the centre of the cell but with nodes at the cell walls. These acoustic modes cannot be induced by pressure changes at the cell walls by heat conduction from the walls, and the background signal, caused by wall absorption, is therefore greatly reduced.

The sensitivity can be further enhanced by frequency modulation of the laser and by intracavity absorption techniques. With the spectraphone inside the laser cavity the photoacoustic signal of nonsaturating transitions is increased by a factor q due to the q-fold laser intensity inside the resonator.

3. Intracavity Absorption

When the absorbing sample is placed inside the laser cavity (see Fig. 9) the detection sensitivity can be considerably enhanced, in favourable cases by several orders of magnitude. Four different effects can be utilized to achieve this "amplified" sensitivity:

Assume the reflectivities of the two resonator mirrors to be $R_1 = 1$ and $R_2 = 1 - T_2$ (mirror absorption is neglected). At a laser output power P_{out} the power inside the cavity is $P_{in} = q \cdot P_{out}$ with $q = 1/T_2$. For $\alpha \cdot L \ll 1$ the laser power $\Delta P(\omega)$ absorbed at the frequency ω in the absorption cell (length L) is

$$\Delta P(\omega) = q \cdot \alpha(\omega) \cdot L \cdot P_{out} \tag{12}$$

If this absorbed power can be directly measured, for example through the resulting pressure increase in the absorption cell (see section 2) or through the laser induced fluorescence (see section 1) the signal will be q times larger than for the case of single pass absorption outside the cavity. With $T_2 = 0.02$ (a figure which can be readily realized in practice) the enhancement factor becomes $q = 50$, as long as saturation effects can be neglected and provided the absorption is sufficiently small that it does not noticeably change the laser intensity. This q-fold amplification of the sensitivity can be also understood from the simple fact, that a laser photon travels on the average q-times back and forth between the resonator mirrors before it leaves the

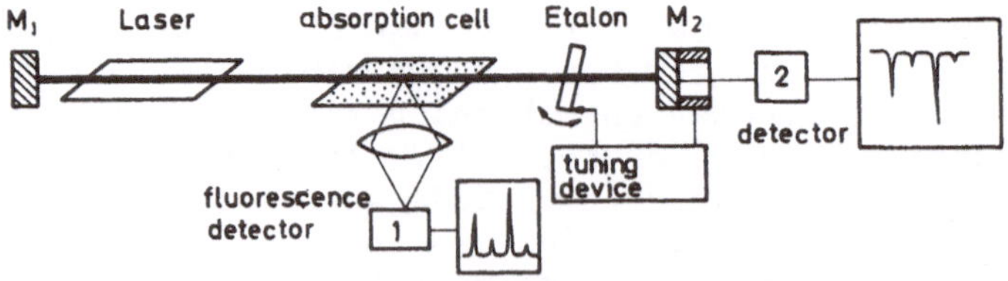

Fig. 9. Intracavity absorption.

resonator. It therefore has a q-fold chance to be absorbed in the sample.

Another way of detecting intracavity absorption with very high sensitivity relies on the dependence of laser output power on absorption losses inside the laser resonator (detector D_2 in Fig. 9). At constant pump power close above threshold already minor changes of the intracavity losses may result in drastic changes of the laser output. Under steady state conditions the laser intensity essentially depends on the pump power and reaches a value I_s where the saturated gain G equals the total losses. If additional losses $\Delta\gamma = \alpha(\omega)\cdot L$ are introduced by the absorbing sample, the relative change of the laser output power is

$$\frac{\Delta P}{P} = \frac{G_o}{G_o - \gamma} \cdot \frac{\Delta\gamma}{\gamma + \Delta\gamma} \tag{13}$$

where G_o is the unsaturated gain.

Compared with the single pass absorption outside the resonator where the relative intensity change is $\Delta I/I = -\alpha\cdot L = -\Delta\gamma$ this represents a sensitivity enhancement by a factor

$$Q = \frac{G_o}{G_o - \gamma}\,\frac{1}{\gamma + \Delta\gamma} \times \frac{G_o/\gamma}{G_o - \gamma} \qquad \text{for } \Delta\gamma \ll \gamma \tag{14}$$

At pump powers far above threshold the unsaturated gain G_o is large compared with the losses γ and eq. (14) reduces to

$$Q \simeq 1/\gamma \qquad \text{for } G_o \gg \gamma \tag{15}$$

If the resonator losses are mainly due to the transmission T_2 of the output mirror, the enhancement factor Q becomes $Q = 1/\gamma = 1/T_2 = q$, which is equal to the enhancement of the previous detection method.

Close above threshold, however, G_o is only slightly larger than γ and the denominator in eq. (14) becomes very small, which means that the enhancement factor Q may reach very large values. At first sight it might appear that the sensitivity could be made arbitrarily large for $G_o \to \gamma$. However, there are experimental as well as fundamental limitations which restrict the maximum achievable value of Q. The increasing instability of the laser output, for instance, limits the detection sensitivity when the threshold is approached. Close at threshold, the spontaneous radiation which is emitted into the solid angle accepted by the detector cannot be neglected. It represents a constant background intensity, nearly independent of γ, which puts a principal upper limit for the relative change $\Delta I/I$ and thus for the sensitivity.

Even larger enhancement factors Q can be achieved with lasers oscillating simultaneously on several competing modes. A detailed calculation (see for instance ref. [18]) yields for the relative intensity change of the absorbing mode:

$$\frac{\Delta I}{I} = \frac{G_o}{G_o - \gamma} \frac{\Delta\gamma}{\gamma + \Delta\gamma} (1 + K \cdot N) \tag{16}$$

where $K(0 \leq K \leq 1)$ is a measure of the coupling strength of the absorbing mode to the N oscillating modes.

With the intracavity absorption technique spurious amounts of molecular constituents have been detected [19,20] or molecular transitions of very small oscillator-strength could be measured [21].

4. Ionization Spectroscopy

Ionization spectroscopy monitors the absorption of photons on a molecular transition $E_i \rightarrow E_k$ by detecting the ions or electrons produced by some means, while the molecule is in its excited state E_k. The necessary ionization of the excited molecule may be performed in various ways:

a) Photo-ionization.

$$M^*(E_k) + h \cdot \nu_2 \rightarrow M^+ + e^- + E_{kin} \tag{17}$$

The ionizing photon may come either from the same laser which has excited level E_k or from a separate light source, which can be another laser or even an incoherent source (see Fig. 10).

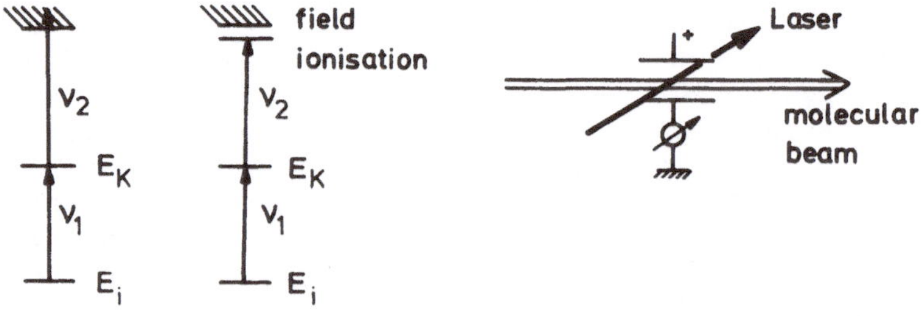

Fig. 10. Ionization spectroscopy.

b) Collision-induced ionization. If the excited level E_k is not too far from the ionization limit, the molecule may be ionized by thermal collisions with other atoms. If E_k lies above the ionization limit of the collision partners A, Penning ionization [22] becomes an efficient process which proceeds as

$$M^*(E_k) + A \rightarrow M + A^+ + e^- \tag{18}$$

c) Field-ionization. If the excited level E_k lies close below the ionization limit, the molecule $M^*(E_k)$ can be ionized by an external electric dc-field. This method is particularly efficient if the excited level is a long lived highly excited Rydberg-state.

With a proper design the collection efficiency δ for the ionized electrons or ions can reach 100%. If the electrons or ions are accelerated to several keV and detected by electron multipliers or channeltrons also a detection efficiency of 100% can be achieved.

The following estimation illustrates the possible sensitivity of ionization spectroscopy: Let N_k be the density of excited molecules in level E_k, P_{kI} the probability per sec that a molecule in level E_k is ionized, and n_a the number of photons absorbed per s on the transition $E_i \rightarrow E_k$. If R_k is the total relaxation rate of level E_k besides the ionization rate (spontaneous transitions plus collision induced deactivation) the signal rate in counts per s is given by

$$\begin{aligned} S &= N_k\, P_{kI} \cdot \delta \cdot \eta = n_a \frac{P_{kI}}{P_{kI}+R_k} \cdot \delta \cdot \eta \\ &= N_i \cdot n_L \cdot \sigma_{ik} \cdot \Delta x \frac{P_{kI}}{P_{kI}+R_k} \cdot \delta \cdot \eta \end{aligned} \tag{19}$$

In case of photo-ionization the ionization probability P_{kI} depends on the intensity of the ionizing radiation. Using intense lasers the ionization rate can be made large compared to the relaxation rate R_k. For the ideal case of $\delta = \eta = 1$ and $P_{kI} \gg R_k$ eq. (19) shows that a signal rate $S = n_a$ can be achieved which equals the absorption rate n_a of photons on the transition $E_i \rightarrow E_k$. This implies that single absorbed photons can be detected with an overall efficiency close to unity. In the experimental practice there are, of course, additional losses and sources of noise which limit the detection efficiency to a somewhat lower level. However, for all absorbing transitions $E_i \rightarrow E_k$, where the upper level E_k can be readily ionized, the ionization spectroscopy is the most sensitive detection technique and superior to all other methods discussed so far.

V. SUB-DOPPLER LASER SPECTROSCOPY

The absorption spectra of many molecules show such a large density of lines that many lines overlap within their Doppler-width. This implies that individual rotational lines, fine-structure or hyperfine-structure splittings often cannot be resolved with Doppler-limited spectroscopy. Recorded with insufficient resolution such spectra appear quasicontinuous, concealing finer details of the spectrum. Doppler-free techniques are therefore required to resolve all individual lines and to take full advantage of the small linewidth of single mode lasers. In this section we will illustrate such Doppler-free techniques by some examples.

1. Laser Spectroscopy in Collimated Molecular Beams

When a collimated molecular beam (beam axis in z-direction) is crossed perpendicularly with a monochromatic laser beam (frequency ω, wavevector $\underset{\sim}{k} = (k,0,0)$ propagating into the x-direction), the absorption probability depends for each molecule on its velocity component v_x (Fig. 11a). The centrefrequency of a molecular transition, which is ω_o in the rest-frame of the moving molecule, is Doppler-shifted to a frequency ω_o' with

$$\omega_o' = \omega_o - \underset{\sim}{k}\cdot\underset{\sim}{v} = \omega_o - k\, v_x, \quad k = |\underset{\sim}{k}| \qquad (20)$$

Only those molecules with velocity components v_x in the interval $dv_x = \delta\omega_n/k$ around $v_x = (\omega - \omega_o)/k$ essentially contribute to the absorption of the monochromatic laser wave, because these molecules are Doppler-shifted into resonance with the laser frequency ω within the natural linewidth $\delta\omega_n$ of the absorbing transition.

When the laser beam in the x-z-plane passes along the x-direction through the molecular beam, its intensity decreases as

$$I(\omega) = I_o \cdot \exp\left[- \int_{x_1}^{x_2} \alpha(\omega,x)dx\right] \qquad (21)$$

For small absorptions ($\alpha\cdot\Delta x \ll 1$) the spectral profile of the absorbed intensity $dI = I_o - I(x)$ is

$$dI(\omega) = I_o \cdot \int \alpha(\omega,x)dx$$

$$= I_o \cdot \int_{-v.\sin\varepsilon}^{+v.\sin\varepsilon} dv_x \cdot \int_{x_1}^{x_2} n(v_x,x)\cdot\sigma(\omega,v_x)dx \qquad (22)$$

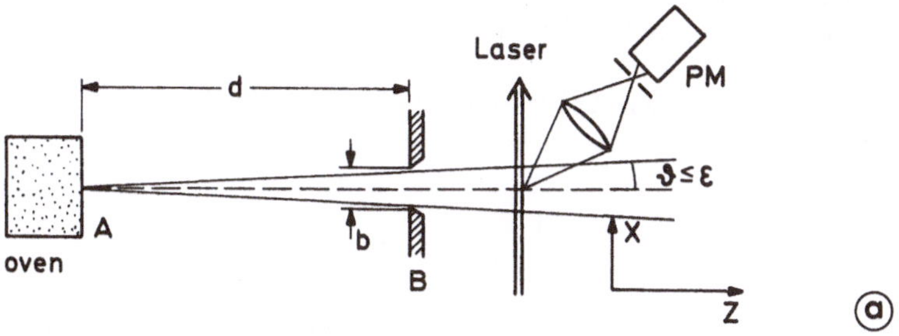

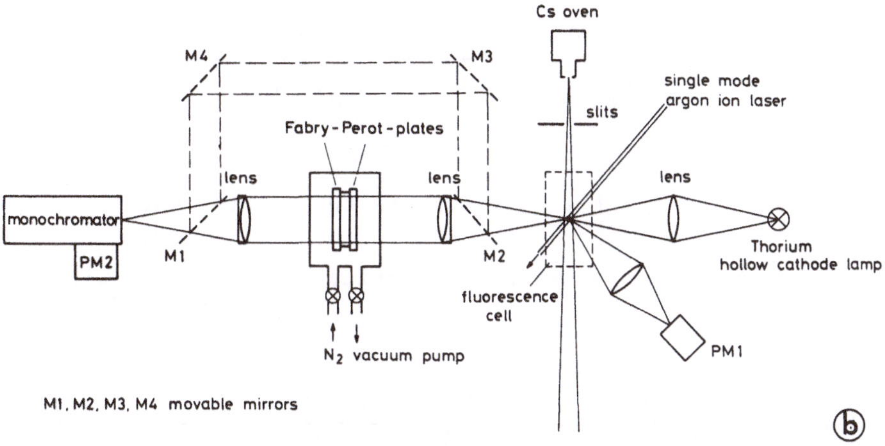

Fig. 11 Laser spectroscopy in a collimated molecular beam.
(a) Schematic diagram; (b) Experimental arrangement.

The absorption cross section $\sigma(\omega, v_x)$ describes the absorption of a monochromatic wave of frequency ω by a molecule with a velocity component v_x. Its spectral profile is represented by a Lorentzian

$$\sigma(\omega, v_x) = \sigma_o \frac{(\gamma/2)^2}{(\omega - \omega_o - kv_x)^2 + (\gamma/2)^2} . \qquad (23)$$

The density $n(v_x, x)\, dv_x$ of molecules with velocity components v_x in the interval dv_x at a point (x,z) in the molecular beam with distance $r = (z^2 + x^2)^{1/2}$ from the beam orifice A is given by

$$n(v_x, x)\, dv_x = C \frac{v_x^2\, z}{r^3}\, e^{-\beta^2 v_x^2 N^2/x^2}\, dv_x \; ; \qquad (24)$$

where the relations $v_x/v = x/r \rightarrow dv_x = (x/r)\cdot dv$ have been used and $\beta = (m/2KT)^{-1/2}$ gives the most probable velocity.

Inserting (20), (23) and (24) into (22) yields the absorption profile

$$dI(\omega) = e(\sigma_o\cdot Z_o\cdot c^3/\omega_o^3)\cdot \int_{-\infty}^{+\infty} d\omega_o \cdot$$

$$\int_{x_1}^{x_2} \frac{\Delta\omega_o^2}{x_o^3}\cdot \frac{(\gamma/2)^2\cdot \exp[\frac{-\beta^2\Delta\omega_o c^2}{\omega_o^2}(1+\frac{Z^2}{x^2})]}{(\omega_o - \omega - \Delta\omega_o)^2 + (\gamma/2)^2}\, dx \qquad (25)$$

where $\Delta\omega_o = \omega_o - \omega_o' = k\cdot v_x = v_x\cdot\omega/c$ is the Doppler shift.

The integration over x can be carried out and gives with $x_{1,2} = \pm\, r\cdot\sin\varepsilon$

$$dI(\omega) = I_o\cdot a\cdot \int \frac{e^{-\beta^2\cdot c^2\cdot(\omega_o - \omega')^2/(\omega_o^2\sin^2\varepsilon)}}{(\omega - \omega')^2 + (\gamma/2)^2}\, d\omega' \qquad (26)$$

with $a = c\cdot\sigma_o(\omega_o/2d\cdot c)^2$. This represents a Voigt profile, i.e., a convolution product of a Lorentzian function with halfwidth γ and a Doppler function. The Doppler-width is, however, reduced by a factor sin ε, which equals the collimation ratio of the beam. The collimation of the molecular beam therefore reduces the Doppler-width of the absorption lines to a width

$$\Delta\omega_D^r = \Delta\omega_D\cdot\sin\varepsilon \qquad \text{with } \Delta\omega_D = \frac{2\omega_o}{c\cdot\beta}(\ln 2)^{1/2} \qquad (27)$$

where $\Delta\omega_D$ is the corresponding Doppler-width in a gas at thermal equilibrium. Figure 12 illustrates the achievable spectral resolution by comparing a section of the Cs_2 absorption spectrum obtained in a cesium vapour cell (upper curve) and in a collimated cesium beam (lower spectrum). Both spectra were taken with a single mode argon laser which could be continuously tuned around $\lambda = 476{,}5$ nm. The absorption was monitored through the total fluorescence intensity $I_{Fl}(\lambda_L)$ as a function of laser wavelength λ_L (excitation spectroscopy, see section IV, 1). The different lines correspond to rotational-vibrational transitions

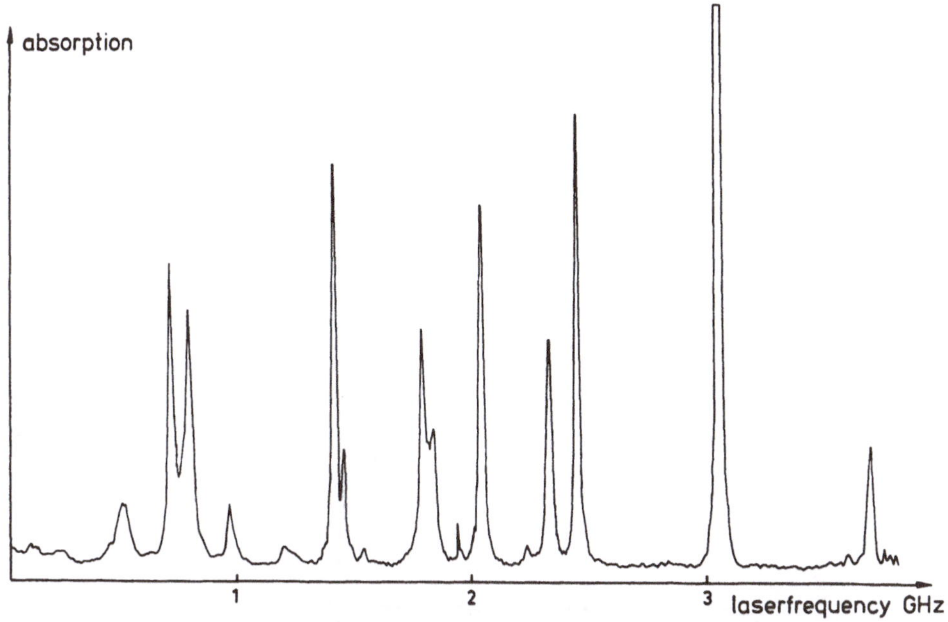

Fig. 12(a). Section of the excitation spectrum of the Cs_2-molecule around $\lambda = 476{,}5$ nm. a) in a cesium vapour cell and b) in a collimated molecular beam.

$(v'',J'') \rightarrow (v',J')$ between two different electronic states of the Cs_2-molecule [23].

Particularly for polyatomic molecules with their complex visible absorption spectra the reduction of the Doppler-width is essential for the resolution of single lines. This is illustrated by a section from the excitation spectrum of the NO_2-molecule, excited with a single mode tunable argon laser around $\lambda = 488$ nm. For comparison the same section of the spectrum is shown in the upper trace, as obtained with Doppler-limited laser spectroscopy in a NO_2-cell (Fig. 12b).

2. Laser Spectroscopy in Fast Ionbeams

In the examples considered so far, the laser beam was crossed perpendicularly with the molecular beam, and the reduction of the Doppler-width was achieved through the limitation of the maximum velocity components v_x by geometrical apertures. One often calls this reduction of the transverse velocity components therefore "geometrical cooling". In fast ionbeams a reduction of the Doppler-width can be achieved in another arrangement where the laser beam travels collinear with a fast ion- or atom beam (Fig. 13) and the reduction of the longitudinal velocity distribution $n(v_z)$

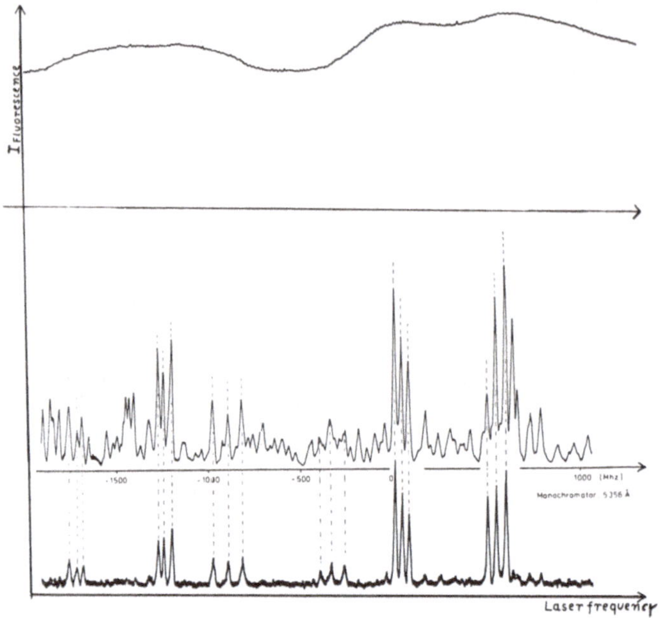

Fig. 12(b). Comparison of NO_2 excitation spectra taken with a single mode argon laser, tunable around λ = 514,5 nm in a cell at 0.1 torr pressure (a) and in a collimated NO_2-beam. The excitation spectrum (b) was taken by monitoring the total undispersed fluorescence $I_{FL}(\lambda_L)$ while in (c) only one vibrational fluorescence band, filtered through a monochromator, was detected as a function of the laser wavelength.

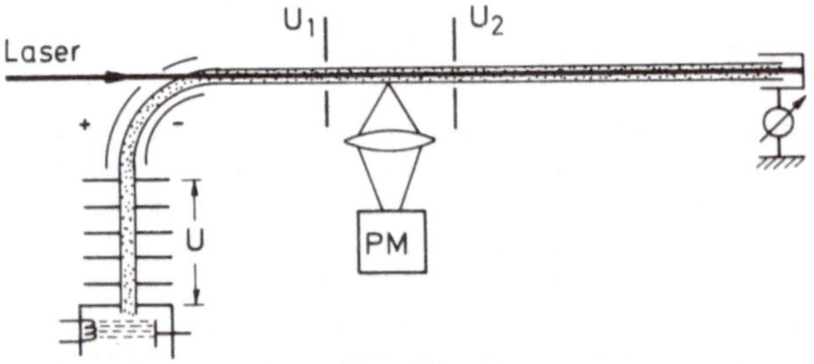

Fig. 13. Laser spectroscopy in fast ion beams.

is achieved by an acceleration voltage ("acceleration cooling") [24]:

Assume two ions start from the ion source with different thermal velocities $v_{10} = 0$ and $v_{20} > 0$. After being accelerated by the voltage U their kinetic energies are

$$E_1 = e\,U = \frac{m}{2} v_1^2$$
$$E_2 = \frac{m}{2} v_{20}^2 + e\,U = \frac{m}{2} v_2^2 \tag{28}$$

Subtracting the first from the second equation yields

$$v_2^2 - v_1^2 = v_{20}^2 \rightarrow \Delta v = v_2 - v_1 = v_{20}^2/2v. \tag{29}$$

This means, that the initial velocity spread v_{20} has decreased to

$$\Delta v = v_{20}\cdot\left(\frac{v_{20}^2\ m}{8\ e\ U}\right)^{1/2} = v_{20}\left(\frac{\Delta E_{th}}{4\cdot e\cdot U}\right)^{1/2} \tag{30}$$

Example:

$$E_{th} = 0.1\ \text{eV}; \qquad e\cdot U = 10\ \text{keV} \rightarrow \Delta v = 1.5 \quad 10^{-3}\cdot v_{20}$$

This reduction of the velocity spread results from the fact that energies rather than velocities are added. For $e\cdot U >> E_{th}$ the velocity change is mainly determined by U but hardly by the fluctuations of the initial thermal velocity. This implies, however, that the acceleration voltage has to be extremely well stabilized to fully utilize this "acceleration cooling".

A definite advantage of this parallel arrangement is the longer interaction zone between the two beams because the laser induced fluorescence can be collected by a lens from a pathlength ΔZ, of several cm, compared to a few mm in the perpendicular arrangement (Fig. 13). A further advantage is the possibility of Doppler-tuning. The absorption spectrum of the ions can be scanned across a fixed laser frequency ω simply by tuning the acceleration voltage U_1 [25]. An absorption line at ω_o is Doppler-tuned into resonance with the laser field, at frequency ω, if

$$\omega = \omega_o - k\cdot v_z = \omega_o\left(1 - \left(\frac{2cU}{mc^2}\right)^{1/2}\right) \tag{31}$$

This allows to use high intensity fixed frequency lasers, such as the argon laser, which have a high gain and even allows to place the interaction zone inside the laser cavity.

3. Saturation Spectroscopy

Saturation spectroscopy is based on the selective saturation of an inhomogeneously broadened molecular transition by optical pumping with a monochromatic tunable laser. The population density $n_i(v_z)dv_z$ of molecules in the absorbing state E_i is selectively depleted of molecules with velocity components

$$v_z \pm dv_z = (\omega_o - \omega \pm \delta\omega)/k \tag{32}$$

in the interval dv_z, because these molecules are Doppler-shifted into resonance with the laser-frequency ω and are excited from E_i to the higher level E_k with $\omega_o = (E_k - E_i)$. The monochromatic laser therefore "burns a hole" into the population distribution $n_i(v_z)$ of the absorbing state and produces simultaneously a peak at the same velocity component v_z in the upper state distribution $n_k(v_z)$ (Fig. 14).

Due to this population depletion the absorption coefficient $\alpha(\omega) = \Delta n(v_z)\sigma_{ik}(\omega_o - \omega - v_z/k)$ is no longer constant but decreases with increasing laser intensity. The absorbed intensity

$$dI = -\alpha(I)\cdot I_o = \alpha_o\cdot I_o - \alpha_1\cdot I_o^2 + \ldots \tag{33}$$

is no longer proportional to the incident intensity but depends in a nonlinear relation on I_o. Spectroscopic techniques based on saturation effects are therefore often called "nonlinear spectroscopy" [26].

We now briefly discuss the basic concepts and experimental arrangements of saturation spectroscopy and we show that this technique allows essentially Doppler-free spectral resolution. Assume a monochromatic laser wave $E = E_o\cdot\cos(\omega t - kz)$ propagates in the z-direction through a gaseous sample of molecules with a thermal velocity distribution. If the molecules can be described

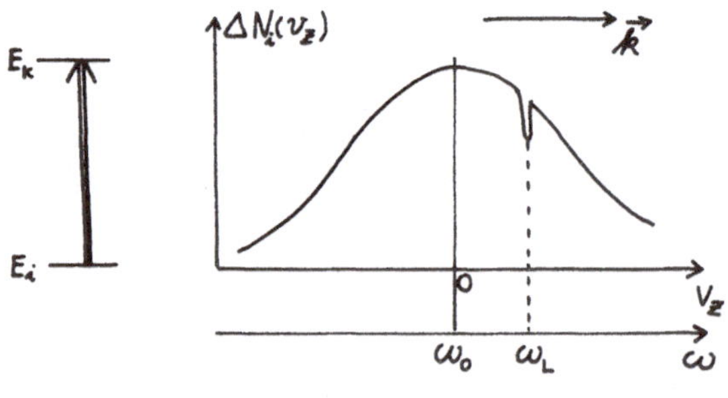

Fig. 14. Hole burning by a monochromatic laser wave, propagating in the +z-direction.

as a two level system with energy levels E_i, E_k with equal relaxation probabilities γ, we can calculate the transition probability W_{ik} for absorption of a photon $\hbar\omega = (E_k - E_i)$ [27],

$$W_{ik} = \frac{S\cdot(\gamma/2)^2}{(\omega_o - \omega - kv_x)^2 + \gamma^2(1 + S)/2} \; ; \; \gamma = \gamma_i + \gamma_k \qquad (34)$$

The saturation parameter $S = (p\cdot E/\hbar)^2/\gamma^2$ represents the ratio of induced absorption rate $(p\cdot E/\hbar)^2/\gamma$ to the relaxation rate γ. Equation (34) describes a Lorentzian line profile with the saturated halfwidth $\gamma_s = \gamma\cdot(1 + S)^{1/2}$.

Due to the transitions induced by the pump-laser the population difference $\Delta n = n_i - n_k$ decreases from its unsaturated value $\Delta n^o(v)$ to a saturated value $\Delta n^s(v)$, which can be obtained as the steady state solution of the rate equation:

$$\frac{\partial}{\partial t}(\Delta n(v)) + \gamma(\Delta n(v) - \Delta n^o(v) = W_{ik}\cdot\Delta n(v) \qquad (35)$$

For $\partial\Delta n(v)/\Delta t = 0$ this yields

$$\Delta n^s(v) = \Delta n^o(v)(1 - \frac{S\cdot(\gamma/2)^2}{(\omega_o - \omega - k\cdot v)^2 + (\gamma/2)^2(1 + S)} \qquad (36)$$

Inserting the Maxwellian velocity distribution

$$\Delta n^o(v) = C\cdot\Delta N^o\cdot e^{-\beta^2 v_z^2} \qquad (37)$$

for the unsaturated population distribution ($\Delta N_o = \int\Delta n^o(v_z)\cdot dv_z$ and $\beta^2 = (m/2KT)$) we obtain the velocity distribution of the saturated population difference

$$\Delta n_s(v_z)\, dv_z = C\Delta N^o\cdot(1 - \frac{(\gamma_s/2)\cdot S}{(\omega_o - \omega - k\cdot v_z)^2 + (\gamma_s/2)^2})\cdot e^{-\beta^2 v_z^2}\, dv_z$$

$$(38)$$

which represents a Gaussian line profile with a hole of width $\gamma_s = \gamma(1 + S)^{1/2}$, centred at the velocity component $v_z = (\omega_o - \omega)/k$ ("Bennet-hole", see Fig. 14).

In order to detect this "Bennet-hole", which has been burnt into the population distribution $n_i(v_z)$ by the pump-wave, a second

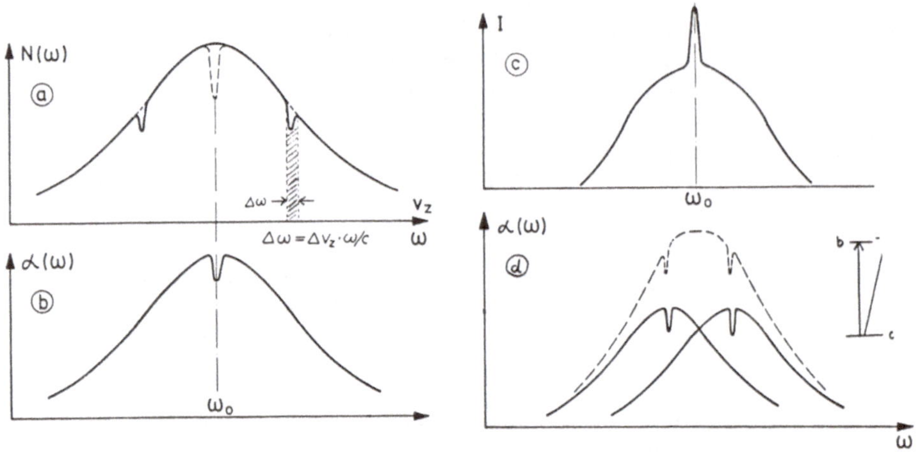

Fig. 15. a) Bennet holes in the population distribution $n_i(v_z)$ burnt by a standing wave with $\omega \neq \omega_o$.
b) Lamb-dip at the centre of Doppler-broadened molecular absorption line.
c) Lamb-peak in the output power of a laser with intra-cavity absorption cell.
d) Resolution of two closely spaced Doppler-broadened transitions by Lamb-dip spectroscopy.

lightwave, called the probe-wave, has to be sent through the sample to probe the population depletion in the hole. This probe-wave may be either provided by the same laser as the pump-wave, or two different lasers may be utilized.

In a simple arrangement the pump-wave is reflected by a mirror back into the sample. Since the reflected wave $E = E_o \cos(\omega t + kz)$ has the opposite wavevector $-\underset{\sim}{k}$, it interacts with another group of molecules centred around $-v_z$. As long as $\omega \neq \omega_o$ two different holes at $v_z = \pm(\omega_o - \omega)/k$ are burnt into the population distribution $n_i(v_z)$ (Fig. 15), which merge together at the line centre $v_z = 0$ for $\omega \to \omega_o$ (dotted curve). The superposition of the two counterpropagating waves

$$E = E_o \cos(\omega t - kz) + E_o \cos(\omega t + kz) = 2E_o \cos \omega t \cos kz$$

represents a standing wave field. The crucial point is now, that the absorption in this field (which means the total absorption of both counter running waves) has a minimum for $\omega = \omega_o$! In this case both waves interact with the same molecules, which are thus exposed to twice the intensity. This means that the saturation parameter S is twice as large and according to Eq.(38) the depletion of $\Delta n(v_z)$ becomes larger for $v_z = 0$ than for $v_z \neq 0$.

To be more quantitative, let us calculate the absorption coefficient $\alpha(\omega)$ in the standing wave field. Using the relation

$\alpha(\omega) = \Delta n \cdot \sigma(\omega)$ we obtain:

$$\alpha_s(\omega) = \int \Delta n_s(v_z) \cdot [\sigma(\omega_o - \omega - kv_z) + \sigma(\omega_o - \omega + kv_z)] dv_z \tag{39}$$

where the saturated population difference $\Delta n(v_z)$ is according to eq. (36) given by

$$\Delta n_s(v_z) = \Delta n_o(v_z) \; [1 - \frac{(\gamma_s/2)\, S_o}{(\omega_o - \omega - kv_z)^2 + (\gamma_s/2)^2} - \frac{(\gamma_s/2)\cdot S_o}{(\omega_o - \omega + kv_z)^2 + (\gamma_s/2)^2}] \tag{40}$$

Inserting (40) into (39) we obtain for the absorption coefficient in the weak field approximation ($S << 1$)

$$\alpha_s(\omega) = \alpha_o(\omega) \; [1 - \frac{S_o}{2} (1 + \frac{(\gamma_s/2)^2}{(\;\; - \omega_o)^2 + (\gamma_s/2)^2}] \tag{41}$$

Since the unsaturated absorption coefficient is

$$\alpha_o(\omega) \;\; = \;\; C\, N_o \cdot \exp\, [\ln 2(\omega - \omega_o)^2/\Delta\omega_D^2]$$

eq. (41) represents a Doppler-profile, modified by a dip at the centre (Fig. 15b) which is called Lamb-dip, after W.E. Lamb, who first described this effect in the approximation of weak saturation [28]. The Lamb-dip profile is Lorentzian with a halfwidth γ_s (FWHM). At the line centre $\alpha_s(\omega_o)$ decreases to $\alpha_o(\omega_o)\cdot(1 - S_o)$; far off resonance to $\alpha_o(\omega)\cdot(1 - S_o/2)$. This is because for $\omega = \omega_o$ the molecules are saturated by twice the intensity, while for $\omega - \omega_o >> \gamma$ both fields interact with different molecules.

The narrow Lamb-dips in the Doppler-broadened absorption coefficient $\alpha(\omega)$, as seen by a monochromatic standing wave, can be used to resolve closely spaced absorption lines, which would be completely masked in Doppler-limited spectroscopy. Fig. 15d gives an example for two transition, between hyperfine components of two molecular states, which are separated by less than their Doppler-width. Although the Doppler-broadened line profiles completely overlap, their Lamb-dips can be clearly resolved.

The detection of these narrow resonances can be realized with different experimental arrangements. A typical example for a possible experimental setup is shown in Fig. 16. The output-beam

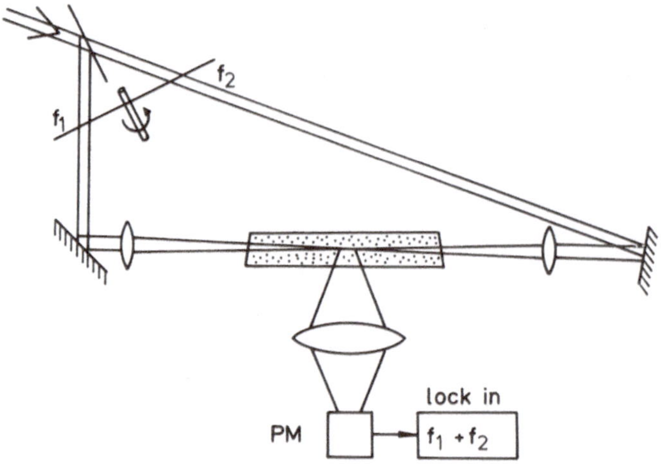

Fig. 16. Experimental setup for saturation spectroscopy with external sample cell.

from a tunable laser is split by the beam splitter BS into a strong pump beam and a weak probe beam, which pass through the absorbing sample into opposite directions. The attenuation of the probe beam is measured as a function of the laser frequency ω. To enhance the sensitivity, the probe beam is again split into two parts. One beam passes the region of the sample which is saturated by the pump beam. The difference of the two probe beam outputs yields the saturation signal.

In cases where the saturation is very small, the direct attenuation of the probe beam is difficult to detect and the small Lamb-dips may be nearly buried under the noise of the Doppler-broadened background. Sorem and Schawlow [29] have demonstrated a very sensitive "intermodulated fluorescence technique", where pump beam and probe beam are chopped at two different frequencies f_1 and f_2. Assume the intensities of the two beams to be $I_1 = I_o(1 + \cos 2\pi \cdot f_1 \cdot t)$ and $I_2 = I_o(1 + \cos 2\pi \cdot f_2 \cdot t)$. The intensity of the laser induced fluorescence is then

$$I_{Fl} = C \cdot n_s \cdot (I_1 + I_2) \tag{39}$$

where n_s is the saturated population density of the absorbing state and the constant C includes the transition probabilities and the collection efficiency of the fluorescence detector. According to (35) the saturated population density at the centre of an absorption line is $n_s = n_o \cdot (1 - a(I_1 + I_2))$. Inserting this into (39) gives

$$I_{Fl} = C\, n_o(I_1 + I_2) - a\, n_o(I_1 + I_2)^2 \tag{40}$$

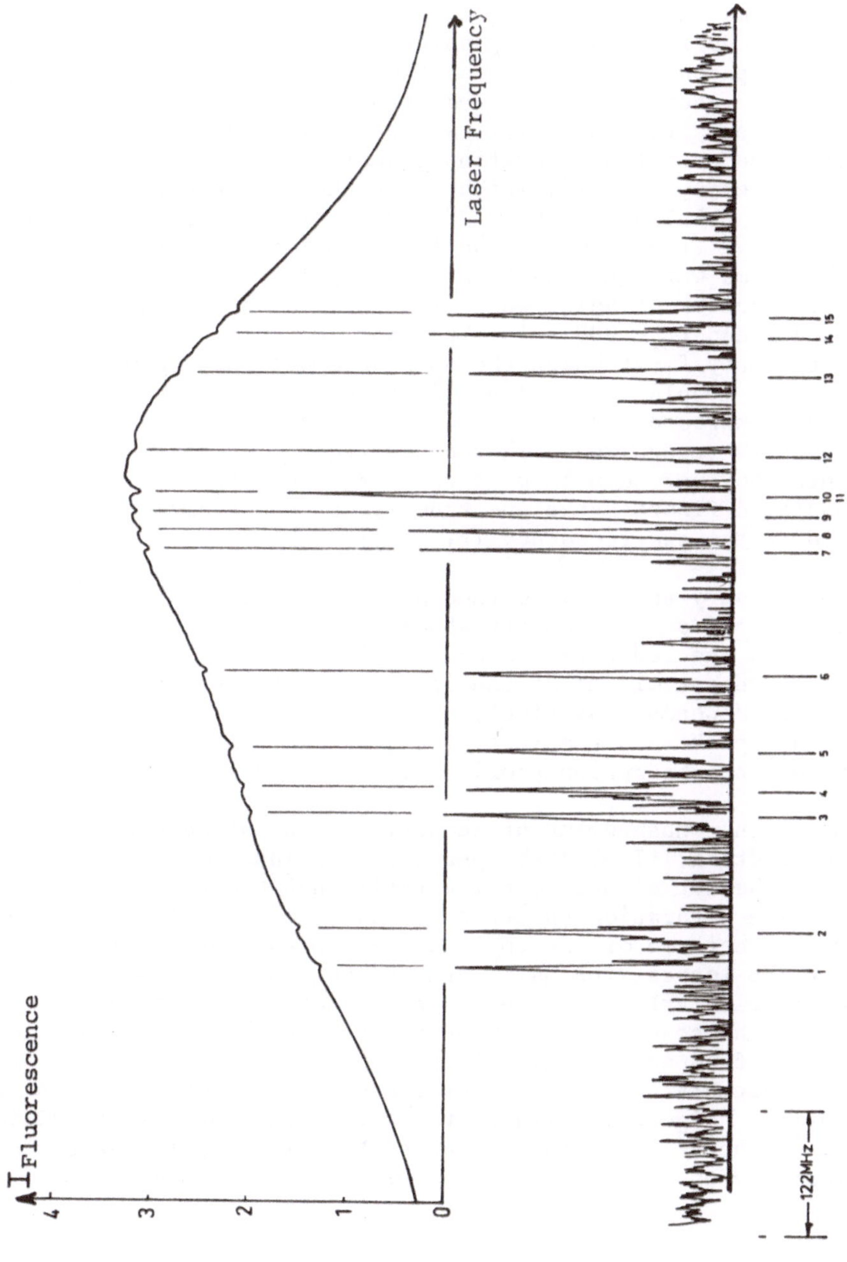

Fig. 17. Saturation spectrum of an I_2-transition, monitored with intermodulated fluoresence-technique.

which shows that the fluorescence intensity contains linear terms, modulated at the chopping frequencies f_1 and f_2 respectively, and quadratic terms with modulation frequencies $(f_1 + f_2)$ and $(f_1 - f_2)$ respectively. While the linear terms represent the normal laser induced fluorescence with a Doppler-broadened line profile, the quadratic terms describe the saturation effect, because they depend on the decrease of the population density $n_i(v_z = 0)$ due to the simultaneous interaction of the molecules with both fields. When the fluorescence is monitored through a lock-in amplifier, tuned to the sum frequency $f_1 + f_2$, the linear background is suppressed and only the saturation signals are detected. This is demonstrated by Fig. 17, which shows the 15 hyperfine-components of the rotational line $(v'' = 1, J'' = 98) \rightarrow (v' = 58, J' = 99)$ in the $X^1\Sigma_g^+ B^3\Pi_{uo-}$ transition of the iodine molecule I_2 [30]. The two laser beams were chopped by a rotating disc with two rows of a different number of holes which interrupted the beams at $f_1 = 600\ s^{-1}$ and $f_2 = 900\ s^{-1}$. The upper spectrum was monitored at the frequency f_1 at which the pump beam was chopped while the probe was not modulated. The nonlinear Lamb-dips on the Doppler-broadened background, which is caused by the linear terms in (40) can be clearly recognized. The centrefrequencies of the hfs-components, however, can be obtained more accurately from the intermodulated fluorescence spectrum (lower spectrum) which was monitored at the sum frequency $(f_1 + f_2)$.

Another very sensitive method of monitoring saturation signals is based on intracavity absorption. If the sample is placed inside the laser resonator, the Lamb-dips in the absorption profiles represent minima of the intracavity losses. Because the laser output depends sensitively on the internal losses it will show a sharp peak when the laser frequency is tuned across the Lamb-dip of the absorption profile (see Fig. 18 and 15c).

Due to the enhancement of sensitivity in intracavity absorption (see section III.3) these peaks in the laser output generally have a much better signal to noise ratio than the Lamb-dips obtained with saturation spectroscopy outside the laser resonator. Furthermore, because of the nonlinear dependence of the laser output on the losses, the halfwidth of the peaks may be smaller than γ_n, particularly when the laser is operated close above threshold. The sensitivity may be further enhanced by modulation of the laser frequency. This allows sensitive lock in-detection and yields signals which represent the derivative of the Lorentzian line profile. Figure 19 illustrates this method by the saturation spectrum of the same I_2-transition as shown in Fig. 17, obtained by intracavity absorption technique with frequency modulation [31].

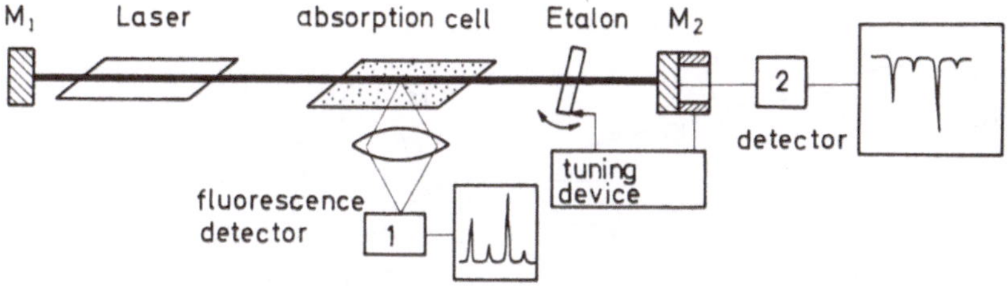

Fig. 18. Intracavity saturation spectroscopy.

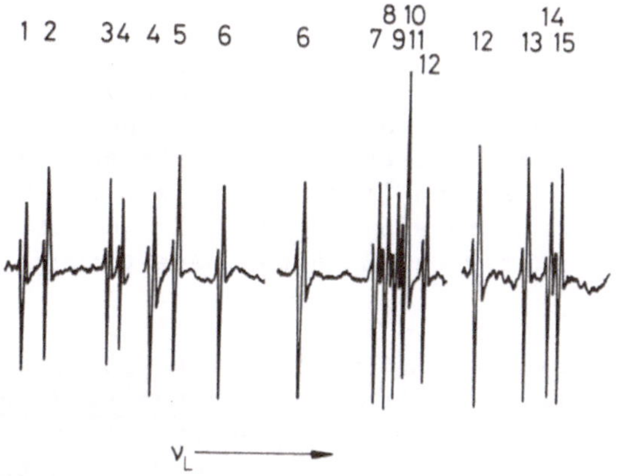

Fig. 19. Intracavity saturation spectrum of an I_2-transition obtained with modulation of the laser frequency (from ref. 31).

4. Polarization Spectroscopy

While saturation spectroscopy monitors the decrease of absorption of a probe beam, caused by a pump wave, which has selectively depleted the absorbing level, the signals in polarization spectroscopy come mainly from the change of refractive index, induced by a polarized pump wave. This very sensitive Doppler-free spectroscopic technique has many advantages over the conventional saturation spectroscopy and will certainly gain increasing attention, in high resolution molecular spectroscopy. We will therefore discuss the basic principle and some of its experimental modifications in more detail [32].

The basic idea of polarization spectroscopy can be understood in a simple way (Fig. 20):

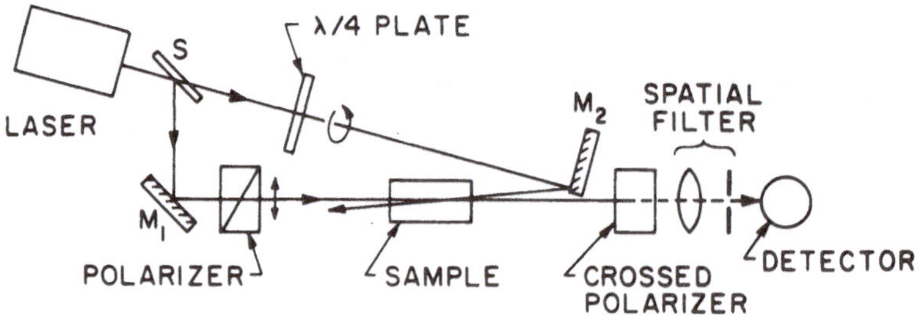

Fig. 20. Experimental arrangement used for polarization spectroscopy (from ref. 33).

The output from a monochromatic tunable laser is split into a weak probe with intensity I_1 and a stronger pump beam with intensity I_2. The probe beam passes through a linear polarizer P_1, the sample cell and a second linear polarizer P_2 which is nearly crossed with P_1. At a crossing angle $(\pi/2 - \theta)$ with $\theta \ll 1$, only the component $E_t = E_o \cdot \sin\theta \approx E_o \cdot \theta$ can pass through P_2 and reaches the detector D.

After having passed a λ/4-plate which produces a circular polarization, the pump beam travels in opposite direction through the sample cell. When the laser frequency is tuned to a molecular transition (J", M") → (J', M'), molecules in the lower level (J", M") can absorb the pump wave. The quantum number M, which describes the projection of J onto the direction of light propagation, follows the selection rule $\Delta M = +1$ for right circularly polarized light, which means that $M'' \rightarrow M' = M'' + 1$. Due to saturation the degenerate M-sublevels of the rotational level J" become partially or completely depleted. The degree of depletion depends on the pump intensity I_2, the absorption cross section $\sigma(J'', M'' \rightarrow J', M')$ and on possible relaxation processes which may repopulate the level (J", M"). From Fig. 21 it can be seen that in case of P- or R-transitions ($\Delta J = +1$ or -1) not all of the M-sublevels are pumped. For example from levels with $M'' = J''$ no R-transitions with $\Delta M = +1$ and no P-transitions with $\Delta M = -1$ are possible. This implies that the pumping process produces an unequal saturation and with it a nonuniform population of the M sublevels, which is equivalent to an anisotropic distribution for the orientations of the angular momentum vector $\underset{\sim}{J}$.

Such an anisotropic sample becomes birefringent for the incident linearly polarized probe beam, and the plane of polarization is slightly rotated. This effect is quite analogous to the Faraday effect where the nonisotropic orientation of J is caused by an external magnetic field. For polarization

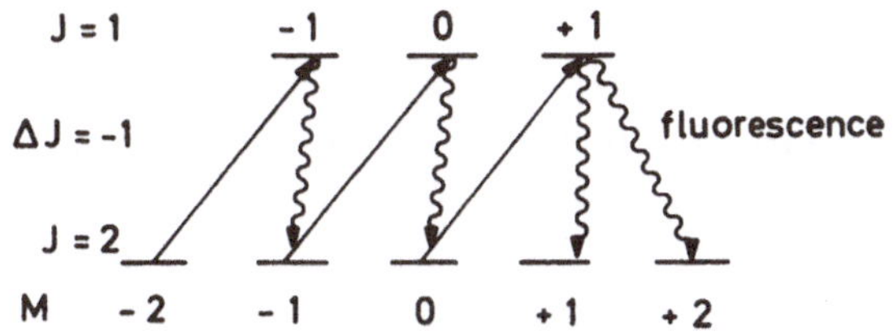

Fig. 21. Optical pumping of rotational levels, degenerate in M_y by circularly polarized light.

spectroscopy no magnetic field is needed. Different from the Faraday effect where all molecules are oriented, here only those molecules show this nonisotropic orientation, which interact with the monochromatic pump wave. As has been already discussed in the previous section, this is the subgroup of molecules with velocity components

$$v_z \pm \Delta v_z = (\omega_o - \omega)/k \pm \delta\omega/k$$

where Δv_z is determined by the homogeneous line width.

For $\omega \neq \omega_o$ the probe wave which passes in the opposite direction through the sample, interacts with a <u>different</u> group of molecules in the velocity interval $v_z \pm \Delta v_z = -(\omega_o - \omega \pm \delta\omega)/k$, and will be therefore not influenced by the pump wave. If, however, the laser frequency ω coincides with the centre frequency ω_o of the molecular transition within its homogeneous line width $\delta\omega$ (i.e., $\omega = \omega_o \pm \delta\omega \rightarrow v_z = 0 \pm \Delta v_z$) both waves can be absorbed by the same molecules and the probe wave experiences a birefringence due to the nonisotropic M-distribution of the absorbing molecules. Only in this case the plane of polarization of the probe wave is slightly rotated by $\Delta\theta$ and the detector D receives a Doppler-free signal every time the laser frequency ω is tuned across the centre of a molecular absorption line.

Let us now discuss the generation of this signal in a more quantitative way following the presentation in [33]: The linearly polarized probe wave

$$E = E_o \cdot e^{i(\omega t - k^+ z)}, \quad \underset{\sim}{E}_o = \{E_{ox}, 0, 0\}$$

can always be composed of a right- and a left-circularly polarized component. While passing through the sample the two components experience different absorption coefficients α^+ and α^- and different refractive indices n^+ and n^- due to the nonisotropic saturation caused by the right circularly polarized pump wave. After a pathlength L through the pumped region of the sample the two components are

$$E_o^+ = E_o^+ \cdot e^{i[\omega t - k^+ \cdot L + i(\alpha^+/2)L]}; \quad 2E_o^+ = E_{ox} + iE_{oy}$$

$$E_o^- = E_o^- - e^{i[\omega t - k^- \cdot L + i(\alpha^-/2)\cdot L]}; \quad 2E_o^- = E_{ox} - iE_{oy} \tag{41}$$

Due to the differences $\Delta n = n^+ - n^-$ and $\Delta\alpha = \alpha^+ - \alpha^-$ caused by the nonisotropic saturation, a phase difference

$$\Delta\phi = (k^+ - k^-)\cdot L = (\omega L/c)\cdot(n^+ - n^-) \tag{42}$$

has developed between the two components and also a small amplitude difference

$$\Delta E = (E_o/2)\cdot[e^{-(\alpha^+/2)L} - e^{-(\alpha^-/2)L}] \tag{43}$$

If both components are again superimposed at $z = L$ after having passed through the sample cell, an elliptically polarized wave comes out with a major axis which is slightly rotated against the x-axis. The y-component of this elliptical wave is

$$E_y = -i(E_o/2)\, e^{i[k^+ - k^- + i(\alpha^+ - \alpha^-)/2]L} \cdot e^{ib} \cdot e^{i(\omega t + \phi)} \tag{44}$$

where the term $i\cdot b$ with $b \ll 1$ in the exponent takes into account, that the windows of the sample cell may have a small birefringence, which introduces an additional ellipticity. In all practical cases the differences $\Delta\alpha$ and Δk are small:

$$(\alpha^+ - \alpha^-)\cdot L \ll 1 \quad \text{and} \quad (k^+ - k^-)\cdot L \ll 1$$

and we can expand the first exponential factor. If the transmission axis of the second polarizer P_2 is close to the y-axis ($\theta \ll 1$), we obtain for the transmitted amplitude:

$$E_t = E_o\cdot[\theta + ib + (\omega\cdot L/2c)\cdot(n^+ - n^-) + i(\alpha^+ - \alpha^-)\cdot L/4]\, e^{i(\omega t + \phi)} \tag{45}$$

The differences $\Delta\alpha = \alpha^+ - \alpha^-$ in absorption coefficients and $\Delta n = n^+ - n$ in refractive indices are due to the different degrees of M-sublevel depopulations experienced by the right- and left-circularly polarized probe components respectively. Although each coefficient α^+ and α^- itself shows a Doppler-broadened spectral profile with a Lamb-dip at the centre, the

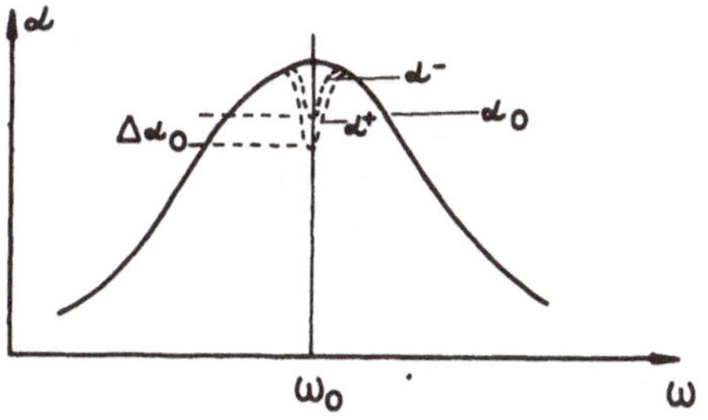

Fig. 22. Lamb-dips in the Doppler-broadened transition induced by pump and probe wave and the polarization signal as the difference of the Lamb-dips.

difference just exhibits the small difference between these Lamb-dips (see Fig. 22). The spectral profile of $\Delta\alpha$ is therefore a Lorentzian

$$\Delta\alpha = \frac{\Delta\alpha_0}{1 + x^2} \quad \text{with } x = (\omega_0 - \omega)/\gamma \tag{46}$$

where $\Delta\alpha_0$ is the maximum difference at the centre $\omega = \omega_0$. Since absorption and dispersion are related by the Kramers-Kronig dispersion relations we obtain a dispersion shaped profile for Δn:

$$\Delta n = \Delta\alpha_0 \cdot \frac{c}{\omega_0} \cdot \frac{x}{1 + x^2} \tag{47}$$

The transmitted intensity is $I_T = E_T E_T^*$. Taking into account that even perfectly crossed polarizers show a residual transmission $I_0 \cdot \xi$ with $\xi << 1$ due to imperfect extinction, we obtain the spectral line profile of the transmitted intensity I_T from eq. (44-47).

$$I_T = I_0 \, [\xi + \theta^2 + b^2 - \frac{1}{2}\theta \cdot \Delta\alpha_0 \cdot L \frac{x}{1 + x^2} + \frac{b}{2} \Delta\alpha_0 \cdot L \cdot \frac{1}{1 + x^2}$$

$$+ \frac{1}{4}(\Delta\alpha_0 \cdot L)^2 \frac{1}{1 + x^2}] \tag{48}$$

The transmitted intensity contains a constant background term $I_0 \cdot (\xi + \theta^2 + b^2)$ which is caused: 1) by the finite transmission $I_0 \cdot \xi$ of the crossed polarizers at $\theta = 0$; 2) by the birefringence of the cell windows, described by the term $I_0 \cdot b^2$ and 3) by the finite uncrossing angle θ of P_2. If the birefringence of the windows can be kept small (for example by squeezing the windows slightly to compensate the birefringence which is mainly caused by the pressure difference of 1 atm at both sides of the windows) the third and the fifth term in (48) can be neglected and the signal consists of

1. a constant term $(\xi + \theta^2)\cdot I_o$

2. a dispersion shaped signal $(\theta\cdot\Delta\alpha_o\cdot L/2)\ x/(1 + x^2)$ and a Lorentzian term $(\Delta\alpha_o\cdot L/2)^2/(1 + x^2)$.

Since $\Delta\alpha_o\cdot L << 1$, the selection of the uncrossing angle θ allows to make either the dispersion term dominant (larger θ) or the Lorentzian term ($\theta = 0$). Fig. 23 shows as example a section from the polarization spectrum of the cesium molecule Cs_2, recorded with $\theta = 0$. In this experiment the term $(\xi + b^2)$ was smaller than 10^{-6}.

Note that the main contribution to the signal comes from the rotation of the plane of polarization of the linearly polarized probe wave, and only to a minor extent from the change of absorption $\Delta\alpha$. In (48) $\Delta\alpha$ appears because $\Delta n = n^+ - n^-$ has been replaced by $\Delta\alpha$ using the dispersion relations. It is the difference in phaseshifts for the two probe components E^+ and E^- rather than the slightly different amplitudes which gives the major part of the signal. This is the reason why this method is much more sensitive than the saturation method, as will be shown below.

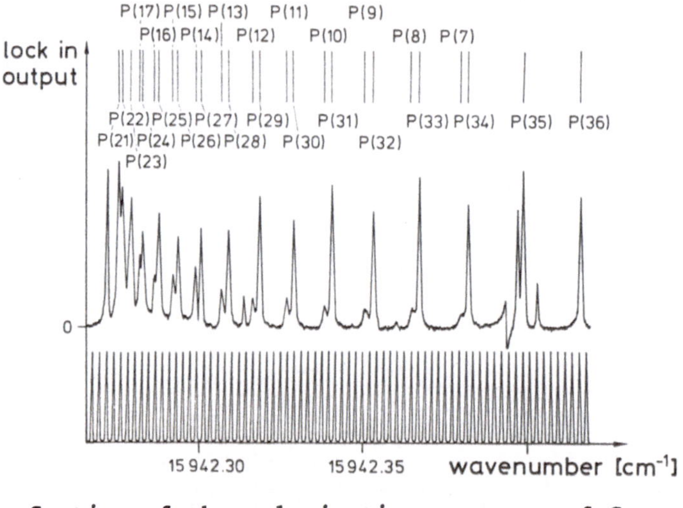

Fig. 23. Section of the polarization spectrum of Cs_2 around $\lambda = 476{,}5$ nm recorded with $\theta = 0$ and circularly polarized pump wave.

Instead of a circularly polarized pump wave also a linearly polarized one can be used with the plane of polarization inclined by 45° against that of the probe. Assume the pump wave is polarized in x-direction. The probe wave then can be composed of two linearly polarized components E_x and E_y and the saturation by the pump will cause a difference $\alpha_x - \alpha_y$ and $n_x - n_y$. Analogous to the derivation above we obtain in this case for the probe intensity, transmitted through the polarizer P_2:

$$I_T = I_o \left[\xi + \theta^2 + b^2 + \frac{1}{2}\theta\Delta\alpha_o \cdot L \frac{1}{1 + x^2} + \frac{1}{2}b \cdot \Delta\alpha_o L \cdot \frac{x}{1 + x^2} + \frac{1}{4}(\Delta\alpha_o \cdot L)^2 \frac{1}{1 + x^2}\right] \qquad (49)$$

which shows that the dispersion term and the Lorentzian term are just interchanged compared to (48). For b = 0 (no birefringence of the cell windows) and $\theta \neq 0$ (polarizers not completely crossed) we obtain pure Lorentzian profiles.

In order to estimate the expected magnitudes of the polarization signals in (48) let us consider the difference $\Delta\alpha = \alpha^+ - \alpha^-$ in the absorption coefficients for the right- and left-hand circularly polarized probe wave components. The absorption of a circularly polarized wave tuned to a rotational transition $J \rightarrow J_1$ is due to the sum of all allowed transition $M_J \rightarrow M_{J1}$ with $\Delta M = \pm 1$ between the (2J + 1) degenerate sublevels M" in the lower level J" and the $(2J_1 + 1)$ sublevels in the upper level J_1,

$$\alpha^+ - \alpha^- = \sum_M n_M(\sigma^+_{JJ_1M} - \sigma^-_{JJ_1M}) \qquad (50)$$

where $\sigma^{\pm}_{JJ_1M}$ is the absorption cross section for the transition $(J, M) \rightarrow (J_1, M + 1)$ resp. $(J, M) \rightarrow (J_1, M - 1)$.

The M-dependence of the cross sections σ_{JJ_1M} can be expressed in terms of Clebsch-Gordon coefficients $C(J, J_1|M, M_1)$ and a reduced matrix element $\tilde{\sigma}_{JJ_1}$ which is independent of M and describes the total rotational transition $J \rightarrow J_1$ [34]. The explicit evaluation yields for a circularly polarized pump

$$\sigma^{\pm}_{JJ_1M} = \begin{cases} \tilde{\sigma}_{J,J+1} \cdot (J \pm M + 1) \cdot (J \pm M + 2) & \text{for } J_1 = J + 1 \\ \tilde{\sigma}_{J,J} \cdot (J \pm M) \cdot (J \pm M + 1) & \text{for } J_1 = J \\ \tilde{\sigma}_{J,J-1} \cdot (J \pm M) \cdot (J M + 1) & \text{for } J_1 = J - 1 \end{cases} \qquad (51)$$

The total cross section

$$\sigma_{JJ_1} = \frac{1}{2J+1} \sum_M \sigma_{J,J_1M}$$

for the transition $J \rightarrow J_1$ is independent of the kind of polarization. Inserting (51) and evaluating the sum over M yields

$$\sigma_{JJ_1} = \frac{1}{3} \tilde{\sigma}_{JJ_1} \begin{matrix} (J+1)\cdot(2J+3) & & \Delta J = +1 \\ J\ (J+1) & \text{for} & \Delta J = 0 \\ J\ (2J-1) & & \Delta J = -1 \end{matrix} \quad (52)$$

The unsaturated level population of a sublevel M is

$$n^o_M = N_o/(2J+1) \quad (53)$$

Without the saturating pump wave we obtain by inserting (52) and (53) into (50)

$$\Delta\alpha^o = \alpha^+ - \alpha^- = 0$$

Due to saturation by the pump wave with intensity I_2 the population of the M-sublevels decreases and the absorption of the probe wave by molecules in sublevel M decreases according to (41) to

$$\alpha_M^s = \alpha_M^o(1 - S_M) \quad (54)$$

The saturation parameter S_M at the line centre ω_o depends on the absorption cross section σ_{JJ_1M} of the pump wave, on the number $n_p = I_2 \hbar\omega$ of pump photons incident on the sample per cm^2 and s, on the saturation-broadened homogeneous linewidth γ_s and on the relaxation rate $R[S^{-1}]$ which tries to refill the depleted level M. In Fig. 24 the M-dependence of σ_{JJ_1M} is plotted for right-hand and left-hand circular polarization (right diagrams) and for linear polarization (left diagrams). These diagrams illustrate that saturation by a circularly polarized pump wave results in larger differences $(\sigma^+ - \sigma^-)$ for P- and R-lines than for Q-lines, while a linearly polarized pump wave favours Q-lines in the detected probe transmission.

Putting the relations (50-54) together we can express the difference $\alpha^+ - \alpha^-$ at the line centre ω_o by the unsaturated absorption coefficient $\alpha_o = N_J{}^o \sigma_{JJ_1}$, the saturation parameter $S_o(\omega_o)$ and a numerical factor $C^*_{JJ_1}$ which stands for the sum over the Clebsch-Gordon coefficients and which is tabulated in [33]. The final result is

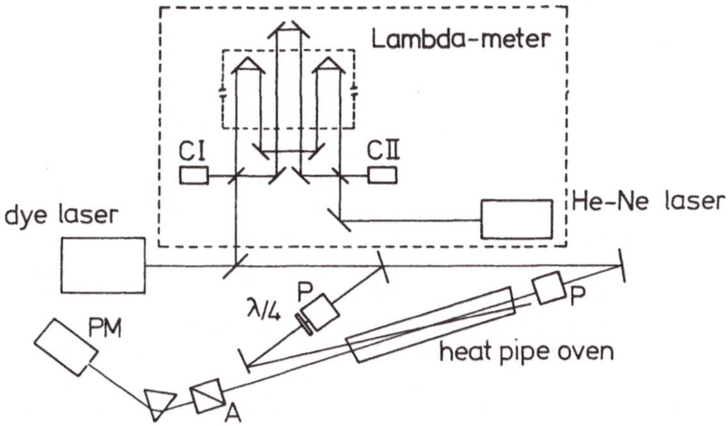

Fig. 24. Experimental arrangement for polarization spectroscopy with travelling Michelson interferometer (wave meter) for absolute wavelength measurements.

$$\alpha^+ - \alpha^- = \alpha^o \, S_o \cdot C^*_{JJ_1} \tag{55a}$$

This is often written in the form

$$\alpha^+ - \alpha^- = \alpha^o \cdot C^*_{JJ_1} \, (I_2/I_S) \tag{55b}$$

where $I_S = S_o \cdot I_2$ represents that pump intensity which causes a saturation parameter $S_o = 1$ at the line centre.

Let us briefly summarize the advantage of polarization spectroscopy, discussed in the previous sections:

1. With the other sub-Doppler-techniques it shares the advantage of high spectral resolution which is mainly limited by the residual Dopplerwidth due to the finite angle between pump beam and probe beam. This limitation corresponds to that imposed on linear spectroscopy in collimated molecular beams by the divergence angle of the molecular beam. The time of flight broadening can be reduced if pump and probe beam are less tightly focussed.

2. The sensitivity is by 2-3 orders of magnitude larger than that of saturation spectroscopy. It is only surpassed at very low sample pressures by that of the intermodulated fluorescence technique (section V, 3).

3. The possibility to distinguish between P, R and O-lines is a particular advantage for the assignment of complex molecular spectra.

4. The dispersion profile of the polarization signals allows a stabilization of the laser frequency to the line centre without any frequency modulation. The large achievable signal to noise ratio assures an excellent frequency stability. The wavelength of such a stabilized single mode laser can be measured with high accuracy, using interferometric techniques, based on a combination of Fabry-Perot-interferometers [35] or a travelling Michelson interferometer [36]. Since the laser wavelength is stabilized onto the centre of a molecular line, the wavelength of this line is known to the same accuracy which reaches 5×10^{-6} nm in the visible range.

5. The combination of optical-optical double resonance techniques and polarization spectroscopy open the way to detailed studies of perturbed excited molecular states.

Polarization spectroscopy has been successfully applied to the sensitive detection of molecular spectra such as Cs_2, NO_2, and others. Figure 24 shows the experimental arrangement, used in our laboratory. Figure 25 illustrates a polarization-spectrum of the P-branch of Cs_2 at the band head of the $0 \rightarrow 0$ vibrational band in a $^1\Pi_u \leftarrow X^1\Sigma_g$-transition. The small dispersion shaped signal belongs to a Q-line from another band.

Part of the laser beam is sent through a 120 cm long confocal Fabry-Perot-interferometer which provides frequency marks every 62 MHz. For absolute wavenumber measurements a travelling Michelson interferometer [36] is used which measures the laser wavelength, stabilized to the centre of a molecular line, to within 2×10^{-4} Å, corresponding to a frequency accuracy of about 20 MHz. Measuring more than 800 Q-, P- and R-lines in the 0-0-band yields the Fortrat-diagram of Fig. 25. The scattering of the experimental points is much less than the diameter of the plotted dots in the expanded insert. A least squares fit to the measured line position gives the rotational constants of the Cs_2-molecule in the $X^1\Sigma_g$- and the $^1\Pi_u$-state with an accuracy of about 10^{-5} [37]. Besides a recently published work [38] on the same band with Doppler-limited resolution this is the first complete resolution of all rotational lines in a band in the visible Cs_2-spectrum.

5. Optical-Optical Double Resonance

The sub-Doppler techniques, discussed in the previous

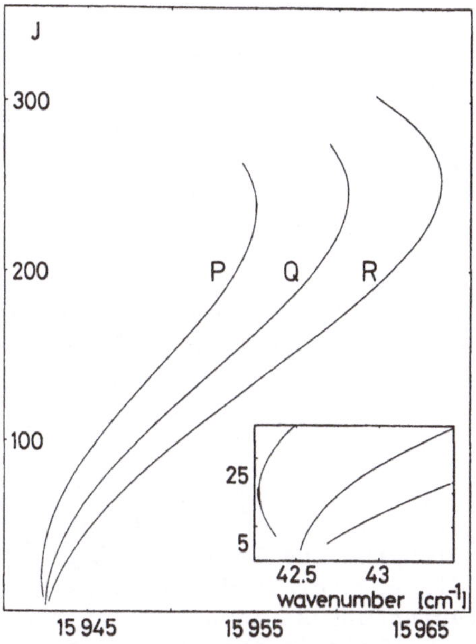

Fig. 25. Fortrat diagram and section of the polarization spectrum of Cs_2 around 15942 cm^{-1}, together with frequency markers ($\Delta\nu$ = 62 MHz) circularly polarized pump, crossing angle $\theta = 0$.

sections can be used in combination with optical double resonance methods to simplify complex molecular spectra and to investigate the level structure of excited molecular electronic states.

The optical-optical double resonance is based on the simultaneous interaction of a molecule with two different laser waves, a pump wave and a probe wave, which pass in opposite directions through the sample [39]. Assume the pump laser is stabilized onto a molecular transition $E_i \rightarrow E_k$. Due to optical pumping the lower level population N_i is partly depleted and the upper level population N_k increases. Chopping of the pump wave intensity therefore results in corresponding modulations of N_i and N_k. When the wavelength of the probe wave is tuned across the molecular absorption spectrum, the absorption of the probe beam will monitor this population modulation whenever the probe wavelength coincides with a molecular transition $E_i \rightarrow E_n$ or $E_k \rightarrow E_m$ (Fig. 26). The modulation phase is opposite for both cases and phase sensitive detection therefore allows to distinguish between transitions $E_i \rightarrow E_m$ or $E_k \rightarrow E_m$.

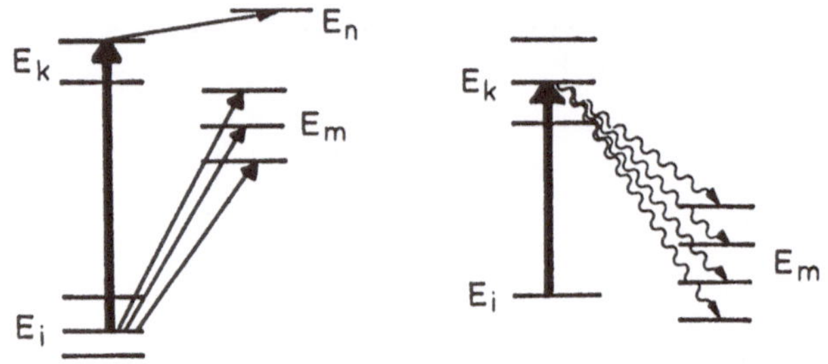

Fig. 26. Schematic diagram for the comparison of laser induced fluorescence and optical-optical double resonance spectroscopy.

Provided the pump transition $E_i \rightarrow E_k$ has been assigned, the common lower level E_i for all probe transitions $E_i \rightarrow E_n$ is known. From the measured wavelengths the level spacings of the upper levels E_n can be immediately obtained. In a way this double resonance technique represents an inversion of the laser induced fluorescence technique. While in the latter method a single upper level is selectively populated and the lower level diagram is deduced from the fluorescence spectrum, in the former method a lower level is selectively depleted and the upper level spectrum is obtained from the modulated absorption spectrum of the probe laser.

Combined with saturation spectroscopy [40] or polarization spectroscopy [41] the OODR-technique allows Doppler-free resolution and substantially reduces the number of lines. This is of particular importance in complex spectra and in case of perturbed upper states. However, if the experiments are performed in gas cells, the wavelength of the single mode pump laser may simultaneously overlap with several Doppler-broadened molecular absorption lines. This causes saturation dips in the population distributions $N_i(v_z)$ of several different levels E_i, where the dips are located at different velocities v_z within the Doppler-profile. Although these dips with width γ_s can be resolved, it is often not easy to decide, which of the dips is at the centre of the Doppler-broadened absorption profile. Furthermore, collisions may redistribute the selectively depleted population and therefore the dips can be transferred to neighbouring levels. Also the fluorescence emitted from the modulated upper level N_k, and terminating at various lower levels will cause a slight population-modulation of the lower levels. All these effects increase the number of double resonance signals and may impede the unambiguous assignment. In case of complex spectra it is therefore desirable to avoid collisions as well as the simultaneous interaction of several molecular transitions with the pump wave.

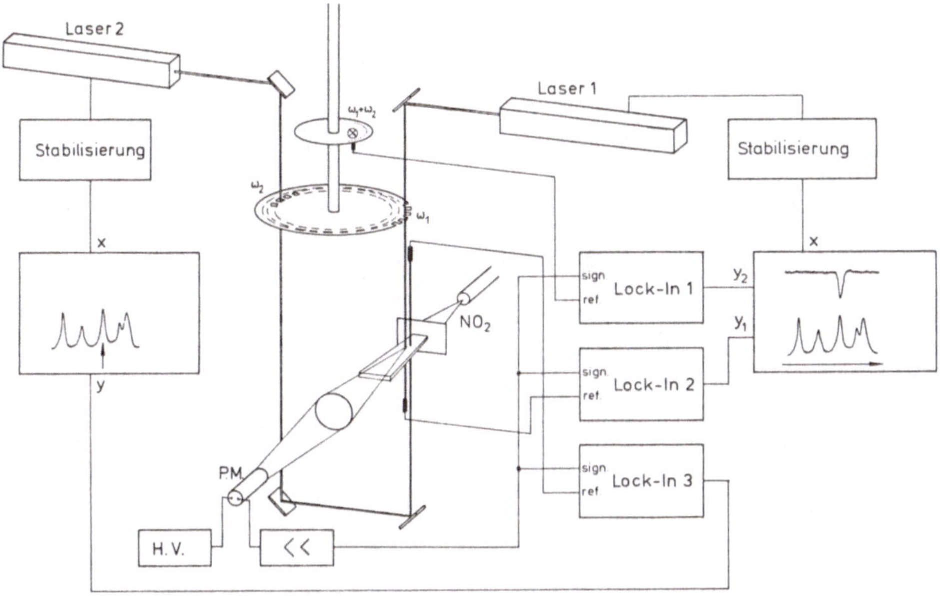

Fig. 27. Experimental arrangement used for OODR in a collimated molecular NO_2-beam.

A possible solution is the use of a collimated molecular beam. Figure 27 shows the experimental arrangement, where the beams from two tunable single mode argon lasers cross the molecular NO_2-beam perpendicularly at two different positions z_1 and z_2. The absorbing transitions are monitored through the laser induced fluorescence. In order to avoid the superposition of the fluorescence spectra from both lasers, an intermodulated fluorescence technique is used (see section V, 3): The two laser beams are chopped at two different frequencies and the OODR-signal is monitored through the laser induced fluorescence at the sum frequency. Figure 28 illustrates the results by showing a small section of the NO_2-excitation spectrum around $\lambda = 488$ nm (lower spectrum). While the pump laser is stabilized on a line indicated by the arrow, the probe laser is tuned. The OODR-spectrum (upper part) demonstrates, that two transitions in the lower spectrum share a common lower level.

The corresponding splitting of the two upper levels, reached from the same lower level, must be caused by perturbations, because the rotational level spacing is orders of magnitude larger. Different fine-structure or hyperfine-structure levels cannot be populated from the same lower level because the selection rules $\Delta S = 0$ and $\Delta I = 0$ hold for electric dipole transitions in unperturbed spectra for the electronic spin S and the nuclear spin I. These perturbations are probably spin-orbit interactions [42] which may simultaneously allow transitions with $\Delta K = 0$ and

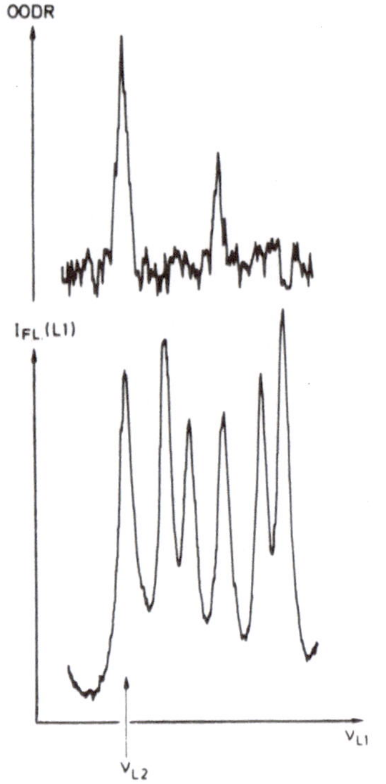

Fig. 28. Section of the linear excitation spectrum of NO_2 around λ = 488 nm and double resonance signals.

$\Delta K = 2$ (K = angular momentum projection quantum number in symmetric tops).

The application of this OODR-technique to a wider spectral range, covered by a dye laser, will be very helpful to separate unperturbed and perturbed lines and to gain insight in the level structure of perturbed upper electronic states. Such experiments are presently underway in our laboratory.

VI. LIFETIME MEASUREMENTS OF SELECTIVELY EXCITED MOLECULAR LEVELS

Besides accurate measurements of molecular energy levels one of the goals of molecular spectroscopy is the determination of transition probabilities, the investigation of transient phenomena and molecular dynamics of free molecules and the study of collisional energy transfer between excited molecules and other collision partners. The capability of pulsed or mode locked lasers to deliver intense short light pulses with tunbale wavelength makes them ideal instruments for the experimental investigations of such problems. If the molecules are irradiated in a collimated molecular beam at sufficiently low pressure, collision processes can be eliminated and the dynamics of the free molecule can be studied. The linewidth of the absorbing transition is determined by the residual Doppler-width, the natural linewidth, the time of flight through the laser beam, or the Fourier limited laserwidth $\Delta\nu \approx 1/\Delta T$ due to the finite duration ΔT of the laser pulse.

The laser pulses can be either generated by flashlamp pumped or nitrogen laser pumped dye lasers or by cw-lasers, where the laser beam is transmitted through a Pockels cell which forms rectangular pulses of variable width and controlled repetition frequency.

Subnanosecond pulses can be generated by mode locking techniques [43]. Of particular convenience is the mode locking of argon lasers by an internal acousto-optic [44] and the synchroneous pumping of cw-dye lasers [45]. Such a system delivers a regular train of short pulses with a tunable wavelength. The pulsewidth depends on the spectral gain profile of the dye laser which is determined by the selecting elements inside the dye laser cavity. Output-pulses below 1 ps have been measured [46]. The repetition rate is determined by the transit time of a light pulse through the resonator. Synchroneous pumping is obtained if these transit times are equal for the dye resonator and the pump laser resonator.

Figure 29 illustrates the experimental arrangement for measuring the lifetimes of selectively excited molecular levels. The selectivity can be achieved either by reduction of the absorption linewidth (excitation in a collimated molecular beam) or by monitoring the laser induced fluorescence through a monochromator. If the excitation is performed with Doppler-limited spectral resolution in a vapour cell, the monochromator still can select a single fluorescence line from a selected upper level, even if several upper levels are simultaneously excited by the laser pulse.

Figure 30 shows two successive laser pulses (left side) and the exponential decay of the fluorescence, obtained with excitation by a mode locked argon laser and time correlated single photon counting techniques [47]. Such measurements yield the lifetimes of individual molecular levels and allow to determine the dependence of the transition moment on the internuclear distance [48]. In polyatomic molecules the primarily excited level may decay by radiationless transitions and the fluorescence may be emitted from several, nearly degenerate levels with different radiative lifetimes. This produces nonexponential decay curves in the measured fluorescence intensity $I_{Fl}(t)$ [49].

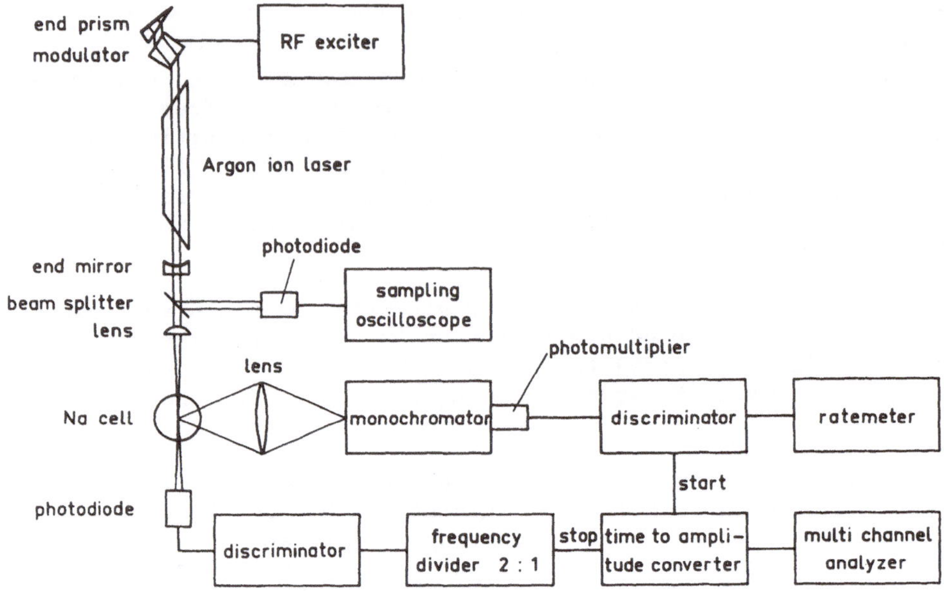

Fig. 29. Experimental arrangement for lifetime measurements with mode locked laser pulses and single photon counting techniques.

A very important field in time resolved spectroscopy is the study of fast relaxation processes. The pump and probe technique where a short pump pulse excites a molecule and a probe pulse with variable delay interrogate it at a known time after the excitation, has become widely used. The two pulses may either come from the same laser where the delay is realized by optical delay paths, or two different lasers can be used with delayed trigger synchronization. Vibrational relaxation of molecules in liquids occurs with typical decay times of 10^{-12} s and a picosecond pulse source is therefore required to study the dynamics of this process [50]. The energy relaxation time T_1 and the dephasing time T_2 can greatly differ [51]. Frequency doubling of mode locked dye lasers allows to cover the whole ultraviolet spectral range, interesting for photochemistry. Since the coupling of electronic motion and nuclear motion generally increases with excitation energy, the study of perturbations, such as intersystem crossing, will certainly profit from the recent development of powerful, ultrashort light pulses in the UV.

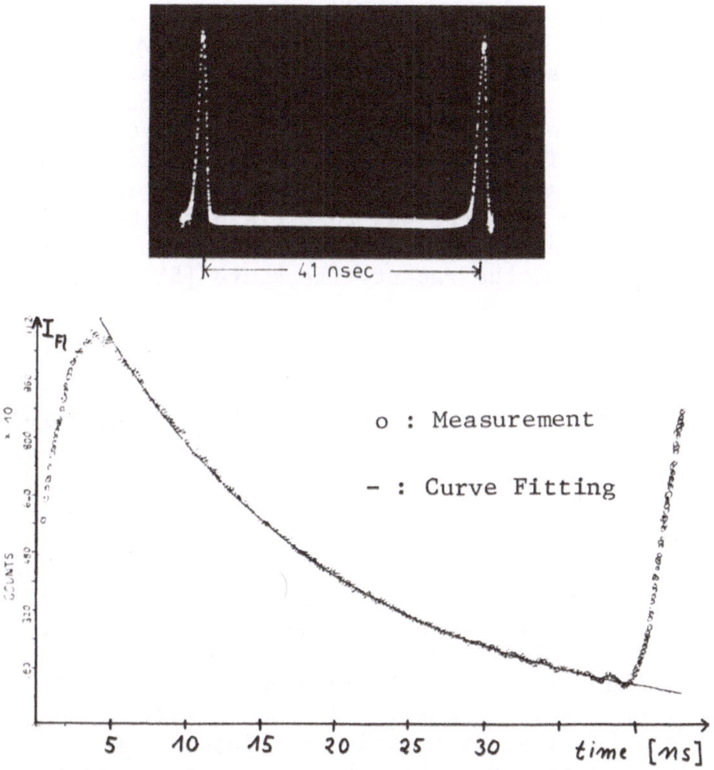

Fig. 30. Two successive pulses from a train of mode locked argon laser pulses, and the exponential decay of a selectively excited (v', J')-level in the $D^1\Pi_u$-state of the NaK-molecule.

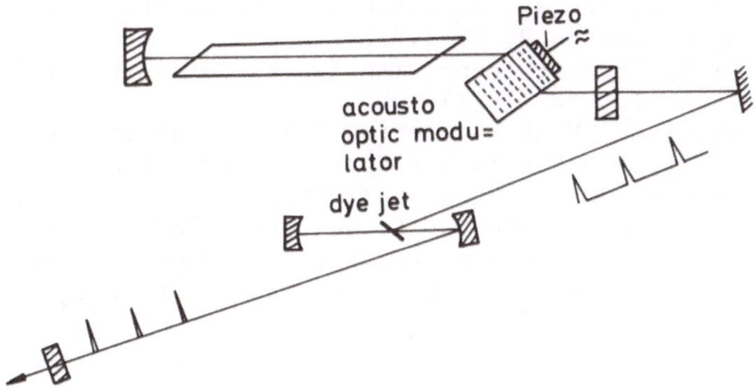

Fig. 31. Synchroneous mode locking of a cw dye-laser.

VII. CONCLUSION

These lectures could only give a short survey on some aspects of high resolution laser spectroscopy. The examples, selected for illustration, have been mainly taken from experiments performed in Kaiserslautern, because the author is of course most familiar with his own work. This does not imply any priority since many other laboratories have performed much more work in this field. The main goal of the lectures was to explain the basic ideas and experimental techniques. For further details the reader is referred to text books [52] and the recent literature [53, 54].

REFERENCES

1. P.W. Smith, "Mode-Selection in Lasers" in "Laser Devices and Applications, I.P. Kaminow and A.E. Siegman, eds., IEEE Press, New York (1973).
2. H.G. Häfele, "Spin Flip Raman Lasers", Appl. Phys. 5, 97 (1974).
3. I. Melngailis and A. Mooradian, "Tunable Semiconductor Diode Lasers" in "Laser Applications in Optics and Spectroscopy" J. Jacobs, M. Sargent, M. Scully and J. Scott, eds., Addison-Wesley Comp., New York (1975).
4. F.B. Dunning, "Tunable Ultraviolet Generation by Sum Frequency Mixing", Laser Focus 14, 72 (May 1978).
5. R.L. Byer, "Optical Parametric Oscillators" in "Quantum Electronics", H. Rabin, L. Tang, eds., Academic Press, New York (1976).
6. U. Wilke, W. Schmidt, "Tunable Coherent Radiation Source from 185 to 880 nm", Appl. Phys. 18, 177 (1979).
7. L.F. Mollenduen and D.H. Olson, J. Appl. Phys. 46, 3109 (1975).
7a. J.J. Hsieh, J.A. Rossi, J.P. Donnelly, Appl. Phys. Lett. 28, 709 (1976).
8. J.F. Butler, "Update on Tunable Diode Lasers", Electro-Opt. Syst. Design 9, 33 (1977).
9. K.W. Nill, "Spectroscopy with Tunable Diode Lasers", Laser Focus 13, 32 (1977).
9a. A. Mooradian, "High Resolution Tunable Infrared Lasers" in: "Very High Resolution Spectroscopy", R.A. Smith, ed., Academic Press, London (1976).
10. H. Welling, G. Litfin and R. Beiyung, "Tunable Infrared Lasers Using Colour Centres", in Spectroscopy III, J.L. Hall and J.C. Carlsten, eds., Springer, Heidelberg (1977).
11. F.P. Schäfer, ed., Dye Lasers, 2nd edition, Springer-Verlag, Heidelberg (1978).
12. B. Steyer, F.P. Schäfer, "Stimulated and Spontaneous Emission from Laser Dyes in the Vapour Phase", Appl. Phys. 7, 113 (1975).

13. A.L. Bloom, "Modes of a Laser Resonator Containing Tilted Birefringent Plates", J. Opt. Soc. Am. 64, 447 (1974).
13a. See information sheets of the manufacturers Coherent Radiation and Spectra Physics.
14. H.W. Schröder, H. Dux and H. Welling, "Single Mode Operation of cw Dye Lasers", Appl. Phys. 7, 21 (1975).
14a. H. Gerhardt, A. Timmermann, "High Resolution Dye Laser Spectrometer for Measurements of Isotope and Hyperfine Structure", Opt. Comm. 21, 343 (1977).
14b. I.V. Hertel and A. Stamatovic, "Spatial Hole Burning and Oligo-Mode Distance Control in cw Dye Lasers, IEEE J. Quant. Electr. QE-11, 210 (1975).
14c. G. Marowsky and K. Kaufman, IEEE J. Quant. Electr. QE-12, 207 (1976).
14d. H.W. Schröder, L. Stein, D. Frölich, B. Fugger and H. Welling, Appl. Phys. 14, 377 (1977).
15. H. Rabin, H. Tang, eds., "Quantum Electronics", Academic Press, New York (1976).
16a. M.J. Colles and C.R. Pidgeon, "Tunable Lasers", Reports Progr. Phys. 38, 329 (1975).
16b. R.S. McDowell, "High Resolution Infrared Spectroscopy with Tunable Lasers", in "Advances in Infrared and Raman Spectroscopy", Vol. 5, R.J. Clark and R.E. Hester, eds., Heydon, London (1978).
17. Yo-Han Pao, "Opto-acoustic Spectroscopy and Detection", Academic Press, New York (1977).
18. W. Brunner, H. Paul, "On the Theory of Intracavity Absorption", Opt. Commun. 12, 252 (1974).
19. G.H. Atkinson, A. Laufer and M. Kurylo, "Detection of Free Radicals by an Intracavity Dye Laser Technique", J. Chem. Phys. 59, 350 (1973).
20. T.W. Hansch, A.L. Schawlow, P. Toschek, "Ultrasensitive Response of a cw Dye Laser to Selective Extinction", IEEE J. Quant. Electr. OE-8, 802 (1972).
21. E.N. Antonow, V.G. Koloshmikov, V.R. Mironenko, "Quantitative Measurements of Small Absorption Coefficients in Intracavity Absorption Spectroscopy Using a cw Dye Laser", Opt. Commun. 15, 99 (1975).
22. H. Hotop, F. Illenberger, H. Morgner and A. Nichaus, Chem. Phys. Lett. 10, 493 (1971).
23. G. Höning, M. Czajkowski, M. Stock and W. Demtröder, "High Resolution Laser Spectroscopy of Cs_2", J. Chem.Phys. 1 Sept. 1979 (in print).
24. S.L. Kaufman, Opt. Commun. 17, 309 (1976).
25. M. Dufay and M.L. Gaillard, "High Resolution Studies in Fast Ion Beams" in "Laser Spectroscopy III", J.L. Hall and J.L. Carlsten, eds., Springer-Verlag, Berlin, Heidelberg (1977).
26. V.S. Letokhov, V.P. Chebotayev, "Nonlinear Laser Spectroscopy", Springer-Verlag, Berlin (1977).

27. A. Maitland, M.H. Dunn, "Laser Physics", North Holland Publ. Comp., Amsterdam (1969).
28. W.E. Lamb, Phys. Rev. 134A, 1429 (1964).
29. M.S. Sorem, A.L. Schawlow, Opt. Commun. 5, 148 (1972).
30. J. Foth, Diplomthesis, Kaiserslautern (1977).
31. J. Foth, F. Spieweck, Phys. Rev. Lett., Sept. 1979 (in print).
32. C. Wieman, T.W. Hänsch, Phys. Rev. Lett. 36, 1170 (1976).
33. R.E. Teets, F.V. Kowalski, W.T. Hill, N. Carlson, T.W. Hänsch, Proc. Soc. Photo-optical Instr. Engineers 113, "Advances in Laser Spectroscopy", San Diego (1977), p.80 ff.
34. M.E. Rose, "Elementary Theory of Angular Momentum", John Wiley & Sons Inc., New York (1957).
35. R.L. Byer, J. Paul and M.D. Duncan, "A Wavelength Meter" in "Laser Spectroscopy III", J.L. Hall and J.L. Carlsten, eds., Springer, Berlin (1977).
36. F.V. Kowalski, R.E. Teets, W. Demtröder and A.L. Schawlow, "An Improved Wavemeter for cw-Lasers", J. Opt. Soc. Am., 68, 1611 (1978).
37. M. Raab, G. Höning, R. Castell, W. Demtröder, "Doppler-free Polarization Spectroscopy of the Cs_2-Molecule at λ = 6270 Å", J. Chem. Phys. Lett., September 1979 (in print).
38. A.I. Kobyliansky, A.N. Kulikov and L.V. Gurvich, Chem. Phys. Lett. 62, 198 (1979).
39. V.P. Chebotayev, "Three Level Laser Spectroscopy" in "High Resolution Laser Spectroscopy", K. Shimoda, ed., Topics in Appl. Phys. Vol. 13, Springer, Berlin (1976).
40. M.E. Kaminsky, R.T. Hawkins, F.V. Kowalski and A.L. Schawlow, Phys. Rev. Lett. 36, 671 (1976).
41. R. Teets, R. Feinberg, T.W. Hänsch, A.L. Schawlow, Phys. Rev. Lett. 37, 683 (1976).
42. J.C.D. Brand, P.H. Chiu, J. Mol. Spectrosc. 75, 1 (1979).
43. S.L. Shapiro, ed., "Ultrashort Light Pulses", in "Topics in Appl. Phys. Vol. 18, Springer-Verlag, Berlin (1977).
44. B. Ellis and A.K. Walton, "A Bibliography on Optical Modulators", Infrared Phys. E11, 85 (1971).
45. C.K. Chan, S.O. Sari, R.E. Foster, J. Appl. Phys. 47, 1139 (1976).
46. J. Kohl, H. Klingenberg and D.v. der Linde, "Picosecond and Subpicosecond Pulse Generation in Synchroneously Pumped Mode Locked cw Dye Lasers", Appl. Phys. 18, 279 (1979).
47. L.J. Cline Love and L.A. Shaver, "Time Correlated Single Photon Technique: Fluorescence Lifetime Measurements", Analyt. Chemistry 48, 364 A (1976).
48. W. Demtröder, W. Stetzenbach, M. Stock, J. Witt, J. Mol. Spectrosc. 61, 382 (1976).
49. F. Paech, R. Schmiedl and W. Demtröder, J. Chem. Phys. 63 4369 (1975).

50. A. Lauberau, A.A. Seilmeier, W. Kaiser, "A New Technique to Measure Ultrashort Vibrational Relaxation Times in Liquid Systems", Chem. Phys. Lett. 36, 232 (1975).
51. A. Lauberau, G. Wochner and W. Kaiser, Opt. Commun. 17, 91 (1976).
52. W. Demtröder, "Laser Spectroscopy, Basic Concepts and Instrumentation", Springer-Series in Chem. Phys. Vol. 5, in print, Springer-Verlag, Berlin (1979).
53. See for instance the Proceedings of the various Conferences on Laser Spectroscopy:
a) Laser Spectroscopy; Proceedings of the 1st Int. Conf. Vale 1973; R.G. Brewer, A. Mooradian, Academic Press, New York (1974).
b) Laser Spectroscopy; Proceedings of the 2nd Int. Conf., Megeve, 1975; S. Haroche, J.C. Pebay-Peyroula, T.W. Hänsch, S.E. Harris, Springer (1975).
c) Proceedings of the Int. Conf. on Tunable Lasers and Applications; A. Mooradian, T. Jaeger, P. Stokseth, Springer Series in Optical Sciences, Vol. 3, Springer, Berlin, Heidelberg, New York (1976).
d) Laser Spectroscopy III; Proceedings of the 3rd Int. Conf. Jackson Lake, 1977; J.L. Hall, J.L. Carlsten (eds.), Springer Series in Optical Sciences, Vol. 7, Springer, Berlin, Heidelberg, New York (1977).
e) Laser Spectroscopy IV; Proceedings of the 4th Int. Conf., Rottach Egern 1979; H. Walther, ed., Spring, Berlin, (1979).
54. a) H. Walther, ed., "Laser Spectroscopy of Atoms and Molecules", Topics in Applied Physics, Vol.2, Springer, Berlin, Heidelberg, New York (1976).
b) K. Shimoda, ed., "High Resolution Laser Spectroscopy", Topics in Applied Physics, Vol.13, Springer, Berlin, Heidelberg, New York (1976).
c) R.A. Smith, ed., "Very High Resolution Spectroscopy", Academic Press, London, New York (1976).

SPECTRAL PROPERTIES OF ATOMIC AND MOLECULAR SYSTEMS

J. M. Combes

Université de Toulon
83130 La Garde, France

R. Seiler

Freie Universität Berlin
Institut für theoretische Physik
Fachbereich Physik, Arnimalle 3, 1000 Berlin 33

1. INTRODUCTION TO SPECTRAL PROPERTIES OF ATOMIC AND MOLECULAR HAMILTONIANS

These first two chapters are intended to give a survey of presently known methods and results in N-body Schrödinger operator theory with applications to the specific situation where interactions are of the Coulomb type. They will provide the mathematical basis for the subsequent rigorous analysis of the Born-Oppenheimer approximation. We assume that the reader is familiar with the notions of self-adjoint operator, spectral decomposition, Hilbert space structure and other elementary concepts of Functional Analysis e.g. compact operator theory, which can be found in ref. [1] or [2, Vol. 1 and 2].

1.1 Kinematics of N-body non-relativistic systems

Consider an assembly of N particles labelled by i{1,2,...N}. We denote by $(m_i, \underset{\sim}{\xi}_i, \underset{\sim}{\Pi}_i)$ the mass parameter, and position and momentum coordinate vectors of particle i with respect to a fixed reference frame of R^3. We want to introduce new phase space variables which are more convenient for many purposes in particular the description of cluster properties of this system. As a simple example recall the well-known reduction to the centre-of-mass system for the Hydrogen atom. Here one introduces the relative or "internal" position and momentum coordinates

$$X_1 = \xi_1 - \xi_2$$

$$P_{12} = \frac{m_2\Pi_1 - m_1\Pi_2}{m_1 + m_2}$$

and "centre-of-mass" position and momentum coordinates

$$\xi_{12} = \frac{m_1\xi_1 + m_2\xi_2}{m_1 + m_2}$$

$$\Pi_{12} = \Pi_1 + \Pi_2 .$$

The virtues of these new coordinates are also well-known and we list some of them without calculation. First of all this new set is "canonical" in the sense that internal and external phase space coordinates have vanishing Poisson bracket, whereas internal or external position and momentum variables have a Poisson bracket equal to one. Then once quantized these new variables will satisfy the usual canonical commutation relations. Second, consider the quadratic forms corresponding to the moment of inertia and "total" kinetic energy of the system:

$$I = m_1\xi_1^2 + m_2\,\xi_2^2$$

$$H_o^{Tot} = \frac{\Pi_1^2}{2m_1} + \frac{\Pi_2^2}{2m_2} .$$

These forms are still diagonalized by the new choice of coordinates as

$$I = M\,\xi_{12}^2 + \mu_{12}\,X_{12}^2$$

$$H_o = \frac{\Pi_{12}^2}{2M} + \frac{P_{12}^2}{2\mu_{12}} \tag{1}$$

where $M = m_1 + m_2$ (total mass) and $\mu_{12}^{-1} = m_1^{-1} + m_2^{-1}$ (reduced mass).

Before analysing the consequences of these facts in the quantum description of the system let us generalize this construction of internal and external canonical coordinates to N-particle systems. This can be done inductively as follows: consider a decomposition $D = \{C_1, C_2, \ldots C_k\}$ of the N-particle system into k disjoint clusters C_i. For each cluster C choose a set of "relative" Jacobi coordinates X_C, $P_C \in R^{3(n_C-1)}$, where n_c is the number of particles in cluster C, and denote by

$$\xi_C = \sum_{i \in C} m_i \, \xi_i / \sum_{i \in C} m_i$$

where $\Pi_C = \sum_{i \in C} \Pi_i$ the position and momenta of the centre of mass of C; then (ξ_C, X_C, Π_C, P_C) is a complete set of variables for cluster C.

Now consider the centres of mass as new particles with total mass $m_C = \sum_{i \in C} m_i$ and define relative coordinates for them $X_D, P_C \in R^{3(k-1)}$.

The centre of mass of the clusters is of course the centre of mass of the system so that one ends up with canonical coordinates

$$(\xi_{CM}, \Pi_{CM} \; ; \; X_D, P_D \; ; \; X_{C_1}, P_{C_1} \; ; \; \dots \; ; \; X_{C_k}, P_{C_k}) \tag{2}$$

In the above formula ξ_{CM} and Π_{CM} are the position and momentum coordinates of the centre of mass of the system. We will call (X_D, P_D) (resp. X_{C_1}, P_{C_1})... (X_{C_k}, P_{C_k}) external (resp. internal) Jacobi coordinates associated with the cluster decomposition of D. The sets $X = (X_D, X_{C_1}, \dots X_{C_k})$ and $P = (P_D, P_{C_1}, \dots, P_{C_k})$ constitute "relative" Jacobi phase space coordinates for the full system. Notice that given D and for $k \geq 2$ or $n_C > 2$ for some $C \in D$ a set of relative Jacobi coordinates associated in this way with D is not unique.

Sets constructed in this way are canonical (in the sense of Poisson brackets). The relations (1) can be generalized by induction showing, e.g., the Huygens formula

$$I = I_{CM} + I_D + \sum_{i=1}^{k} I_{C_i} ,$$

where the meaning of each of these moments of inertia is obvious. For the total kinetic energy

$$H_o = \sum_{i=1}^{N} \frac{\Pi_i^2}{2m_i}$$

one has a decomposition

$$H_o^{Tot} = \frac{\Pi_{CM}^2}{2M} + H_{o,D} + \sum_{i=1}^{k} H_{o,C_i} . \tag{3}$$

Here $H_{o,C}$ is the kinetic energy of particles in cluster C in the rest frame $\Pi_C = 0$; the term $H_{o,D}$ is the kinetic energy of the centres of mass of all $C \in D$ in the rest frame $\Pi_{CM} = 0$.

Example: Consider the molecule H_2^+. The above construction provides the following possible set of Jacobi coordinates associated with D = (AB,e) (see Fig. 1):

$$\underset{\sim}{X}_{AB} = \underset{\sim}{\xi}_A - \underset{\sim}{\xi}_B$$

$$\underset{\sim}{P}_{AB} = \frac{m_B \underset{\sim}{\Pi}_A - m_A \underset{\sim}{\Pi}_B}{m_A + m_B}$$

$$\underset{\sim}{X}_e = \underset{\sim}{\xi}_e - \frac{m_A \underset{\sim}{\xi}_A + m_B \underset{\sim}{\xi}_B}{m_A + m_B}$$

$$\underset{\sim}{P}_e = \frac{(m_A + m_B)\underset{\sim}{\Pi}_e - m(\underset{\sim}{\Pi}_A + \underset{\sim}{\Pi}_B)}{m_A + m_B + m}$$

$$\underset{\sim}{\xi}_{CM} = \frac{m_A \underset{\sim}{\xi}_A + m_B \underset{\sim}{\xi}_B + m \underset{\sim}{\xi}_e}{m_A + m_B + m}$$

$$\underset{\sim}{\Pi}_{CM} = \underset{\sim}{\Pi}_A + \underset{\sim}{\Pi}_B + \underset{\sim}{\Pi}_e$$

Here A,B,e label respectively the two protons (with masses $m_A = m_B$) and the electron (with mass m). With this choice the kinetic energy can be written as

$$H_o^{Tot} = \frac{\Pi_{CM}^2}{2M} + \frac{P_e^2}{2\mu_e} + \frac{P_{AB}^2}{2\mu_{AB}} , \tag{4}$$

with $M = m_A + m_B + m$, $\mu_e^{-1} = m^{-1} + (m_A+m_B)^{-1}$, $\mu_{AB}^{-1} = m_A^{-1} + m_B^{-1}$.

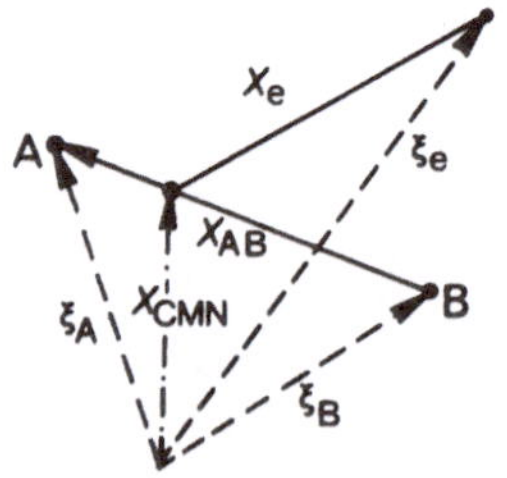

Fig. 1. Laboratory and centre-of-mass-of-the-nuclei system.

Remark: In view of the analysis of diatomic systems let us mention at this point that the above choice of variables is standard only for one electron. If there are N electrons most authors choose instead of Jacobi coordinates the following "canonical" set of phase space variables (assuming here two different nuclear masses m_A and m_B):

In addition to $(\underset{\sim}{X}_{AB},\underset{\sim}{P}_{AB})$ as above and the total centre of mass variables $(\underset{\sim}{\xi}_{CM},\underset{\sim}{\Pi}_{CM})$ we define (see Fig. 1)

$$\underset{\sim}{X}_i = \underset{\sim}{\xi}_i - \frac{m_A\underset{\sim}{\xi}_A + m_B\underset{\sim}{\xi}_B}{m_A + m_B}$$

$$\underset{\sim}{P}_i = \underset{\sim}{\Pi}_i - \frac{1}{m_A + m_B + N}(\underset{\sim}{\Pi}_A + \underset{\sim}{\Pi}_B + \sum_{j=1}^{N} \underset{\sim}{\Pi}_j),$$

$i = 1,2,\ldots N$, that is the relative position coordinates of particle i with respect to the centre of mass of A and B and their conjugate variables.* This set is canonical by construction so, e.g. the Poisson bracket of $\underset{\sim}{P}_i$ and $\underset{\sim}{X}_j$ is zero for $i \neq j$. Furthermore one still has a nice additivity property of the energy in the following sense:

$$H_o^{Tot} = \frac{\Pi_{CM}^2}{2M_T} + \frac{1}{2m}(\sum_{i=1}^{N} P_i^2) + \frac{\rho}{2\mu_{AB}}(\sum_{i=1}^{N} \underset{\sim}{P}_i)^2 + \frac{1}{2\mu_{AB}} P_{AB}^2 \qquad (5)$$

where $M_T = m_A + m_B + Nm$, $M = m_A + m_B$ and $\rho = \mu_{AB}/M$. This generalizes (4) but here the "isotropic" term $\frac{\rho}{2\mu_{AB}}(\sum_{i=1}^{N} \underset{\sim}{P}_i)^2$ appears explicitly.

1.2 Hilbert space structure for N-body quantum systems

Quantum mechanical states of the N particle assembly described above are usually represented by elements of the Hilbert space

$$\mathcal{H}^{Tot} = L^2(\mathbb{R}^{3N})$$

$$= \{\phi, \int |\phi(\underset{\sim}{\xi}_1, \ldots \underset{\sim}{\xi}_N)|^2 \, d\underset{\sim}{\xi}_1 \ldots d\underset{\sim}{\xi}_N < \infty\}.$$

Consider now a set of Jacobi coordinates associated with a cluster decomposition D. Up to a constant the volume element $d\underset{\sim}{\xi}_1,\ldots d\underset{\sim}{\xi}_N$ is proportional to $d\underset{\sim}{\xi}_{CM}\, d\underset{\sim}{X}_D\, d\underset{\sim}{X}_{C_i}\ldots d\underset{\sim}{X}_{C_k}$ since Jacobi coordinates

* For N = 1 it is just the set described in the above example.

are obtained from the $\{\xi_i\}$ through a bijective linear transformation. It is easy to prove that the set of finite linear combinations of product elements having the form $h_{CM}(\xi_{CM})h_D(X_D)\ h_{C_1}(X_{C_1}\ \ldots\ h_{C_k}(X_{C_k})$, where all h's are square integrable in their respective variables, is dense in $L^2(R^{3N})$. So one can write

$$H^{Tot} = H_{CM} \otimes H_D \otimes H_{C_1} \ldots \otimes H_{C_k} \tag{6}$$

where the meaning of each factor in (6) is obvious. Notice that H_{CM}, the Hilbert space of states for the total centre of mass, does not depend on D; so the same is true for

$$H = H_D \otimes H_{C_1} \ldots \otimes H_{C_k} \tag{7}$$

Elements of H describe states of the N particle system in the centre of mass reference frame; they depend only on relative Jacobi position variables.

The position and momentum operators are represented as usual, i.e., multiplication by the corresponding functions for position observables $f(\xi_1,\ldots,\xi_N)$ and application of the "Pseudo-differential" operators $g(-i\hbar\nabla_{\xi_1},\ldots,-i\hbar\nabla_{\xi_N})$ for momentum observables $g(\Pi_1,\ldots,\Pi_N)$ (here $\hbar$ is the Planck constant and ∇_ξ denotes the gradient operator with respect to the coordinate ξ). One major consequence of the fact that Jacobi variables associated with the cluster decomposition D are canonical, is that the operator corresponding to one of these phase space Jacobi variables acts only on the factor of the decomposition (6) having its own label (CM, D or some $C \in D$); to be correct one should write them as the corresponding multiplication operator on this factor times the identity on the other factors. In particular the correct mathematical expression for the decomposition of the kinetic energy operator (3) would be (denoting by the same symbol observables and the corresponding operators)

$$H_o^{Tot} = H_{o,CM} \otimes I_{H_D} \otimes I_{H_{C_1}} \ldots \otimes I_{H_{C_k}} + \ldots$$

$$\ldots + I_{H_{CM}} \otimes I_{H_D} \otimes \ldots \otimes H_{o,C_k}. \tag{8}$$

We will not use this cumbersome notation systematically but it is necessary to mention it in order to be able to use one important consequence.

Namely, under some extra assumptions on the spectra $\sigma(A_1)$ and $\sigma(A_2)$ of two closed operators A_1 and A_2 (see [2] Vol.4) acting on two Hilbertspaces H_1 and H_2 then

$$\sigma(A_1 \otimes I_{H_2} + I_{H_1} \otimes A_2) = \sigma(A_1) + \sigma(A_2) \quad (9)$$

where $A_1 \otimes I_{H_2} + I_{H_1} \otimes A_2$ acts on $H_1 \otimes H_2$.

The result holds in particular for self-adjoint operators; in the case of (8) it is of little interest but we will use it in more delicate situations. Another important consequence of the tensorial structure associated with the choice of a canonical set of Jacobi variables is that whenever we deal with isolated systems (no external forces) it will be possible to work only in the "relative" Hilbert space H (7) since observables which are functions of relative Jacobi variables act on it and not on the factor H_{CM}. In particular this is the case for the kinetic energy in the total centre of mass reference frame

$$H_o = H_o^{Tot} - \frac{\Pi_{CM}^2}{2M} \quad (10)$$

which, according to (8), is just represented by the operator

$$H_o = H_{o,D} \otimes I_{H_{C_1}} \cdots \otimes I_{H_{C_k}} + \cdots + I_{H_D} \otimes I_{H_{C_1}} \cdots$$
$$\cdots \otimes H_{o,C_k}$$

1.3 N-particle Schrödinger Hamiltonians

We consider now the situation where the particles interact via two-body potentials $V_{ij}(\xi_i - \xi_j)$. The Hamiltonian of such an isolated system in the centre of mass reference frame is

$$H = H_o + V \quad (11)$$

$$V = \sum_{1 \le i < j \le N} V_{ij}. \quad (12)$$

Notice that V is a function of relative Jacobi variables only so the operator associated with the observable H acts on H and not on H_{CM}. We want to find conditions ensuring existence of a spectral decomposition, solution of the time-dependent Schrodinger equation etc. If the functions V_{ij} are real then H is formally symmetric; i.e., $H \subset H^*$ (in the sense of extension of unbounded operators [1]); we need to verify $H = H^*$ which requires a knowledge of D(H), the domain of H. This is usually done for nice potentials (see the definition below) in a perturbative way. One first studies the domain of the kinetic energy operator H_o and then shows that the perturbation operator V does not change it if the singularities of the functions V_{ij} are not too strong. The mathematical basis of this so-called "regular perturbation theory" approach is a classical theorem of Kato and Rellich:

Proposition 1 ([1])

Let A be a self-adjoint operator on a Hilbert space $\mathcal{H}$ and B a linear (possibly unbounded) operator on $\mathcal{H}$ such that $\mathcal{D}(B) \supset \mathcal{D}(A)$ and there exist positive constants a,b such that $\forall \phi \in \mathcal{D}(A)$

$$||B\phi|| \leq a||A\phi|| + b||\phi|| \tag{13}$$

Then if B is symmetric and a < 1 the operator A + B with domain D(A) is self-adjoint.

An operator B satisfying (13) will be called A-bounded. The lower bound of the set of a's such that (13) holds will be called the A-bound of B (we will see later that it can be zero even in some cases where B is unbounded).

The domain of H_0 is defined in a natural way through Fourier transformation $\mathfrak{F}$ since for $\phi \in C_0^\infty(R^{3(N-1)})$

$$\tilde{\mathfrak{F}}(H_0\phi)(\underset{\sim}{P}) = \hbar^2 H_0(\underset{\sim}{P})(\tilde{\mathfrak{F}}\phi)(\underset{\sim}{P}) \tag{14}$$

where $\tilde{\mathfrak{F}}\phi$ is the Fourier transform of ϕ and $H_0(\underset{\sim}{P})$ is the function defined by (3) and (10) of $\underset{\sim}{P} = (\underset{\sim}{P}_D, \underset{\sim}{P}_{C_1}, \ldots \underset{\sim}{P}_{C_k}) \in R^{3(N-1)}$. The natural domain of definition for H_0 is then the set of elements such that the right hand side of (14) is in $\mathcal{H}$, i.e.,

$$\mathcal{D}(H_0) = \{\phi \in L^2(R^{3(N-1)}),\ H_0(\underset{\sim}{P})(\tilde{\mathfrak{F}}\phi)(\underset{\sim}{P}) \in L^2(R^{3(N-1)})\} \tag{15}$$

The proof that multiplication by $H_0(\underset{\sim}{P})$ is a self-adjoint operator is a trivial exercise and the statement that H_0 as defined by (14) and (15) is self-adjoint follows readily from the unitarity of the Fourier transform. One can also characterize the domain of H_0 using the Sobolev spaces

$$\mathcal{H}^m(R^n) = \{\phi,\ \mathcal{D}^\alpha\phi \in L^2(R^n)\ \forall\, \alpha,\ |\alpha| \leq m\} \tag{16}$$

where m is a positive integer, $\alpha = (\alpha_1, \ldots, \alpha_n)$ is a n-dimensional index and $|\alpha| = \sum_{i=1}^{n} \alpha_i$. The Sobolev norm is defined by

$$||\phi||_{\mathcal{H}^m(R^n)} = \Big(\sum_{|\alpha|\leq m} ||\mathcal{D}^\alpha\phi||^2_{L^2(R^n)}\Big)^{\frac{1}{2}}$$

and one then has ([1], [2]),

$$\mathcal{D}(H_o) = \mathcal{H}^2(R^{3(N-1)}). \tag{17}$$

For further purposes let us mention some of the so-called "Sobolev inequalities" stating that functions $\phi \in \mathcal{H}^m(R^n)$ can't have too strong singularities:

<u>Proposition 2</u> ([2] Vol.1, 3)

Let $L^p(R^n) = \{\phi, \int|\phi(X)|^p\, d^nX < \infty\}$ and $\mathcal{H}^m(R^n)$ as above. Then if

$$\frac{1}{p} = \frac{1}{2} - \frac{m}{n} \geq 0$$

one has $\mathcal{H}^m(R^n) \subset L^p(R^n)$, and there exists a positive constant C such that

$$||\phi||_{L^p(R^n)} \leq C||\phi||_{\mathcal{H}^m(R^n)} \tag{18}$$

If $\frac{1}{2} - \frac{m}{n} \leq 0$ one can take $p = \infty$, i.e., ϕ is bounded pointwise by a constant times its Sobolev norm. Let us prove this last result in the case n = 3, m = 2 which is particularly relevant for us:

By Fourier transform it is enough to show that

$$||\tilde{\mathcal{F}}\phi||_{L^1(R^3)} \leq C||\phi||_{\mathcal{H}^2(R^3)}.$$

Now $||\phi||_{\mathcal{H}^2(R^3)}$ is equivalent to the L^2 norm of $(1 + P^2)(\tilde{\mathcal{F}}\phi)(P)$ since Fourier transform exchanges differentiation and multiplication. Let $u = \tilde{\mathcal{F}}\phi$, then

$$\begin{aligned}\int|u(\underset{\sim}{P})|d^3\underset{\sim}{P} &= \int(1 + P^2)^{-1}(1 + P^2)|u(\underset{\sim}{P})|d^3\underset{\sim}{P} \\ &\leq \left(\int(1 + P^2)^{-2}d^3\underset{\sim}{P}\right)^{\frac{1}{2}}\left(\int(1 + P^2)^2|u(\underset{\sim}{P})|^2d^3\underset{\sim}{P}\right)^{\frac{1}{2}}.\end{aligned}$$

Notice that one can easily deduce from this inequality that $\forall\, X \in R^3$

$$|\phi(\underset{\sim}{X})| \leq C(||\phi||_{L^2(R^3)} + ||\Delta\phi||_{L^2(R^3)}), \tag{19}$$

where Δ denotes the ordinary Laplacian on R^3. Using the way the different quantities on each side of (19) transform under the scaling $\phi(X) \to \phi(\lambda\underset{\sim}{X})$ $\lambda \in R^+$, one can deduce from (19) that $\forall\, a > o$, there exists a,b > o such that

$$|\phi(\underset{\sim}{X})| \leq a||\Delta\phi||_{L^2(R^3)} + b||\phi||_{L^2(R^3)} \tag{19b}$$

(see, e.g., [2] Vol. 2).

This is particularly interesting in view of the applications of the Kato-Rellich theorem to Schrödinger Hamiltonians. Consider first the case N = 2. Let the potential $V = V_{12}$ be a real function such that

1) $V \in L^2_{loc}(R^3)$

2) $V(X) \to o$ as $X \to \infty$. (20)

By (19) it is obvious that $D(V) \supset D(H_o)$.

For a given $\varepsilon > o$ let us split

$$V = V_1^{(\varepsilon)} + V_2^{(\varepsilon)},$$

where $V_1^{(\varepsilon)} \in L^2(R^3)$ and $|V_2^{\varepsilon}(\underset{\sim}{X})| < \varepsilon \quad \forall\, \underset{\sim}{X} \in R^3$. Then

$$||V_2^{(\varepsilon)}\phi||_{L^2(R^3)} \leq \varepsilon||\phi||_{L^2(R^3)}$$

and by (20)

$$||V_1^{(\varepsilon)}\phi||_{L^2(R^3)} \leq ||\phi||_{L^\infty(R^3)}\, ||V_1^{(\varepsilon)}||_{L^2(R^3)}$$

$$\leq a'||\Delta\phi||_{L^2(R^3)} + b'||\phi||_{L^2(R^3)}$$

with $a' = a||V_1^{(\varepsilon)}||_{L^2(R^3)}$

$b' = b||V_1^{(\varepsilon)}||_{L^2(R^3)}$.

Since a can be chosen arbitrarily small one can apply Proposition 1 with the conclusion that $H = H_o + V$ is self-adjoint with domain $H^2(R^3)$.

Before turning to the self-adjointness problem for N-particle Hamiltonians with two-body potentials of the above type let us make the following remarks.

Remarks

1) Using the explicit form of the kernel of $(H_o + 1)^{-1}$ in the 3-dimensional case

$$(H_o + 1)^{-1}(\underset{\sim}{X},\underset{\sim}{Y}) = (4M)^{-1} e^{-|\underset{\sim}{X}-\underset{\sim}{Y}|}/|\underset{\sim}{X} - \underset{\sim}{Y}|$$

one can easily check that if $f \in L^2(R^3)$ then the operator $f(H_o + 1)^{-1}$ is Hilbert-Schmidt. By a limiting procedure this implies that for a potential of the above type $V(H_o + 1)^{-1}$ is a compact operator. Such operators are called H_o-compact.

2) Let $(\phi_n)_n$ be a sequence in $L^2(R^3)$ weakly convergent to zero and such that in addition for some $C < \infty$

$$||\phi_n||_{H^2(R^3)} \leq C.$$

Then necessarily ϕ_n has support away from the origin for large n in the sense, that if χ_R denotes the characteristic function of the sphere of radius R in R^3 then

$$||\chi_R\phi_n||_{L^2(R^3)} \to 0 \quad (R \to \infty) . \tag{21}$$

This follows from compactness of $\chi_R(H_o + 1)^{-1}$, boundedness of $((H_o + 1)\phi_n)_n$ in $L^2(R^3)$, the equality $\chi_R\phi_n = \chi_R(H_o + 1)^{-1}(H_o + 1)\phi_n$ and the fact that any strongly convergent subsequence of $(\chi_R\phi_n)_n$ necessarily has limit zero, since $(\phi_n)_n$ converges weakly to zero.

We now turn to the main result of this paragraph.

Proposition 3:

Assume $V_{ij} \in L^2_{loc}(R^3)$ and $V_{ij}(x) \to o$ as $x \to \infty$ for all pairs i,j. Then H is self-adjoint with domain $H^2(R^{3(n-1)})$.

Proof:

Let Δ_{ij} denote the Laplacian with respect to the 3-dimensional variable $X_{ij} = \xi_i - \xi_j$. Then $\forall\, a > o \quad \exists\, b > o$ such that

$$||V_{ij}\phi||_{L^2(R^{3(N-1)})} \leq a||\Delta\phi||_{L^2(R^{3(N-1)})} + b||\phi||_{L^2(R^{3(N-1)})}$$

by the above arguments (one just has to integrate out the extra Jacobi variables in a decomposition D having (ij) as a cluster.) Now $H_o = -\frac{1}{2}\Delta_{ij}\mu_{ij}^{-1} + H'_{o,ij}$ where $H'_{o,ij}$ is a positive operator commuting with $-\Delta_{ij}$ (see (3)); from this it follows easily that

$$2\mu_{ij}^{-1}||\Delta_{ij}\phi||_{L^2(R^3)} \leq ||H_o\phi||_{L^2(R^3)}$$

Hence V_{ij} is H_o-bounded with zero relative bound and accordingly V too; this completes the proof of the Kato-Rellich theorem.

1.4 Spectral properties of Schrödinger Hamiltonians

One expects the spectrum of H to be the union of a discrete spectrum (bound-states) and a continuum (scattering states). We will give in this paragraph a mathematical formulation of these facts using some simple and very useful recent "geometrical" techniques. Since perturbation of the continuous spectrum involves a rather elaborate theory, we will content ourselves here with the analysis of the essential spectrum, which in most situations of interest, in particular those investigated here, is the continuous spectrum with possibly some embedded eigenvalues.

Definition:

Let A be a closed linear operator on $\mathcal{H}$. The discrete spectrum of A, denoted by $\sigma_d(A)$ is the set of isolated eigenvalues of A with finite multiplicities. The essential spectrum, denoted by $\sigma_{ess}(A)$, is

$$\sigma_{ess}(A) = \sigma(A)\backslash\sigma_d(A)$$

Definition:

The Weyl spectrum of A, denoted by $\sigma_w(A)$ is

$$\sigma_w(A) = \{\lambda \in C,\ \exists \text{ sequence } (\phi_n)_n \subset \mathcal{D}(A) \text{ s.t.} \quad ||\phi_n|| = 1,\ (\phi_n)_n \text{ converges weakly to zero and } ||(A - \lambda I)\phi_n|| \to 0\,\} \tag{22}$$

Proposition 4 ([4]):

Assume $\sigma(A) \neq C$ and let $\rho(A) = C\backslash\sigma(A)$ (resolvent set). Then

i) $\sigma_w(A) \subset \sigma_{ess}(A)$

ii) The boundary of $\sigma_{ess}(A)$ is contained in $\sigma_w(A)$

iii) $\sigma_w(A) = \sigma_{ess}(A)$ if and only if each connected component of the complement of $\sigma_w(A)$ contains a point of $\rho(A)$.

According to iii) one has $\sigma_w(A) = \sigma_{ess}(A)$ for any self-adjoint operator A since any connected component of $\complement\sigma_w(A)$ necessarily contains some point with Im $z \neq 0$ hence in $\rho(A)$.

Let us show now invariance of the essential spectrum of H_o under potentials satisfying (20) in the case N = 2 where $\sigma_{ess}(H_o) = R^+$. A general theorem ([1]) asserts that the essential spectrum is invariant under a relatively compact perturbation. We prefer to use here a direct "geometrical" proof in view of the analysis for general N.

Proposition 5:

Let V satisfy (20). Then $\sigma_{ess}(-\Delta + V) = R^+$.

Proof:

Let $\lambda \in R^+$ and let $(\phi_n)_n$ be a Weyl sequence as in (22) such that $||(-\Delta - \lambda I)\phi_n|| \to o$. One has $V\phi_n = V\chi_R\phi_n + V(I - \chi_R)\phi_n$ where χ_R is the characteristic function of the sphere of radius R. Now $V\chi_R\,\phi_n$ tends strongly to zero as $n \to \infty$ by an argument already used in the Remark 1-2, §1.3. Furthermore $V(I - \chi_R)\phi_n \to o$ uniformly in n as $R \to \infty$. Using these facts it is easy to extract a subsequence $(\phi_{k(n)})_N$ converging strongly to zero. Accordingly $||(-\Delta + V - \lambda)\phi_{k(n)}|| \to o$ so that $\lambda \in \sigma_{ess}(-\Delta + V)$. Conversely one can use the same type of argument to show that $\lambda \in \sigma_{ess}(-\Delta + V) \Rightarrow \lambda \in \sigma_{ess}(-\Delta)$.

This argument does not work anymore for N > 2 since $V(I - \chi_R) \to o$ as $R \to \infty$ does not hold anymore. However it is still true that $\chi_R(H_o + 1)^{-1}$ is compact; this says that any sequence $(\phi_n)_n$ bounded in $H^2(R^{3(N-1)})$ and converging weakly to zero in $L^2(R^{3(N-1)})$ ultimately has support outside the sphere S_R of radius R for any R (one uses the same arguments as in Remark 1-2, §1.3). One expects the states ϕ_n to split into a sum of cluster states ϕ_n^D over all 2-cluster decompositions D. The support of ϕ_n^D (see Fig. 2) is in this region of configuration space where particles in different clusters of D are far from each other so that the effective Hamiltonian in this region is

$$H_D = H - V_D \qquad (23)$$

$$V_D = \sum_{\substack{i \in C_1 \\ j \in C_2}} V_{ij}.$$

Thus a Weyl sequence for H in the sense of (22) gives Weyl sequences for at least one H_D and one can repeat the argument used for N = 2. The precise mathematical formulation of these facts

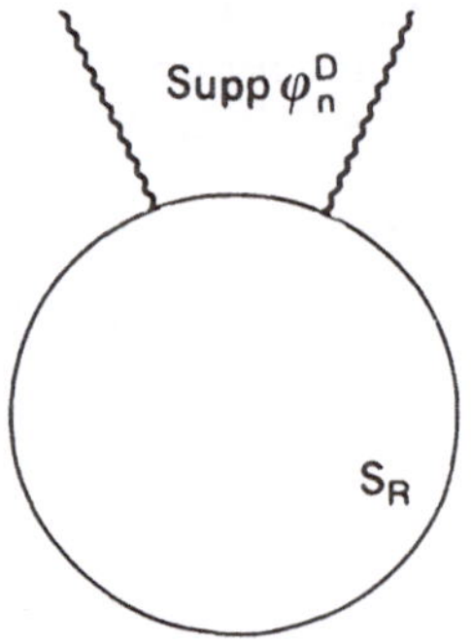

Fig. 2. Decomposition of configuration space.

is due to Enss and we summarize it here. The result is the so-called HVZ theorem ([5], [6], [7]).

Theorem 6:

Let the two-body potentials V_{ij} satisfy (20). Then

$$\sigma_{ess}(H) = [\Lambda,\infty), \tag{24}$$

where

$$\Lambda = \inf_{D=\{C_1,C_2\}} (\inf \sigma(H_{C_1}) + \sigma(H_{C_2})). \tag{25}$$

Proof:

We refer to Simon [8] for the construction of a set of functions $J_D(\underset{\sim}{X})$, $\underset{\sim}{X} \in R^{3(N-1)}$ $D = (C_1,C_2)$, homogeneous of degree zero such that

i) $$\sum_{D=(C_1,C_2)} J_D^2 = 1$$

ii) $$J_D(\underset{\sim}{X}) = o \quad \text{if} \quad \min_{\substack{i\in C_1\\ i\in C_2}} |\xi_i - \xi_j| \le \frac{1}{\sqrt{2}} N^{-\frac{1}{2}}(N-1)^{-3/2}|\underset{\sim}{X}| \tag{26}$$

where $|\underset{\sim}{X}| = \sum_{1\le i<j\le N} |\xi_i - \xi_j|^2.$

So functions J_D live in regions where particles in different clusters of D are far from each other and $V_D J_D = 0\ (1/|\underset{\sim}{X}|)$ as

$|\underset{\sim}{X}| \to \infty$. We assume furthermore that they are smooth; by homogeneity one has $\nabla J_D = 0\,(1/|\underset{\sim}{X}|)\;(|\underset{\sim}{X}| \to \infty)$.

Consider now a Weyl sequence $(\phi_n)_n$ such that $(\phi_n)_n$ converges weakly to zero in $L^2(R^{3(n-1)})$, $||\phi_n||_{L^2(R^{3(N-1)})} = 1$ and

$$||(H - \lambda)\phi_n||_{L^2(R^{3(N-1)})} \to 0.$$

We make a splitting $\phi_n(1 - \chi_R)\phi_n + \chi_R\phi_n$ where χ is a smooth characteristic function of some neighbourhood of zero and $\chi_R(X) = \chi(X/R)$; then $(\chi_R\phi_n)_n$ tends strongly to zero as $n \to 0$. So one can extract a subsequence such that $\chi_n\,\phi_{k(n)} \to 0$.

Let $\phi_n^D = J_D(1 - \chi_n)\phi_{k(n)}$.

Then $V_D\phi_n^D$ tends to zero as $n \to \infty$ since $V_D J_D(1 - \chi_n) \to o$. Furthermore since $\sum||\phi_n^D||^2 \to 1$ then for some subsequence still denoted by ϕ_n^D and some D one has for all n $||\phi_n^D|| \geq a > o$. Now notice that since $J_D \leq 1$ one has

$$||(H_D - \lambda)\phi_n^D|| \leq ||(H - \lambda)\phi_{k(n)}|| + ||V_D\,\phi_n^D|| + ||[H_o, J_D(1 - \chi_n)]\phi_{k(n)}||.$$

The last term decays like $1/n$ so that finally $||(H_D - \lambda)\phi_n^D|| \to o$; since obviously $\phi_n^D \to o$ weakly then $(\phi_n^D)_n$ is after normalization a Weyl sequence for H_D, i.e. $\lambda \in \sigma_{ess}(H_D)$.

So finally

$$\sigma_{ess}(H) \subset \bigcup_D \sigma_{ess}(H_D) \tag{27}$$

To prove the converse statement let $D = (C_1^{\cdot}, C_2)$, $\lambda \in \sigma(H_D)$ and $(\phi_n)_n$ such that $||\phi_n|| = 1$ and

$$||(H_D - \lambda)\phi_n|| \to o$$

(the existence of such a sequence is the Weyl criterion for the full spectrum; one does not require $\phi_n \to o$ weakly). Let $a_D \in R^{3(N-1)}$ have the direction $\underset{\sim}{X}_D = \underset{\sim}{X}_{C_1} - \underset{\sim}{X}_{C_2}$ and $T(\underset{\sim}{a})$ denote the translation operator

$$(T(\underset{\sim}{a})\phi)(\underset{\sim}{X}) = \phi(\underset{\sim}{X} - \underset{\sim}{a}), \tag{28}$$

so that $T(\underset{\sim}{a}_D)$ separates the centres of mass of C_1 and C_2 without changing internal variables and $T(a_D)H_D T^{-1}(a_D) = H_D$. For each n pick N(n) so large that

$$||V_D T(N(n)\underset{\sim}{a}_D)\phi_n|| \to o \quad \text{and let} \quad \psi_n = T(N(n)\underset{\sim}{a}_D)\phi_n.$$

Then $||(H - \lambda)\psi_n|| \to o$, i.e. $\lambda \in \sigma(H)$. Now by (9) and $H_D = H_{o,D} + H_{C_1} + H_{C_2}$ one has:

$$\sigma(H_D) = \sigma_{ess}(H_D) = [\inf \sigma(H_{C_1}) + \sigma(H_{C_2}), \infty)$$

and the proof of (24) is completed.

Remarks

1) The above analysis extends to situations with fixed centres of forces lying on a bounded region by introducing a fictitious particle of infinite mass at the origin.

2) The physical meaning of the HVZ theorem is intuitively clear in terms of scattering states. Essentially it says that $\sigma_{ess}(H)$ is the energy spectrum of states formed of at least two non interacting clusters, i.e. the spectrum of states describing clusters infinitely separated from each others. This is exactly how scattering states should look like.

3) For N = 2 there are known criteria for infinitude or finiteness of the number of eigenvalues below the essential spectrum ([9]). Namely one has finiteness (resp. infinitude) for potentials decaying more (resp.less) rapidly than $|\underset{\sim}{X}|^{-2}$. One also has the celebrated Lippman-Schwinger bounds on the number of bound-states in terms of the potential only. The situation is drastically more complicated for N-body systems due to the difficulty to control the position of Λ; quite complicated phenomena like the Efimov effect [10] can give strongly unstable results under a small change of potentials or coupling constants (see [9] for a general discussion).

Let us mention, however, that for atomic systems it is now proved that the number of bound-states is infinite for atoms and positive ions ([11], [12]) whereas it is finite for negative ions ([13]).

1.5 Application: Introduction to spectral properties of electronic Hamiltonians

Let us consider a diatomic molecule with nuclei A,B and N electrons. The Hamiltonian of such a system is (see (5))

$$H = \frac{1}{2\mu_{AB}} P_{AB}^2 + \frac{\rho}{2\mu_{AB}} \left(\sum_{i=1}^{N} P_i\right)^2 + \frac{1}{2}\sum_{i=1}^{N} P_i^2 + \frac{Z_A Z_B}{|\underset{\sim}{X}_{AB}|}$$

$$- \sum_{i=1}^{N} \frac{Z_A}{|\underset{\sim}{\xi}_i - \underset{\sim}{\xi}_A|} - \sum_{i=1}^{N} \frac{Z_B}{|\underset{\sim}{\xi}_i - \underset{\sim}{\xi}_B|} + \sum_{1 \le i < j \le N} \frac{1}{|\underset{\sim}{\xi}_i - \underset{\sim}{\xi}_j|} \tag{29}$$

Here Z_A, Z_B denote the nuclear charges; we adopt the convention $\hbar = m = 1$. We can reexpress the distances between electrons and nuclei in terms of the relative coordinates defined in §1.1. Let

$$\sigma_A = \frac{m_A}{m_A + m_B}, \quad \sigma_B = \frac{m_B}{m_A + m_B}; \quad \text{then}$$

$$\underset{\sim}{\xi}_i - \underset{\sim}{\xi}_A = \underset{\sim}{X}_i - \sigma_A \underset{\sim}{X}_{AB} \quad \text{and} \quad \underset{\sim}{\xi}_i - \underset{\sim}{\xi}_B = \underset{\sim}{X}_i + \sigma_B \underset{\sim}{X}_{AB}$$

The electronic Hamiltonian is obtained from (29) by letting all nuclear masses tend to infinity, i.e.,

$$H_{el} = \frac{1}{2}\sum_{i=1}^{N} P_i^2 - \sum_{i=1}^{N} \frac{Z_A}{|\sigma_A \underset{\sim}{X}_{AB} - \underset{\sim}{X}_i|} - \sum_{i=1}^{N} \frac{Z_B}{|\sigma_B \underset{\sim}{X}_{AB} + \underset{\sim}{X}_i|}$$

$$+ \frac{Z_A Z_B}{|\underset{\sim}{X}_{AB}|} + \sum_{1\ i\ j\ N} \frac{1}{|\underset{\sim}{X}_i - \underset{\sim}{X}_j|} \tag{30}$$

This operator acts on $\mathcal{H} \approx L^2(R^{3(N+1)}) = \mathcal{H}_{AB} \otimes \mathcal{H}_{el}$ where $\mathcal{H}_{el}$ is the Hilbert space of electronic states, $\mathcal{H}_{el} \approx L^2(R^{3N})$. Notice that $\underset{\sim}{X}_{AB}$ commutes with the momentum operators of electrons so that one can fix the inter-nuclear vector $\underset{\sim}{X}_{AB}$ and consider the so-called "clamped" nuclei Hamiltonian $H(\underset{\sim}{X}_{AB})$ on H_{el} where now $\underset{\sim}{X}_{AB}$ is considered as a parameter.

In mathematical terms H_{el} is the "direct integral" of clamped nuclei Hamiltonians

$$H_{el} = \int^{\oplus} H(\underset{\sim}{X}_{AB})\, d^3\underset{\sim}{X}_{AB}.$$

It acts on $\mathcal{H}$ decomposed itself as such an integral

$$\mathcal{H} = \int^{\oplus} \mathcal{H}(\underset{\sim}{X}_{AB})\, d^3\underset{\sim}{X}_{AB}$$

where $\mathcal{H}(\underset{\sim}{X}_{AB}) \approx \mathcal{H}_{el}\ \forall\ \underset{\sim}{X}_{AB}$.

The spectrum of the clamped nuclei Hamiltonian $H(\underset{\sim}{X}_{AB})$ can be analysed with the HVZ theorem (notice the proof given above extends easily to this type of N-particle Hamiltonian with external forces). So it has the following structure

$$\sigma(H(\underset{\sim}{X}_{AB})) = \{E_o(\underset{\sim}{X}_{AB}), \ldots, E_n(\underset{\sim}{X}_{AB}) \ldots\} \cup [\Lambda(\underset{\sim}{X}_{AB}), \infty) \tag{31}$$

where $E_n(\underset{\sim}{X}_{AB})$ are isolated eigenvalues with finite multiplicities and $\Lambda(\underset{\sim}{X}_{AB})$ is the bottom of the essential spectrum. For electronic systems of this type it is well-known ([12]) that $\Lambda(\underset{\sim}{X}_{AB})$ is the lowest eigenvalue of an (N - 1) electron clamped nuclei Hamiltonian. Notice that the HVZ theorem and hence the structure (31) remains true if one reduces the system with respect to its symmetries (rotations around the internuclear axis, reflections in planes containing the internuclear axis, permutations of electrons and rotations and reflections interchanging A and B in case $Z_A = Z_B$). In a given symmetry sector the bottom of the essential spectrum is then given by the lowest eigenvalue of the (N - 1) electron system with "compatible" symmetry ([12]).

It turns out that the spectral values of $H(\underset{\sim}{X}_{AB})$ depend only on $r = |\underset{\sim}{X}_{AB}|$. This comes from the fact that if O is a rotation in R^3 and $\underset{\sim}{U}(O)$ denotes the corresponding unitary transformation on $\mathcal{H}_{el}$ then

$$H(O(\underset{\sim}{X}_{AB})) = U(O)H(\underset{\sim}{X}_{AB})\, U^{-1}(O). \tag{32}$$

Since a unitary transformation leaves the spectrum of a self-adjoint operator invariant one obtains

$$\begin{aligned} E_n(O(\underset{\sim}{X}_{AB})) &= E_n(\underset{\sim}{X}_{AB}) \qquad \forall\, O \\ &= E_n(r) \end{aligned}$$

and $\Lambda(O(\underset{\sim}{X}_{AB})) = \Lambda(\underset{\sim}{X}_{AB}) = \Lambda(r)$.

We will analyse later continuity properties of $E_n(r)$, $\Lambda(r)$ as functions of r. Varying r one obtains the so-called electronic curves as depicted in Fig. 3. We content ourselves for the moment with the following remarks about their local structures:

1) Since the potentials in (30) satisfy a uniform Kato-Rellich estimate (13) it is easy to show that electronic curves are bounded below ([1]).

2) One expects a Coulomb like behaviour at $\underset{\sim}{X}_{AB} = 0$ due to the singularity of the term $Z_A Z_B/r$.

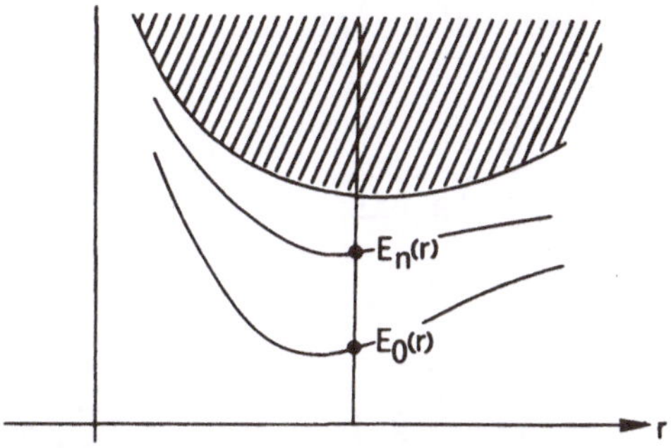

Fig. 3. Spectrum of the clamped nuclei Hamiltonian.

3) The asymptotic behaviour of the electronic curves is what intuition says it should be. Namely when $r \to \infty$ one expects the system to split into separated atoms and accordingly that $E_n(r)$ converges to sums of two atomic bound-state energies. To show this fact we are going to use an extension of the geometrical ideas used in the proof of the HVZ theorem [15].

Let D be a cluster decomposition of the molecule into two atoms

$$\bar{A} = \{A, C_A\} \quad , \quad \bar{B} = \{B, C_B\}$$

$$C_A \cup C_B = \{1, 2, \ldots, N\} \quad , \quad C_A \cap C_B = \emptyset .$$

Let H_D be the "separated atoms" Hamiltonian

$$H_D = H_{\bar{A}} + H_{\bar{B}} \tag{33}$$

where, e.g.,

$$H_{\bar{A}} = \frac{1}{2m} \left(\sum_{i \in C_A} P_i^2 \right) - Z_A \sum_{i \in C_A} \frac{1}{|\underset{\sim}{X}_i|} + \sum_{i,j \in C_A} \frac{1}{|\underset{\sim}{X}_i - \underset{\sim}{X}_j|} .$$

We also define $H_D(\underset{\sim}{X}_{AB})$ as follows: Let $T_D(\underset{\sim}{X}_{AB})$ be the translation operator on $\mathcal{H}_{el}$ which moves the clusters C_A and C_B away from each other by $\underset{\sim}{X}_{AB}$ about the centre of mass of nuclei A and B

$$\begin{aligned} \underset{\sim}{X}_i &\to \underset{\sim}{X}_i - \sigma_A \underset{\sim}{X}_{AB} \qquad i \in C_A \\ \underset{\sim}{X}_j &\to \underset{\sim}{X}_j - \sigma_B \underset{\sim}{X}_{AB} \qquad j \in C_B . \end{aligned} \tag{34}$$

Then let

$$H_D(\underset{\sim}{X}_{AB}) = T_D(\underset{\sim}{X}_{AB}) H_D T_D^{-1}(\underset{\sim}{X}_{AB}). \tag{35}$$

Notice that

$$H(\underset{\sim}{X}_{AB}) = H_D(\underset{\sim}{X}_{AB}) + V_D(\underset{\sim}{X}_{AB}), \tag{36}$$

where (see (23))

$$V_D(\underset{\sim}{X}_{AB}) = \sum_{\substack{i \in C_A \\ i \in C_B}} \frac{1}{|\underset{\sim}{X}_i - \underset{\sim}{X}_j|} - \sum_{i \in C_A} \frac{Z_B}{|\sigma_B \underset{\sim}{X}_{AB} + \underset{\sim}{X}_i|}$$

$$- \sum_{j \in C_B} \frac{Z_A}{|\sigma_A \underset{\sim}{X}_{AB} - \underset{\sim}{X}_j|}.$$

Notice also that by unitary equivalence

$$\sigma(H_D(\underset{\sim}{X}_{AB})) = \sigma(H_D) = \sigma(H_{\bar{A}}) + \sigma(H_{\bar{B}}). \tag{37}$$

Our aim is to show that in some sense $H(\underset{\sim}{X}_{AB})$ approaches for large internuclear separations $\underset{\sim}{X}_{AB}$ the "direct sum" of separated atoms Hamiltonians $H_D(\underset{\sim}{X}_{AB})$. Let us prove first [14], [15].

Proposition 7:

Let σ_d be the set of limit points of $\sigma_d(H(\underset{\sim}{X}_{AB}))$:

$$\sigma_d = \{ \lim_{|\underset{\sim}{X}_{AB}| \to \infty} E(\underset{\sim}{X}_{AB}), \ E(\underset{\sim}{X}_{AB}) \in \sigma_d(H(\underset{\sim}{X}_{AB})) \}$$

Then one has the inclusion

$$\sigma_d \subset \bigcup_{\{\bar{A}, \bar{B}\}} \{\sigma_d(H_{\bar{A}}) + \sigma_d(H_{\bar{B}})\} \cup \Lambda_3, \tag{38}$$

where

$$\Lambda_3 = \inf_{\{\bar{A}, \bar{B}, C\}, C \neq \emptyset} \{\sigma_d(H_{\bar{A}}) + \sigma_d(H_{\bar{B}}) + \sigma_d(H_C)\}$$

Proof:

Let us introduce, as in the proof of the HVZ theorem, a smooth partition of unity on the configuration space of the two nuclei and N electron system (see (26)); there the sum runs now from 1 to N + 2 where nuclei A (resp. B) is labelled N + 1 (resp. N + 2) so that

$$|\underset{\sim}{X}| \geq |\underset{\sim}{X}_{AB}| = r. \tag{39}$$

Notice that for fixed value of $\underset{\sim}{X}_{AB}$ the functions $\{J_D^2(\cdot,\underset{\sim}{X}_{AB})$, $D = \{C_1 C_2\}\}$ form a partition of unity on the configuration space of the N electrons; remark also that the partitions D considered here involve also those where the two nuclei belong to the same cluster C_1 or C_2.

Consider now an electronic curve E(r) such that $E(r) < \Lambda(r)$. We devote to Chapter 2 the proof that E(r) is a continuous function. Consider a normalized eigenfunction

$$(H(\underset{\sim}{X}_{AB}) - E(r))\ell(\underset{\sim}{X}_{AB}) = o \tag{40}$$

Denoting by $J_D(\underset{\sim}{X}_{AB})$ the function of electronic coordinates $J_D(\cdot,\underset{\sim}{X}_{AB})$ one has

$$\sum_D ||J_D(\underset{\sim}{X}_{AB})(H_{AB}) - E(r)\ell(\underset{\sim}{X}_{AB})||^2 = 0.$$

Now there are two types of D's. First let $D = (C_A, C_B)$; then

$$J_D(\underset{\sim}{X}_{AB})H(\underset{\sim}{X}_{AB}) = H_D(\underset{\sim}{X}_{AB})J_D(\underset{\sim}{X}_{AB}) + V_D(\underset{\sim}{X}_{AB})J_D(\underset{\sim}{X}_{AB})$$
$$- 2\underset{\sim}{\nabla}J_D(\underset{\sim}{X}_{AB})\cdot\underset{\sim}{\nabla} + \Delta J_D(\underset{\sim}{X}_{AB})$$

where $\underset{\sim}{\nabla}$ and Δ denote respectively the gradient and Laplacian on the electronic configuration space. By the homogeneity of J_D one has $|\underset{\sim}{\nabla}J_D(\underset{\sim}{X}_{AB})| = O(1/r)$ and $|\underset{\sim}{\nabla}J_D(\underset{\sim}{X}_{AB})| = O(1/r^2)$. By construction of J_D one also has $|V_D(\underset{\sim}{X}_{AB})J_D(\underset{\sim}{X}_{AB})| = O(1/r)$; in these estimates (39) plays of course an essential role. Summing up, we get from (40) setting $\ell_D(\underset{\sim}{X}_{AB}) = (J_D\ell)(\underset{\sim}{X}_{AB})$

$$||(H_D(\underset{\sim}{X}_{AB}) - E(r)\ell_D(\underset{\sim}{X}_{AB})|| = O(\tfrac{1}{r}) \tag{41}$$

for all $D = (C_A, C_B)$.

Second, consider now the decompositions D where A and B belong to the same cluster C_1. Then the electrons belonging to the cluster C_2 are separated by a distance $O(1/|\underset{\sim}{X}|) = O(1/r)$ from nuclei A and B and thus their interactions with the nuclei is $O(1/r)$; by arguments similar to those used above and in the HVZ theorem one has then an estimate like (41) but where now $H_D(X_{AB})$ is the clamped nuclei Hamiltonian for an ionized molecule, namely the one obtained by extracting the electrons of C_2.

Now let us assume inductively that

$$\lim_{r\to\infty} \Lambda(r) = \inf_{\substack{(\bar{A},\bar{B},C)\\ C\neq\phi}} (\inf \sigma_d(H_{\bar{A}}) + \inf \sigma_d(H_{\bar{B}})$$

$$= \Lambda_3$$

i.e. a sum of bound-state energies for a pair of atoms containing less than N electrons. This is true for N = 1 since in this case $\Lambda(r) = 0$ by Proposition (5).

Now let E be a limit point of a sequence $E(r_n)$, $r_n \to \infty$; since $E(r) < \Lambda(r)$ $\forall$ r one has $E \leq \Lambda_3$. If $E = \Lambda_3$ there is nothing to prove. If $E < \Lambda_3$ we know from $||\sum_D \ell_D(\underset{\sim}{X}_{AB})||^2 = 1$ that for some subsequence and <u>at least</u> one D one has $||\ell_D(\underset{\sim}{X}_{AB,k(n)})|| > a > 0$ together with

$$||(H_D(\underset{\sim}{X}_{AB,k(n)}) - E)\ \ell_D(\underset{\sim}{X}_{AB,k(n)})|| = 0(\frac{1}{r_{k(n)}}). \qquad (42)$$

Assume first D is a cluster decomposition for which A and B belong to the same cluster; then $\sigma(H_D(\underset{\sim}{X}_{AB})) \subset [\Lambda(r) + 0(1/r), \infty) = [\Lambda_3 + o(1), \infty)$ so that (42) is impossible.

So necessarily $D = (\bar{A},\bar{B})$ and by (37) and the Weyl criterion for the full spectrum one has $E \in \sigma_d(H_{\bar{A}}) + \sigma_d(H_{\bar{B}})$ Q.E.D.

<u>Remarks</u>

1) If one takes into account the fact that electronic curves are continuous (see Ch. 2) and the discreteness of

$$\bigcup_{D=(\bar{A},\bar{B})} \{\sigma_d(H_{\bar{A}}) + \sigma_d(H_{\bar{B}})\}$$

below Λ_3, the intermediate value theorem for continuous functions implies modulo (38) that electronic curves are asymptotic to sums of atomic bound-state energies.

2) The proof of the proposition also says something about the asymptotic behaviour of electronic wave-functions at large inter-nuclear distances. In fact by (42) $(\ell_D(\underset{\sim}{X}_{AB}))$ converges to an eigenvector of $H_D(\underset{\sim}{X}_{AB})$; these eigenvectors have by (33) and (35) the structure

$$\sum_{j,j'} T_D(\underset{\sim}{X}_{AB})\ \ell_{\bar{A}}^{(j)} \otimes \ell_{\bar{B}}^{(j)}$$

where $\ell_{\bar{A}}^{(j)}$ and $\ell_{\bar{B}}^{(j)}$ are atomic bound state wave-functions. In other words once projected in the right region of configuration space the molecular wave-function $\ell(\underset{\sim}{X}_{AB})$ looks like a tensor product of atomic bound state wave-functions translated away from each other by a distance $\underset{\sim}{X}_{AB}$. Notice that

$$\ell(\underset{\sim}{X}_{AB}) = \sum_D J_D^2(\underset{\sim}{X}_{AB})\ \ell(\underset{\sim}{X}_{AB})$$

$$= \sum_D J_D(\underset{\sim}{X}_{AB})\ \ell_D(\underset{\sim}{X}_{AB}). \tag{43}$$

It is a priori embarrassing that one does not have instead of (43) a behaviour

$$\lim_{r\to\infty} ||\ell(X_{AB}) - \sum_D \ell_D(X_{AB})|| = 0. \tag{44}$$

However one can recover (44) from (43) in the case $E < \Lambda_3$ by noticing that since atomic wave-functions have exponential fall-off one can show easily

$$J_{D'}(X_{AB})\ell_D(X_{AB}) \to 0 \qquad (r \to \infty)$$

if $D \neq D'$. Since $\sum_D J_D^2(X_{AB}) = 1$ one gets (44) and finally the limit

$$\lim_{r\to\infty} ||\ell(X_{AB}) - \sum_D \sum_{j,j'} \alpha_D^{jj'}(\underset{\sim}{X}_{AB})\ T_D(\underset{\sim}{X}_{AB})\ \ell_{\bar{A}}^{(j)} \otimes \ell_{\bar{B}}^{(j)}|| = 0$$

(where the coefficients $\alpha_D^{jj'}(\underset{\sim}{X}_{AB})$ have in general no limit).

3) The "geometrical" proof of the Proposition can easily be symmetry adapted (see [15]), i.e., it remains valid in any symmetry sector.

To conclude this chapter we now prove a converse statement to (38):

Proposition 8:

Let σ be the set of limit points of $\sigma(H(\underset{\sim}{X}_{AB}))$

$$\sigma = \{\lim_{r\to\infty} E(\underset{\sim}{X}_{AB}),\ E(\underset{\sim}{X}_{AB}) \in \sigma(H(\underset{\sim}{X}_{AB})) .$$

Then $\sigma \supset \sigma_d(H_{\bar{A}}) + \sigma_d(H_{\bar{B}})$ for any pair of atoms $\bar{A},\bar{B}$.

Proof

Let $E \in \sigma(H_{\bar{A}}) + \sigma(H_{\bar{B}})$ and $\ell_{\bar{A}} \otimes \ell_{\bar{B}}$ be a corresponding eigenstate for H_D, $D = (\bar{A}, \bar{B})$. Then $\ell(\underset{\sim}{X}_{AB}) = T_D(\underset{\sim}{X}_{AB})\ \ell_{\bar{A}} \otimes \ell_{\bar{B}}$ is an eigenstate for

$H_D(\underset{\sim}{X}_{AB})$ with eigenvalue E.

Now one has by (35) and (36)

$$||(H(\underset{\sim}{X}_{AB}) - E)\ell(\underset{\sim}{X}_{AB})|| = ||V_{AB}(\underset{\sim}{X}_{AB})\ell(\underset{\sim}{X}_{AB})|| \tag{45}$$

It is a simple exercise using the exponential fall-off of atomic wave-functions $\ell_{\bar{A}}$, $\ell_{\bar{B}}$ to show that the quantity (45) is $O(1/r)$. Assume then that $E \not\in \sigma$; then $||(H(X_{AB}) - E)\ell(X_{AB})|| \geq a||\ell(X_{AB})||$ for a constant $a > o$ and all X_{AB} with $r > r_o$. This obviously gives a contradiction.

Remarks

1) The above analysis can be used to prove rigorously the validity of the 1/r expansion of electronic curves. Details of such an analysis can be found in [15].

2) We will see later that electronic wave-functions $\ell(x)$ are twice-differentiable in the nuclear variables x. It is straightforward to show "geometrically" that such derivatives converge to those of their "atomic" asymptotic limits.

3) Our geometrical analysis carries over to polyatomic molecules (see also [15]).

2. THE BORN-OPPENHEIMER APPROXIMATION

2.1 Introduction

The Born-Oppenheimer approximation scheme is one of the most important tools in molecular physics. It is based on the smallness of the ratio electronic mass to nuclear mass $\kappa^4 = \mu^{-1}$ typically of the order 10^{-3}.

The two main results of the 1927 article by Born and Oppenheimer are [16]: The Schrödinger equation for the total molecular system can be replaced by another Schrödinger equation for the nuclei only. The electrons give rise to an effective potential. The error introduced by the substitution is of order κ^8. The second main result is their formula for the nuclear energies,

$$E(\kappa) = E_o + \kappa^2 \nu(n + \tfrac{1}{2}) + \kappa^4 \frac{\ell(\ell+ 1)}{r_o^2} + \kappa^6 \ldots \tag{1}$$

The first term E_o is due to electronic motion in the clamped nuclei approximation. The second term comes from the vibration of the nuclei with frequency ν around the minimum r_0 of the effective potential. The third term is due to the rotation of the total molecule with angular momentum ℓ. The importance of the above formula (1) is that it accounts for the three orders of magnitude typically encountered in molecular spectroscopy.

The article by Born and Oppenheimer mentioned is preceded by an article by Born and Heisenberg of 1924 with the same title [17]. The authors try to explain the three orders of magnitude in molecular spectra. Although the authors understood the physical mechanism i.e., the relevance of electronic motion versus nuclear vibration and rotation they were unsuccessful; the Schrödinger equation was still missing at that time.

The concept of adiabatic wave functions (see below) is of great importance in the Born-Oppenheimer approximation scheme. It was used at about the same time as Born and Oppenheimer by Kronig [17] and Condon [18] in the discussion of the fine structure in molecular spectra. Very rapidly the Born-Oppenheimer approximation scheme was used to analyse detailed questions in molecular spectroscopy. The method was so successful and became so popular in physics and chemistry that the lack of mathematical foundation has not really penetrated in the realms of interest of many scientists working in molecular physics. However, with the development of experimental techniques the precise nature of approximations entering the relevant theory are also getting more attention; we hope therefore that our analysis of the Born-Oppenheimer scheme is not merely an esoteric mathematical game.

From a conceptual point of view the Born-Oppenheimer approximation described in the following is a method for solving the complicated many body problem by approximation, with a precise estimate on the error introduced. The method is motivated by classical intuition for the nuclear motion. Hence, the success of this approximation proves the validity of the classical picture of a molecule as a rigid body, the shape being described by the minima of the potential energy surfaces (see below).

The Born-Oppenheimer scheme involves two steps:

Step one: Reduction of the eigenvalue problem for the molecular Schrödinger operator to a WKB problem for a nuclear Schrödinger operator with an effective potential reflecting the electronic degrees of freedom.

Step two: Discussion of the WKB problem.

Note that the construction of the nuclear Schrödinger operator is crucially dependent on the energy range one is interested in. The nuclear Schrödinger operator gets more complicated for larger energy ranges, as will be explained below. Furthermore the construction requires a detailed analysis of the molecular Schrödinger operator in the limit $\kappa^4 = \mu^{-1} = o$. This is the so called clamped nuclei approximation; it plays the role of zeroth order approximation in the Born-Oppenheimer scheme.

The lectures on the Born-Oppenheimer approximation are organized as follows: In the second section of this chapter we discuss the clamped nuclei Schrödinger operator. This is an extension of what has been presented in the first chapter. After that we explain the reduction of the WKB problem in section 3, the analysis of which is presented in the last chapter.

2.2 The Clamped Nuclei Approximation

The Schrödinger operator in the clamped nuclei approximation, or the electronic Hamiltonian, has been introduced earlier (see §1.4); it is given by

$$H(\underset{\sim}{X},\alpha) = \sum_{\ell=1}^{N} \tfrac{1}{2}P_\ell^2 - \sum_{\ell=1}^{N} \frac{\alpha Z_A}{|\sigma_A \underset{\sim}{X} - \underset{\sim}{X}_\ell|} - \sum_{\ell=1}^{N} \frac{\alpha Z_B}{|\sigma_B \underset{\sim}{X} + \underset{\sim}{X}_\ell|} + \sum_{i<k} \frac{\alpha}{|\underset{\sim}{X}_i - \underset{\sim}{X}_k|} + \frac{\alpha Z_A Z_B}{|\underset{\sim}{X}|} . \tag{2}$$

To simplify notation we write $\underset{\sim}{X}$ for $\underset{\sim}{X}_{AB}$, r for $r_{AB} = |X_{AB}|$ and H_o^{el} for the kinetic part in the clamped nuclei Hamiltonian $H(\underset{\sim}{X},\alpha)$. If the α-dependence is unimportant we write simply $H(\underset{\sim}{X})$. Sometimes it is useful to discard the nuclear-nuclear interaction in $H(\underset{\sim}{X})$. We define therefore the electronic part by

$$H^{el}(\underset{\sim}{X}) = H(\underset{\sim}{X}) - \frac{\alpha Z_A Z_B}{r} . \tag{3}$$

In fact $H(\underset{\sim}{X})$ is an operator-valued function of the nuclear configuration $\underset{\sim}{X}$. For every $\underset{\sim}{X}, H(\underset{\sim}{X},\alpha)$ is self-adjoint with domain of definition $\mathcal{D}$ independent of X. $\mathcal{D}$ equals the natural domain of the kinetic term H_o^{el}. Hence $\mathcal{D}$ is the set of square integrable electronic wave functions $\phi(\underset{\sim}{x}_1,\ldots,\underset{\sim}{x}_N)$ with square integrable first and second derivatives.

The electron-nuclei and electron-electron interaction belong to the class of dilation analytic potentials [18]. Most of the results of this section will be correct for all potentials in this class. For simplicity only the special case with Coulomb interactions will be considered here: for the more general case we refer to the literature [14]. The main object of this section is the analysis of regularity properties of eigenvalues and eigenfunctions of $H(\underset{\sim}{X})$ in the variable $\underset{\sim}{X}$.

There are two groups of results. The first one is based on the fact that $H(\underset{\sim}{X},\alpha)$ is unitarily equivalent to the family of operators $r^{-2}H(\hat{X},\alpha r)$ (see below). From this analyticity and convexity properties of eigenvalues in the distance $r > o$ will follow. The second group of results is founded on the differentiability of the resolvent $(H(\underset{\sim}{X}) - Z)^{-1}$ in $\underset{\sim}{X}$ up to second order. It provides second order differentiability of eigenvalues and also of eigenfunctions, if there is no degeneracy (see ref. [1] Theorem 6.8 and following discussion).

Regularity of eigenvalues and eigenfunctions of $H(X)$ is of utmost importance for the second step in the Born-Oppenheimer approximation, the harmonic approximation and for estimates of the error terms introduced by the approximation procedure.

Now we explain the unitary equivalence of $H(\underset{\sim}{X},\alpha)$ and $r^{-2}H(\hat{X},\alpha r)$, where $\hat{X}$ denotes the direction $\hat{X} = r^{-1}\underset{\sim}{X}$.

Let $D(r)$ be the unitary scaling operator by the length of $\underset{\sim}{X}$. Then from the definition of $H(\underset{\sim}{X},\alpha)$ follows

$$D(r)\, H(\underset{\sim}{X},\alpha)\, D(r)^{-1} = r^{-2}H(\hat{X},\alpha r) \,. \tag{4}$$

Hence the spectrum of $H(\underset{\sim}{X})$ is up to a factor identical with the spectrum of $H(\hat{X},\alpha r)$. The essential point of this construction is the following: the family of operators $H(\hat{X},\alpha r)$ is analytic in its second argument, i.e., for every ϕ in the domain D (independent of α) and every electronic wave function ψ the scalar product $\langle\psi,H(\underset{\sim}{X},\alpha)\phi\rangle$ is analytic in αr (in the terminology of Kato this is type A analyticity).

The proof of this fact is simple: Every $\phi \in \mathcal{D}$ can be written in the form $\phi = (H_o^{el} - Z)^{-1}\chi$ for Z in the resolvent set $\rho(H_o^{el})$. Hence the scalar product takes the form

$$\begin{aligned}\langle\psi,H(\hat{X},\alpha r)\phi\rangle = \langle\psi,H_o^{el}\phi\rangle + \alpha r \langle\psi, \{- \sum_{\ell} \frac{Z_A}{|\sigma_A\hat{X} - \underset{\sim}{X}_\ell|} \\ - \sum_{\ell} \frac{Z_B}{|\sigma_B\hat{X} + \underset{\sim}{X}_\ell|} + \sum_{i<k} \frac{1}{|\underset{\sim}{X}_i - \underset{\sim}{X}_k|} \\ + \frac{Z_A Z_B}{r} \}\, (H_o^{el} - Z)^{-1}\chi\rangle.\end{aligned} \tag{5}$$

Since $\phi \in \mathcal{D}$ the first term is well defined and αr-independent. The second term is well defined because the Coulomb potentials in $H(\underset{\sim}{X},\alpha)$ are small perturbations of the kinetic term, and so all the factors in (5) are well defined. This proves explicitly that all the scalar products are linear in αr, and in particular they are analytic.

Every discrete eigenvalue of a family of operators analytic in a parameter is analytic in this parameter (see ref. [1] page 386). Hence every discrete eigenvalue $E(r)$ - they are functions of the length r only, and not of the direction $\hat{X}$ as shown previously in §1.5 - is analytic in r provided $\underset{\sim}{X} \neq 0$ [14], [19].

Now we come to the concavity statements [20]. Due to Weyl's minimax principle (e.g.,[1], [2]) the sum of the first n eigenvalues F_n of $H(\hat{X},\alpha r)$ is given by

$$r^2 F_n(r) = r^2 \sum_{\ell=1}^{n} E_\ell(r) = \inf_{\{\phi_1,\ldots,\phi_n \in \mathcal{D},\ \text{orthonormal}\}} \sum_{\ell=1}^{n} \langle \phi_\ell, H(\hat{X},\alpha r)\phi_\ell \rangle \qquad (6)$$

when E_ℓ denotes the ℓ-th repeated discrete eigenvalue of $H(\underset{\sim}{X})$ below the essential spectrum. As we just have seen, the infimum runs over a family of linear functions in r, hence $r^2F_n(r)$ is concave in r (see Figure 4) on every connected subset of the positive real numbers R_+ for which all the discrete eigenvalues $E_\ell(r), \ell = 1,\ldots,n$ exist.

For diatomic molecules with only one electron, a related result on the groundstate energy $E_o^{el}(r)$ of $H^{el}(X)$ is known: $E_o^{el}(r)$ is monotone in r (see ref. [21] and [22]). Let us sketch the argument: due to the Hellmann-Feynman theorem (see below) one has

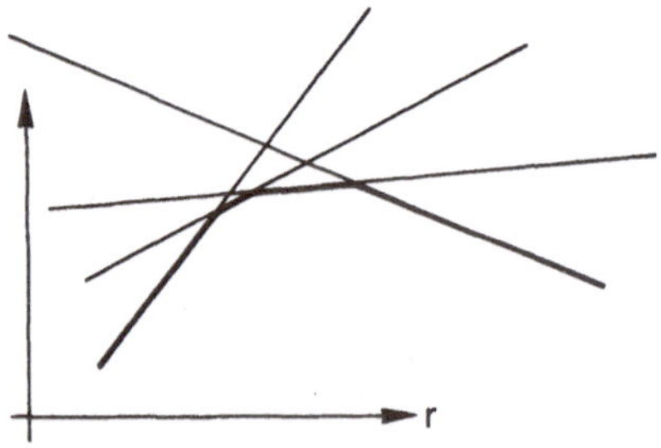

Fig. 4. Concavity of $r^2F_n(r)$; every straight line is characterized by a set of orthonormal wave functions $\{\phi_\ell, \ell = 1,\ldots,n\}$.

$$\frac{\partial E_o^{el}}{\partial r} = \langle \mathcal{C}_o(\underset{\sim}{X}), \frac{\partial}{\partial r} \frac{\alpha Z_A}{|\underset{\sim}{X} - \underset{\sim}{x}_1|} \mathcal{C}_o(\underset{\sim}{X}) \rangle$$

where $\mathcal{C}_o(\underset{\sim}{X})$ is the non degenerate ground state wave function of

$$\tfrac{1}{2}P^2 - \frac{\alpha Z_A}{|\underset{\sim}{X} - \underset{\sim}{x}_1|} - \frac{\alpha Z_B}{|\underset{\sim}{x}_1|} .$$

With the notation explained in Fig. 5 one gets

$$\frac{\partial E_o^{el}}{\partial r} = - \alpha Z_A \int d\underset{\sim}{x}_1 |\mathcal{C}_o(\underset{\sim}{X};\underset{\sim}{x}_1)|^2 \frac{\cos\theta}{|\underset{\sim}{X} - \underset{\sim}{x}_1|^2}$$

One expects the density $|\mathcal{C}_o(\underset{\sim}{X},\underset{\sim}{x}_1)|^2$ to be localized mainly on the B-side of the plane π through A perpendicular to $\underset{\sim}{X}$. Since the function $\cos\theta|\underset{\sim}{X} - \underset{\sim}{x}_1|^{-2}$ changes sign under reflection at π the integral is negative. The argument breaks down for more than one electron because the localization of the electron density gets complicated due to the delicate balance between electron-nuclear attraction and electron-electron repulsion. The mathematical formulation of the localization of $|\mathcal{C}_o(\underset{\sim}{X},\underset{\sim}{x}_1)|^2$ has been given in refs. [21] and [22] which basically use a correlation inequality, and a refined subharmonicity argument, respectively.

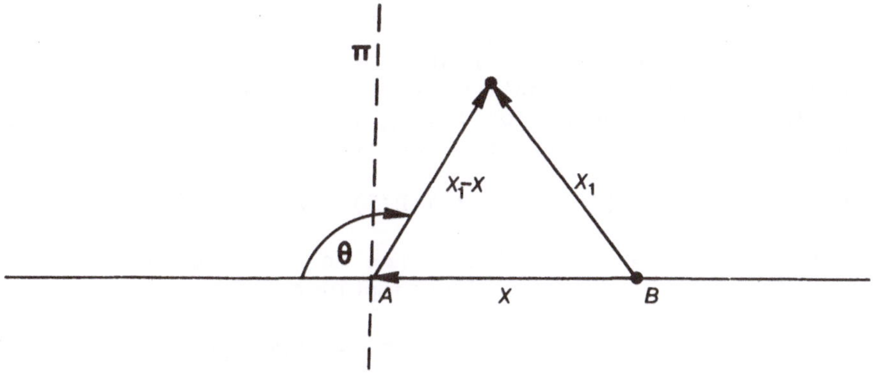

Fig. 5. The electronic wave function is concentrated mainly to the right of the plane π.

Before entering into the second group of results on the spectrum and eigenfunctions of $H(\underset{\sim}{X},\alpha)$ let us mention the infinitesimal version of the scaling relation (4), the virial theorem. The sum of the first n eigenvalues follows from equation (4),

$$F_n(r) = \frac{1}{r^2} \sum_{\ell=o}^{n} <\psi_\ell, H_o^{el} + \alpha r \{- \sum_\ell \frac{Z_A}{|\sigma_A \underset{\sim}{X} - \underset{\sim}{x}_\ell|}$$

$$- \sum_\ell \frac{Z_B}{|\sigma_B \underset{\sim}{X} - \underset{\sim}{x}_\ell|} + \sum_{i<k} \frac{1}{|\underset{\sim}{X}_i - \underset{\sim}{X}_k|} + \frac{Z_A Z_B}{r} \psi_\ell>$$

where ψ_ℓ is the ℓ-th repeated eigenfunction of $H(\hat{X},\alpha r)$. By differention with respect to r one gets the generalized virial equation

$$\frac{\partial F_n}{\partial r} = \frac{1}{r} (- 2F_n + \Sigma_\ell <V>_\ell) , \tag{7}$$

where we denoted by $<V>_\ell$ the expectation value of the potential part in $H(\underset{\sim}{X})$ in the ℓ-th repeated eigenfunction of $H(\underset{\sim}{X})$. Due to linearity of equation (7) the analogous equation holds for every single discrete eigenvalue

$$\frac{\partial E_n}{\partial r} = \frac{1}{r}(- 2E_n + <V>_n). \tag{8}$$

To start the second group of results we shall show continuity of the resolvent $(H(\underset{\sim}{X}) - Z)^{-1}$ in norm for every Z in the resolvent set $\rho(H(\underset{\sim}{X}))$. In such a case $H(\underset{\sim}{X})$ is called norm resolvent continuous (n-r-continuous).

For the proof it is convenient to remember that for our case of self-adjoint operators continuity of the resolvent holds for any $Z \in \rho(H(\underset{\sim}{X}))$ if it holds for one $Z_o \in \rho(H(\underset{\sim}{X}))$. This follows from the identity between the two resolvents

$$(H(\underset{\sim}{X}) - Z)^{-1} = \{1 - (Z - Z_o)(H(\underset{\sim}{X}) - Z_o)^{-1}\}^{-1}(H(\underset{\sim}{X}) - Z_o)^{-1}$$

and the fact that the number one is not in the spectrum of $(Z - Z_o)(H(\underset{\sim}{X}) - Z_o)^{-1}$ due to our assumptions.

For a Z_o with real part sufficiently negative the Neumann series

$$(H(\underset{\sim}{X}) - Z_o)^{-1} = (H_o^{el} - Z_o)^{-1} \sum_{n=o}^{\infty} [V(\underset{\sim}{X})(H_o^{el} - Z_o)^{-1}]^n$$

converges in norm. It suffices therefore to prove continuity of $V(\underset{\sim}{X})(H_o^{el} - Z)^{-1}$.

$V(X)$ is the sum of four terms, the interaction V_A of the electrons with nucleus A, the interaction V_B with nucleus B, the electron-electron interaction and the nuclear-nuclear interaction. We show that $V_A(\underset{\sim}{X})(H_o^{el} - Z_o)^{-1}$ is norm continuous: the analogous argument then holds for V_B. The last two terms are trivial. By definition $V_A(\underset{\sim}{X}) = T(\sigma_A \underset{\sim}{X}) V_A(0) T(\sigma_A \underset{\sim}{X})^{-1}$ where $T(.)$ denotes the translation operator acting on electronic wave functions (see §1). $T(a)$ is strongly continuous in a (this follows easily from the Lebesgue dominance theorem). Since translations commute with the kinetic operator H_o^{el} the following identity holds:

$$V_A(\underset{\sim}{X})(H_o^{el} - Z)^{-1} = T(\sigma_A \underset{\sim}{X}) V_A(0)(H_o^{el} - Z)^{-1} T(-\sigma_A \underset{\sim}{X}) \tag{9}$$

$$= T(\sigma_A \underset{\sim}{X}) V_A(0)(H_o^{el} - Z)^{-1}(T(-\sigma_A \underset{\sim}{X}) - 1)$$

$$+ T(\sigma_A \underset{\sim}{X}) V_A(0)(H_o^{el} - Z)^{-1}.$$

Consider the limit $\underset{\sim}{X} \to o$. The last term is a product of $T(\sigma_A \underset{\sim}{X}) \overset{s}{\to} 1$ with a compact operator $V_A(0)(H_o^{el} - Z_o)^{-1}$ since Coulomb potentials are H_o^{el}-compact perturbations. The product converges in norm to $V_A(0)H_o^{el} - Z_o)^{-1}$ due to standard results (see, e.g.,1], [2] or the appendix of [19]). By the same argument, the first term on the r.h.s. of (9) converges in norm to zero. This concludes the proof of norm resolvent continuity of $H(\underset{\sim}{X})$.

Now we want to show that $(H(\underset{\sim}{X}) - Z)^{-1}$ is differentiable in $\underset{\sim}{X}$ up to second order for every $Z \in \rho(H(\underset{\sim}{X}))$. The method of proof is already seen from the first derivative. It has to be shown that every term of the following identity makes sense

$$i\partial_\mu \, (H(\underset{\sim}{X}) - Z)^{-1} = - i(H(\underset{\sim}{X}) - Z)^{-1}\{\partial_\mu V(X)\}(H(X) - Z)^{-1} \tag{10}$$

By definition of $V(\underset{\sim}{X})$ an identity holds

$$i\partial_\mu V(\underset{\sim}{X}) \equiv [\sum_{\ell=1}^{N} P_{\ell,\mu}, \sigma_A V_A(\underset{\sim}{X}) - \sigma_B V_B(\underset{\sim}{X})] + i\partial_\mu \, \alpha \frac{Z_A Z_B}{r} . \tag{11}$$

The last term on the r.h.s. of (11) is just a number and can be disregarded for this discussion. The first term, the commutator of the total electronic momentum with $\sigma_A V_A - \sigma_B V_B$, has to be inserted in (10). This leads to

$$i\partial_\mu (H(\underset{\sim}{X}) - Z)^{-1} = - (H(\underset{\sim}{X}) - Z)^{-1} [\sum_\ell P_{\ell,\mu}, \sigma_A V_A(\underset{\sim}{X}) - \sigma_B V_B(\underset{\sim}{X})] (H(\underset{\sim}{X}) - Z)^{-1} + \ldots .$$

Next consider the second term in the commutator,

$$\{(H(\underset{\sim}{X}) - Z)^{-1}(\sigma_A V_A(\underset{\sim}{X}) - \sigma_B V_B(\underset{\sim}{X})\} \sum_{\ell=1}^{N} P_\ell \, (H(\underset{\sim}{X}) - Z)^{-1}.$$

The resolvent to the right maps the Hilbert space of wave functions onto the domain of definition of the kinetic operator H_o^{el}. On $\mathcal{D}$ the total electronic operator is well defined. As we have already seen several times, the term in the curly bracket is bounded: the first term in the commutator is treated analogously (see [14] for details). This proves n.r.-differentiability of $H(\underset{\sim}{X})$ for $\underset{\sim}{X} \neq 0$.

The formal reason for differentiability of the resolvent is the following: The linear growth in the electronic momentum is compensated by the quadratic term in the kinetic energy which is part of $H(\underset{\sim}{X})$ in the resolvent.

A more detailed analysis shows that differentiability up to second order is linked with the kinetic term in $H(\underset{\sim}{X})$ being of second order in the momenta.

The differentiability of $H(\underset{\sim}{X})$ implies a useful result on the total projectors defined below. Let E be a discrete eigenvalue of $H(\underset{\sim}{X}_o)$ and Γ a circle in the resolvent set $\rho(H(\underset{\sim}{X}_o)$ around E with no other point of the spectrum of $H(\underset{\sim}{X}_o)$ included. The total projector is defined by

$$P(\underset{\sim}{X}_o) = - \frac{1}{2\pi i} \int_\Gamma dz (H(\underset{\sim}{X}_o) - Z)^{-1}$$

and projects on all the eigenfunctions of $H(\underset{\sim}{X}_o)$ with eigenvalue E. It follows now from the above result that $P(\underset{\sim}{X})$ is twice differentiable in $\underset{\sim}{X}$ at $\underset{\sim}{X}_o$.

To get information on eigenfunctions let $\phi_1, \ldots, \phi_d$ be an orthonormal basis of the range of $P(\underset{\sim}{X}_o)$. Then the wave functions

$$\phi_\ell(\underset{\sim}{X}) = P(\underset{\sim}{X})\phi_\ell, \quad \ell = 1, \ldots, d$$

span the range of $P(\underset{\sim}{X})$ for $\underset{\sim}{X}$ in a neighbourhood of $\underset{\sim}{X}_o$. By construction $\phi_\ell(\underset{\sim}{X})$ is twice differentiable.

Notice that the argument of differentiability of $H(\underset{\sim}{X})$ extends to $\underset{\sim}{X} = o$ if we replace $H(\underset{\sim}{X})$ by $H^{el}(\underset{\sim}{X})$. The singularities in $H(\underset{\sim}{X})$

at $\underset{\sim}{X} = o$ originate only from the nucleus-nucleus interaction. Since however total projectors and eigenfunctions of $H(\underset{\sim}{X})$ are not affected by the change of a constant, the results about differentiability are correct for arbitrary nuclear configuration including the point $\underset{\sim}{X} = o$.

The results on regularity of eigenvalues and total projectors of $H(\underset{\sim}{X})$ have several consequences. As a first one we mention the Hellman-Feynman theorem. Let E be a discrete eigenvalue of $H(\underset{\sim}{X}_o)$ with total projector $P(\underset{\sim}{X}_o)$. Let furthermore $E_\ell(r)$, $\ell = 1,\ldots,d$, be the set of eigenvalues of $P(\underset{\sim}{X})\ H(\underset{\sim}{X})\ P(\underset{\sim}{X})$ for $\underset{\sim}{X} \sim \underset{\sim}{X}_o$. The Hellman-Feynman theorem asserts that the derivatives $\{\partial E\ /\partial r(r_o)\}$ are just the eigenvalues of the finite rank operator $P(\underset{\sim}{X}_o)\ \partial H/\partial r\ (\underset{\sim}{X}_o)\ P(\underset{\sim}{X}_o)$; here $\underset{\sim}{X}$ is a function of r and the angles Ω.

To prove this statement we introduce the family of adiabatic operators $U(r)$, $r \sim r_o$, with the properties

1) $U(r)$ is unitary, twice differentiable in r and $U(r_o) = 1$.

2) $U(r)\ P(X_o = r_o\hat{X}_o)U(r)^{-1} = P(r\hat{X}_o)$.

For the construction of such an operator we refer to chapter II.4.4 of ref. [1]. By construction the family of operators $P(r\hat{X}_o)H(r\hat{X}_o)P(r\hat{X}_o)$ has the same spectrum as $P(\underset{\sim}{X}_o)U^{-1}(r)H(r\hat{X}_o)U(r)P(\underset{\sim}{X}_o)$. The latter is essentially a family of self-adjoint matrices because $P(\underset{\sim}{X}_o)$ projects on a finite dimensional space. At this point we use a result on matrix valued functions which is in fact not difficult to prove (see ref. [1], chapter II.5.4). Let $A(t)$ be family of self-adjoint matrices differentiable in $r \sim r_o$. Then the eigenvalues are differentiable in r. The derivatives of the eigenvalues at r_o are given by the eigenvalues of $\partial A/\partial r(r_o)$. Application of this result to our case with $A(r) = P(\underset{\sim}{X}_o)U^{-1}(r)H(r\hat{X}_o)U(r)P(\underset{\sim}{X}_o)$ together with the identity

$$\frac{\partial}{\partial r} P(\underset{\sim}{X}_o)U^{-1}(r)H(r\hat{X}_o)U(r)P(\underset{\sim}{X}_o)\Big|_{r=r_o} = P(\underset{\sim}{X}_o)\ \frac{\partial H}{\partial r}(r_o\hat{X}_o)P(\underset{\sim}{X}_o) \tag{12}$$

proves the theorem.

The next application of the regularity properties concerns the slope of the electronic eigenvalues $E^{el}(r)$ at $r = o$. We want to prove that $\partial E/\partial r(r)$ for every discrete eigenvalue of $H^{el}(X)$ vanishes at zero [23]. For non degenerate eigenvalues, as for instance the groundstate energy, this is a simple consequence of differentiability and rotational covariance (see §1.5). The general case is slightly more complicated.

Instead of $H^{el}(r\hat{X}_o)$ let us look at the family of operators $A(r) = T(ar\hat{X}_o)\, H^{el}(r\hat{X}_o) T(ar\hat{X}_o)^{-1}$ where T denotes the unitary translation operator in electronic wave functions and a is a number to be chosen later. Of course A(r) and $H^{el}(r\hat{X}_o)$ have the same spectrum. Due to the Hellman-Feynman theorem the derivative of an electronic energy curve $E^{el}(r)$ at r = o is given by an eigenvalue of $P(o)\, \partial A/\partial r(o)\, P(o)$ where P(o) is an appropriately chosen total projector of $A(o) = H^{el}(o)$. A short computation yields

$$\frac{\partial A}{\partial r}(o) = - \alpha \sum_{\ell=1}^{N} \frac{Z_A(\sigma_A + a) - Z_B(\sigma_B - a)}{|x_\ell|^3} (\hat{X}_o, x_\ell).$$

If we now choose $a = (Z_A + Z_B)^{-1} (Z_B\sigma_B - Z_A\sigma_A)$ this expression vanishes. Hence the slope of all electronic energy curves at r = o vanishes.

As a last application of the results on regularity we prove the following identity

$$\text{Trace}\ P_1(\underset{\sim}{X})\frac{\partial}{\partial X_\mu} P_2(\underset{\sim}{X}) = \frac{\text{Trace}\ P_1(\underset{\sim}{X})\ \frac{\partial H}{\partial X_\mu}\ P_2(\underset{\sim}{X})}{E_1(r) - E_2(r)} . \tag{13}$$

$P_i(\underset{\sim}{X})$ denotes the total projector for discrete eigenvalue $E_i(\underset{\sim}{X})$, i = 1,2. The identity follows by differentiation of o = Trace $P_1(\underset{\sim}{X})\ H(\underset{\sim}{X})\ P_2(\underset{\sim}{X})$ and o = Trace $P_1(\underset{\sim}{X})\ P_2(\underset{\sim}{X})$. The formula (13) is useful for discussing the validity of the Born-Oppenheimer approximation: notice that it holds also in the limit $|\underset{\sim}{X}| \to o$.

After this discussion of the clamped nuclei approximation we are ready to look into the concepts of adiabatic wave functions and nuclear Schrödinger operators.

In the following we shall make use of the fact that the clamped nuclei Schrödinger operator is bounded below. In fact $H(\underset{\sim}{X})$ is bounded below by a constant independent of $\underset{\sim}{X}$ because the Coulomb interaction of electrons with nuclei are H_o^{el} bounded with relative bound independent of $\underset{\sim}{X}$. A bound based on such an argument is very crude. There are much better bounds available [24]; they will however not be used in the following.

2.3 Adiabatic Wave Functions and the Nuclear Schrödinger Operator

In this chapter we reduce the eigenvalue problem for the molecular Schrödinger operator to an eigenvalue problem for the nuclei only. An effective potential reflects the electronic degrees of freedom. As it will be shown in the next chapter this method gives rise to an asymptotic expansion of the energy eigenvalues in terms of $\kappa = \sqrt[4]{\mu^{-1}}$ to order κ^6.

Now we describe the construction of the nuclear Schrödinger operator. The first step is the definition of the space of adiabatic wave functions. Let $E(r)$ be a real valued smooth function of r with $E(r) \in \rho(H(\underset{\sim}{X}))$ so that $\sigma(H(\underset{\sim}{X}))$ below $E(r)$ is only discrete. Define the projector $P(\underset{\sim}{X})$ on the space of electronic wave functions as the spectral projectors of $H(\underset{\sim}{X})$ for the interval $(-\infty, E(r))$,

$$P(\underset{\sim}{X}) = -\frac{1}{2\pi i}\int_{\Gamma} dZ\ (H(\underset{\sim}{X}) - Z)^{-1}$$

where Γ is a closed curve about the real axis to the left of $E(r)$.

By construction $P(\underset{\sim}{X})$ has finite dimensional range and is twice differentiable in $\underset{\sim}{X}$. It gives rise to a projector on the "big space" of molecular wave functions by direct integration, $P = \int d\underset{\sim}{X}\ P(\underset{\sim}{X})$, i.e. for every test function $\phi(\underset{\sim}{X},\underset{\sim}{X}_1,\ldots,\underset{\sim}{X}_N)$ with compact support $(P\phi)(\underset{\sim}{X},\underset{\sim}{X}_1,\ldots\underset{\sim}{X}_N) = P(\underset{\sim}{X})\phi(\underset{\sim}{X},\underset{\sim}{X}_1,\ldots\underset{\sim}{X}_N)$ where $\phi(\underset{\sim}{X}\ldots)$ is interpreted as an electronic wave function only. Obviously P is uniformly bounded by one on this dense set of functions, hence it is a well defined operator on arbitrary molecular wave functions. Notice that the dimension n of range $P(\underset{\sim}{X})$ is $\underset{\sim}{X}$-independent.

In the following an important role will be attributed to the adiabatic Hamiltonian defined by

$$H^{AD} = P\ H\ P \tag{15}$$

with domain $\mathcal{D}(H^{AD}) = P\ \mathcal{D}(H) \otimes Q\ ,\ P + Q = 1$.

Recall that $\mathcal{D}(H)$ is just the Sobolev space $\mathcal{H}^2$ of twice differentiable molecular wave functions. In order that definition (15) is sensible it is necessary that $P\ \mathcal{D}(H) \subset \mathcal{D}(H)$. Let $\phi \in \mathcal{D}(H)$ and $\phi(\underset{\sim}{X})$ be the electronic wave function for fixed $\underset{\sim}{X}$. Now $P(\underset{\sim}{X})\ \phi(\underset{\sim}{X})$ is twice differentiable in $\underset{\sim}{X}$ due to the differentiability of $P(\underset{\sim}{X})$ and $\phi(\underset{\sim}{X})$. The derivatives of $P(\underset{\sim}{X})$ up to order two are uniformly bounded in $\underset{\sim}{X}$ [14]. Hence the inclusion holds.

The operator H^{AD} is self-adjoint. This is a consequence of the following:

<u>Proposition 1</u>:

Let A be self-adjoint with domain $\mathcal{D}(A) \subset \mathcal{H}$ and P an orthogonal projector so that PAQ + QAP is infinitely A-small in the sense of Kato [1,2] and $P\mathcal{D}(A) \subset \mathcal{D}(A)$. Then PAP is self-adjoint on $P\mathcal{D}(A) \otimes Q\mathcal{H}$.

Proof:

PAQ + QAP is symmetric on $\mathcal{D}(A)$. Hence A − (PAQ + QAP) is self-adjoint on $\mathcal{D}(A)$ and equals PAP + QAQ. This implies the statement of the proposition.

In order to apply this result we just have to verify that PHQ + QHP is H bounded with relative bound zero: let ϕ be an arbitrary element in $\mathcal{D}(H)$. Then

$$||PHQ\phi||^2 = \frac{\kappa^4}{2} \int d\underset{\sim}{X} ||P(\underset{\sim}{X})\ p^2 Q(\underset{\sim}{X})\phi(\underset{\sim}{X})||^2$$

$$\leq \frac{\kappa^4}{2} \int d\underset{\sim}{X} ||P(\underset{\sim}{X})(\nabla_{\underset{\sim}{X}} Q(\underset{\sim}{X}))\nabla_{\underset{\sim}{X}}\phi(\underset{\sim}{X})||^2 + $$

$$+ ||P(\underset{\sim}{X})(-\Delta_{\underset{\sim}{X}} Q(\underset{\sim}{X}))\phi(\underset{\sim}{X})||^2$$

We have neglected in H the isotopic term $\rho(\Sigma p_\ell)^2$. However this does not change the argument because $P(\Sigma p_\ell)^2 Q$ is bounded. Since $\nabla_X Q(\underset{\sim}{X})$ and $\Delta_X Q(\underset{\sim}{X})$ are uniformly bounded in $\underset{\sim}{X}$ as operators on electronic wave functions (see section 2) we end up with

$$||PHQ\phi||^2 \leq \lambda\kappa^4[||\nabla_{\underset{\sim}{X}}\phi||^2 + ||\phi||^2], \quad \lambda \text{ constant.}$$

But ∇_X is H bounded with relative bound zero. An analogous argument proves the bound for the second term QHP. Hence the lemma is applicable and H^{AD} self-adjoint.

Next we shall prove that the operator $U(E) = PHQ\ (H - E)^{-1}\ QHP$, $E < \mathcal{E} = \inf_r \mathcal{E}(r)$ is H^{AD} bounded with relative bound zero. $(H - E)^{-1}$ denotes the restriction of the operator to the subspace $Q\mathcal{H}$, the orthogonal complement of $P\mathcal{H}$. By the previous result QHQ is self-adjoint on $Q\mathcal{D} \otimes \mathcal{H}^{AD}$. The spectrum $\sigma(QHQ)$ is strictly bounded below by $\inf \mathcal{E}(r)$. This follows from the inequality

$$\langle\phi, QHQ\phi\rangle \geq \int d\underset{\sim}{X} \langle\phi(\underset{\sim}{X}), Q(\underset{\sim}{X})H(\underset{\sim}{X})Q(\underset{\sim}{X})\phi(\underset{\sim}{X})\rangle$$

$$\geq \mathcal{E}\ ||\phi||^2$$

$$\mathcal{E} = \inf \mathcal{E}(r),$$

which holds for every $\phi \in \mathcal{D}(H)$. Hence $E \notin \sigma(QHQ)$ and the range of $(H - E)^{-1}$ equals $Q\mathcal{D}(H)$. On those functions PHQ is well defined and therefore $PHQ(H - E)^{-1}Q$ bounded. Furthermore QHP is H^{AD} bounded with relative bound zero by the same argument as we used in the proof of the last proposition.

Now we are ready to formulate the following equivalence statement:

Proposition 2:

For every $E < \mathcal{E}$ there is a linear isomorphism I between the kernel of $H - E$ and the kernel of $H^{AD} - U(E) - E$ given by

$$I : \phi \rightarrow P\phi$$

$$I^{-1} : P\chi \rightarrow P\chi - Q(H - E)^{-1} QHP\chi. \tag{16}$$

Proof:

Let ϕ be an eigenfunction of H with discrete eigenvalue $E < \mathcal{E}$; obviously we get the two equations

$$\begin{aligned} H^{AD}\phi + PHQ\phi &= EP\phi \\ QHP\phi + QHQ\phi &= EQ\phi. \end{aligned} \tag{17}$$

The second equation can be used to compute $Q\phi = - Q(H - E)^{-1}QHP\phi$ and if this expression is inserted in (17) it follows immediately that $P\phi$ belongs to the kernel of $H^{AD} - U(E) - E$. The kernel of the linear map I is zero because otherwise there would be a normalized eigenvector $\phi = Q\phi$ with $\langle\phi, H\phi\rangle < \mathcal{E}$. But this contradicts the inequality $\langle\phi(\underset{\sim}{X}), H(\underset{\sim}{X})\phi(\underset{\sim}{X})\rangle \geq \mathcal{E}\, \langle\phi(\underset{\sim}{X}), \phi(\underset{\sim}{X})\rangle$, which holds by assumption. Finally the map I is onto if we can show that the explicit formula for (16) for the inverse is correct. To prove that, consider a vector $P\chi$ in the kernel of $H^{AD} - U(E) - E$ and define ϕ by

$$\phi = P\chi - Q(H - E)^{-1}QHP\chi.$$

Then we have to compute $(H - E)\phi$ and prove that this expression vanishes. We will show

$$0 = P(H - E)\phi = Q(H - E)\phi \text{ , but}$$

$$\begin{aligned} P(H - E)\phi &= P(H - E)P\chi - P(H - E)Q(H - E)^{-1}QHP\chi \\ &= (H^{AD} - E)P\chi - U(E)P\chi = 0. \end{aligned}$$

Furthermore we get

$$Q(H - E)\phi = Q(H - E)P\chi - Q(H - E)Q(H - E)^{-1}QHP\chi$$

and this expression vanishes. Hence the mapping I is surjective. Notice that all vectors are always in just the domains and spaces they have to be in in order that everything is well defined.

Due to the aforementioned equivalence the Schrödinger equation for the molecular system can be replaced by the equation

$$(H^{AD} - U(E)\chi = E\chi , \quad \chi \in P\,\mathcal{D} \subset \mathcal{H}^{AD}.$$

Notice that U is a non local E-dependent potential. The great importance of looking at the eigenvalue problem for H in this way is that U(E) is of order κ^8 and positive,

$$U(E) = \frac{\kappa^8}{4} P(p^2 + \rho(\Sigma p_\ell)^2)Q(H - E)^{-1}Q(p^2 + (\Sigma p_\ell)^2)P.$$

The positivity of U(E) is related to the fact that the spectrum of QHQ is bounded below by $\mathcal{E}$. Furthermore U(E) is monotonically increasing with E and will in general diverge for E approaching the lower limit of the spectrum σ(QHQ).

For the following discussion it is useful to identify the subspace of adiabatic wave functions $\mathcal{H}^{AD}$ with a direct sum of n copies of nuclear wave function spaces.

The unitary identification map is constructed as follows:

Let $\{e_i|\ i = 1,\dots,n\}$ be an orthonormal basis of range P(0). Due to the adiabatic theorem (see [1] pages 369, 375) there is a twice differentiable family of unitary operators $U_{\hat{X}_o}(r)$ so that $\{e_i(r\hat{X}_o) = U_{\hat{X}_o}(r)e_i\}$ is an orthonormal basis of range $P(rX_o)$. Let U be the representation of rotations on electronic wave functions. By the covariance relations (see §1.5) the following is an orthonormal basis of range $P(\underset{\sim}{X})$ $\{e_i(\underset{\sim}{X} = r\hat{X}) = U(O_{\hat{X}_o}(\hat{X}))U_{\hat{X}_o}(r)e_i$: $O_{\hat{X}_o}(\hat{X})$ is the standard family of rotations mapping $\hat{X}_o$ into $\hat{X}\}$. Since $e_i(r\hat{X}_o)$ is in the domain of angular momenta operators, $e_i(\underset{\sim}{X})$ is twice differentiable in $\underset{\sim}{X} \neq o$.

The mapping J is now defined by

$$\phi = P\,\phi \to J\phi = \{<e_i(\underset{\sim}{X}),\phi(\underset{\sim}{X})> \ ; \ i = 1,\dots,n\}. \tag{18}$$

By construction J is unitary and maps $P\,\mathcal{D}(H)$ onto a direct sum of Sobolev spaces

$$J: P\,\mathcal{D}(H) \to \mathcal{H}^2 \otimes \dots \otimes \mathcal{H}^2.$$

This follows from the definition (18) of J.

The adiabatic Schrödinger operator under conjugation with J transforms into

$$H^{AD} \rightarrow JH^{AD}J^{-1}$$
$$= \tfrac{1}{2}\kappa^4 (P^2 \otimes 1 + T_{eff}) + V_{eff} ,$$

where we have introduced the effective kinetic term T_{eff} and the effective potential V_{eff} defined by

$$T_{eff,k\ell} = \langle e_k(\underset{\sim}{X}),[-\Delta_{\underset{\sim}{X}} + \rho(\Sigma p_\ell)^2]e_\ell(\underset{\sim}{X})\rangle - 2\langle e_k(\underset{\sim}{X}),i\underset{\sim}{\nabla}_{\underset{\sim}{X}} e_\ell(X)\rangle P$$

$$V_{eff,k\ell} = E_\ell(r = |\underset{\sim}{X}|)\delta_{k\ell}.$$

The domain of definition $\mathcal{D}(JH^{AD}J^{-1}) = \mathcal{H}^2 \otimes \ldots \otimes \mathcal{H}^2$. Notice that the second term in T_{eff} is only nontrivial for $k \neq \ell$.

By conjugation the non-local potential U(E) turns into an operator which is bounded below by zero and above by $\kappa^8 D_2$,

$$0 \leq JU(E)J^{-1} \leq \kappa^8 D_2$$

where D_2 is a second order differential operator defined by

$$D_2(E) = \frac{1}{2(\mathcal{E} - E)} \{-\alpha^2\Delta_{\underset{\sim}{X}} + \beta^2\} \otimes 1$$

$$\alpha^2 = \sup_{\underset{\sim}{X}} 4 \sum_{i=o}^{n} ||\underset{\sim}{\nabla} e_i(\underset{\sim}{X})||^2 \tag{19}$$

$$\beta^2 = \sup_{\underset{\sim}{X}} \sum_{i=o}^{n} (||\Delta_{\underset{\sim}{X}} e_i(\underset{\sim}{X})|| + \rho||(\Sigma p_\ell)^2 e_i(\underset{\sim}{X})||)^2.$$

The second term on the right hand side of (19) is finite because the square of the total electronic momentum can be estimated by $H^{el}(\underset{\sim}{X})$. Furthermore, due to the regularity properties mentioned previously, the constants α and β are finite. D_2 is of course self-adjoint on the same domain as $JH^{AD}J^{-1}$.

We have now transformed the problem of solving the molecular Schrödinger equation for small κ into a problem of n nuclear wave functions only, n = dim $P(\underset{\sim}{X})$. With the notation $H^{NAD} = H^{AD} - \kappa^8 D_2$ for the Schrödinger operator including the non-adiabatic term $\kappa^8 D_2$ we get the upper and lower bound

$$H^{NAD}(E) \leq H^{AD} - U(E) \leq H^{AD}. \tag{20}$$

The non-local operator $H^{AD} - U(E)$ can be estimated by two Schrödinger operators with local potentials.

The inequalities (20) allow the control of the spectrum of H below $\mathcal{E}$ up to order κ^6. Notice that $H^{NAD}(E)$ and $H^{AD} - U(E)$ are analytic families of operators in E, monotonically decreasing and concave. From this follows that all the eigenvalues of those operators are decreasing and the sum of the first m are concave.

The spectrum of H is geometrically given by the intersections of the spectrum of $H^{AD} - U(E)$ as a function of E with the straight line $f(E) = E$ (see Fig. 6).

Hence there is for every pair of discrete eigenvalues (E^{AD}, E^{NAD}), $E^{AD} < \mathcal{E}$, $E^{NAD} \in \sigma(H^{NAD}(E^{AD}))$ an eigenvalue $E \in \sigma_d(H)$ with $E^{NAD} \leq E \leq E^{AD}$. In the following section it will be shown how this can be used to get asymptotic series of E in κ up to order κ^6 from the corresponding expansions for E^{NAD} and E^{AD}.

To close this section we shall mention the well-known results of Brattsev [25] and Epstein [26]. Consider the Born-Oppenheimer Schrödinger operator for dim $P(\underset{\sim}{X}) = n = 1$ defined by

$$H^{BO} = \tfrac{1}{2}\kappa^4 P^2 + V_{eff}. \tag{21}$$

Then the groundstate $E_o^{BO} = \inf \sigma(H^{BO})$ is a lower bound for the molecular ground state E_o. This follows from the following chain of inequalities.

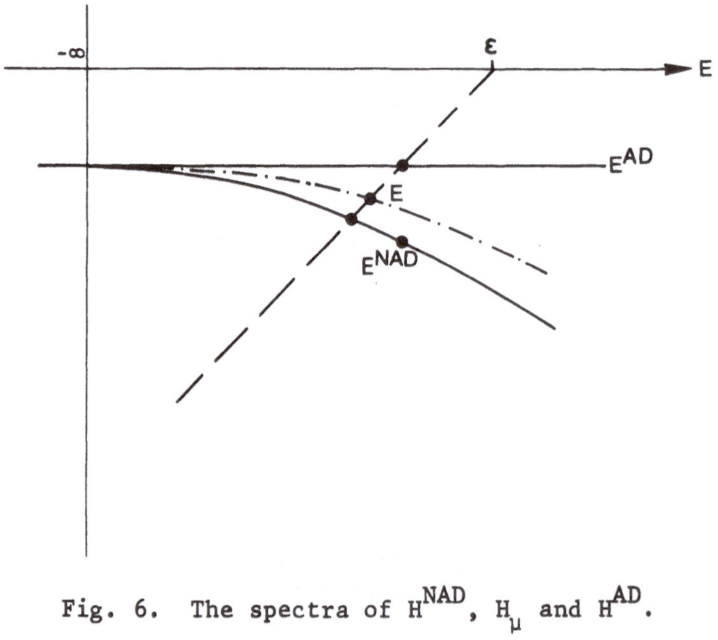

Fig. 6. The spectra of H^{NAD}, H_μ and H^{AD}.

$$\langle\phi,H\phi\rangle \geqq \langle\phi,(\tfrac{1}{2}\kappa^4 p^2 + H^{el})\phi\rangle \geqq \langle\phi,(\tfrac{1}{2}\kappa^4 p^2 + V_{eff})\phi\rangle$$

$$\geqq E^{BO} \langle\phi,\phi\rangle$$

The first inequality follows from positivity of the isotopic term in H. The others from Weyl's minimax principle.

The groundstate E_o^{AD} of H^{AD} gives an upper bound on the correct groundstate energy E_o. This follows from positivity of the non-local potential U(E) (20). Hence we get $E_o^{BO} \leq E_o \leq E_o^{AD}$.

It would of course be most useful to get a generalization of this lower bound to arbitrary eigenvalues of the molecular Hamiltonian.

3. ASYMPTOTIC EXPANSIONS FOR MOLECULAR ENERGIES

To conclude this mathematical analysis of the Born-Oppenheimer approximation we now have to study asymptotic expansions for the molecular levels for small values of κ.

We will only discuss here the eigenvalues of H^{AD}; for them the $\kappa = o$ limit corresponds to the classical limit problem as we will see later. Our analysis can be easily extended to the levels of $H^{NAD}(E)$ for any fixed E.

As is well-known the classical limit problem falls into the domain of singular perturbation theory since the small parameter appears as a coefficient of the higest order differential operator. We present here some methods which nevertheless are based on standard perturbation theory and which apply to the various situations appearing in diatomic problems. To our knowledge this approach to the classical limit is new; we will only give here a brief description, details can be found in [27], [28].

The adiabatic levels of the first electronic curve

Let us start with the simplest situation where the dimension n of P(x) is one. Then after reduction to the sector with angular momentum ℓ the operator $JH^{AD} J^{-1}$ (see ch. 2) is transformed into the Sturm-Liouville operator

$$h = \frac{\kappa^4}{2} \left(- \frac{d^2}{dr^2} + \frac{\ell(\ell + 1)}{r^2} + t^{(o)}(r)\right) + V(r), \qquad (1)$$

where $t^{(o)}$ is the momentum independent part of the effective kinetic energy term T_{eff} and thus contains the isotopic term and the term coupling the nuclei and electrons; $V(r)$ is the effective potential which in this case is simply the electronic groundstate $E_o(r)$. The operator h acts onto $L^2(R^+)$ and has a Dirichlet boundary condition at $r = o$.

Let us assume first that E_o has, as in Fig. 3, a single minimum V_o at r_o. Since one expects that near the classical limit $\kappa = o$, the lowest eigenstates of h will be strongly localized near the minimum r_o, one has to look with a "microscope" around this point with a resolution power which, according to Born and Oppenheimer, should be κ. In mathematical terms we perform the canonical transformation

$$(D(\kappa)\phi)(r) = \kappa^{\frac{1}{2}}\phi(r_o + \kappa(r - r_o)).$$

Since $D(\kappa)$ is a unitary transformation from $L^2(R^+)$ to $L^2(-\kappa^{-1}r_o,\infty)$ the point spectra of h and $D(\kappa)\, h\, D^{-1}(\kappa)$ are identical. It is convenient to write

$$D(\kappa)\, hD^{-1}(\kappa) = V_o + \kappa^2\, \tilde{h}(\kappa) \tag{2}$$

with

$$\tilde{h}(\kappa) = -\frac{d^2}{dr^2} + W(\kappa,r) + \kappa^2 \frac{\ell(\ell + 1)}{|r_o + \kappa(r - r_o)|^2} + \kappa^2 t^{(o)}(\kappa,r),$$

where

$$W(\kappa,r) = \kappa^{-2}[V(r_o + \kappa(r - r_o)) - V_o] \tag{3}$$

$$t(\kappa,r) = t(r_o + \kappa(r - r_o)).$$

Eigenvalues of h are related to those of $\tilde{h}$ by

$$E_n = V_o + \kappa^2\, \mu_n(\kappa)$$

according to (2). Notice that as $\kappa \to o$ $W(\kappa,r)$ converges pointwise to $r^2V''(r_o)/2$, and tends to $\kappa^{-2}[V(+\infty) - V_o]$ as $r \to +\infty$. In fact the convergence to an harmonic oscillator potential is uniform on intervals of the form $(-\alpha\kappa^{-1},\alpha\kappa^{-1})$ for some $\alpha > o$. This is sufficient to prove the following facts ([27]).

1) $\tilde{h}(\kappa)$ converges in norm resolvent sense to the harmonic oscillator Hamiltonian on $L^2(R)$:

$$h_o = -\frac{d^2}{dr^2} + r^2 \frac{V''(r_o)}{2}$$

2) As a consequence (see [1]) eigenvalues of h_o are stable in the sense that for each eigenvalue $\mu_n = (2n + 1)\sqrt{V''(r_o)/2}$ of h_o there exist one and only one eigenvalue $\mu_n(\kappa)$ of $\tilde{h}(\kappa)$ such that $\lim_{\kappa\to o} \mu_n(\kappa) = \mu_n$.

3) Let us consider the normalized real eigenvector of $\tilde{h}(\kappa)$ associated with this eigenvalue:

$$\tilde{h}(\kappa)\phi_n(\kappa) = \mu_n(\kappa)\phi_n(\kappa)$$

Then $||\phi_n(\kappa) - \phi_n|| \to o$ as $\kappa \to o$ where ϕ_n is the nth Hermite function.

These results are enough to obtain the so called harmonic approximation for eigenvalues of h. In order to obtain higher order approximations a regularization procedure based on a uniform WKB-like estimate is necessary:

$$\sup_r \left| e^{-a\int_{-\kappa^{-1}r_o}^{\infty} (W(\kappa,r))^{\frac{1}{2}}} \phi_n(\kappa,r)\right| < C \tag{4}$$

for some positive constants a and c uniformly in κ with $o < \kappa < \kappa_o$. Then the formal Rayleigh-Schrödinger perturbation series are asymptotic to $\mu_n(\kappa)$ to any order; here of course the analyticity properties of electronic quantities (discussed in §2) are essential to justify the expansions of $W(\kappa,r)$ and $t(\kappa,r)$ in powers of κ. For example, using (2), one obtains to order 4 for the nth eigenvalue $E_n(\kappa)$ of h

$$E_n(\kappa) = V_o + \kappa^2(2n + 1)\sqrt{\frac{V''(r_o)}{2}} + \kappa^4\{\frac{\ell(\ell + 1)}{r_o^2} + \delta_n\} + o(\kappa^6) \tag{5}$$

where $\delta_n = t^o(r_o) + \beta_n$, with β_n the anharmonic correction.

One can also show that nuclear wave functions have asymptotic expansions; in particular the normalized real eigenvector of h, ψ_n, with eigenvalue $E_n(\kappa)$, satisfies

$$\lim_{\kappa\to o} ||\psi_n(\kappa,r) - \kappa^{-\frac{1}{2}}r^{-1}\phi_n(\kappa^{-1}(r - r_o))|| = o .$$

Multiminima problems

These results obviously are insufficient if the potential V has many minima. This happens in some diatomic molecules with $Z_A >> Z_B$ such as HgH. In fact one expects that if the potential V has say, two minima V_1 and V_2 at r_1 and r_2 with $V_2 > V_1$ then there will not only be eigenstates of h localized around r_1 with energies converging to V_1 and having asymptotic expansions of the form (5), but such states should also exist for the second minimum at r_2 provided this minimum is not embedded in the continuum, that is $V_2 \leq \lim_{r\to\infty} V(r)$. This can be shown using the following arguments; let us consider a point b, $r_1 < b < r_2$ and the operators h^D and h^N obtained from h (see (1)) by imposing a Dirichlet or Neumann boundary condition at b. In the sense of ordering of self-adjoint operators ([1], [27]) one has then

$$h_N \leq h \leq h_D,$$

which implies an ordering of their respective eigenvalues (Dirichlet-Neumann bracketing)

$$\mu_m^N \leq \mu_m \leq \mu_m^D , \tag{6}$$

where m labels the m-th eigenvalue, counted from the lowest one and taking multiplicities into account. The reason for the introduction of these differential operators is that they both split into direct sums

$$h^{N,D} = h_1^{N,D} \otimes h_2^{N,D}$$

where $h_1^{N,D}$ (resp. $h_2^{N,D}$) is the differential operator (1) with Neumann or Dirichlet boundary conditions at b on $L^2(o,b)$ (resp. $L^2(b,\infty)$). Furthermore, the above analysis of the single minimum situation applies as well to $h_i^{N,D}$. So h^D and h^N have two sets of eigenvalues $\mu_{n,i}(\kappa)$, n = 0,1,..., i = 1,2, having asymptotic expansions in κ near $\kappa = o$ of the form (5), and these expansions are the same for Neumann or Dirichlet boundary conditions at b. In other words, the differences $\mu_{n,i}^D(\kappa) - \mu_{n,i}^N(\kappa)$ have zero asymptotic expansion. In fact using (4) one obtains using Green's formula

$$\langle h^D\phi_{n,i}^D , \phi_{n,i}^N\rangle - \langle\phi_{n,i}^D , h^N\phi_{n,i}^N\rangle$$

$$= (\phi_{n,i}^D)'(b^+)\phi_{n,i}^N(b^+) - (\phi_{n,i}^D)'(b^-)\phi_{n,i}^N(b^-)$$

that for some a > o

$$\mu_{n,i}^D - \mu_{n,i}^N = o(e^{-a\kappa^{-2}}) .$$

This information together with (5) is not however sufficient to prove our expectations, since the n-th levels of h^D and h^N do not necessarily have the same asymptotic expansions. This ambiguity can be resolved using the following mathematical facts. For some positive constant c large enough, the differences $(h^{D,N} + c)^{-1} - (h + c)^{-1}$ are self-adjoint rank-one operators. As proved in [27] or as a consequence of the Weinstein-Aronzjan formula, rank-one perturbations of a self-adjoint operator shift the eigenvalues in such a way that the perturbed ones intertwine the unperturbed ones, whereas eigenvalues of multiplicity d with $d > 1$ remain eigenvalues of the perturbed operator with multiplicity d or $d - 1$. The case $d = 2$ can occur for h^D or h^N due to the structure (7). However, one can exclude the possibility that the eigenvalues $\mu(\kappa)$ have $d = 2$ by a non-crossing rule for these energy levels as functions of κ; this is due to the fact that Sturm-Liouville operators have non-degenerate spectra under the regularity properties of $V(r)$ and $t^o(r)$ which have been described in §2. Finally we obtain rigorous prescriptions for the location of energy levels of h with respect to those of h^N and h^D. An illustration is given in Fig. 7 where we have shown in particular what happens around crossing points of Neumann and Dirichlet levels. It is interesting to notice the spectacular "repulsion" effect for the levels $E(\kappa)$ in the neighbourhood of such points showing very explicitly the non-crossing mechanism, which in turn implies jumps of $E(\kappa)$ between Neumann-Dirichlet brackets of different minima. Summing up, one can now assert that the energy levels of the molecule are, up to exponentially small corrections $o(e^{-a\kappa^{-2}})$ due to the tunnel effect, those of two independent minima and that these levels have asymptotic expansions of the usual Born-Oppenheimer type.

An additional technique is necessary when V_2 is embedded in the continuum. Then one expects that this second minimum will produce not bound-states but resonances. Actually for small values of κ one can show [28] that there exist such resonances; their energies have a real part which has an asymptotic expansion of the Born-Oppenheimer type whereas their imaginary part is exponentially small $o(e^{-a\kappa^{-2}})$. The basic tool needed for this extension of our bound-state analysis is the so-called "exterior scaling method" [28] introduced by B. Simon.

Simultaneous consideration of many electronic curves

Let us finally consider the situation where $\dim P(\underset{\sim}{X}) = n > 1$ so that the adiabatic Hamiltonian is now a matrix-Schrödinger operator. For this more general problem we also have to consider the momentum-dependent part of T_{eff} which gives, after angular momentum reduction, an additional term $t^{(1)}(r)\frac{d}{dr}$ with only

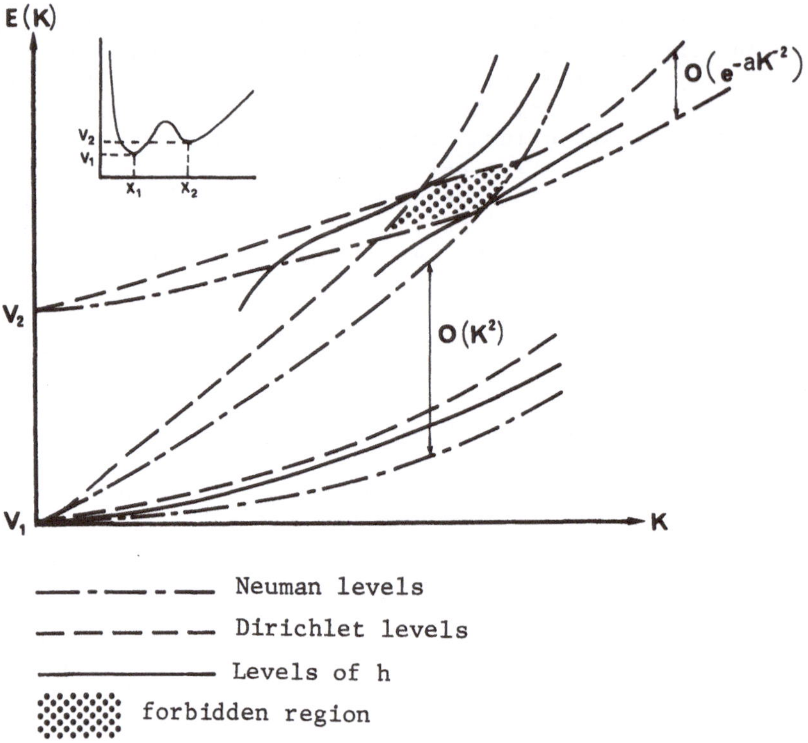

Fig. 7. Neumann-Dirichlet bracketing for multiminima potentials

non-diagonal components in the Schrodinger matrix. Such terms will give corrections of order κ^6 to the energy levels of the individual curves, whereas the non-diagonal components of $t^{(o)}(r)$ will give corrections of order κ^8. This can be seen in two different ways, either by treating the non-diagonal terms as a perturbation of the diagonal part which is a direct sum of Hamiltonians of type (1) for which we know existence of perturbative expansions for both energies and eigenvectors; or by generalizing the scaling arguments used for single levels and introducing the transformation

$$(D(\kappa)\phi)_j(r) = \kappa^{\frac{1}{2}}\phi_j(r_j + \kappa(r - r_j)), \quad j = 0,1, \ldots,$$

where r_j is the position of the minimum V_j of $E_j(r)$, assuming for a moment single minima not embedded in the continuum. Then H^{AD} is unitarily equivalent in a given angular momentum sector to $V + \kappa^2 h(\kappa)$ with

$$V_{ij} = V_j \, \delta_{ij} \quad \text{and} \quad h_{jj}(\kappa) = h_j(\kappa),$$

and where $h_j(\kappa)$ has the form (3) for $i \neq j$

$$h_{ij}(\kappa) = [\kappa t_{ij}^{(1)}(r_i + \kappa(r - r_i)) \frac{d}{dr}$$

$$+ \kappa^2 t_{ij}^{(o)}(r_i + \kappa(r - r_i))]T(1 + \kappa(r_i - r_j)) ,$$

and T(a) is the translation operator by a.

From this it is easy to obtain by a formal expansion in κ the first correction terms due to the non-diagonal parts: for potential curves having many minima, or minima in the continuum, the same modifications as for the single curve should be made, namely Neumann-Dirichlet bracketing with exterior complex scaling in the case of resonances.

ACKNOWLEDGEMENTS

Research supported by the Délégation Générale à la Recherche Scientifique et Technique and the Humboldt Gesellschaft.

REFERENCES

1. T. Kato, Perturbation Theory of Linear Operators, Berlin-Heidelberg-New York, Springer Verlag, (1966).
2. M. Reed, B. Simon, Methods of Modern Mathematical Physics, Vols. 1-4, Academic Press, (1970-79).
3. F. Treves, Basic linear partial differential equations, Academic Press (1975).
4. P. Deift, W. Hunziker, B. Simon, E. Vock, Pointwise bounds on eigenfunctions and wave packets in N-body quantum systems IV, Commun. Math. Phys. 64, 1 (1978).
5. W. Hunziker, Helv. Phys. Acta 39, 451 (1966), CMP 11, 19-55 (1968).
6. C. Van Winter, H.J. Brascamp, The n-body problem with spin orbit and Coulomb interaction, Comm. Math. Phys. 11, 19 (1968).
7. G.M. Zhislin, Investigation of the spectrum of the Schrödinger operator for a many particle system, Tr.Mosk.Mat.Obs.9, 81-128 (1960).
8. B. Simon, Geometric methods in multiparticle quantums systems, Commun. Math. Phys. 55, 259-274 (1977).
9. B. Simon, On the infinitude or finiteness of the number of bound-states of an N-body quantum systems, Helv.Phys. Acta 43, 607 (1970).
10. V.N. Effimov, Energy levels from resonant two-body forces in a three-body system, Phys. Lett. 33B, 563 (1970).
11. G.M. Zhislin, A.G. Sigalov, On the spectrum of the energy operator for atoms with fixed nucleus, Amer. Math. Soc. Tr. Serv. Vol.91, 297-310.

12. E. Balslev, Spectral theory of Schrodinger operators of many-body systems with permutation and rotation symmetries, Ann. Phys. 73, 49 (1972).
13. J.M. Combes, T.O. Connor, Finiteness of the number of bound states for negative ions, in preparation.
14. J.M. Combes, R. Seiler, Regularity and asymptotic properties of the discrete spectrum of electronic Hamiltonians, Int. J. Quant. Chem. 14, 213 (1978).
15. J. Morgan, B. Simon, Behaviour of molecular potential energy curves for large nuclear separations, Preprint, to appear in Int. J. Quant. Chem. (1980).
16. M. Born and R. Oppenheimer, Zur Quantentheorie der molekeln, Annalen der Physik 84, 457 (1927).
17. M. Born and W. Heisenberg; Zur Quantentheorie der molekeln, Annalen der Physik 74, 1 (1924).
18. E.U. Condon, A theory of intensity distributions in band systems, Phys. Rev. 28, 1182 (1926).
19. P. Aventini, R. Seiler, On the electronic spectrum of the diatomic molecular ion, Commun. Math. Phys. 41, 119 (1975).
20. H. Narnhofer and W. Thirring, Convexity properties for Coulomb systems, Acta Physica Austriaca 41, 281 (1975). E.R. Davidson, J. Chem. Phys. 36, 3527 (1962).
21. E. Lieb and B. Simon, Monotonicity of the electronic contribution to the Born-Oppenheimer energy, J. Phys. BL 537 (1978).
22. M. and T. Hoffmann-Osterhoff; to appear in Journal of Physics A (1979).
23. R.A. Buckingham, Trans. Faraday Soc. 54, 453 (1958).
24. P. Hertel, E. Lieb, W. Thirring, Lower bound to the energy of complex atoms; H. Grosse, P. Hertel, W. Thirring, Lower bounds to the energy levels of atomic and molecular systems, Acta Phys. Austr. 49, 89 (1978).
25. V.F. Brattsev, Dokl. Akad. Nauk SSSR 160, 570 (1965).
26. S. Epstein, Ground-state energy of a molecule in the adiabatic approximation, J. Chem. Phys. 44, 836 (1966) and 44, 4062 (1966).
27. J.M. Combes, P. Duclos, R. Seiler, Dirichlet-Neumann decoupling and Krein's formula I, to appear.
28. J.M. Combes, P. Duclos, R. Seiler, Dirichlet-Neumann decoupling and Krein's formula II, to appear.

THEORY OF ELECTRON MOLECULE COLLISIONS

P. G. Burke

Queen's University, Belfast BT7 1NN, Northern Ireland
and
Science Research Council, Daresbury Laboratory
Warrington WA4 4AD, England

1. INTRODUCTION

In the last few years considerable progress has been made in the theory of electron molecule collisions. This is firstly because of an ever increasing need for electron molecule collision cross sections in many applications in astrophysics, in the physics of the upper atmosphere, in laboratory plasma physics and in laser physics. Secondly because recent experiments measuring electron molecule collision cross sections are providing very detailed information to test theory, and finally because the ready availability of powerful computing facilities is making it possible to obtain numerical results from new theories to an extent impossible a few years ago.

We divide these lectures into seven sections. After this introductory section in which we summarize very briefly the scope of the lectures we consider in section 2 the collision of an electron with a molecule in which the nuclei are constrained to fixed locations in space. This approach called the fixed-nuclei approximation has gained favour in recent years, since firstly it gives a considerable simplification in the equations which are to be solved, and secondly it has been found to give reliable results for many transitions when the scattering amplitude is appropriately averaged over the nuclear coordinates. We then consider in section 3 how the theory must be modified to take account of the rotational motion. In this, we lay particular emphasis on the frame transformation theory method in which different representations are used in different regions of configuration space and the resultant solutions matched on the boundaries of these regions. This approach enables the simplicity of the fixed-nuclei approximation to be largely retained even at low energies when the time of the collision is of the order of or less than the rotational period of the molecule. In section 4 we consider new developments involving the use of L^2 integrable trial functions in collisions. Important amongst these are the R-matrix and T-matrix methods. The reasons for these developments are two fold. Firstly they avoid in a natural way single centre expansion approaches which are very slowly convergent for all but the simplest molecules and secondly they enable standard molecular structure codes to be simply adapted for use in the collision. In section 5 we consider the extensions of the theory to take account of vibrational excitation and dissociative attachment. In this we take particular note of the important role resonances play in such processes and of the recent use of the R-matrix method in a completely ab-initio calculation for e - N_2 scattering. Next in section 6 we consider the special problems raised by the long range potential seen by an electron incident on a polar molecule. We show that, as well as giving rise to cross sections which are strongly peaked at small angles, the potential can bind the electron into negative ion states. We also consider recent experiments and their theoretical interpretations which show strong threshold peaks. Finally in the last section we consider methods which have been developed to describe the scattering of electrons by molecules at intermediate and high energies. Important amongst these methods is the continuum multiple scattering method which has been used to study the scattering of electrons over a very broad energy range and has shown that shape resonances can give rise to enhanced vibrational excitation cross sections at intermediate energies.

2. FIXED-NUCLEI APPROXIMATION

In this section we consider the approximation in which the nuclei are held fixed during the collision. Thus the Hamiltonian contains only electronic terms. We will use a frame of reference to describe the collision which is rigidly attached to the molecule. This is called the molecular frame or body-fixed frame. We will see that this approximation gives good results for collision cross sections when the time of the collision is short compared with typical rotation and vibration times.

The Schrödinger equation describing the collision of an electron with an N electron molecule can be written using atomic units and assuming that relativistic terms in the Hamiltonian can be neglected (Burke and Sinfailam, 1970, Chung and Lin, 1978, Lane, 1980).

$$(H_{e\ell} - E)\psi = (-\frac{1}{2}\nabla^2 + H_T + V - E)\psi = 0 \qquad (1)$$

where $-\frac{1}{2}\nabla^2$ is the kinetic energy operator for the scattered electron, H_T is the target electronic Hamiltonian and V is the potential interaction between the electron and the target. In the case of a diatomic molecule this is defined by

$$V(\underline{r}_1..\underline{r}_N, \underline{r} ; R) = \sum_{i=1}^{N} \frac{1}{|\underline{r} - \underline{r}_i|} - \frac{Z_A}{|\underline{r} - \underline{r}_A|} - \frac{Z_B}{|\underline{r} - \underline{r}_B|} \tag{2}$$

where Z_A and Z_B are the nuclear charges located at $\underline{r}_A$ and $\underline{r}_B$, R is the internuclear distance and we have denoted the target electron coordinates by $\underline{r}_1,..,\underline{r}_N$ and the scattered electron coordinate by $\underline{r}$.

We now expand the total collision wavefunction in the form

$$\psi = \mathscr{A} \sum_{i=1}^{n} \phi_i(1,2...N)\, F_i(N+1) + \sum_{i=1}^{m} \xi_i(1,2...N+1)a_i \tag{3}$$

where the ϕ_i are target eigenstates and possibly pseudostates included to represent the polarizability of the target by the incident electron, F_i are functions representing the scattered electron and ξ_i are short range correlation functions multiplied by coefficients a_i. These correlation functions are needed if we assume that the F_i are orthogonal to the orbitals in ϕ_i and can also be included to help represent electron-electron correlation effects. The whole wavefunction depends parametrically on the internuclear distances, and is antisymmetrized by $\mathscr{A}$ with respect to exchange of the space and spin coordinates of any pair of electrons.

Assuming that we know the wavefunctions ϕ_i and ξ_i (e.g., usually an SCF wavefunction is used for the target ground state) we can obtain coupled equations for the unknown functions F_i and for the coefficients a_i by substituting ψ into the Kohn-variational principle.[1] This leads to the following coupled equations

$$\begin{aligned} &\langle\phi_i \mid (H_{e\ell} - E)\psi\rangle = 0 \qquad i = 1,2...n \\ &\langle\xi_i \mid (H_{e\ell} - E)\psi\rangle = 0 \qquad i = 1,2...m \end{aligned} \tag{4}$$

where the integration in the first n equations is carried out over the space and spin coordinates of the N target electrons, and the integration in the last m equations is carried out over the space and spin coordinates of all N + 1 electrons.

In order to proceed further we need to make some assumptions about the form of the wavefunction for the scattered electron. One approach which has been widely used is to make a single-centre expansion of this function. While this can lead to convergence difficulties near the nuclei, it is appropriate at large distances where the electron molecule interaction is weak and nearly central.

For simplicity we restrict our consideration to a linear molecule. Then due to axial symmetry the total electronic angular momentum projected onto the molecular axis $\Lambda = \lambda_i + m_\ell$ is a good quantum number, where λ_i is the projection of the angular momentum of the target and m_ℓ is that for the scattered electron. Further because the Hamiltonian is spin-independent we can couple the spin of the target with the spin of the scattered electron to form an eigenstate of the total spin angular momentum operator $\underline{S}^2$ and its z component S_z. Accordingly it is appropriate to adopt a coupling scheme which is diagonal in $\Lambda S M_S$ and eq.(3) can be rewritten as

$$\psi^{\Lambda S M_S} = \sum_{i=1}^{n} \sum_{\ell=0}^{\infty} \Phi_{i\ell}^{\Lambda S M_S}(1,..,N,\hat{\underline{r}}_{N+1}\ \sigma_{N+1})\ r_{N+1}^{-1}\ f_{i\ell}^{\Lambda S}(r_{N+1})$$

$$+ \sum_{i=1}^{m} \xi_i^{\Lambda S M_S}(1,2...,N+1)\ a_i^{\Lambda S} \tag{5}$$

The channel functions in this equation are defined by

$$\Phi_{i\ell}^{\Lambda S M_S}(1,..,N,\ \hat{\underline{r}}_{N+1},\ \sigma_{N+1})$$

$$= \sum_{M_{S_i} m_{s_i}} \phi_i^{\lambda_i S_i M_{S_i}}(1,2...,N)\ Y_\ell^{m_\ell}(\hat{\underline{r}}_{N+1}) \tag{6}$$

$$\times\ \eta_{1/2}^{m_{s_i}}(N+1)\ (S_i M_{S_i}\ \tfrac{1}{2}\ m_{s_i} | S_i\ \tfrac{1}{2}\ S\ M_S)$$

where the $Y_\ell^{m_\ell}$ are spherical harmonics, the $\eta_{1/2}^{m_{s_i}}$ are electron spin functions and $(S_i\ M_{S_i}\ \frac{1}{2}\ m_{s_i} | S_i\ \frac{1}{2}\ S\ M_S)$ are Clebsch Gordan coefficients where we have chosen the z axis to lie along the internuclear axis. In the case of a non-linear molecule the spherical harmonics must be replaced by more general angular functions belonging to the appropriate irreducible representation of the symmetry group of the molecule (Burke et al, 1972).

Projecting the Schrödinger equation $(H_{e\ell} - E)\psi^{\Lambda SM_S} = 0$ onto the channel functions $\Phi_{i\ell}^{\Lambda SM_S}$ and onto the symmetry adapted correlation functions $\xi_i^{\Lambda SM_S}$ gives the following infinite set of coupled integro-differential equations

$$\left(\frac{d^2}{dr^2} - \frac{\ell(\ell+1)}{r^2} + k_i^2\right) f_{i\ell}^{\Lambda S}(r) = 2 \sum_{i'=1}^{n} \sum_{\ell'=0}^{\infty} (V_{i\ell,i'\ell'}^{\Lambda S} + W_{i\ell,i'\ell'}^{\Lambda S} + X_{i\ell,i'\ell'}^{\Lambda S})\, f_{i'\ell'}^{\Lambda S}(r) \tag{7}$$

where for notational simplicity we have not explicitly shown the dependence of these equations on the internuclear coordinate R. Also if the molecule is homonuclear the summation over ℓ and ℓ' runs only over even or odd values.

In these equations the wave numbers k_i are defined by

$$k_i^2 = 2(E - E_i) \tag{8}$$

where E_i is the electronic energy of the ith target state at the given value of R. The direct potential matrix $V_{i\ell,i'\ell'}^{\Lambda S}$ is defined by

$$V_{i\ell,i'\ell'}^{\Lambda S}(r_{N+1}) = \langle \Phi_{i\ell}^{\Lambda SM_S}(1,\ldots,N,\hat{\underline{r}}_{N+1}\sigma_{N+1})|V(\underline{r}_1\underline{r}_2\ldots\underline{r}_N\underline{r}_{N+1};R)|\Phi_{i'\ell'}^{\Lambda SM_S}(1,2\ldots N,\hat{\underline{r}}_{N+1}\sigma_{N+1})\rangle \tag{9}$$

where the integration in this equation is taken over all (N+1) electron coordinates except the radial coordinate of the (N+1)th electron. The exchange potential matrix $W_{i\ell,i'\ell'}^{\Lambda S}$ is defined by

$$W^{\Lambda S}_{i\ell,i'\ell'} f^{\Lambda S}_{i'\ell'}(r_{N+1}) = - N \langle \Phi^{\Lambda SM_S}_{i\ell}(1,\ldots,N,\hat{\underline{r}}_{N+1}\sigma_{N+1})$$

$$|(H_{e\ell} - E)| \quad \Phi^{\Lambda SM_S}_{i'\ell'}(1,\ldots,N-1,N+1,\hat{\underline{r}}_N\sigma_N) f^{\Lambda S}_{i'\ell'}(r_N)\rangle \tag{10}$$

where the coordinates of the scattered electron are exchanged with the coordinates of one of the target electrons. Finally the correlation potential $X^{\Lambda S}_{i\ell,i'\ell'}$ arises from the elimination of the m equations involving the correlation functions in eq. (5). This can be most simply done following Feshbach (1958, 1962) by diagonalizing $H_{e\ell}$ in the space of the correlation functions giving a set of eigenfunctions $\xi^{\Lambda SM_S}_{\lambda}$ and corresponding eigenvalues ε_λ. We then find that

$$X^{\Lambda S}_{i\ell,i'\ell'} f^{\Lambda S}_{i'\ell'}(r_{N+1}) = - \sum_{\lambda} \langle \Phi^{\Lambda SM_S}_{i\ell}(1,\ldots,N,\hat{\underline{r}}_{N+1}\sigma_{N+1})$$

$$|(H_{e\ell}-E)|\xi^{\Lambda SM_S}_{\lambda}\rangle (\varepsilon_\lambda - E)^{-1} \langle \xi^{\Lambda SM_S}_{\lambda}|(H_{e\ell}-E)|$$

$$\Phi^{\Lambda SM_S}_{i'\ell'}(1,\ldots,N;\hat{\underline{r}}_{N+1}\sigma_{N+1}) f^{\Lambda S}_{i'\ell'}(r_{N+1})$$

$$- N \Phi^{\Lambda SM_S}_{i'\ell'}(1\ldots, N-1, N+1, \hat{\underline{r}}_N\sigma_N) f^{\Lambda S}_{i'\ell'}(r_N)\rangle \tag{11}$$

Although considerable simplifications are possible in the expressions for the potentials given in eqs. (9), (10) and (11) if we assume that the basis orbitals used in eq. (5) are all orthogonal we will not persue this aspect here. However it is convenient at this stage to make some remarks about their general behaviour and representation in recent calculations.

We define the static potential for an electron incident on the molecule in its ground state as

$$V_S(\underline{r},R) = \langle \phi_1(1,\ldots,N)|V(\underline{r}_1\ldots\underline{r}_N,\underline{r};R)|\phi_1(1,\ldots,N)\rangle \tag{12}$$

where the integration is carried out over the space and spin coordinates of the N target electrons. This potential can be expanded about the centre of gravity of the molecule giving

$$V_S(\underline{r},R) = \sum_\lambda V_\lambda(r)\, P_\lambda(\cos\theta)$$

$$\underset{r\to\infty}{\sim} -\frac{\mu}{r^2} P_1(\cos\theta) - \frac{Q}{r^3} P_2(\cos\theta) + \ldots \tag{13}$$

where μ and Q are the dipole and quadrupole moments of the molecule and where θ is the polar angle of the vector $\underline{r}$. If we substitute eq.(13) into eq.(9) we obtain in the case of a molecule in a Σ state

$$V^{\Lambda S}_{1\ell,1\ell'}(r) = \left(\frac{2\ell' + 1}{2\ell + 1}\right)^{1/2} \sum_\lambda (\ell'0\lambda0|\ell'\lambda\ell0)$$

$$(\ell' m_\ell \lambda 0|\ell'\lambda\ell m_\ell)\, V_\lambda(r) \tag{14}$$

for the direct potential matrix elements coupling the ground state channels. In a similar way we can obtain corresponding expressions for the direct potential due to any state of the molecule. It follows from this analysis that outside the molecule, that is where the charge distribution of the molecule is effectively zero, the direct potential has long range r^{-2}, r^{-3} and higher components arising from the permanent dipole and quadrupole moments of the particular electronic states.

In addition to these potentials, which are diagonal in i, i.e., i = i' in eq.(9), we also have off-diagonal potentials. The most important of these couple electronic states between which optical dipole transitions are allowed. The asymptotic form of the corresponding potentials is

$$V^{\Lambda S}_{i\ell,i'\ell'}(r) \underset{r\to\infty}{\sim} a^{\Lambda S}_{i\ell,i'\ell'}\, r^{-2}, \text{ optically allowed transitions} \tag{15}$$

where the coefficient $a^{\Lambda S}_{i\ell,i'\ell'}$ is proportional to the oscillator strength of the transition from state i to state i'. Castillejo et al. (1960) showed that the result of coupling such electronic states to the ground electronic state is to give in second-order an effective r^{-4} polarization potential seen by an electron incident on this state. Of course the full polarization potential of the ground state can only be obtained by coupling all allowed target eigenstates including continuum states. Although this infinity of states cannot be included explicitly in expansion (3) or (5) Damburg and Karule (1967) showed that their effect, at least so far as the long-range r^{-4} component of the polarization potential is concerned, can be represented by suitably chosen pseudostates. In the case of a linear molecule one pseudostate is needed to represent α_{11} and one to represent $\alpha_{\perp}$, the parallel and perpendicular components of the polarizability. It follows that the static potential defined by eq.(13) is augmented by a polarization potential

$$V_p(\underline{r},R) \underset{r\to\infty}{\sim} -\frac{\alpha_0}{2r^4} - \frac{\alpha_2}{2r^4} P_2(\cos\theta) \tag{16}$$

where α_0 and α_2 are defined by

$$\begin{aligned} \alpha_0 &= \frac{1}{3}(\alpha_{11} + 2\alpha_{\perp}) \\ \alpha_2 &= \frac{2}{3}(\alpha_{11} - \alpha_{\perp}) \end{aligned} \tag{17}$$

The use of pseudostates to represent the polarizability has been widely used in electron atom scattering (e.g., Le Dourneuf et al, 1977) but in the case of molecules pseudostates have so far only been used for e - H_2 scattering (Schneider, 1978). The more usual procedure has been to represent the polarization by a parametrized potential of the form

$$V_p(\underline{r};R) = -\left(\frac{\alpha_0}{2r^4} + \frac{\alpha_2}{2r^4} P_2(\cos\theta)\right)C(r) \tag{18}$$

where C(r) is a cut-off factor

$$C(r) = 1 - \exp\left(-\left(\frac{r}{r_c}\right)^6\right) \tag{19}$$

and where r_c is a parameter which is usually adjusted to give the best agreement with experiment.

Turning now to the exchange potential we see from eq.(10) that it involves an integration over the radial functions $f_{i'\ell'}(r)$. We can write it in the form

$$W^{\Lambda S}_{i\ell,i'\ell'} f^{\Lambda S}_{i'\ell'}(r) = \sum_{\alpha} \int_0^{\infty} K^{\Lambda S\alpha}_{i\ell,i'\ell'}(r,r') f^{\Lambda S}_{i'\ell'}(r')dr' \tag{20}$$

where the exchange kernels $K^{\Lambda S\alpha}_{i\ell,i'\ell'}(r,r')$ can be expressed in terms of the radial orbitals $u^{\beta}_{\ell}(r)$ arising from the single-centre expansion of the occupied target molecular orbitals labelled β. A typical term in eq.(20) involving the electron-electron repulsion term in $H_{e\ell}$ is

$$y_\lambda(u^{\beta_1}_{\ell_1} f;r)\, u^{\beta_2}_{\ell_2}(r) = \left(\int_0^\infty u^{\beta_1}_{\ell_1}(r')\, f(r')\, \frac{r^\lambda_<}{r^{\lambda+1}_>}\, dr'\right) u^{\beta_2}_{\ell_2}(r) \tag{21}$$

where $r_<$ and $r_>$ are the smaller and greater of r and r'. We see immediately that the exchange potential is non-local, which causes difficulty in the solution of eq.(7), and it has the range of the occupied target orbitals which vanish exponentially. The last property means that for any choice of target eigenstates and pseudostates we can choose a radius r = a beyond which the exchange potential is negligible and the direct potential can be represented by an expansion.

$$V^{\Lambda S}_{i\ell,i'\ell'}(r) = \sum_{\lambda=1}^{\lambda_{max}} a^{\Lambda S\lambda}_{i\ell,i'\ell'}\, r^{-\lambda-1} \qquad r > a \tag{22}$$

We will see later that this result has important implications for methods of solution of eq.(7); in particular it is one of the main motivations for the development of the R-Matrix method.

The difficulties caused by the inclusion of the non-local exchange potential in the solution of eqs.(7) is beyond the scope of the present lectures. Here we just mention the main methods of including exchange. These can be classified as follows:

(i) Representation of exchange by a local exchange potential

(ii) Representation of exchange by orthogonalization to bound orbitals

(iii) Exact numerical solution of the integro-differential equations by iterative or by non-iterative methods.

Considerable effort has been given to the first of these since the coupled integro-differential equations are then reduced to much simpler coupled differential equations. One of the most successful and widely used approaches has been the free-electron gas model introduced by Hara (1967) (see also discussions by Riley and Truhlar (1975), Baille and Darewych (1977) and Morrison and Collins (1978)). Basically the method is based on the picture introduced by Slater (1960) in which the total wavefunction is assumed to be made up of plane waves which are antisymmetrized in accordance with the Pauli exclusion principle and the exchange energy obtained by summing all states up to the Fermi level. An alternative local exchange potential based on a semiclassical approximation has also been derived by Riley and Truhlar (1975).

While these local exchange potentials have had considerable success they do not ensure that the continuum orbital representing the scattered electron is orthogonal to the bound orbitals which are fully occupied which is a requirement of the exact solution. The imposition of this constraint, which gives an inhomogeneous term in the coupled differential equations, was considered by Burke and Chandra (1972). It was shown in that paper and in subsequent papers (e.g., Chandra and Temkin, 1976a,b) that this constraint did indeed represent a substantial part of the exchange interaction.

Recently Collins et al. (1979) have considered the possibility of including both a local exchange potential and imposing an orthogonality constraint. Their results for the e-LiH Σ eigenphase sum is shown in figure 1, where they are compared with exact static exchange calculations obtained by solving eqs.(7) by an iterative method. A low energy resonance which appears when only the local exchange potential is included is removed by orthogonalization and the resultant phase shifts including orthogonalization and local exchange are in excellent agreement with the exact results.

Both the above approaches for including exchange are of course approximations to the exact numerical solution of eq.(7) which, at least for simple molecules is the ultimate aim of our theory. In recent years considerable progress has been made in the solution of these integro-differential equations (see review by Buckley and Burke, 1979) and accurate static exchange results have now been obtained for H_2 and N_2 (e.g., Collins et al. 1978). However the problems of the exact inclusion of exchange within the single-centre framework considered in this section are formidable, and it is probable that in the future the emphasis will move towards including exchange in a multi-centre formalism similar to those discussed in a later section of these lectures.

Finally we note that the correlation potential defined by eq.(11) is also non-local and short ranged. In this case the range is determined by the range of the correlation functions. Thus the radius $r = a$ defined in eq.(22) can be generalized so that both the exchange potential and the correlation potential are negligible beyond $r = a$. So far this type of potential has not been included in the single-centre framework considered in

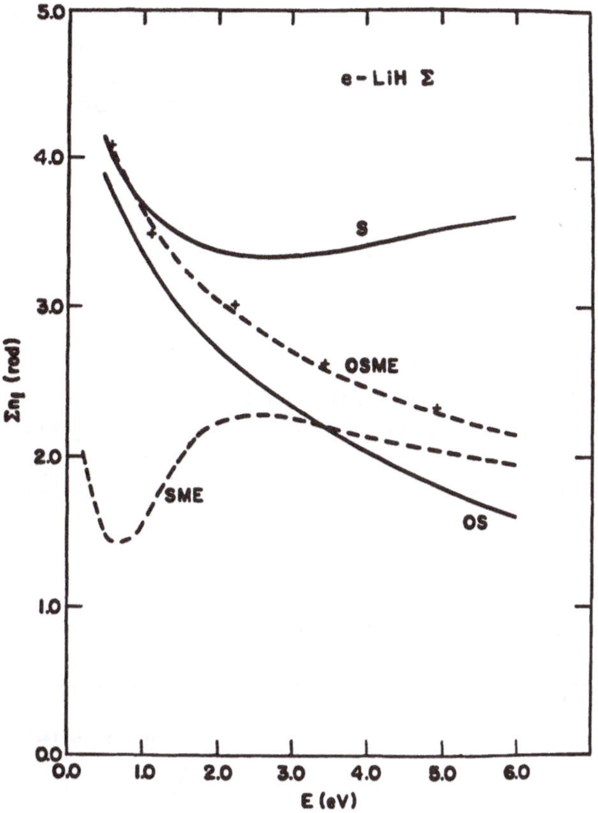

Figure 1 The Σ eigenphase sum for e-LiH scattering. Curves S, static; OS, orthogonalized static; SME, static with model exchange; OSME, orthogonalized static with model exchange; crosses, exact static exchange calculations. (from Collins et al. 1979).

this section.

The S-matrix and cross section can be obtained from the asymptotic form of the solutions of eq.(7) for each Λ and S. We look for regular solutions satisfying the asymptotic boundary conditions

$$f^{\Lambda S}_{i\ell,i'\ell'}(r) \underset{r\to\infty}{\sim} k_i^{-1/2}(\sin(k_i r - \frac{1}{2}\ell\pi)\delta_{ii'}\delta_{\ell\ell'} + \cos(k_i r - \frac{1}{2}\ell\pi)\, K^{\Lambda S}_{i\ell,i'\ell'}),\ k_i^2 > 0 \text{ (open channels)}$$

$$\underset{r\to\infty}{\sim} 0,\ k_i^2 < 0 \text{ (closed channels)}, \qquad (23)$$

where the second pair of subscripts $i'\ell'$ distinguish the n_a linearly independent solutions where n_a is the number of open channels. We can combine these real solutions to form solutions with ingoing and outgoing wave boundary conditions defining the S-matrix. We find in matrix notation that

$$\underline{S}^{\Lambda S} = \frac{\underline{1} + i\,\underline{K}^{\Lambda S}}{\underline{1} - i\,\underline{K}^{\Lambda S}} \tag{24}$$

where $\underline{S}^{\Lambda S}$ is a unitary symmetric $n_a \times n_a$ matrix for each Λ and S. The total cross section obtained after averaging over all incident beam directions relative to the molecular axis can be obtained following a similar analysis to Blatt and Biedenharn (1952). We find that

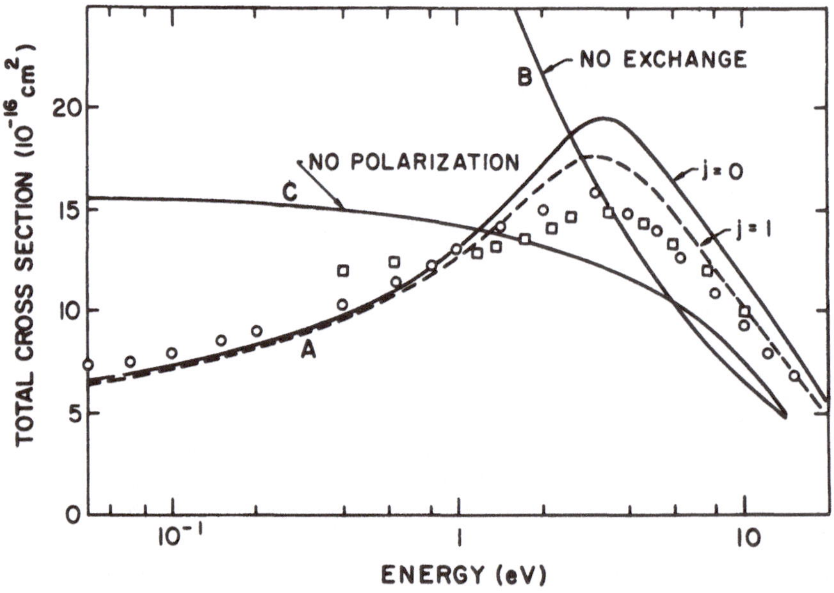

Figure 2 The total cross section for e-H_2 scattering. Theoretical curve from Henry and Lane (1969): curves A, exchange and polarization (—— , j=0, and ---, j=1); B, no exchange; C, no polarization. Circles, measurements of Golden et al. (1966); squares, measurements of Ramsauer and Kollath (1930). (from Golden et al. 1971).

$$\sigma(i \rightarrow i') = \frac{\pi}{k_i^2} \sum_S \frac{(2S+1)}{2(2S_i+1)} \sum_{\Lambda \ell \ell'} |S^{\Lambda S}_{i\ell,i'\ell'} - \delta_{ii'}\delta_{\ell\ell'}|^2 \qquad (25)$$

For each value of Λ and S the coupled integro-differential equations (7) must be solved, retaining enough values of ℓ and ℓ' to give cross sections which have converged to the required accuracy. Then sufficient values of Λ and S must be considered to obtain the total cross section.

We conclude this section by considering a few representative results. First as an indication of the importance of including both exchange and polarization effects in low energy electron molecule scattering we show in figure 2 a calculation for e - H_2 scattering carried out by Henry and Lane (1969). They solved equations analogous to eq.(7) but formulated in the laboratory frame as discussed in the next section. It is clear

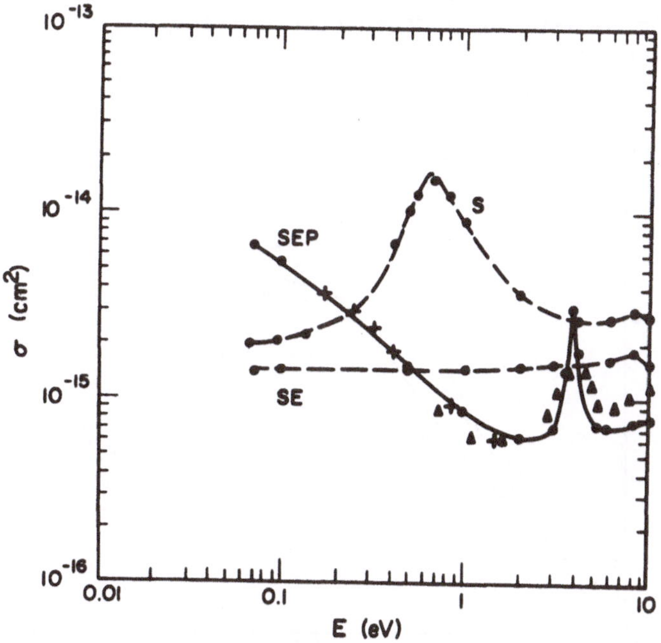

Figure 3 The total cross section for e-CO_2 scattering. Theoretical curves from Morrison et al. (1977): S, static; SE, static plus local exchange; SEP, static plus local exchange plus polarization. Crosses, measurements of Ramsauer and Kollath (1927), Kollath (1932); triangles, measurements of Bruche (1927), Brode (1933) (from Morrison et al. 1977).

from this figure that both effects have to be included to obtain agreement with experiment.

In figure 3 we compare the total cross section for e - CO_2 scattering calculated by Morrison et al. (1977) with experiment. Again the importance of including both exchange (a free-electron-gas local exchange potential) and polarization (a cut-off potential given by eqs.(18) and (19)) is evident. Although the cut-off parameter in the polarization potential was adjusted so that the position of the $^2\Pi_u$ resonance at 3.8 eV agrees with experiment, the general shape of the final cross section is in excellent agreement with the normalized experimental cross section over the full energy range shown. These calculations also throw light on the problems of convergence of the single centre expansions used in eqs.(5) and (13). It was found that in order to obtain fully converged results in the worst $^2\Sigma_g$ case as many as 32 channels had to be retained, while because of the nuclear contribution in eq.(2) 40 terms in eq.(13) were required. Clearly the search for methods which do not involve a single centre expansion is essential if results for more complex molecules are to be routinely calculated particularly if exchange is included exactly.

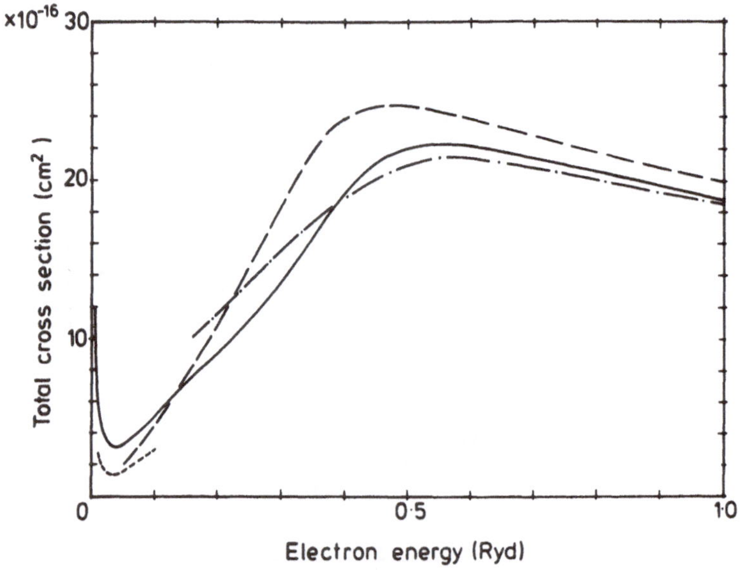

Figure 4 The total cross section for e-CH_4 scattering. Solid line, calculations of Gianturco and Thompson (1976); dash-dot line, measurements of Brode (1925); dashed line, measurements of Ramsauer and Kollath (1930). (from Gianturco and Thompson, 1976).

In figure 4 we compare the total cross section for e-CH_4 scattering calculated by Gianturco and Thompson (1976) with experiment. In this case exchange was included by orthogonalizing to the bound orbitals, and polarization by a cut-off potential given by eqs.(18) and (19). The Ramsauer-Townsend minimum at low energies and the broad T_2 symmetry (mainly d-wave) resonance at higher energies are both correctly described. This calculation shows that the single centre expansion method with model exchange and potential polarization potentials is capable of yielding relatively good results for non-linear molecules.

3. FRAME TRANSFORMATION THEORY - INCLUSION OF ROTATIONAL MOTION

We now return to a more detailed discussion of the region of validity of the fixed-nuclei approximation. When the spacings between the rotational levels of the molecule, which are important in the collision, are negligible compared with rate of change of the corresponding S-matrix elements coupling these levels (or alteratively if the time of collision is short compared with the rotation time) then it has been shown (eg Bottcher, 1969, Burke and Sinfailam 1970, Chang and Fano 1972) that there is an orthogonal transformation which connects the S-matrix calculated in the fixed-nuclei approximation with the S-matrix defined in the laboratory frame in which the molecule rotates. Further Chang and Fano (1972) have pointed out that even if this condition is not satisfied, the fixed-nuclei approximation can always be used in an internal region, where the electronic terms dominate the nuclear rotational terms in the Hamiltonian, and the laboratory frame used only in the external region. The wavefunction in these two regions can then be connected on the surface separating these regions by a frame transformation.

In order to explore this further we consider the expansion of the electron molecule collision wavefunction in the laboratory frame. Following Arthurs and Dalgarno (1960) we couple the angular momentum ℓ of the scattered electron with the rotational angular momentum j of the target molecule to form an eigenstate of the total angular momentum operator $\underline{J}^2$ and its z component J_z which are conserved in the collision. Thus instead of eq.(5) we now have

$$\psi^{JM_JSM_S} = \sum_{ij\ell} \Phi_{ij\ell}^{JM_JSM_S}(1,..,N,\hat{\underline{r}}_{N+1}\sigma_{N+1})\, r_{N+1}^{-1}\, f_{ij\ell}^{JS}(r_{N+1}) \qquad (26)$$

where the channel functions are defined by

$$\phi^{JM_JSM_S}_{ij\ell}(1,\ldots,N,\hat{\underline{r}}_{N+1}\sigma_{N+1}) = \sum_{\substack{m_j m_\ell \\ M_{S_i} m_{s_i}}} \phi_i^{S_i M_{S_i}}(1,2\ldots N)$$

$$Y_\ell^{m_\ell}(\hat{\underline{r}}_{N+1})\, Y_j^{m_j}(\hat{R})\, \eta_{1/2}^{m_{s_i}}(N+1)(\ell m_\ell j m_j|\ell j\, JM_J) \tag{27}$$

$$(S_i M_{S_i} \tfrac{1}{2} m_{s_i} |\, S_i \tfrac{1}{2}\, S\, M_S)$$

where $Y_j^{m_j}(\hat{R})$ are the rotor wavefunctions for the molecule which for simplicity we assume is in a Σ state. We note that since the correlation functions do not contribute at the radius where the frame transformation is usually applied we have not included them in expansion (26).

We can derive coupled integro-differential equations for the radial functions $f^{JS}_{ij\ell}$ by projecting the Schrödinger equation

$$(H_{rot} + H_{e\ell} - E)\,\psi^{JM_JSM_S} = 0 \tag{28}$$

onto the channel functions. In eq.(28) H_{rot} is the rotational part of the molecular Hamiltonian, whose eigenvalues are $Bj(j+1)$ if we disregard rotational stretching and, $H_{e\ell}$ is defined in eq.(1). We obtain the equations

$$\left(\frac{d^2}{dr^2} - \frac{\ell(\ell+1)}{r^2} + k^2_{ij}\right) f^{JS}_{ij\ell}(r) = 2 \sum_{i'j'\ell'} (V^{JS}_{ij\ell,i'j'\ell'} + W^{JS}_{ij\ell,i'j'\ell'})\, f^{JS}_{i'j'\ell'}(r) \tag{29}$$

which must be compared with eq.(7). The wave number k_{ij} now depends on the rotational state j through the relation

$$k^2_{ij} = 2(E - E_i - Bj(j+1)). \tag{30}$$

The direct potential is $V^{JS}_{ij\ell,i'j'\ell'}$ while the exchange potential $W^{JS}_{ij\ell,i'j'\ell'}$ like the correlation potential is zero at the radius where the frame transformation is usually applicable.

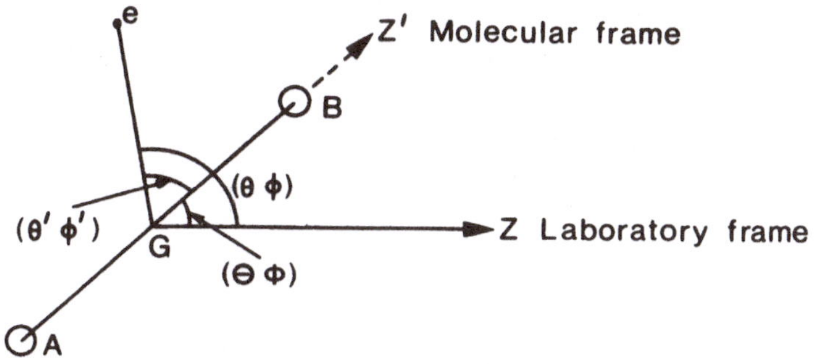

Figure 5 Frames used in electron molecule scattering.

To relate eqs.(7) and (29) we must compare the coupling schemes used in deriving these equations. To do this we introduce the co-ordinates defined in figure 5. The direction of the molecular axis is $\underline{R} \equiv (\Theta,\Phi)$ with respect to the z axis defined in the laboratory frame, and the colliding electron has primed angular co-ordinates (θ',ϕ') with respect to the molecular axis and unprimed angular co-ordinates (θ,ϕ) with respect to the laboratory axis.

The angular basis functions corresponding to eqs.(6) and (27) are respectively (Chang and Fano, 1972)

$$X_{JM_J}^{\ell m_\ell \eta}(\hat{\underline{r}},\hat{\underline{R}}) = \left(Y_\ell^{m_\ell}(\theta',\phi')\, D^{(J)}_{m_\ell M_J}(\Theta,\Phi) + \eta\, Y_\ell^{-m_\ell}(\theta',\phi')\, D^{(J)}_{-m_\ell M_J}(\Theta,\Phi) \right) \frac{2J+1}{8\pi(1+\delta_{m_\ell 0})} \tag{31}$$

and

$$\mathscr{Y}_{JM_J}^{\ell j}(\underline{r},\underline{R}) = \sum_{m_\ell m_j} Y_\ell^{m_\ell}(\theta,\phi)\, Y_j^{m_j}(\Theta,\Phi)\, (\ell m_\ell j\, m_j|\ell j J M_J) \tag{32}$$

where we have again restricted consideration to Σ states, where the D-functions are Wigner symmetric top functions, and where η is a parity quantum number which combines with J to yield the inversion parity of the form $I = \eta(-1)^J$. It is now straightforward to show that the functions X and $\mathscr{Y}$ are connected by an orthogonal transformation given by

$$X^{\ell m_\ell \eta}_{JM_J} = \sum_j \mathscr{Y}^{\ell j}_{JM_J} \; U^{\ell j \eta}_{j m_\ell} \tag{33}$$

and

$$\mathscr{Y}^{\ell j}_{JM_J} = \sum_{m_\ell} X^{\ell m_\ell \eta}_{JM_J} \; \tilde{U}^{\ell J \eta}_{m_\ell j} \tag{34}$$

where

$$\tilde{U}^{\ell J \eta}_{m_\ell j} = (-1)^{J+m_\ell-j} \; (\ell - m_\ell J m_\ell | \ell \; J \; j \; o) \; \frac{1 + \eta(-1)^{J-j-\ell}}{[2(1 + \delta_{m_\ell o})]^{1/2}} \tag{35}$$

The generalization of the transformation for non-Σ states has been given by Chang and Fano (1972).

The essential distinction between eqs.(7) and (29) is now clear. In the derivation of eqs.(7) H_{rot} is omitted and thus the corresponding equations are diagonal in Λ and S. If however H_{rot} had been included there would have been matrix elements of this operator coupling channels with different values of $\Lambda = m_\ell$ defined by

$$\langle X^{\ell m_\ell \eta}_{JM_J} | H_{rot} | X^{\ell m'_\ell \eta}_{JM_J} \rangle = B \sum_j \tilde{U}^{\ell J \eta}_{m_\ell j} \; j(j+1) \; U^{\ell J \eta}_{j m'_\ell} \quad , \tag{36}$$

If these matrix elements are important it would be necessary to solve eqs.(29) where H_{rot} is diagonal rather than eqs.(7). However if these matrix elements can be neglected then eqs.(7) and (29) are equivalent and we would then choose to solve eqs.(7) which has fewer channels coupled together.

This equivalence implies that there is an orthogonal transformation relating the solutions of eqs.(7) and (29) in the absence of H_{rot}. We introduce solutions of eqs.(29) which are regular at the origin and which satisfy the asymptotic boundary conditions

$$f^{JS}_{ij\ell,i'j'\ell'}(r) \underset{r\to\infty}{\sim} k_{ij}^{-1/2}(\sin (k_{ij} r - \tfrac{1}{2}\ell\pi)\delta_{ii'}\delta_{jj'}\delta_{\ell\ell'} +$$

$$\cos(k_{ij} r - \tfrac{1}{2}\ell\pi) \; K^{JS}_{ij\ell,i'j'\ell'}), \; k^2_{ij} > 0 \text{ (open channels)}$$

$$\underset{r\to\infty}{\sim} 0, \; k^2_{ij} < 0 \text{ (closed channels)} \tag{37}$$

If we now assume that expansions (5) and (26) are identical apart from the coupling scheme (i.e., we include the same target states etc.) then

$$f^{JS}_{ij\ell,i'j'\ell'}(r) = \sum_{m_\ell} U^{\ell J\eta}_{jm_\ell} \; f^{m_\ell S}_{i\ell,i'\ell'}(r) \; \tilde{U}^{\ell' J\eta}_{m_\ell j'} \tag{38}$$

It follows immediately by considering this equation in the limit of large r that the S-matrix in the two representations are related by

$$S^{JS}_{ij\ell,i'j'\ell'} = \sum_{m_\ell} U^{\ell J\eta}_{jm_\ell} \; S^{m_\ell S}_{i\ell,i'\ell'} \; \tilde{U}^{\ell' J\eta}_{m_\ell j'} \tag{39}$$

and the corresponding rotational excitation cross section can be obtained from the expression

$$\sigma(ij \rightarrow i'j') = \frac{\pi}{k^2_{ij}} \sum_S \frac{2S+1}{2(2S_i+1)} \sum_{J\ell\ell'} (2J+1) \times \left| S^{JS}_{ij\ell,i'j'\ell'} - \delta_{ii'}\delta_{jj'}\delta_{\ell\ell'} \right|^2 . \tag{40}$$

It follows that if the matrix elements of H_{rot} can be neglected compared with the incident kinetic energy and the potential, then rotational excitation cross sections can be obtained by solving eq.(7) and then using eqs.(39) and (40).

If the incident kinetic energy is low, or the S-matrix elements are rapidly varying, which is the situation close to a narrow resonance, then the matrix elements of H_{rot} can no longer be neglected and eq.(39) is invalid. However as we have already mentioned Chang and Fano (1972) pointed out that the potential interaction will dominate H_{rot} for sufficiently small r. Thus there will always be a region near the origin where eq.(7) can be used. However at larger values of r the potential interaction becomes small compared with H_{rot} and eqs.(29) must be used. This is illustrated in figure 6 where A denotes the inner region where the molecular frame can always be used and B denotes the region where eqs.(29) must be used at low energies or close to narrow resonances.

At this point we remember that there is another division of configuration space that we have considered where the direct potential achieves its asymptotic form defined by eq.(22), and

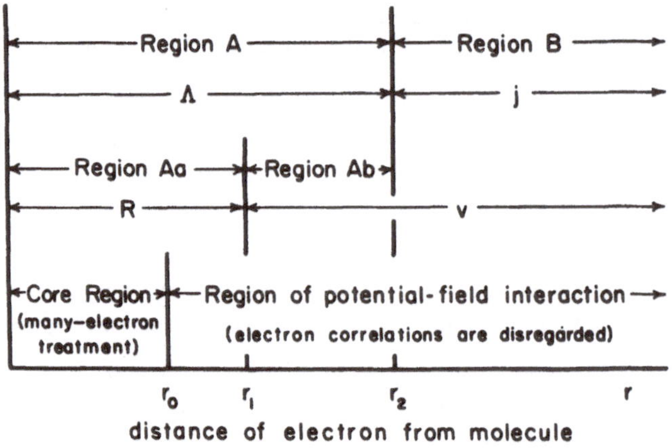

Figure 6 Regions of configuration space where different representations are appropriate (from Chang and Fano, 1972).

where the exchange, and correlation potentials are negligible. For molecules in their ground states we expect the region where exchange and correlation effects are important to be < 10 a.u., while the quadrupole and polarization potentials at such distances are of the order of 10^{-3} a.u. (~ 100 cm^{-1}) which lies between the rotational and vibrational energies. This is illustrated in figure 6 where region A is shown to contain the core region where exchange and correlation effects must be taken into account in a many electron treatment of the scattering.

This discussion then suggests the following method of solution in situations where H_{rot} is important. For $r < r_0$ in figure 6 we solve eq.(7) retaining the direct, exchange and correlation potentials. This enables the solution matrices $\underline{f}^{\Lambda S}$ defined by eq.(23) and their derivations $d\underline{f}^{\Lambda S}/dr$ to be determined at $r = r_0$. From these solutions we can then construct the R-matrix (Wigner 1946a,b, and Wigner and Eisenbud 1947) defined by

$$\underline{f}^{\Lambda S}(r_0) = \underline{\mathscr{R}}^{\Lambda S} \cdot \left(r \frac{d\underline{f}^{\Lambda S}}{dr} - \underline{b}\, \underline{f}^{\Lambda S} \right)_{r = r_0} \tag{41}$$

where $\underline{b}$ is an arbitrary constant matrix. The importance of the R-matrix in this context is that it provides a complete description of the action of the potentials in the region $r < r_0$ on the solution for $r > r_0$. We then transform the R-matrix at $r = r_0$ to the laboratory frame using the orthogonal transformation matrix

$U^{\ell J\eta}_{mj\ell}$. Finally we solve eqs.(29) in the region $r > r_0$ using the boundary conditions at $r = r_0$ defined by the transformed R-matrix. In this last step the direct potential is given by its asymptotic form eq.(22) and the exchange and correlation potentials, which are negligibly small, are omitted.

A detailed study of the application and validity of frame transformation theory to e-CO scattering has recently been carried out by Le Dourneuf et al.(1979). Using a model potential for this problem suggested by Crawford and Dalgarno (1971) they solved eq.(7) in an internal region and eq.(29) in the corresponding external region and they then considered three matching procedures involving transforming either the K-matrix defined by eq.(23), the R-matrix defined by eq.(41) or the M-matrix defined by

$$\underline{M}(r_0) = \underset{\sim}{k}^{\ell+1/2}\ \underline{K}^{-1}(r_0)\ \underset{\sim}{k}^{\ell+1/2} \qquad (42)$$

This matrix introduced by Ross and Shaw (1961) is an analytic function in the neighbourhood of thresholds, since the branch cuts in $\underline{K}$ have been removed by the kinematical factors $\underset{\sim}{k}^{\ell+1/2}$. Their results for the ratio of the $j = 2 \rightarrow 3$ cross section at 0.005 eV using frame transformation theory to the exact cross section calculated by solving eq.(29) for all r plotted as a function of the position where the frame transformation is applied are shown in figure 7. From this figure it is clear that transforming the M-matrix is by far the best procedure. While transformation of the R-matrix is satisfactory at small r, it breaks down at larger r because poles in this quantity become closely spaced in energy and cause difficulties. On the other hand the transformation on the K-matrix does not converge at small r. In conclusion these results strongly indicate that the frame transformation on the M-matrix can be used to give reliable results for rotational excitation cross sections from fixed nuclei calculations in an internal region.

Finally we remark that the principles of frame transformation theory have been combined with quantum defect theory by Fano (1970, 1975) and collaborators and used as a method of parameterizing molecular ion collision and photoionization data. The internal region in this case is characterized by eigenchannels defined in the body-fixed frame which are parameterized by an orthogonal transformation matrix $U_{i\alpha}$ and eigen quantum defects μ_α. These eigenchannel parameters then serve as boundary conditions for the solution of the problem in the external region in the laboratory frame where only the Coulomb potential is retained.

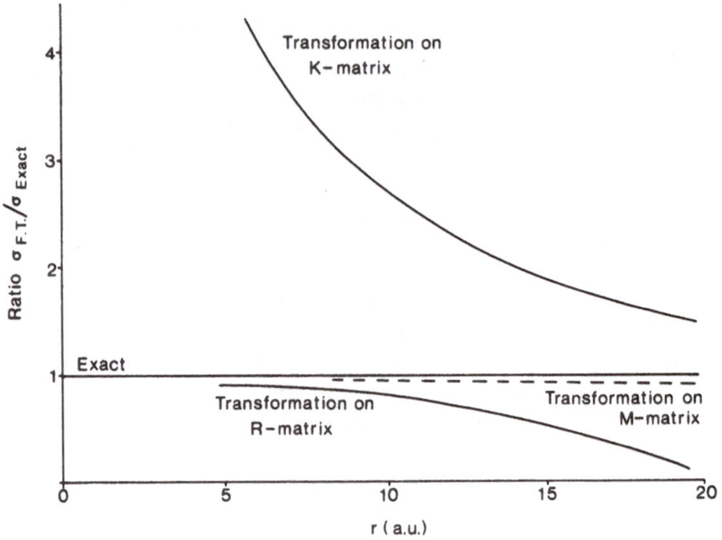

Figure 7 Application of frame transformation theory to e-CO scattering. The ratio $\rho = \sigma_{calc}/\sigma_{exact}$ for the $j = 2 \rightarrow 3$ transition at 0.005 eV is plotted as a function of the frame transformation point (from Le Dourneuf et al. 1979).

This approach has been used to analyse the angular distribution of photoelectrons from H_2 by Dill (1972), to study the photo-absorption of Rydberg levels of molecular hydrogen by Atabek et al. (1974) and by Jungen and Atabek (1977) and to analyse the dissociative recombination in NO by Lee (1977).

4. L^2 METHODS

4.1 Introduction

We have seen that a basic difficulty with the single-centre expansion approach described in section 2 is its slow convergence for all molecules except the lightest. In molecular structure calculations this problem is of course overcome by using a multi-centre expansion. It is thus of interest to see if bound state techniques and if possible computer codes can be used in collision calculations. This is the motivation behind the development of L^2 methods described in this section.

The first discussions of the use of L^2 methods in electron scattering calculations were made by Temkin (1966), Hazi and Taylor (1970) and Krauss and Mies (1970). They noted that if the collision wavefunction was expanded as follows

$$\psi_k = \sum_j \chi_j(1,2....,N+1)\ a_{kj} \tag{43}$$

where the χ_j are L^2 discrete basis functions (i.e., they are square integrable) and where the a_{kj} are determined by diagonalizing the electronic part of the Hamiltonian (using a standard molecular structure code)

$$\langle \psi_k \mid H_{e\ell} \mid \psi_j \rangle = E_k\ \delta_{kj} \tag{44}$$

then the ψ_k provided an accurate representation of the wavefunction at the energy E_k in the region where the χ_j are non-zero. Since in a typical case most of the E_k lie in the continuous spectrum of $H_{e\ell}$ this enables the collision wavefunction to be determined at these energies in an "internal region". An approximation phase shift can then be determined by projecting ψ_k onto the appropriate channel function $\Phi_{i\ell}$ defined by eq.(6) yielding a radial function

$$r^{-1}_{N+1}\ R_{ki\ell}\ (r_{N+1}) = \langle \Phi_{i\ell} \mid \psi_k \rangle \tag{45}$$

where the integral is taken over all electronic coordinates except the radial coordinate of the scattered electron. The phase shift in this channel is then determined by fitting $R_{ki\ell}(r)$ to the appropriate combination of spherical Bessel functions near the boundary of the "internal region".

This approach has been used by McCurdy et al. (1976) to obtain phase shifts for e-N_2 scattering and we present their results for the $p\pi$ and $d\pi$ phase shifts in figure 8. The results are seen to be in reasonable agreement with the static exchange calculations of Burke and Sinfailam (1970) who solved eq.(7) directly. The main differences are probably attributable to the different target wavefunctions which were used, the problems of convergence of the single centre expansion and the inclusion of exchange in the calculation of Burke and Sinfailam and the neglect of coupling between angular momenta in the asymptotic region (low ℓ spoiling) in the calculations of McCurdy et al. using eq.(45).

The results of McCurdy et al. exemplify the difficulties of this simple L^2 approach. These are that the method gives the phase shift only at a discrete set of energies in the continuum. To obtain the phase at other energies it is necessary either to interpolate or to vary some parameter in the basis functions to move the energies E_k defined by eq.(44). Further, the definition of the "internal region" is ill-defined leading to difficulties in the extraction of the phase shift. Finally the low ℓ spoiling approximation is by no means always applicable and certainly it

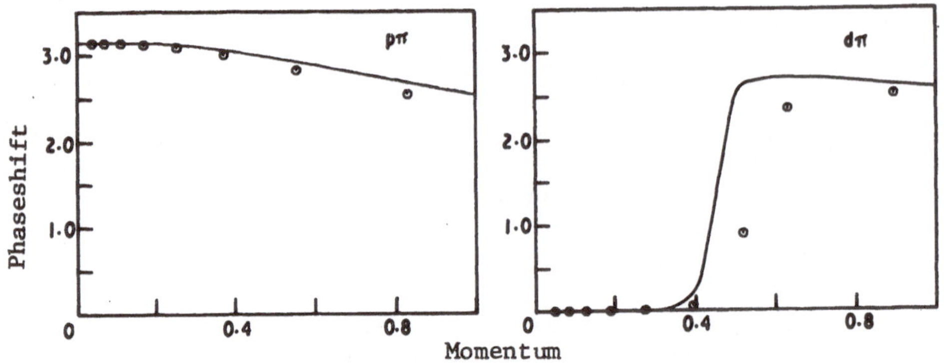

Figure 8 The pπ and dπ phase shifts for $e-N_2$ scattering. Solid line, static exchange calculations by Burke and Sinfailam (1970); circles, L^2 calculations by McCurdy et al. (1976). (From McCurdy et al, 1976).

is of doubtful validity when there are strong long range potentials coupling the channels as is the case for polar molecules. Consequently in the last few years interest has turned to two approaches the R-matrix and the T-matrix methods which avoid these difficulties.

However, before discussing the R-matrix and T-matrix methods we mention that L^2 methods have recently been extensively applied to the photoionization of atoms and molecules (Reinhardt, 1979) and to the calculation of resonance positions and widths (e.g., Hazi, 1979).

4.2 R-Matrix Method

This method was first introduced by Wigner (1946a,b) and Wigner and Eisenbud (1947) in a study of nuclear reactions, and extended to electron atom scattering by Burke and Hibbert (1969) Burke et al. (1971) and Burke and Robb (1975). Recently it has been developed further and applied extensively to electron molecule scattering by Schneider (1975a,b), Schneider and Hay (1976a,b), Morrison and Schneider (1977), Schneider (1978), Burke et al. (1977), and Buckley et al. (1979).

The method automatically takes advantage of the division of configuration space into an internal and an external region suggested by frame transformation theory. This division of space is illustrated in figure 9. In the internal region, where

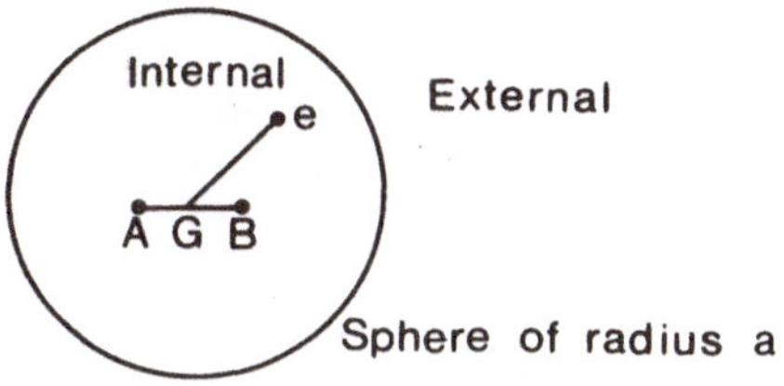

Figure 9 Division of configuration space in the R-Matrix method.

exchange and correlation effects are important, the wavefunction is expanded using the fixed-nuclei approximation in terms of a multicentre discrete basis in the molecular frame of reference. In the external region, where only the long-range components of the direct potential remain, a single-centre expansion using only a few partial waves is appropriate and either the molecular frame or the laboratory frame is used depending on the relative importance of H_{rot}. The link between these two regions is provided by the R-matrix on the surface defined by eq.(41). The basic problem of the R-matrix method is to calculate this matrix from the expansion of the wavefunction in the internal region.

In the following discussion we describe the R-matrix approach of Burke et al. (1977) which uses a multicentre STO basis. The work of Schneider and collaborators uses an approach based on prolate spheroidal co-ordinates which has some advantages for diatomic molecules, but which cannot be so easily extended to polyatomic molecules. We expand the wave function describing the electron molecule system in an internal region bounded by a sphere of radius r = a in the form

$$\psi_k^{\Lambda SM_S} = \sum_{ij} \Phi_i^{\lambda_i SM_S}(1,\dots,N,\sigma_{N+1})\, \psi_{ij}^{m_i}(\underline{r}_{N+1})\, c_{ijk}^{\Lambda SM_S} + \sum_i \xi_i^{\Lambda SM_S}(1,2\dots,N+1)\, c_{ik}^{\Lambda SM_S} \tag{46}$$

which replaces eq.(5). We see that the channel function in eq.(46) now do not involve the angular co-ordinates of the scattered electron but are still eigenstates of $\underline{S}^2$ and S_z. The molecular orbitals used in the definition of the target eigenstates and pseudostates and in the correlation functions are expanded in terms of STO's centred on A and B which have effectively vanished by the boundary of the internal region. On the other hand the continuum molecular orbitals $\psi_{ij}^{m_i}$ are expanded in terms of the same set of STO's on A and B but also include STO's centred on the centre of gravity G of the molecule which are of longer range and may be non-zero on the boundary. We will see that it is these STO's on G which provide the connection with the single-centre expansion used in the external region.

The coefficients $c_{ijk}^{\Lambda SM_S}$ and $c_{ik}^{\Lambda SM_S}$ are determined by diagonalizing the operator $H_{e\ell} + L_b$

$$\langle\psi_k^{\Lambda SM_S}|(H_{e\ell} + L_b)|\,\psi_{k'}^{\Lambda SM_S}\rangle = E_k\,\delta_{kk'} \tag{47}$$

where the integrals in this equation are carried out over the internal region and L_b is a surface projection operator introduced by Bloch (1956) which ensures that $H_{e\ell} + L_b$ is Hermitian. It is defined by

$$L_b = \frac{1}{2}\sum_{i\ell}|\,\Phi_{i\ell}^{\Lambda SM_S}\rangle\,\delta(r - a)\left(\frac{d}{dr} - \frac{b_{i\ell} - 1}{r}\right)\langle\Phi_{i\ell}^{\Lambda SM_S}| \tag{48}$$

where the channel functions used here are those defined in eq.(6) which involves a projection onto the angular co-ordinates of the scattered electron. We can now write the Schrödinger equation (1) as

$$(H_{e\ell} + L_b - E)\,\psi^{\Lambda SM_S} = L_b\,\psi^{\Lambda SM_S} \tag{49}$$

which has the formal solution

$$\psi^{\Lambda SM_S} = (H_{e\ell} + L_b - E)^{-1}\,L_b\,\psi^{\Lambda SM_S} \tag{50}$$

The inverse operator in this equation can be expanded in terms of the eigenfunctions defined by eq.(47) giving

$$|\psi^{\Lambda SM_S}\rangle = \sum_k \frac{|\psi_k^{\Lambda SM_S}\rangle\,\langle\psi_k^{\Lambda SM_S}|L_b\,|\psi^{\Lambda SM_S}\rangle}{E_k - E} \tag{51}$$

We now project this equation onto the channel function $\Phi_{i\ell}^{\Lambda SM_S}$ and evaluate it on the boundary r = a. If we define the radial functions

$$r_{N+1}^{-1} f_{i\ell}(r_{N+1}) = \langle \Phi_{i\ell}^{\Lambda SM_S} | \psi^{\Lambda SM_S} \rangle \tag{52}$$

where the integration in this equation is carried out over the coordinates of all N + 1 electrons except the radial co-ordinate of the scattered electron, then we obtain immediately eq.(41) where the R-matrix is given by

$$\mathscr{R}_{i\ell,i'\ell'}(E) = \sum_k \frac{\gamma_{i\ell k}\,\gamma_{i'\ell' k}}{E_k - E} \tag{53}$$

The reduced width amplitudes $\gamma_{i\ell k}$ are defined in terms of the value of continuum orbitals on the boundary. Remembering that we have used a 3-centre expansion for these orbitals which can be written as

$$\psi_{ij}^{m_i} = \psi_{ij}^{m_i}(\underline{r}_A) + \psi_{ij}^{m_i}(\underline{r}_B) + \sum_\ell r^{-1} v_{ij\ell}(r)\, Y_\ell^{m_i}(\hat{\underline{r}}) \tag{54}$$

where only the last term centred on G contributes on the boundary, then

$$\gamma_{i\ell k} = \left(\frac{1}{2a}\right)^{1/2} \sum_j v_{ij\ell}(a)\, c_{ijk}^{\Lambda SM_S} \tag{55}$$

An important point to remember is that the number of ℓ values that need to be retained on G is appropriate to the external region, and not to represent the nuclear singularities which are now represented by the orbitals centred on the nuclei. It has been found in practise that for homonuclear diatomic molecules such as N_2 at most two ℓ values and perhaps 6-8 basis orbitals on G for each value of ℓ are required to give accurate results for energies up to 10-15 eV.

The most time consuming part of this calculation is setting up and diagonalizing $H_{e\ell} + L_b$ which must be carried out once for each set of symmetries ΛSM_S. This can be done by relatively small modifications of standard multicentre integral codes by subtracting the tail from these integrals involving STO's on G which extend beyond the boundary and calculating and adding on the Bloch surface terms. These modifications have recently been made to the ALCHEMY code by Kendrick (1980). Once the R-matrix is determined then the coupled differential equations need to be solved in the external region for each energy where the cross section is required.

Results obtained for e-H_2 and e-N_2 scattering in the static exchange approximation (where just the ground electronic state is retained in expansion (46)) by Morrison and Schneider (1977) and by Buckley et al.(1979) are in good agreement. Schneider (1978) has also reported calculations on H_2 in which the polarizability was represented by pseudostates. Finally an alternative way of including the effect of polarizability has been explored by Schneider and Hay (1976a,b) for F_2. They used the F_2 SCF orbitals rather than F_2 orbitals in representing the target, so allowing for part of the relaxation occurring during the collision. A similar approach has been used by Schneider et al. (1979b) for vibrational excitation in e-N_2 scattering and these results will be discussed in the next section.

4.3 T-Matrix Method

Another L^2 method, this time in which the T-matrix is expanded in a discrete basis has been proposed by Rescigno et al. (1974a,b, 1975a).

The original method starts from the Lippmann-Schwinger integral equation for the transition matrix

$$T = U + U\, G_o^+\, T \tag{56}$$

where U is twice the non-local potential interaction between the electron and the target molecule and G_o^+ is the free-particle Green's function with outgoing wave boundary conditions. We now expand U in a finite basis set of square integrable functions ϕ_α taken in the applications to be Gaussians, giving

$$U^t(\underline{r},\underline{r}') = \sum_{\alpha\beta} |\,\phi_\alpha(\underline{r})\rangle \langle\phi_\alpha(\underline{r})\,|\,U(\underline{r},\underline{r}')|\,\phi_\beta(\underline{r}')\rangle \langle\phi_\beta(\underline{r}')| \tag{57}$$

Inserting this truncated potential U^t into eq.(56) gives a matrix equation for the T-matrix which can be solved yielding

$$\underline{T}^t = (\underline{1} - \underline{U}^t\, \underline{G}_o^+)^{-1}\, \underline{U}^t \tag{58}$$

where the matrix $\underline{G}_o^+$ is defined by

$$\langle\phi_\alpha|G_o^+(E)|\phi_\beta\rangle = \lim_{\varepsilon\to 0+} \int d^3k \frac{\langle\phi_\alpha \mid \underline{k}\rangle \langle\underline{k} \mid \phi_\beta\rangle}{q^2 - k^2 + i\varepsilon} \tag{59}$$

and where $E = 1/2\ q^2$ and $|k\rangle$ and $|k'\rangle$ are the plane wave states

$$|k\rangle = (2\pi)^{-3/2}\, e^{i\,\underline{k}_i\cdot\underline{r}} \tag{60}$$

The scattering amplitude for a transition from state $\underline{k}_i$ to $\underline{k}_f$ is then given by

$$f(k_i \to \underline{k}_f) = -\,2\pi^2 \sum_{\alpha\beta} \langle\underline{k}_f|\alpha\rangle \langle\alpha|T^t(E)|\beta\rangle \langle\beta|\underline{k}_i\rangle \tag{61}$$

where the matrix elements involving the plane wave states can be evaluated analytically for Gaussian basis functions (Ostlund, 1975 and Levin et al. 1979).

This theory has been used to obtain static exchange results for H_2, N_2, CO and F_2 by Rescigno et al. (1974a,b, 1975a), Fliflet et al. (1978) and Rescigno (1979). In addition the polarizability was included for H_2 by Klonover and Kaldor (1977, 1978, 1979a,b) using second-order perturbation theory. However, it has been found that errors due to truncation of the potential can only be made small, particularly at energies below 5 eV where the long range potentials are important, by the use of large basis sets and this can lead to problems of linear dependence. Although the inclusion of a variational correction using the Kohn method (Fliflet and McKoy, 1978a,b) helps to overcome this difficulty, it has proved better to use a modification of this method based on a variational principle from the outset. The method adopted has been widely used in nuclear physics (Lovelace, 1964) and involves writing the potential in the separable form

$$U^S = \sum_{\alpha\beta} U|\alpha\rangle\ (U^{-1})_{\alpha\beta}\ \langle\beta|U \tag{62}$$

Substituting into the Lippmann-Schwinger equation then gives

$$\langle\underline{k}_f|T^S|\underline{k}_i\rangle = \sum_{\alpha\beta} \langle\underline{k}_f|U|\alpha\rangle \left[(U - U\,G_o\,U)^{-1}\right]_{\alpha\beta} \langle\beta|U|\underline{k}_i\rangle \tag{63}$$

which can be shown to satisfy the Schwinger variational principle (Adkhikari and Sloan, 1975). An important feature of this new approach is that the expansion functions ϕ_α only appear in conjunction with the potential, which allows the use of trial functions which do not have the correct asymptotic form. This is of course in contrast to the Kohn variational method. The method involves two new types of matrix element $\langle\underline{k}|U|\alpha\rangle$ and

$\langle\alpha|U\ G_o\ U|\beta\rangle$. The first can be evaluated in closed form when Gaussians are used and the second can be approximated by closure. In practice, single centre expansions are used for evaluating the matrix elements. Results involving only 3 s type basis functions, 1 p type basis function and 1 d type basis function have recently been obtained by Watson et al (1979) for H_2 which are in good agreement with other methods and work is now underway on LiH.

As an example of the application of this method we show in figure 10 the angular distribution of $v = 0 \rightarrow 1$ vibrational excitation cross section for e-H_2 scattering calculated by Klonover and Kaldar (1979a) compared with the measurements of Linder and Schmidt (1971). The calculations were carried out with and without the inclusion of the second-order polarization and the vibrational excitation cross sections were obtained using the adiabatic theory of Chase (1956) described in the next section. The calculation used 30 cartesian Gaussian basis orbitals to describe the occupied H_2 orbitals and the excited orbitals which account for the polarization effects, and 61 basis

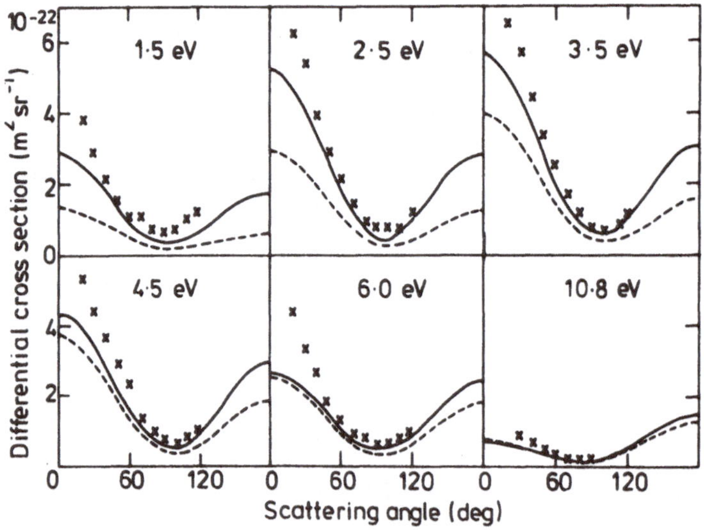

Figure 10 Differential $v = 0 \rightarrow 1$ vibrational excitation cross section for e-H_2 scattering. Solid line, T-matrix calculations including polarization; dotted line, T-matrix calculations omitting polarization both by Klonover and Kaldor (1979a); crosses, measurements of Linder and Schmidt (1971). (From Klonover and Kaldor, 1979a).

orbitals to span the scattered electron orbital, some of the latter orbitals being centred on the centre of gravity. The calculations were based on eq.(58) for the T-matrix. The results show clearly the importance of the polarization effects at low energies in both the forward and backward directions.

In conclusion it is clear that the T-matrix method promises to be a powerful way of calculating electron molecule collision cross sections.

5. VIBRATIONAL EXCITATION AND DISSOCIATIVE ATTACHMENT

It was shown by Massey (1935) that the Born approximation predicts that the cross section for vibrational excitation for non-polar molecules is of the order of 10^{-18} cm^2. However observations of vibrational excitation for e-N_2 scattering near 2 eV made by Schulz (1962, 1964, 1973) showed that the cross section is instead of the order of $10^{-16} cm^2$. The explanation of this result first given by Herzenberg and Mandl (1962) is that the cross section at this energy is dominated by a N_2 resonance, now known to have $^2\Pi_g$ symmetry. The capture of the incident electron into this resonance state means that it stays in the neighbourhood of the molecule for a time of the order of the vibrational time and it thus has a high probability of transferring energy to the vibrational motion. This result has since been found to a very general one for vibration excitation, and for dissociative attachment which also proceeds strongly through resonances, and has been very important in influencing the development of the theory. Nevertheless we will start this section by considering briefly the adiabatic theory of vibrational excitation which is of particular relevance in non-resonant situations.

5.1 Adiabatic Nuclei Approximation

The adiabatic nuclei approximation developed by Chase (1956) to describe the excitation of deformed nuclei can be used to obtain rotational and vibrational excitation cross sections. The basic assumption made in this theory is that the collision time is very much shorter than typical rotation and vibration times and thus the scattering amplitude for the transition can be obtained by taking the appropriate matrix element of the scattering amplitude calculated in the fixed-nuclei approximation. If we define $f_{i'i}(\underline{\beta}, \underline{\Omega}; R)$ as the electronic excitation amplitude describing the scattering of an electron into a laboratory angle $\underline{\Omega}$ by a fixed molecule whose orientation is defined by the angle $\underline{\beta}$ and whose internuclear separation is given by R then the transition amplitude from an initial vibrational-rotational state vj to a final vibrational-rotational state v'j' is given by

$$f_{i'v'j',ivj}(\underline{\Omega}) = \int_0^\infty dR \int d\underline{\beta}\ \chi^*_{v'}(R)\ \psi^*_{j'}(\underline{\beta})\ f_{i'i}(\underline{\beta},\underline{\Omega};R)\ \chi_v(R)\ \psi_j(\underline{\beta}) \tag{64}$$

Here χ_v (R) and $\chi_{v'}$ (R) are the vibrational wavefunctions and $\psi_j(\underline{\beta})$ and $\psi_{j'}(\underline{\beta})$ are the rotational wavefunctions for the relevant electronic state.

The condition for the validity of eq.(64) in so far as rotational excitation is concerned is analogous to the validity of the orthogonal transformation defined by eq.(39): i.e., we saw that eq.(39) was valid if the incident energy was large compared with the rotational splitting of the levels which is true if the time of collision is considerably shorter than the time of rotation. The formal equivalence between these approaches can be demonstrated by carrying out the angular integral in eq.(64) (Chang and Temkin, 1969 and Temkin and Faisal, 1971).

The range of validity of eq.(64) in so far as vibrational excitation is concerned is limited to energies away from threshold which are not dominated by narrow resonances and in such situations it has had extensive use. However Chang and Fano (1972) have suggested that this range can be extended by an extension of the frame transformation approach introduced earlier to describe rotational excitation. If we include the nuclear kinetic energy operator into the Hamiltonian in the molecular frame of reference then the Schrödinger equation (1) is replaced by

$$(T_R + H_{e\ell} - E)\ \psi = 0 \tag{65}$$

where the rotational term in the Hamiltonian H_{rot} has been omitted, and where

$$T_R = -\frac{1}{2\mu}\ \frac{d^2}{dR^2} \tag{66}$$

μ being the reduced mass of the two nuclei. Referring to figure 6 Chang and Fano now assume that region A, where the matrix elements of H_{rot} are negligible compared with other terms in the Hamiltonian, can be sub-divided into two regions Aa and Ab where in region Aa the matrix elements of the nuclear kinetic energy operator may be neglected. In this region eqs.(7) are solved using perhaps the fixed-nuclei R-matrix method for a number of values of the internuclear separation spanning the range of interest in vibrational excitation. The R-matrix (or perhaps more appropriately the M-matrix defined by eq.(42)) is then calculated on the boundary of this region and transformed to the representation defined by the vibrational state $\chi_v(R)$ using

$$\mathcal{R}^{\Lambda S}_{i'\ell'v',i\ell v} = \int_0^\infty \chi^*_{i'v'}(R)\, \mathcal{R}^{\Lambda S}_{i'\ell',i\ell}(R)\, \chi_{iv}(R)\, dR \quad . \qquad (67)$$

Outside region Aa, the coupled "Hybrid equations" considered in detail below, coupling v as well as i and ℓ are solved using

$\mathcal{R}^{\Lambda S}_{i'\ell'v',i\ell v}$ (or $M^{\Lambda S}_{i'\ell'v',i\ell v}$) to define the boundary condition.

Finally if H_{rot} is important in region B, a further frame transformation to include this operator in the laboratory frame of reference would be necessary on the outer boundary of region Ab as described previously.

The remaining question to be asked, if this vibrational frame transformation theory is going to be a practical method of calculation, is whether region Aa where T_R can be neglected includes the region where electron exchange and correlation effects are important. If it does not then the non-local terms in the potential will couple the wavefunction in regions Aa and Ab making this approach difficult to apply. Recent results using a new R-matrix approach for vibrational excitation, to be discussed below, suggest that a suitable inner region Aa cannot be chosen, but the question still remains untested at this time.

Another interesting way of extending the adiabatic theory has recently been proposed by Nesbet (1979). In this approach, called the "energy-modified adiabatic approximation", the S-matrix elements connecting target states χ_v and $\chi_{v'}$ is defined by

$$S_{v'v} = \langle \chi_{v'} | S\,(E - H_i ; R) | \chi_v \rangle \qquad (68)$$

Here $S(E - H_i ; R)$ is the S-matrix calculated in the fixed-nuclei approximation at the internuclear separation R at an energy defined by the operator $H_i = T_R + E_i(R)$. This has the effect of including the internal energy of the target and has been used successfully to describe the observed structure in the vibration excitation in e-N_2 scattering in terms of a parametrized $^2\Pi_g$ resonance potential energy curve.

5.2 Hybrid Theory

The most natural way of including effects of T_R in eq.(65) is to expand the total wavefunction in terms of the eigenfunctions $\chi_{iv}(R)$ of the vibrational motion. This approach was considered by Chandra and Temkin (1976a,b) and by Choi and Poe (1977). In this theory expansion (5) is replaced by

$$\psi^{\Lambda SM_S} = \mathscr{A} \sum_{i\ell v} \Phi_{i\ell}^{\Lambda SM_S}(1,\ldots,N,\hat{\underline{r}}_{N+1}\sigma_{N+1})\, r_{N+1}^{-1}\, f_{i\ell v}^{\Lambda S}(r_{N+1})\, \chi_{iv}(R) \tag{69}$$

where the correlation terms are omitted partly for notational simplicity, and partly because such terms have not yet been included in any application of this theory.

We now substitute eq.(69) into eq.(65) and project onto the channel functions $\Phi_{i\ell}^{\Lambda SM_S} \chi_{iv}$. In simplifying these equations the action of T_R on the functions $\Phi_{i\ell}^{\Lambda SM_S}$ is neglected in accordance with the Born-Oppenheimer approximation. We then obtain the following coupled integro-differential equations

$$\left(\frac{d^2}{dr^2} - \frac{\ell(\ell+1)}{r^2} + k^2_{iv}\right) f_{i\ell v}^{\Lambda S}(r) = 2 \sum_{i'\ell'v'} (V_{i\ell v,i'\ell'v'}^{\Lambda S} + W_{i\ell v,i'\ell'v'}^{\Lambda S})\, f_{i'\ell'v'}^{\Lambda S}(r) \tag{70}$$

where the wave numbers k_{iv} are defined by

$$k_{iv}^2 = 2(E - E_i - \omega(v + 1/2)) \tag{71}$$

the direct potential is defined in terms of the matrix elements of the direct potential in the fixed-nuclei approximation given by eq.(9)

$$V_{i\ell v,i'\ell'v'}^{\Lambda S}(r) = \langle \chi_v(R) |\, V_{i\ell,i'\ell'}^{\Lambda S} \,|\, \chi_{v'}(R) \rangle \tag{72}$$

where $V_{i\ell,i'\ell'}^{\Lambda S}$ depend on R, and the exchange potential can be formally expressed in a similar way in terms of the matrix elements of the exchange potential in the fixed-nuclei approximation given by eq.(10).

The solution of eqs.(70) then proceeds in the usual way. We look for regular solutions satisfying the asymptotic boundary conditions which in analogy with eqs.(23) are:

$$f^{\Lambda S}_{i\ell v,i'\ell'v'}(r) \underset{r\to\infty}{\sim} k_i^{-1/2}(\sin(k_{iv}r - \frac{1}{2}\ell\pi)\,\delta_{ii'}\,\delta_{\ell\ell'}\,\delta_{vv'}$$

$$+ \cos(k_{iv}r - \frac{1}{2}\ell\pi)\,K^{\Lambda S}_{i\ell v,i'\ell'v'}),\ k^2_{iv} > 0 \text{ (open channels)}$$

$$\underset{r\to\infty}{\sim} 0,\ k^2_{iv} < 0 \text{ (closed channels)} \qquad (73)$$

The cross section for vibrational excitation is then given by

$$\sigma(iv \to i'v') = \frac{\pi}{k^2_{iv}} \sum_S \frac{(2S+1)}{2(2S_i+1)} \sum_{\Lambda\ell\ell'} |S^{\Lambda S}_{i\ell v,i'\ell'v'}$$

$$- \delta_{ii'}\,\delta_{\ell\ell'}\,\delta_{vv'}|^2 \qquad (74)$$

when the S-matrix is related to the K-matrix by eq.(24).

The numerical solution of eqs.(70) is formidable since convergence must be obtained both in the number of angular momentum values ℓ and in the vibrational quantum values v retained in the expansion. In order to make them tractable in the case of the scattering of electrons by N_2 in the resonant $^2\Pi_g$ state, Chandra and Temkin (1976a,b) represented the exchange potential by an equivalent orthogonalization procedure discussed earlier. They also include a parametrized polarization potential defined by eq.(18) where the cut-off parameter was adjusted to make the calculated position of the $^2\Pi_g$ resonance agree with experiment. With these approximations and a careful consideration of the variation of the polarizability with internuclear separation (Temkin, 1978 and Kendrick, 1978) recent results are in reasonable accord with experiment (Temkin, 1979a,b).

Finally we remark that the hybrid theory can be combined with the frame transformation theory when the rotational terms in the Hamiltonian are important. In this case the solution of eq.(70) is used to calculate an R-matrix on some boundary as in eq.(41). Coupled differential equations in the laboratory frame analogous to eqs.(29), but involving the vibrational quantum number v are then solved in the outer region.

5.3 Resonance Theory

Although the development and use of the hybrid theory is a substantial step forward and provides a standard against which more approximate methods can be judged, it is very costly in computer time. It is therefore important to develop methods which are guided more by the physical processes which are known to be important. It has been mentioned earlier that the dominant mechanism for vibrational excitation and also for dissociative attachment is through the capture of the incident electron into a resonance state, which leads to an enhanced probability of energy transfer between the incident electron and the nuclei. This is the basis of the resonance theories developed and applied largely by Herzenberg and collaborators over the last two decades (e.g., Bardsley and Mandl, 1968, Herzenberg, 1978).

As an example of a resonance theory which has been particu-particularly successful in describing vibrational excitation and dissociative attachment we show in figure 11 the potential energy curves and nuclear wavefunctions which arise in the "boomerang model" introduced by Herzenberg (1968) and Birtwistle and

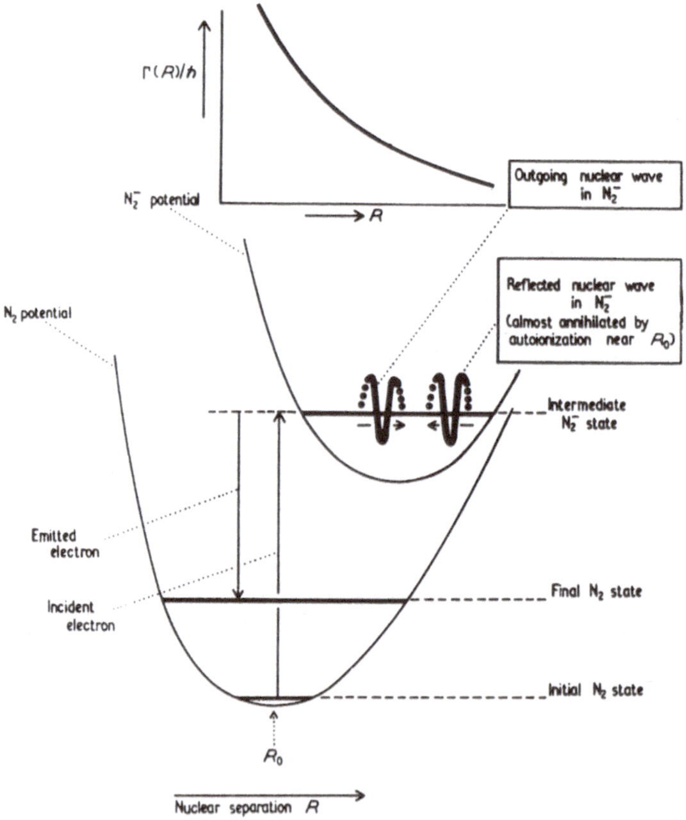

Figure 11 The potential energy curves and nuclear wavefunctions in the "boomerang model" (from Birtwistle and Herzenberg, 1971).

Herzenberg (1971). The incident electron is captured into the resonance state as illustrated in the figure. The nuclei then move apart and are reflected once from the outer turning point of the resonance state before the electron is emitted leaving the molecule in an excited vibrational state. This model is based on the assumption that the dependences of the width $\Gamma(R)$ on the internuclear separation R is such that there can only be a single outgoing and reflected wave as is the case for e-N_2 scattering.

As in the R-matrix method the formulation of the resonance theory starts from a division of configuration space into two regions where the interaction between the scattered electron and the molecule is assumed to be small in the external region (Bardsley et al. 1966). For each value of the internuclear co-ordinate R we define a complete set of resonance states in the internal region which satisfy the equation

$$(H_{e\ell} - W_n(R))\,\psi_n = 0 \tag{75}$$

The resonance states ψ_n satisfy the Siegert (1939) outgoing wave boundary conditions in the external region

$$\psi_n \underset{r\to\infty}{\sim} \phi_i\, e^{ik_{ni} r} \tag{76}$$

where ϕ_i are the target electronic states, the wave numbers k_{ni} are defined in analogy with eq.(8) as

$$k_{ni}^2 = 2(W_n - E_i) \tag{77}$$

and where for notational simplicity here and below we have omitted the angular and spin channels, but these would have to be included as in eqs.(5) and (6) in a detailed theory.

If the total energy W_n is below the first electronic threshold then all the channels are closed, the wave numbers k_{ni} are then pure imaginary and the wavefunction defined by eq.(76) decays exponentially asymptotically. This corresponds to a bound negative ion state of the electron molecule system. On the other hand if the energy is such that some of the channels are open then the Siegert prescription makes the eigenvalues complex. This corresponds to a resonance state of the electron molecule system. We will write in this case

$$W_n(R) = \mathrm{Re}\,W_n(R) - \frac{1}{2}\,i\,\Gamma_n(R) \tag{78}$$

where the resonance widths $\Gamma_n(R)$ are real and greater than zero.

In accordance with the Born-Oppenheimer separation of the electronic and nuclear motion we now expand the total electron molecule wavefunction in the internal region in terms of this complete set of resonance states giving

$$\psi = \sum_n \psi_n (1,2\ldots N+1)\, \xi_n(R,\underline{k}) + \phi_{inc}\, \chi_{iv}(R) \tag{79}$$

where the last term in this equation corresponds to a plane wave incident on the initial state of the target. To obtain equations for the wavefunctions $\xi_n(R,\underline{k})$ describing the nuclear motion we substitute eq.(79) into eq. (65), multiply by the functions ψ_n and integrate over all the electronic co-ordinates within the internal region. If we now assume that ψ_n varies slowly with R in comparison to ξ_n so that the nuclear kinetic energy operator acts only on ξ_n, we obtain with the aid of eq.(75)

$$\left(-\frac{1}{2\mu}\frac{d^2}{dR^2} + W_n(R) - E\right)\xi_{niv}(R,\underline{k}_{iv}) = -\zeta_{niv}(R,\underline{k}_{iv})\,\chi_{iv}(R) \tag{80}$$

where $\chi_{iv}(R)$ is the initial vibrational state of the molecule and $\zeta_{niv}(R,\underline{k}_{iv})$ is the so-called "entry amplitude" into the resonance state described by ψ_n. This is defined by

$$\zeta_{niv}(R,\underline{k}_{iv}) = \langle\psi_n\,|V|\,\phi_i\, e^{i\underline{k}_{iv}\cdot\underline{r}}\rangle \tag{81}$$

where the integral is taken over the internal region and V is the potential interaction between the electron and the molecule defined by eq.(2).

The amplitude for vibrational excitation is now obtained by calculating the decay of the resonance states into the appropriate final state. If we introduce the Green's function $G_n(R,R')$ of the operator on the left hand side of eq.(80) then we obtain

$$\xi_{niv}(R,\underline{k}_{iv}) = -\int_0^\infty dR'\, G_n(R,R')\,\zeta_{niv}(R',\underline{k}_{iv})\,\chi_{iv}(R'). \tag{82}$$

The amplitude for vibrational excitation is then

$$f_{i'v',iv} = -\sum_n \int_0^\infty dR \int_0^\infty dR'\, \chi^*_{i'v'}(R)\,\zeta^*_{ni'v'}(R,\underline{k}_{i'v'})\, G_n(R,R')\,\zeta_{niv}(R',k_{iv})\,\chi_{iv}(R') \tag{83}$$

where the appropriate sum over channels is assumed.

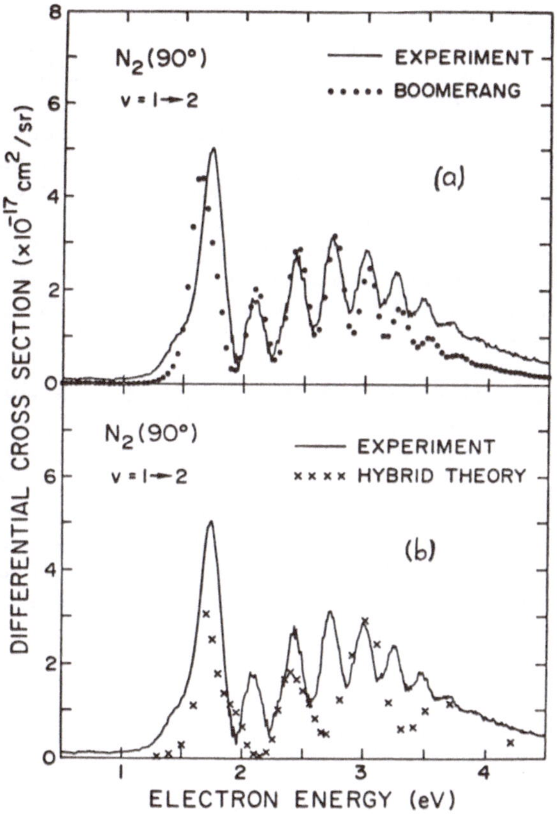

Figure 12 The differential cross section at 90° for $v = 1 \rightarrow 2$ vibrational excitation in e-N_2 scattering. (a) Solid line, measurement of Wong (1978); dots, "boomerang model" calculations of Dubé and Herzenberg (1979). (b) Solid line, measurement as in Figure 12a; crosses, hybrid theory calculations of Chandra and Temkin (1976a,b) (from Dubé and Herzenberg, 1979).

Dubé and Herzenberg (1979) have applied this theory to calculate the vibrational excitation cross section in e-N_2 scattering and we compare their results for the $v = 1 \rightarrow 2$ transition at 90° with measurements made by Wong (1978) in figure 12a. Also shown in figure 12b are the hybrid calculations of Chandra and Temkin (1976a,b) compared with the same measurements. This "boomerang model" theory of Dubé and Herzenberg assumes that there is one resonance state of N_2 with $^2\Pi_g$ symmetry contributing to the cross section and the real and imaginary parts of its potential energy curve are written in terms of six adjustable parameters. In practise the permissible values of all but one of these parameters are fixed down almost completely by an ab-initio L^2 calculation by Krauss and Mies (1970), and the remaining parameter is a "correlation energy" which Krauss and Mies adjusted to get the resonance appearing at the right energy. Having paramaterized the N_2 state the theory is then completely determined. Clearly both the "boomerang model" and the hybrid theory reproduce the observed oscillations in the cross section, which arises from the relaxation of the nuclear wavefunction during the collision. These oscillations would not be obtained in the adiabatic nuclei theory based on eq.(64) which instead would have given one broad peak in the cross section (see in this connection figure 22).

Dissociative attachment can be described by a small extension of this theory. The electron is captured by the molecule from some initial vibrational state into an intermediate resonance state. The nuclei then move apart along the resonance potential energy curve. If they separate to the point where this curve becomes real before the electron autoionizes then dissociative attachment occurs. Otherwise we have vibrational excitation as described above. The cross section for dissociative attachment is obtained by calculating the flux of negative ions at large values of the internuclear co-ordinate giving

$$\sigma_{DA} = \frac{2\pi^2}{k_{iv}} \frac{K}{\mu} \lim_{R\to\infty} |a(R)|^2 \tag{84}$$

where k_{iv} and K are the wave numbers describing the motion of the incident electron and outgoing ion in the centre-of-mass and a(R) is the amplitude of the outgoing wave solution of eq.(80).

This theory of dissociative attachment has had many successful applications. For example in a recent study of dissociative attachment in e-H_2 scattering Wadehra and Bardsley (1978) have successfully explained the effect of vibrational and rotational excitation of the target molecule on the dissociative attachment cross section observed by Allan and Wong (1978). As for vibrational excitation the input to this theory is the real and imaginary parts of the resonance potential energy curve.

5.4 R-Matrix Method

We have seen earlier that the R-matrix method has proved to be a convenient technique for the ab-initio calculation of electron molecule collision cross sections for fixed internuclear separation. It is therefore of interest to see if this approach can be extended to provide an ab-initio treatment of vibrational excitation and dissociative attachment. This problem has recently been discussed by Schneider et al. (1979a).

The approach of Schneider et al. (1979a) is to adopt an expansion for the electron molecule wavefunction which is analogous to the expansion defined by eq.(79) in the resonance theory case except that the basis states are now R-matrix states. Thus we write

$$\psi = \sum_n \psi_n(1,2...N+1)\ \xi_n(R) \tag{85}$$

where the R-matrix states ψ_n are defined by diagonalizing $H_{e\ell} + L_b$ for each fixed value of R as in eq.(47). In order to obtain an equation for the wavefunctions $\xi_n(R)$ describing the nuclear motion we project the Schrödinger equation

$$(T_R + H_{e\ell} + L_b - E)\ \psi = L_b\ \psi \tag{86}$$

onto the R-matrix states. If as before we assume in accordance with the Born-Oppenheimer approximation that the nuclear kinetic energy operator acts only on the $\xi_n(R)$ we obtain the equation

$$\left(-\frac{1}{2\mu}\ \frac{d^2}{dR^2} + E_n(R) - E\right)\xi_n(R) = \langle\psi_n\ |L_b|\ \psi\rangle \tag{87}$$

In comparing this equation with eq.(80) obtained in the resonance theory case we note that the potential on the left hand side is now real since the R-matrix eigenvalues defined by eq.(47) are real and the surface term on the right hand side replaces the "entry amplitude".

In order to obtain the R-matrix coupling the electronic and vibrational channels we now project the total wavefunction defined by eq.(85) onto the appropriate channel functions and evaluate this quantity on the boundary. We obtain

$$\langle\chi_{i'v'}\ \Phi_{i'\ell'}\ |\psi\rangle_{r=a} = \sum_n \langle\chi_{i'v'}\ \Phi_{i'\ell'}|\ \psi_n(1,2...N+1)\ \xi_n(R)\rangle_{r=a} \tag{88}$$

If we now substitute for $\xi_n(R)$ from eq.(87) by introducing the Green's function $G_n(R,R')$ of the operator $T_R + E_n(R) - E$ we obtain

$$\langle\chi_{i'v'}\Phi_{i'\ell'}|\psi\rangle_{r=a} = \sum_{iv\ell} \mathscr{R}_{i'v'\ell',iv\ell}\left[\left(\frac{d}{dr} - \frac{b-1}{a}\right)\langle\chi_{iv}\Phi_{i\ell}\ |\psi\rangle\right]_{r=a} \tag{89}$$

where the generalized R-matrix

$$\mathscr{R}_{i'v'\ell',iv\ell} = \sum_n \langle \chi_{i'v'}(R')\, \gamma_{i'\ell'n}(R') \, G_n(R',R)\, \gamma_{i\ell n}(R)\, \chi_{iv}(R)\rangle \tag{90}$$

and where the reduced width amplitudes are defined by eq.(55).

If dissociative attachment is possible then the Bloch operator L_b in eqs.(86) and (87) must be augmented by an additional term allowing for dissociation into A + B^- type channels. Schneider et al (1979a) show that this leads to additional terms in the generalized R-matrix coupling the electronic and vibrational channels with the dissociating channels.

We can relate eq.(90) with the result obtained using the frame transformation theory of Chang and Fano (1972). If we neglect the kinetic energy operation T_R in the internal region then the Green's function $G_n(R',R)$ can be written

$$G_n(R',R) = \frac{\delta(R' - R)}{E_n(R) - E} \tag{91}$$

and then eq.(90) becomes

$$\mathscr{R}_{i'v'\ell',iv\ell} = \int_0^\infty \chi^*_{i'v'}(R) \sum_n \frac{\gamma_{i'\ell'n}(R)\,\gamma_{i\ell n}(R)}{E_n(R) - E} \chi_{iv}(R)\, dR \tag{92}$$

which reduces to eq.(67) when we use eq.(53). It is easy to see that while eq.(91) is a good approximation to $G_n(R',R)$ for the higher R-matrix levels where $E_n(R) - E$ is large and positive it cannot be a good approximation when $E_n(R) \approx E$ which occurs when the incident energy is close to the resonance R-matrix eigenenergy. However since almost all the computational effort is expended in setting up and diagonalizing $H_{e\ell} + L_b$ for the required set of R values, and very little in determining the Green's function $G_n(R',R)$ it is just as easy and very much more accurate to use eq.(90) rather than eq.(92) to determine the R-matrix.

Once the generalized R-matrix has been determined, the final step in the calculation is to solve the collision problem in the external region. The appropriate equations in this region are the hybrid equations defined by eq.(70) without of course the exchange potential which is zero in this region. These equations are solved subject to the asymptotic boundary conditions given by eq.(73) where the R-matrix provides the starting condition at $r = a$. In this way the K-matrix and consequently the cross section defined by eq.(74) are determined.

This theory has recently been used to calculate the vibrational excitation cross section in e-N_2 scattering by Schneider et al. (1979b) and Le Dourneuf et al. (1979). They represented the target by orbitals obtained by an SCF calculation for N_2^- in its $^2\Pi_g$ state. In this way they included part of the target polarizability. They then set up the Hamiltonian matrix for the scattered electron interacting with this distorted target in the static exchange approximation. The Green's functions $G_n(R',R)$ were determined by diagonalizing $T_R + E_n(R) - E$ in a basis set of functions $\theta_{nq}(R)$ for each n. Finally the solution of the equations in the external region were obtained assuming that the electron molecule interaction was zero in this region.

We show in figure 13 the potential energy curves and the resonance width calculated by Schneider et al. (1979b) and Le

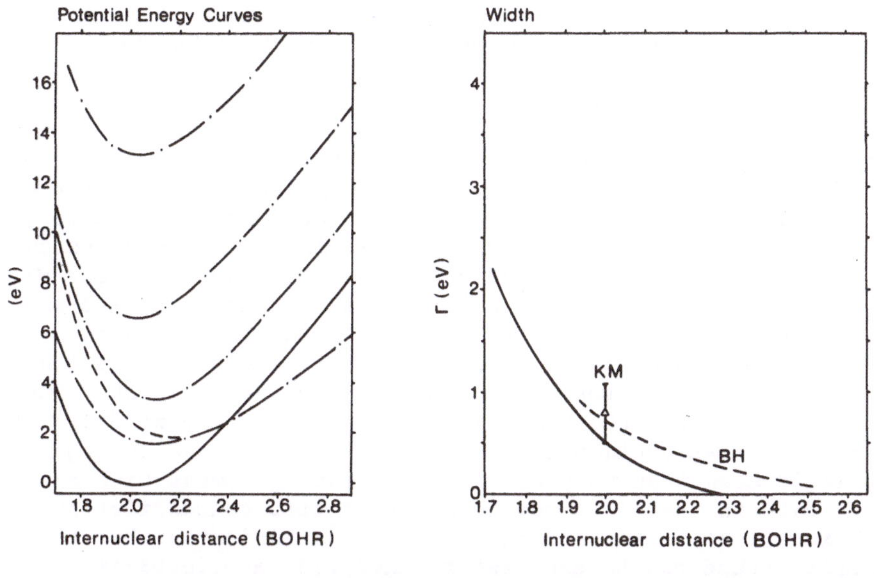

Figure 13 The potential energy curves and the resonance width for the $^2\Pi_g$ state in e-N_2 scattering. The potential energy curves are: solid line, $^1\Sigma_g$ ground state of N_2; dash-dot lines, R-Matrix eigenenergies; dashed line, real part of resonance energy from the R-Matrix calculation. The widths are: solid line, R-matrix calculation; dotted line, calculated by Birtwistle and Herzenberg, (1971); triangle; calculated by Krauss and Mies (1970) (from Le Dourneuf et al. 1979).

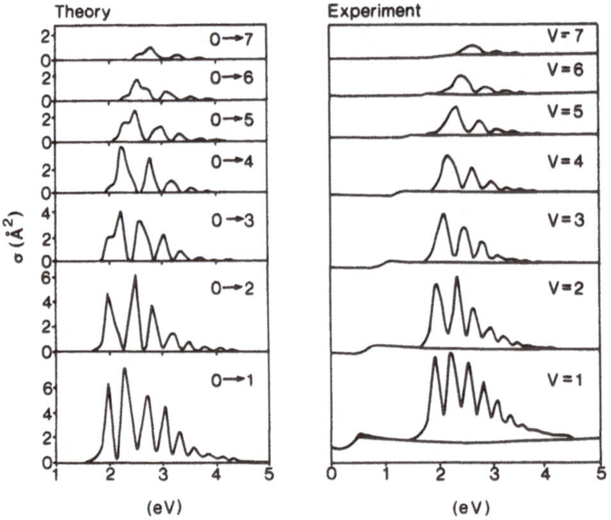

Figure 14 Total vibrational excitation cross sections for $e-N_2$ scattering. Comparison of the R-Matrix calculation of Le Dourneuf et al.(1979) with the experiment of Ehrhardt and Willmann (1967((from Le Dourneuf et al. 1979).

Dourneuf et al.(1979). We see that the real part of the resonance energy obtained by fitting the eigenphase shifts from the R-matrix calculation to the usual resonance formula is close to the second R-matrix eigenenergy at small internuclear distances and is close to the first eigenenergy at large internuclear distances. This shows that there is a close but not a one-to-one relationship between the $W_n(R)$ in eq.(80) and the $E_n(R)$ in eq.(87). The resonance width obtained from these calculations are seen to be in reasonable agreement with earlier work. The results obtained by the R-matrix method for the vibrational excitation cross section in $e-N_2$ scattering is shown in figure 14. It is seen that they are in excellent agreement with experiment. This is the first completely ab-initio calculation for vibrational excitation for this system and indicates that the R-matrix method can be expected to have wide applicability.

In comparing the resonance theory approach with the R-matrix approach we have seen that while they are both basically ab-initio approaches, the former leads itself more naturally to a way of fitting and interpreting the data, while the latter can more easily be used to calculate the cross sections from first principles. In this respect they are complementary and both will undoubtedly have a wide role to play in the future.

6. SCATTERING BY POLAR MOLECULES

Although the theory developed in the previous sections can and has been successfully applied to the collisions of electrons with polar molecules it is of interest to consider these molecules separately. This is because as Massey (1932) pointed out in the first significant study of such systems "... the collision of electrons with a top possessing such a dipole moment may be treated by Born's method whatever the velocity of the electron may be". Although this statement needs some qualification in the light of our present day knowledge, it remains a guiding principle for many transitions where the long range dipole interaction defined by the first term in eq.(13) dominates the cross section. In this section we have only room to give a brief survey of the important features of the collision of electrons with polar molecules and we refer to a comprehensive review by Itikawa (1978) and to a recent paper by Collins and Norcross (1978) for more detailed discussions.

It is useful to commence our discussion by summarizing the cross sections given in the first Born approximation for the scattering of an electron by a rotating point dipole with a potential interaction

$$V_{PD}(r) = -\frac{\mu}{r^2}\cos\theta \tag{93}$$

where μ is the dipole in a.u. (= 2.5418 debye where 1 debye = 10^{-18}esu cm). We find that the differential cross section for a transition from an initial rotational state j to a final rotational state j' is

$$\frac{d\sigma_{j'j}}{d\Omega}(\theta) = \frac{4}{3}\mu^2\frac{j_>}{2j+1}\frac{k'}{k}\frac{1}{k^2+k'^2-2kk'\cos\theta}, \tag{94}$$

where $j_>$ is the larger of j and j', k and k' are the wave numbers for the incident and scattered electron and the selection rule imposed by the cos θ dependence of the potential is that $j' = j \pm 1$. The total cross section obtained by integrating eq.(94) over all scattering angles θ is

$$\sigma_{j'j} = \frac{8\pi}{3}\mu^2\frac{j_>}{2j+1}\frac{1}{k^2}\ln\frac{k+k'}{|k-k'|} \tag{95}$$

Finally the momentum transfer cross section defined by

$$\sigma^{M}_{j'j} = 2\pi \int_0^\infty d\theta \sin\theta\,(1 - \cos\theta)\,\frac{d\sigma_{j'j}}{d\Omega}(\theta) \tag{96}$$

becomes using eq.(94)

$$\sigma^{M}_{j'j} = \frac{8\pi}{3}\,\mu^2\,\frac{j_>}{2j+1}\,\frac{1}{k^2}\left(1 - \frac{(k-k')^2}{2kk'}\,\ell n\,\frac{k+k'}{|k-k'|}\right) \tag{97}$$

We see from eq.(94) that the differential cross section peaks very strongly in the forward direction. This part of the scattering cross section arises from electrons with large impact parameters or large angular momenta corresponding to tail of the potential eq.(93) where it is weak. This region is accurately given by the first Born approximation. However the cross section at large scattering angles contains contributions from small impact parameters or small angular momenta. In this case the electron can penetrate to small r where electron exchange and correlation effects are important and where the Born approximation is no longer applicable. More accurate approaches based on the coupled channel formalisms described in the earlier sections must then be used.

Turning to the total scattering cross section this is dominated in many cases by the forward peak in the differential cross section and hence the Born approximation can often give results which are in reasonable accord with experiment. This result however does not hold true for the momentum transfer cross section where the $(1 - \cos\theta)$ factor in eq.(96) suppresses the forward scattering peak. In this case, the Born approximation can give completely erroneous results in the case of large dipole moments.

Another important aspect of the collision of electrons with polar molecules occurs in the limit when the moment of inertia of the molecule tends to infinity and thus k' tends to k. In this case the differential cross section given by eq.(94) diverges in the forward direction while the total cross section is infinite. The total differential and momentum-transfer cross sections obtained by summing over j' reduce in this limit to

$$\frac{d\sigma}{d\Omega} = \frac{2}{3k^2}\,\mu^2\,\frac{1}{1 - \cos\theta} \tag{98}$$

and to

$$\sigma^M = \frac{8\pi}{3k^2}\,\mu^2 \tag{99}$$

which are independent of j. These results were first obtained in the adiabatic approximation by Altshuler (1957) who considered the elastic scattering of an electron by a fixed dipole. The divergence in the forward direction in eq.(98) arises of course from the contribution from the long range tail of the dipole.

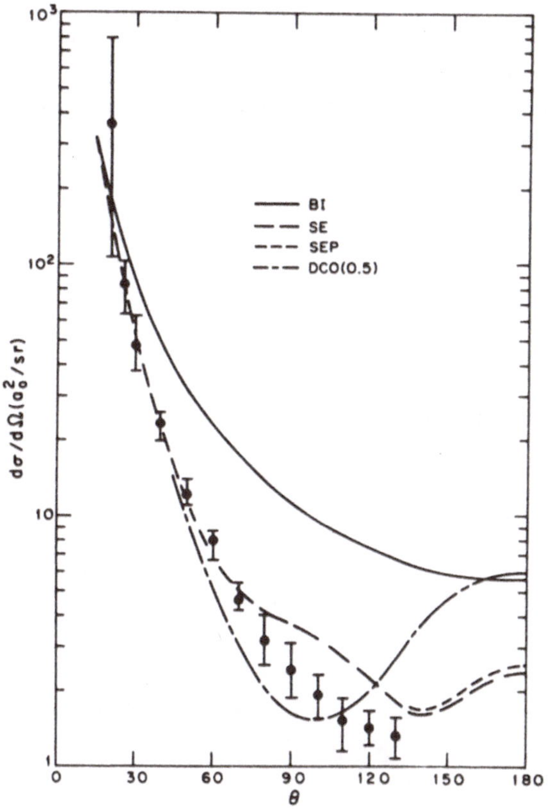

Figure 15 Total differential cross section for e-LiF scattering at 5.44 eV Solid line, the Born approximation; large dashed line, static plus local exchange calculations of Collins and Norcross (1978); small dashed line, static plus local exchange plus polarization calculations of Collins and Norcross (1978); large and small dashed line, cut-off dipole model calculation of Collins and Norcross (1978); circles, experiments of Vuskovic et al.(1978) (from Collins and Norcross, 1978).

Thus a frame transformation theory in which the fixed-nuclei approximation is used in the internal region, and in which the laboratory frame, with the correct moment of inertia of the molecule adopted, is used in the external region, would remove this divergence.

Some of these effects are illustrated in the next three figures. We show in figure 15 the total differential cross section for 5.44 eV electrons scattering from LiF. We see that the Born approximation, while accurate at small angles, breaks down for scattering angles greater than about 30°. However, including just the dipole potential with a cut-off at 0.5 a.u. in the coupled eqs.(7) gives reliable results out to almost 90°. Finally we see that in order to obtain reasonable agreement with experiment in the backward scattering direction both an accurate

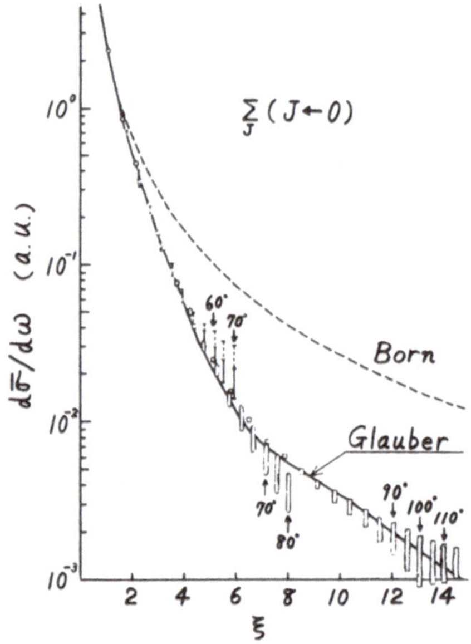

Figure 16 Computed reduced differential cross sections for electron scattering by alkali halides summed over the final rotational states. Born and Glauber approximation calculations for a point dipole potential are compared with close coupling calculations with different choices of the short range potential. Lines, close coupling results for LiF by Collins and Norcross (1978); circles and ovals, close coupling results for CsF by Itikawa (1976); rectangles, close coupling results for KI by Collins and Norcross (1978) (from Shimamura, 1979).

representation of the static potential at short distances and a representation of the exchange potential must be included, although as expected for strongly polar molecules we see that the role of the polarization potential is small.

In figure 16 we show a comparison made by Shimamura (1979) of the reduced differential cross sections for electron scattering from alkali halides summed over all final rotational states computed in the Born approximation the Glauber approximation and by solving the coupled eqs.(29) exactly. The reduced differential cross section plotted in this figure is defined by

$$\frac{d\bar{\sigma}_{j'j}}{d\Omega} = \frac{E}{\mu^4}\frac{d\sigma_{j'j}}{d\Omega} \tag{100}$$

where the incident electron energy E and the dipole moment μ are measured in atomic units. In addition the abscissa in this figure is a reduced momentum transfer defined by

$$\xi = 4\ \mu \sin \theta/2 \tag{101}$$

where μ is measured in atomic units. The calculations in the Glauber approximation were made by Ashihara et al. (1975) for the pure point dipole potential given by eq.(93). This approximation, unlike the Born approximation includes terms from all orders in the potential in the expansion of the scattering amplitude such that unitarity is satisfied. However, like the Born approximation, the resultant expression for the cross section for the dipole potential is easy to evaluate. It is clear from this figure that the Glauber approximation gives very good agreement with the exact solution of the coupled equations over a much larger angular range than the Born approximation. Further, the results indicate that the reduced differential cross sections lies on a single universal curve for scattering angles where the short range potential does not contribute. These results together with figure 15 shows that the first breakdown in the Born approximation for the point dipole potential as the scattering angle increases is due to the breakdown in unitarity and not to the poor representation of the short range part of the potential.

In figure 17 we compare the measured momentum transfer cross section as a function of the dipole moment with a number of calculations. For small dipole moments all calculations are in good agreement and also agree fairly well with experiment. As the dipole moment increases the calculations diverge from each other and, in the absence of experiment in this region, from the probably correct result given by solving eqs.(7) using a range of cut-off dipole potentials. The Born approximation BI obvious-

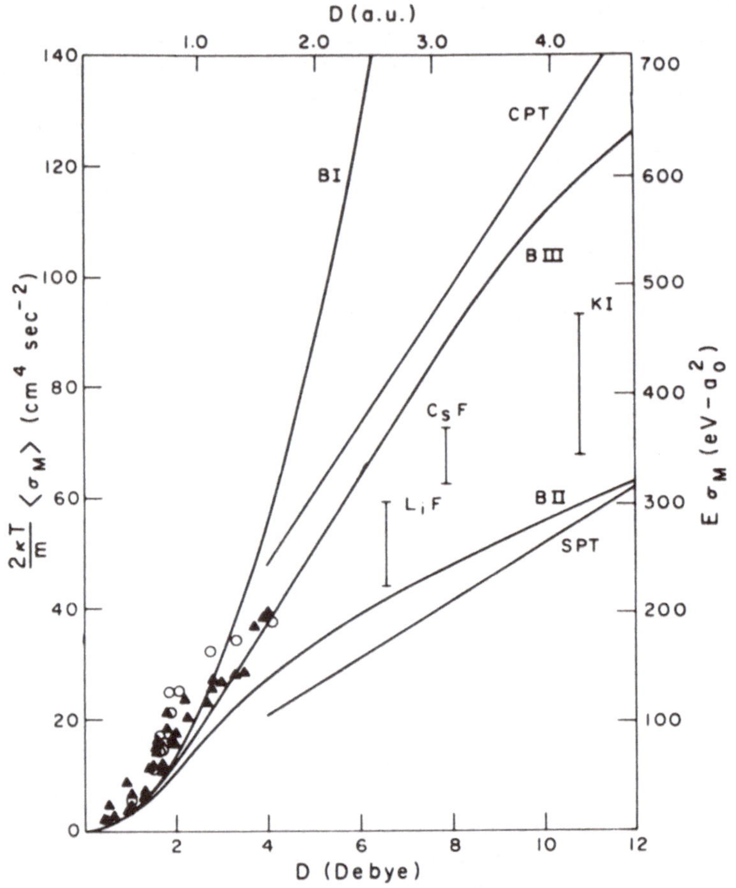

Figure 17 Total momentum transfer cross section as a function of the dipole moment. The experimental data for $\langle \sigma_M \rangle$ are from thermal-energy swarm measurements: Triangles, measurements of Christophorou et al. (1967) and Christophorou and Christodoulides (1969); circles, results of the re-analysis of the measurements of Christophorou and Pittman (1970) by Fabrikant (1977). The calculated results are: BI, first Born approximation; BII and BIII, unitarized Born approximations; CPT, classical perturbation theory of Dickinson (1977); SPT, semi-classical perturbation theory of Mukherjee and Smith (1978) and Hickman and Smith (1978); and Hickman and Smith (1978); lines, close coupling calculations with different choices of the short range potential by Collins and Norcross (1978) (from Collins and Norcross, 1978).

ly diverges the most strongly due to the breakdown in unitarity. Simple unitary corrections suggested by Seaton (1961) were applied by Collins and Norcross (1978) and give the improved BII and BIII results. In fact these corrected Born results appear to be better than the classical perturbation theory result of Dickinson (1977) and the semi-classical perturbation theory result of Mukkerjee and Smith (1978) and Hickman and Smith (1978) although this is probably fortuitous. In conclusion we see that in order to obtain reliable results for the momentum transfer cross section for polar molecules with dipole moments in excess of about 4 debye it is necessary to include unitarity and a reasonable estimate of the short range potential.

The scattering of an electron by a fixed dipole leads to another interesting effect. When the dipole moment exceeds a certain critical value $\mu_c = 0.6393$ a.u. (1.6249 debye) the long range field will support an infinite number of bound states. (We give in table 1 the dipole moments of some typical molecules for illustration). In order to see the reason for this we examine the asymptotic form of the Schrödinger equation for an electron moving in the field of a polar molecule in the fixed-nuclei approximation. We obtain

$$\left(\nabla^2 + k^2 + \frac{2\mu \cos \theta}{r^2}\right)\psi(\underline{r}) = 0 \qquad r > a \qquad (102)$$

TABLE 1

Dipole moments of some molecules of interest in scattering experiments and theoretical studies

Molecule	μ(a.u.)
CO	0.0441
HBr	0.325
$HC\ell$	0.436
HF	0.719
H_2O	0.728
LiH	2.31
CsF	3.10
NaF	3.21
$CaC\ell$	3.54
$KC\ell$	4.04
$CsC\ell$	4.09
KI	4.26

where a is the value of r beyond which the short range exchange and correlation potentials vanish. We now expand $\psi(\underline{r})$ in analogy with our earlier work as

$$\psi(\underline{r}) = \sum_{\ell} r^{-1} u_{\ell}^{m}(r) Y_{\ell}^{m}(\hat{\underline{r}}) \tag{103}$$

Substituting this expansion into eq.(102) and projecting onto the spherical harmonics $Y_{\ell}^{m}(\hat{\underline{r}})$ gives the following set of coupled differential equations for each value of m

$$\left(\frac{d^2}{dr^2} - \frac{\ell(\ell+1)}{r^2} + k^2\right) u_{\ell}^{m}(r) = 2 \sum_{\ell'} V_{\ell\ell'}^{m}(r)\, u_{\ell'}^{m}(r) \tag{104}$$

where the potential $V_{\ell\ell'}^{m}(r)$ is given in analogy with eq.(14) by

$$V_{\ell\ell'}^{m}(r) = \left(\frac{2\ell'+1}{2\ell+1}\right)^{1/2} (\ell'\ 0\ 1\ 0|\ell'\ 1\ \ell\ 0)$$
$$(\ell'\ m\ 1\ 0|\ell'\ 1\ell m)\mu\, r^{-2} = a_{\ell\ell'}^{m}(\mu)\, r^{-2} \tag{105}$$

We now solve eq.(104) by diagonalizing the r^{-2} interaction. We look for the orthogonal matrix $A_{\ell n}$ such that

$$\sum_{\ell\ell'} A_{\ell n} \left[\ell(\ell+1)\delta_{\ell\ell'} + a_{\ell\ell'}^{m}(\mu)\right] A_{\ell' n'} = \lambda_n^m(\lambda_n^m + 1)\delta_{nn'} \, . \tag{106}$$

Equation (104) then becomes

$$\left(\frac{d^2}{dr^2} - \frac{\lambda_n^m(\lambda_n^m+1)}{r^2} + k^2\right) v_n^m(r) = 0 \tag{107}$$

where

$$v_n^m(r) = \sum_{\ell} A_{\ell n}\, u_{\ell}^{m}(r) \tag{108}$$

and we remember that both the eigenvalues λ_n^m and the orthogonal matrix elements $A_{\ell n}$ depend on the dipole moment.

Provided that the eigenvalues $\lambda_n^m(\lambda_n^m + 1)$ satisfy

$$\lambda_n^m(\lambda_n^m + 1) > -\frac{1}{4} \tag{109}$$

then there is nothing unusual in the solution of eq.(107). However if this is not the case λ_n^m becomes complex with values given by

$$\lambda_n^m = -\frac{1}{2} \pm i \operatorname{Im} \lambda_n^m \ . \tag{110}$$

The corresponding zero-energy solution of eq.(107) is then

$$v_n^m(r) = r^{1/2} \sin (\operatorname{Im} \lambda_n^m \ell nr + \delta_n^m) \quad r > a \tag{111}$$

where the phase shift δ_n^m is determined by the boundary condition at $r = a$. Clearly this solution has an infinite number of nodes for $r > a$ which corresponds to an infinity of negative energy bound states supported by the long range "attractive" r^{-2} potential.

Mittleman and von Holdt (1965) have determined the eigenvalues $\lambda^m(\lambda^m + 1)$ and found that for values of the dipole moment larger than μ_c eq.(109) is violated for m=0. This result clearly depends on the assumption that the nuclei are fixed. If rotation is included, which is going to be important for sufficiently large values of r, then the infinite sequence of bound states will be cut off or may move into the continuum to give resonances or virtual states. In the above formalism the inclusion of rotation means that the k^2 values in eq.(104) will depend on the rotational quantum numbers and cannot be diagonalized simultaneously with the r^{-2} potential.

The question concerning the role which the dipole moment plays in low energy electron molecule scattering has also been raised by recent experiments. In a series of papers by Rohr and Linder (1975, 1976), Seng and Linder (1976) and Rohr (1977a,b, 1978a,b) sharply peaked structures were reported close to threshold in the vibrational excitation functions of a number of polar molecules. We show in figure 18 the measured differential cross sections for vibrational excitation of HF, HCl and HBr by electron impact. All curves show a pronounced peak close to threshold and a second broader peak, which is most developed for HBr, at higher energies. We see that while the absolute value of the cross section, including the threshold peak region, increases from HF to HBr, the relative magnitude of the threshold peak, like the dipole moment, is decreasing from HF to HBr. Further information is provided by angular distribution measurements in the threshold resonance region which are found to be flat

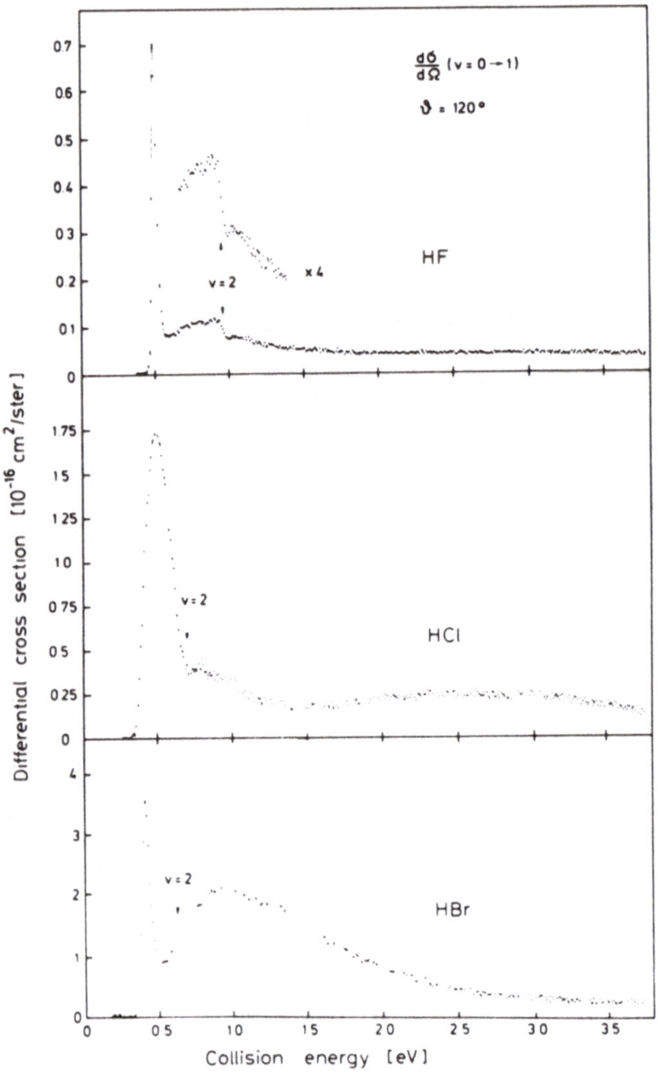

Figure 18 Measured differential cross sections at 120° for v = 0 → 1 vibrational excitation for e-HF, e-HCℓ and e-HBr scattering (from Rohr, 1979).

within the error of experiment suggesting that these structures are mainly s-wave in character.

There are a number of interpretations which have been given for these threshold peaks, some of which are discussed below.

A stabilization calculation by Taylor et al. (1977) and

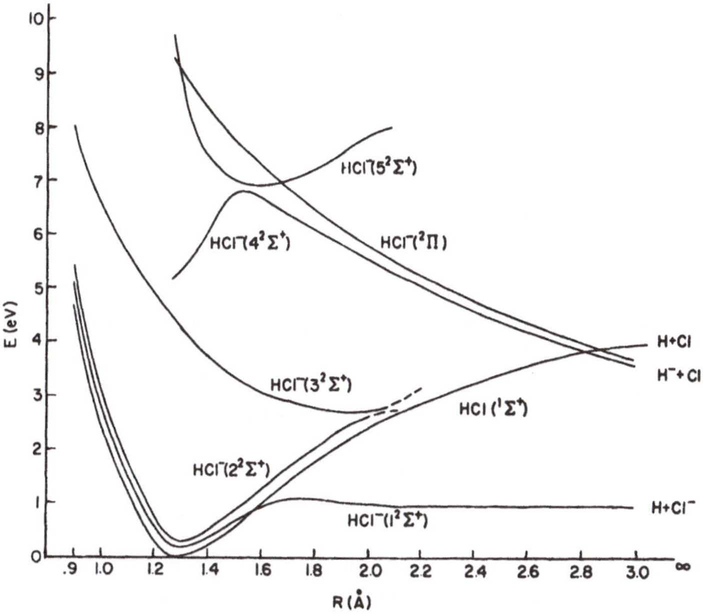

Figure 19 Calculated potential energy curves for $HC\ell$ and $HC\ell^-$ (from Goldstein et al, 1978).

Goldstein et al. (1978) for HCl^- revealed two very low-lying $^2\Sigma^+$ states of HCl^- as shown in figure 19. In addition a third $^2\Sigma^+$ state of HCl^- was found a few eV above the ground state of HCl which might be associated with the broader peak at higher energies seen in figure 18. Since the stabilization calculation involves the diagonalization of the Hamiltonian in a finite L^2 basis as in eq.(44) it provides information on the real part of the energy of the state given by E_k in this equation but not on the imaginary part of the energy. However the calculated potential energy curves, as well as being consistent with the vibrational excitation data, also explain qualitatively the measured dissociative attachment cross sections leading to H^- + Cl and H + Cl^- (Azria et al. 1974, Ziesel et al. 1975 and Abouaf and Teillet-Billy, 1977).

The calculations of Taylor et al.(1977) and Goldstein et al. (1978) were interpreted by Nesbet (1977) in terms of virtual states of the e-HCl system. The concept of a virtual state is illustrated in figure 20 which shows the location of poles in the S-matrix in the complex k plane and in the complex E plane corresponding to a bound state, a virtual state and a resonance state. At a pole in the S-matrix the wavefunction satisfies the outgoing wave boundary condition defined by eq.(76), and thus a virtual state is distinguished from a bound state by the fact that its wavefunction diverges rather than converges exponen-

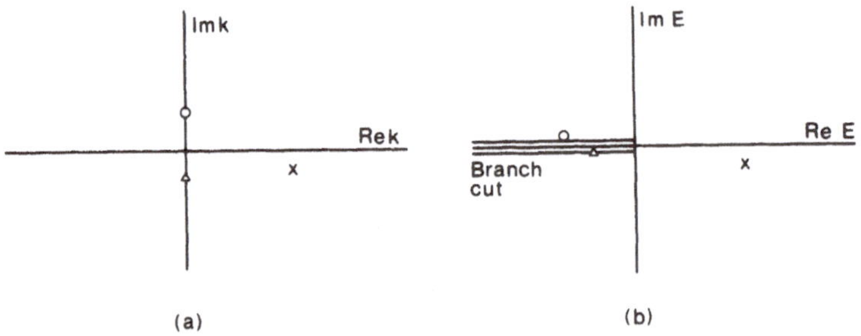

Figure 20 Possible location of poles in the S-matrix in (a) the complex k plane and (b) their corresponding location in the complex E plane. Circle, bound state pole; triangle, virtual state pole; cross, resonance state pole.

tially at large distances. Nesbet argued that the stabilization calculation was trying to reproduce two virtual states of the HCl^- system at the equilibrium nuclear separation distance. Since this calculation can only give eigenvalues on the positive real k axis or on the positive imaginary k axis it is clear from figure 20a that the closest point in the complex k plane that can be reached to a virtual state pole is the point $k = 0$.

Dubé and Herzenberg (1977) fitted the $v = 0 \rightarrow 1$ and $v = 0 \rightarrow 2$ vibrational excitation cross section for e-HCl scattering. They parametrized the logarithmic derivative of the wavefunction or R-matrix, at a radius of 2 a.u. and solved the scattering problem in the outer region with just the dipole potential present. They obtained a good fit with experiment if the short range part of the potential was almost strong enough to support an s-wave bound state. This result is again consistent with the virtual state concept discussed above.

Other models for e-HCl which have been considered are by Gianturco and Rahman (1977), who ascribe the threshold peak to a virtual state caused by the long range dipole potential in contradiction to Dubé and Herzenberg who as we have seen find that the virtual state is caused mainly by the short range potential, by Domcke et al. (1977) who fitted the data to a resonance model involving five parameters and finally by Rudge (1979) who obtained reasonable agreement with experiment by solving the close coupling equations involving vibration and rotation of the molecule with a model potential.

There is still not a definite conclusion from all this work although most of the evidence now points to the threshold peaks being caused by virtual states which are supported mainly by the short range part of the potential. Further evidence for the rather peripheral role played by the dipole potential comes from work by Rohr (1977c, 1979) which showed similar threshold peaks in the case of the polyatomic non-polar molecules CH_4 and SF_6. It is certainly important that accurate calculations, which are now possible in principle using one of the methods described in the previous sections, should be carried out for systems such as e-HCl scattering to finally resolve these questions.

7. SCATTERING AT INTERMEDIATE AND HIGH ENERGIES

In this last section we commence by briefly discussing recent work on models which, while perhaps not as intrinsically accurate as some of the approaches considered in earlier sections, allow the electron molecule electronically elastic cross section to be surveyed rapidly over a broad energy range. Further, since the models are relatively straightforward to apply they can and are being used to study the scattering of electrons from polyatomic molecules which are at present too complex to be treated by other more accurate approaches. Finally we conclude by mentioning work on electronically inelastic electron molecule collisions.

The first model which we will consider is the Continuum Multiple-Scattering Method (CMSM) which was developed primarily by Dill and Dehmer (1974) and recently comprehensively reviewed

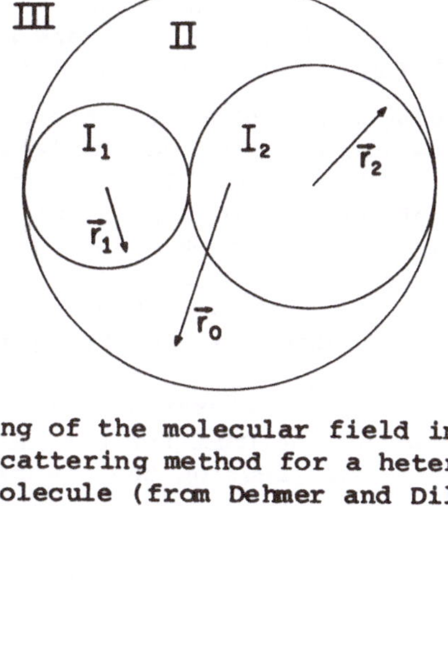

Figure 21 Partitioning of the molecular field in the continuum multiple-scattering method for a heteronuclear diatomic molecule (from Dehmer and Dill, 1979).

by these authors (Dehmer and Dill, 1979). This approach has also been independently considered by Ziesche and John (1976). This approach which was first used in bound state studies (e.g., Johnson, 1973) involves the partition of the molecular field into three regions as illustrated in figure 21. In region I the electron molecule potential is represented by direct and local exchange terms; in region II the potential is approximated by a constant; while in region III a polarization potential is added to the direct and local exchange potentials. Several forms have been explored for the local exchange potential but that due to Hara (1967) seems most effective. Usually the potentials in regions I and III are spherically symmetric although this limitation has been relaxed by Siegel et al (1976) without undue additional effort.

The form assumed for the wavefunction representing the scattered electron in the three regions is as follows. In region I surrounding each nucleus

$$\psi_{I_i} = \sum_{\ell m} a_{\ell m}^{I_i} f_\ell^{I_i}(kr_i)\, Y_\ell^m(\hat{\underline{r}}_i) \quad . \tag{112}$$

In the region between the the inner spheres and the outer sphere

$$\psi_{II} = \sum_{\ell m} a_{\ell m}^{II}\, j_\ell(kr_o)\, Y_\ell^m(r_o) + \sum_i \sum_{\ell m} b_{\ell m}^{II_i}\, n_\ell(kr_i)\, Y_\ell^m(\hat{\underline{r}}_i) \quad . \tag{113}$$

Finally in the outer region

$$\psi_{III} = \sum_{\ell m} \left[a_{\ell m}^{III}\, f_{\ell m}^{III}(kr_o) + b_{\ell m}^{III}\, g_\ell^{III}(kr_o)\right] Y_\ell^m(\hat{\underline{r}}_o) \quad . \tag{114}$$

In these equations f_ℓ and g_ℓ are the regular and irregular solutions in the appropriate regions, and j_ℓ and n_ℓ are the regular and irregular Bessel functions. The coefficients $a_{\ell m}$ and $b_{\ell m}$ are obtained by solving a set of similtaneous equations obtained by requiring that the function and derivatives match on all the spherical boundaries. The phase shifts and cross sections are then obtained by matching ψ_{III} to the asymptotic form defined by eq.(23). If vibrational excitation cross sections are required the calculation is repeated for a number of different internuclear separations and the cross section determined using the adiabatic nuclei approximation defined by eq.(64).

We show in figure 22 the vibrational excitation cross sections calculated using the CMSM for e-N_2 scattering. For the vibrationally elastic cross section the contribution from the

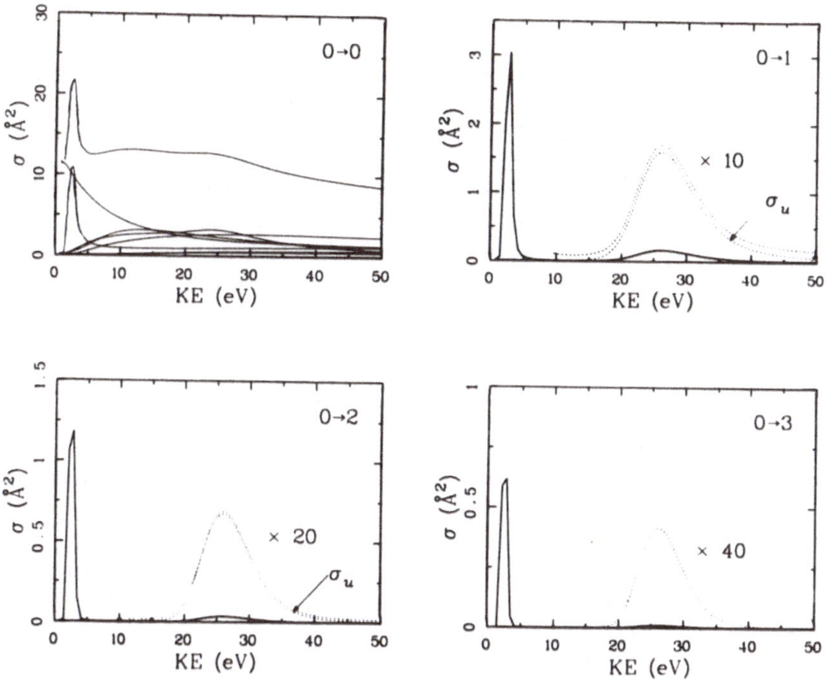

Figure 22 Vibrational excitation cross sections for e-N_2 scattering calculated using the continuum multiple-scattering method (from Dehmer and Dill, 1979).

various symmetry states are also displayed. The well known low energy $^2\Pi_g$ resonance is clearly seen in all the cross sections, although the structure is not given by the adiabatic nuclear approximation used in the CMSM. What is new is the appearance at higher energies of a broad $^2\Sigma_u$ resonance centred at about 26 eV which stands out most prominently in the vibrational excitation cross sections. This feature had been observed previously in experiments by Pavlovic et al. (1972) but the CMSM calculation shows conclusively that it is a shape resonance with mainly f-wave symmetry. The importance of shape resonances in enhancing the vibrational excitation cross section at intermediate energies for many molecules as well as their related role in photoionization of molecules has been emphasised particularly by Dehmer and Dill (1979) and by Dill et al. (1979).

An alternative two-potential approach for intermediate and high energy electronically elastic collisions has recently been proposed by Poe and co-workers (Choi et al. 1979 and Poe, 1979). The essential physical assumptions of this approach are as follows. The scattering can be adequately described as arising from two types of potential; a short-range potential describing the shielded nuclear Coulomb potential and a long-range permanent and polarization potential. The short-range potential's contribution to the scattering is then treated in the impulse approximation, while the long-range potential's contribution is calculated in the Born approximation. Finally these two contributions are summed incoherently. This approach has been applied to e-N_2, e-CO and e-CO_2 scattering and the results over a broad range of energies and angles are in good agreement with the available data. The physical assumptions are seen to be very similar to those of the CMSM and the two approaches can be expected to give similar results at the higher energies although as the energy decreases the CMSM approach will be more reliable since it is non-perturbative.

We have space in these lectures to do no more than mention recent work on electronic excitation. The basic expansion defined by eqs.(3) and (5) will of course enable electronic excitation cross sections to be calculated, however the resultant eqs.(7) have so far proved too complex to solve except in a very few cases. These include calculations by Chung and Lin (1978) who solved these equations for e-H_2 scattering to obtain cross sections for excitation of the $b^3\Sigma_u^+$, $a^3\Sigma_g^+$, $e^3\Sigma_u^+$, $c^3\Pi_u$ and the $^1\Sigma_u^+$ states, and calculations by Robb and Collins (1979) who have obtained cross sections for excitation in e-H_2^+ scattering. Turning to more approximate methods, distorted wave calculations, using L^2 basis functions for the initial and final states, have been made by Rescigno et al.(1975b, 1976) and Fliflet and McKoy (1979) and calculations based on the Born-Ochkur-Rudge approximation have been made by Chung and Lin (1972, 1974) and Chung et al.(1975). However, the overall picture in this case is one where very few calculations have been carried out, and those that have, are very much of an exploratory character. It is clear that electronic excitation and of course ionization while they have been extensively studied experimentally, are very much subjects for the future as far as theoretical work is concerned.

Acknowledgement

These lectures were written and presented immediately following the International Symposium on Electron Molecule Collisions held at the University of Tokyo on 6th and 7th September 1979. It is a pleasure to thank the organizers of this symposium led by Professor Takayanagi Dr. Shimamura and Dr. Matsuzawa and to acknowledge the many valuable discussions with the participants.

REFERENCES

Abouaf, R., and Teillet-Billy, D., 1977, J. Phys. B: Atom. Molec. Phys. 10:2261.
Adkhikari, S.K., and Sloan, I.H., 1975, Phys. Rev., C11:1133.
Allan, M., and Wong, S.F., 1978, Phys. Rev. Lett., 41:1791.
Altshuler, S., 1957, Phys. Rev., 107:114.
Arthurs, A.M. and Dalgarno, A., 1960, Proc. Roy. Soc., A256:334.
Ashihara, O., Shimamura, I., and Takayanagi, K., 1975, J. Phys. Soc. Japan, 38:1732.
Atabek, O., Dill, D., and Jungen, Ch., 1974, Phys. Rev. Lett., 33:123.
Azria, R., Roussier, L., Paineau, R., and Tronc, M., 1974, Rev. Phys. Appl., 9:469.
Baille, P., and Darwych, J.W., 1977, J. Chem. Phys. 67:3399.
Bardsley, J.N., Herzenberg, A., and Mandl, F., 1966, Proc. Phys. Soc., 89:321.
Bardsley, J.N., and Mandl, F., 1968, Rep. Prog. Phys., 31:471.
Birtwistle, D.T., and Herzenberg, A., 1971, J. Phys. B: Atom. Molec. Phys., 4:53.
Blatt, J.M., and Biedenharn, L.C., 1952, Rev. Mod. Phys., 24:258.
Bloch, C., 1957, Nucl. Phys., 4:503.
Bottcher, C., 1969, Chem. Phys. Lett., 4:320.
Brode, R.B., 1925, Phys. Rev. 25:636.
Brode, R.B., 1933, Rev. Mod. Phys., 5:257.
Bruche, E., 1927, Ann. Physik, 83:1065.
Buckley, B.D. and Burke, P.G., 1979, Invited paper at the Symposium on Electron Molecule Collisions, edited by I. Shimamura and M. Matsuzawa (published by the University of Tokyo) p.37
Buckley, B.D., Burke, P.G., and Vo Ky Lan, 1979, Comput. Phys. Commun. 17:175.
Burke, P.G., and Chandra, N., 1972, J. Phys. B: Atom.Molec.Phys., 5:1696
Burke, P.G., Chandra, N., and Gianturco, F.A., 1972, J. Phys. B: Atom. Molec. Phys., 5:2212.
Burke, P.G., and Hibbert, A., 1969, in Abstracts of the Sixth International Conference on the Physics of Electronic and Atomic Collisions (Massachusetts Institute of Technology, Cambridge, Mass. 1969), 367.
Burke, P.G., Hibbert, A., and Robb, W.D., 1971, J. Phys. B: Atom. Molec. Phys., 4:153.
Burke, P.G., Mackey, I., and Shimamura, I., 1977, J. Phys. B: Atom. Molec. Phys., 12:2497.
Burke, P.G., and Robb, W.D., 1975, Adv. Atom. Molec. Phys. 11:143.
Burke, P.G., and Sinfailam, A.L., 1970, J. Phys. B: Atom. Molec. Phys., 3:641.
Castillejo, L., Percival, I.C., and Seaton, M.J., 1960, Proc. Roy. Soc. A254:259.
Chandra, N., and Temkin, A., 1976a, Phys. Rev., A13:188.

Chandra, N., and Temkin, A., 1976b, Phys. Rev., A14:507.
Chang, E.S., and Temkin, A., 1969, Phys. Rev. Lett., 23:399.
Chang, E.S., and Fano, U., 1972, Phys. Rev., A6:173.
Chase, D.M., 1956, Phys. Rev., 104:838.
Choi, B.H. and Poe, R.T., 1977, Phys. Rev., A16:1821; Ibid, 1831.
Choi, B.H., Poe, R.T., Sun, J.C., and Shan, Y., 1979, Phys. Rev. A19:116.
Christophorou, L.G., and Christodoulides, J., 1969, J. Phys. B: Atom. Molec. Phys., 2:71.
Christophorou, L.G., Hurst, G.S., and Hendrick, W.G., 1967, J. Chem. Phys., 45:1081.
Christophorou, L.G., and Pittman, D., 1970, J. Phys. B: Atom. Molec. Phys., 3:1252.
Chung, S., and Lin, C.C., 1972, Phys. Rev. A6:988.
Chung, S., and Lin, C.C., 1974, Phys. Rev. A9:1954.
Chung, S., and Lin, C.C., 1978, Phys. Rev. A17:1874.
Chung, S., Lin, C.C., and Lee, E.T.P., 1975, Phys. Rev. A12:1340.
Collins, L.A., and Norcross, D.W., 1978, Phys. Rev., A18:467.
Collins, L.A., Robb, W.D., and Morrison, M.A., 1978, J. Phys. B: Atom. Molec. Phys., 11:L777.
Collins, L.A., Robb, W.D., and Norcross, D.W., 1979, to be published in Phys. Rev. A.
Crawford, O.H., and Dalgarno, A., 1971, J. Phys. B: Atom. Molec. Phys., 4:494.
Damburg, R., and Karule, E., 1967, Proc. Phys. Soc., 90:677.
Dehmer, J.L., and Dill, D., 1979, to be published in "Electron and Photon Molecule Collisions" edited by V. McKoy, T.N. Rescigno, and B.I. Schneider (Plenum Press, New York).
Dickinson, A.S., 1977, J. Phys. B: Atom. Molec. Phys., 10:967.
Dill, D., 1972, Phys. Rev., A6:160.
Dill, D., and Dehmer, J.L., 1974, J. Chem. Phys., 61:692.
Dill, D., Welch, J., Dehmer, J.L., and Siegel, J., 1979, Phys. Rev. Lett., 43:1236.
Domcke, W., Cederbaum, L.S., and Kaspar, F., 1979, J. Phys. B: Atom. Molec. Phys., 12:L359.
Dubé, L., and Herzenberg, A., 1977, Phys. Rev. Lett., 38:820.
Dubé, L., and Herzenberg, A., 1979, Phys. Rev., 20:194.
Ehrhardt, H., and Willmann, K., 1967, Z. Phys., 204:462.
Fabrikant, I.I., 1977, J. Phys. B: Atom. Molec. Phys., 10:1761.
Fano, U., 1970, Phys. Rev., A2:353.
Fano, U., 1975, J. Opt. Soc. Am., 65:979.
Feshbach, H., 1958, Ann. of Phys., 5:357.
Feshbach, H., 1962, Ann. of Phys., 19:287.
Fliflet, A.W., Levin, D.A., Ma, M., and McKoy, V., 1978, Phys. Rev. A17:160.
Fliflet, A.W., and McKoy, V., 1978a, Phys. Rev., A18:1048.
Fliflet, A.W., and McKoy, V., 1978b, Phys. Rev., A18:2107.
Fliflet, A.W., and McKoy, V., 1979, submitted to Phys. Rev. A.
Gianturco, F.A., and Thompson, D.G., 1976, J. Phys. B: Atom. Molec. Phys. 9:L383.

Gianturco, F.A., and Rahman, N.K., 1977, Chem. Phys. Lett., 48:380.
Golden, D.E., Bandel, H.W., and Salerno, J.A., 1966, Phys. Rev. 146:40.
Golden, D.E., Lane, N.F., Temkin, A., and Gerjuoy, E., 1971, Rev. Mod. Phys., 43:642.
Goldstein, E., Segal, G.A., and Wetmore, R.W., 1978, J. Chem. Phys., 68:74.
Hara, S., 1967, J. Phys. Soc. Japan, 22:710.
Hazi, A.U., and Taylor, H.S., 1970, Phys. Rev. A1:1109.
Hazi, A.U., 1979, to be published in: "Electron and Photon-Molecule Collisions", eds., V. McKoy, T.N. Rescigno, and B.I. Schneider, (Plenum Press, New York).
Henry, R.J.W., and Lane, N.F., 1969, Phys. Rev., 183:221.
Herzenberg, A., 1968, J. Phys. B: Atom. Molec. Phys., 1:548.
Herzenberg, A., 1978, in: "Electronic and Atomic Collisions", Edited by G. Watel, (North Holland Publishing Co., Amsterdam) p.1.
Herzenberg, A., and Mandl, F., 1962, Proc. Roy. Soc., A270:48.
Hickman, A.P., and Smith, F.T., 1978, Phys. Rev. A17:968.
Itikawa, Y., 1976, J. Phys. Soc., Japan, 41:619.
Itikawa, Y., 1978, Phys. Reports, 46:117.
Johnson, K.H., 1973, in "Advances in Quantum Chemistry" edited by P.O. Lowdin, (Academic Press, New York) p.143.
Jungen, Ch., and Atabek, O., 1977, J. Chem. Phys., 66:5584.
Kendrick, J., 1978, J. Phys. B: Atom. Molec. Phys., L601.
Kendrick, J., 1980, Daresbury Laboratory Report.
Klonover, A., and Kaldor, U., 1977, Chem. Phys. Letters, 51:321.
Klonover, A., and Kaldor, U., 1978, J. Phys. B: Atom. Molec. Phys., 11:321.
Klonover, A., and Kaldor, U., 1979a, J. Phys. B: Atom. Molec. Phys., 12:323.
Klonover, A., and Kaldor, U., 1979b, J. Phys. B: Atom. Molec. Phys., 12:L61.
Kollath, R.K., 1932, Ann. Physik, 15:485.
Krauss, M., and Mies, F.H., 1970, Phys. Rev., A1:1592.
Lane, N.F., 1980, Rev. Mod. Phys., 52:29.
Le Dourneuf, M., Vo Ky Lan, and Burke, P.G., 1977, Comm. Atom. Molec. Phys., 7:1.
Le Dourneuf, M., Vo Ky Lan, and Schneider, B.I., 1979, Invited paper at the Symposium on Electron Molecule Collisions. Edited by I. Shimamura and M. Matsuzawa, (published by the University of Tokyo) post deadline contribution.
Lee, C.M., 1977, Phys. Rev. A16:109.
Levin, D.A., Fliflet, A.W., Ma, M., and McKoy, V., 1979, J. Comput. Phys., to be published.
Linder, F., and Schmidt, H., 1971, Z. Naturf., 26a:1603.
Lovelace, C., 1964, Phys. Rev., 135:B1225.
Massey, H.S.W., 1932, Proc. Camb. Phil. Soc., 28:99.
Massey, H.S.W., 1935, Trans. Faraday Soc., 31:556.

McCurdy, C.W., Jr., Rescigno, T.N., and McKoy, V., 1976, J. Phys. B: Atom. Molec. Phys., 9:691.
Mittleman, M.H., and von Holdt, R.E., 1965, Phys. Rev., 140:A726.
Morrison, M.A., and Collins, L.A., 1978, Phys. Rev., A17:918.
Morrison, M.A., Lane, N.F., and Collins, L.A., 1977, Phys. Rev., A15:2186.
Morrison, M.A., and Schneider, B.I., 1977, Phys. Rev., A16:1003.
Mukherjee, D., and Smith, F.T., 1978, Phys. Rev., A17:954.
Nesbet, R.K., 1977, J. Phys. B: Atom. Molec. Phys., 10:L739.
Nesbet, R.K., 1979, Phys. Rev., A19:551.
Ostlund, N.S., 1975, Chem. Phys. Lett., 34:419.
Pavlovic, ., Boness, M.J., Herzenberg, A., and Schulz, G.J., 1972, Phys. Rev., A6:676.
Poe, R.T., 1979, Invited paper at the Symposium on Electron Molecule Collisions. Edited by I. Shimamura and M. Matsuzawa (published by the University of Tokyo) 49.
Ramsauer, C., and Kollath, R.K., 1927, Ann. Physik, 83:1129.
Ramsauer, C., and Kollath, R.K., 1930, Ann. Physik, 4:91.
Reinhardt, W.P., 1979, Comput. Phys. Commun., 17:1.
Rescigno, T.N., 1979, private communication.
Rescigno, T.N., McCurdy, C.W., and McKoy, V., 1974a, Chem. Phys. Lett., 27:401.
Rescigno, T.N., McCurdy, C.W., and McKoy, V., 1974b, Phys. Rev., A10:2240.
Rescigno, T.N., McCurdy, C.W., and McKoy, V., 1975a, Phys. Rev., A11:825.
Rescigno, T.N., McCurdy, C.W., and McKoy, V., 1975b, J. Phys. B: Atom. Molec. Phys., 8:L433.
Rescigno, T.N., McCurdy, C.W., McKoy, V., and Bender, C.F., 1976, Phys. Rev., A13:216.
Riley, M.E., and Truhlar, D., 1975, J. Chem. Phys., 63:2182.
Robb, W.D., and Collins, L.A., 1979, private communication.
Rohr, K., 1977a, J. Phys. B: Atom. Molec. Phys., 10:L399.
Rohr, K., 1977b, J. Phys. B: Atom. Molec. Phys., 10:L735.
Rohr, K., 1977c, J. Phys. B: Atom. Molec. Phys., 10:1175.
Rohr, K., 1978a, J. Phys. B: Atom. Molec. Phys., 11:1849.
Rohr, K., 1978b, J. Phys. B: Atom. Molec. Phys., 11:4109.
Rohr, K., 1979, Invited paper at the Symposium on Electron Molecule Collisions. Edited by I. Shimamura and M. Matsuzawa (published by the University of Tokyo) p.67.
Rohr, K., and Linder, F., 1975, J. Phys., B: Atom. Molec. Phys., 8:L200.
Rohr, K., and Linder, F., 1976, J. Phys. B: Atom. Molec. Phys., 9:2521.
Ross, M.H., and Shaw, G.L., 1961, Ann. Phys., 13:147.
Rudge, M.R.H., 1979, J. Phys. B: Atom. Molec. Phys., submitted.
Schneider, B.I., 1975a, Chem. Phys. Lett., 31:237.
Schneider, B.I., 1975b, Phys. Rev., A11:1957.
Schneider, B.I., and Hay, P.J., 1976a, J. Phys. B: Atom. Molec. Phys., 9:L165.

Schneider, B.I., and Hay, P.J., 1976b, Phys. Rev., A13:2049.
Schneider, B.I., 1978, in "Electronic and Atomic Collisions", Edited by G. Watel, (North Holland Publishing Co., Amsterdam) p.257.
Schneider, B.I., Le Dourneuf, M., and Burke, P.G., 1979a, J. Phys. B: Atom. Molec. Phys., 12:L365.
Schneider, B.I., Le Dourneuf, M., and Vo Ky Lan, 1979b, Phys. Rev. Lett., 43:1926.
Schulz, G.J., 1962, Phys. Rev., 125:229.
Schulz, G.J., 1964, Phys. Rev., 135:A988.
Schulz, G.J., 1973, Rev. Mod. Phys., 45:423.
Seaton, M.J., 1961, Proc. Phys. Soc., 77:174.
Seng, G., and Linder, F., 1976, J. Phys. B: Atom. Molec. Phys., 9:2539.
Shimamura, I., 1979, Invited paper at the Symposium on Electron Molecule Collisions. Edited by I. Shimamura and M. Matsuzawa (published by the University of Tokyo) p.13.
Siegel, J., Dill, D., and Dehmer, J.L., 1976, J. Chem. Phys., 64:3204.
Siegert, A.J.F., 1939, Phys. Rev., 56:750.
Slater, J.C., 1960, Quantum Theory of Atomic Structure, Vols.1 and 2. (McGraw Hill, New York).
Taylor, H.S., Goldstein, E., and Segel, G.A., 1977, J. Phys. B: Atom. Molec. Phys., 10:2253.
Temkin, A., 1966, in "Autoionization". Edited by A. Temkin. (Mono Book Corporation, Baltimore) p.55.
Temkin, A., and Faisal, F.H.M., 1971, Phys. Rev., A3:520.
Temkin, A., 1978, Phys. Rev., A17:1232.
Temkin, A., 1979a, to be published in "Electron and Photon-Molecule Collisions", eds., V. McKoy, T.N. Rescigno, and B.I. Schneider, (Plenum Press, New York).
Temkin, A., 1979b, in "Invited Papers and Progress Reports", XI ICPEAC. Edited by K. Takayanagi. (North Holland Publishing Company).
Vuskovic, L., Srivastava, S.K., and Trajmar, S., 1978, J. Phys. B: Atom. Molec. Phys., 11:1643.
Wadehra, J.M., and Bardsley, J.N., 1978, Phys. Rev. Lett., 41:1791.
Watson, D.K., Lucchese, R.R., McKoy, V., and Rescigno, T.N., 1979, submitted to Phys. Rev. A.
Wigner, E.P., 1946a, Phys. Rev. 70:15.
Wigner, E.P., 1946b, Phys. Rev. 70:606.
Wigner, E.P., and Eisenbud, L., 1947, Phys. Rev., 72:29.
Wong, S.F., 1978, unpublished measurements reproduced by A. Herzenberg in "Electronic and Atomic Collisions" Edited by G. Watal. (North Holland Publishing Company) p.1.
Ziesche, P., and John, W., 1976, J. Phys. B: Atom. Molec. Phys., 9:333.
Ziesel, J.P., Nenner, I., and Schulz, G.J., 1975, J. Chem. Phys., 63:1943.

LIST OF PARTICIPANTS

DIRECTOR:
DR. R.G. WOOLLEY, Cavendish Laboratory, University of Cambridge, Madingley Road, CAMBRIDGE CB3 OHE, United Kingdom.

ORGANIZING COMMITTEE:
PROFESSOR R.S. BERRY, Department of Chemistry, University of Chicago, 5735 South Ellis Avenue, CHICAGO, Illinois 60637, U.S.A.
PROFESSOR J.-M. COMBES, Departément de Mathématiques, Centre Universitaire de Toulon, Château Saint-Michel, 83130 LA GARDE, France.
DR. B.T. SUTCLIFFE, Department of Chemistry, University of York, Heslington, YORK YO1 5DD, United Kingdom.

LECTURERS:
PROFESSOR P.G. BURKE, F.R.S., Daresbury Laboratory, Science Research Council, Daresbury, WARRINGTON WA4 4AD, United Kingdom.
DR. E.B. DAVIES, Mathematical Institute, University of Oxford, 24-29 St. Giles, OXFORD OX1 3LB, United Kingdom.
PROFESSOR DR. W. DEMTRÖDER, Fachbereich Physik, Universität Klaiserslautern, D-675 KAISERSLAUTERN, Pfaffenbergstrasse 95, Postfach 3049, West Germany.
PROFESSOR R.A. HARRIS, Department of Chemistry, University of California, BERKELEY, California 94720, U.S.A.
DR. L. LATHOUWERS, Faculteit der Wetenschappen, Rijksuniversitair Centrum Antwerpen, Middelheimlaan 1, 2020 ANTWERP, Belgium.
PROFESSOR D.H. LEVY, The James Franck Institute, University of Chicago, 5640 Ellis Avenue, CHICAGO, Illinois 60637, U.S.A.
PROFESSOR H. PRIMAS, Laboratory of Physical Chemistry, Swiss Federal Institute of Technology, Universitätstrasse 16, CH-8092 ZÜRICH, Switzerland.
PROFESSOR S.A. RICE, The James Franck Institute, University of Chicago, 5640 Ellis Avenue, CHICAGO, Illinois 60637, U.S.A.
PROFESSOR R. SEILER, Freie Universität Berlin, Institut für Theoretische Physik (WE4), Arnimallee 3, 1000 BERLIN 33, West Germany.
PROFESSOR P. VAN LEUVEN, Dienst Theoretische en Wiskundige Natuurkunde, Rijksuniversitair Centrum Antwerpen, Groenenborgerlaan 171, 2020 ANTWERP, Belgium.

STUDENTS:

Francois G. AMAR, Department of Chemistry, University of Chicago, 5735 South Ellis Avenue, CHICAGO, Illinois 60637, U.S.A.

Sheldon ARONOWITZ, National Aeronautics and Space Administration, Ames Research Center, 239-9 MOFFETT FIELD, California 94035, U.S.A.

Michael BERMAN, Institute of Chemistry, University of Tel-Aviv, Ramat-Aviv, TEL AVIV, Israel.

Vincenzo CARRAVETTA, Laboratorio di Chimica Quantistica, Via Risorgimento 35, 56100 PISA, Italy.

Renzo CIMIRAGLIA, Istituto di Chimica Fisica, Via Risorgimento 35, 56100 PISA, Italy.

Charles W. CLARK, Department of Physics, University of Chicago, 1108 East 58 Street, CHICAGO, Illinois 60637, U.S.A.

Monique COMBESCURE, L.P.T.H.E. - Bâtiment 211, Université de Paris-Sud, 91405 ORSAY CEDEX, France.

Peter D. DACRE, Department of Chemistry, University of Sheffield, SHEFFIELD S3 7HF, United Kingdom.

Jens Peder DAHL, Chemistry Department B, Chemical Physics, Technical University of Denmark, DTH 301, DK-2800 LYNGBY, Denmark.

Esper DALGAARD, Department of Chemistry, University of Aarhus, DK-8000 AARHUS, Denmark.

Ian D. DAWBARN, Mathematical Institute, University of Oxford, 24-29 St. Giles, OXFORD OX1 3LB, United Kingdom.

David A. DOLSON, Department of Chemistry, Indiana University, BLOOMINGTON, Indiana 47405, U.S.A.

Maarten C.A. DONKERSLOOT, Laboratorium voor Organische Chemie, Eindhoven University of Technology, P.O. Box 513, EINDHOVEN, The Netherlands.

Pierre DUCLOS, Departément de Mathématiques, Centre Universitaire de Toulon, Château Saint-Michel, 83130 LA GARDE, France.

John O. EAVES, Joint Institute for Laboratory Astro-physics, University of Colorado, BOULDER, Colorado 80309, U.S.A.

Gregory EZRA, Physical Chemistry Laboratory, University of Oxford, South Parks Road, OXFORD OX1 3QZ, United Kingdom.

Knut FAEGRI, Jr., Department of Chemistry, University of Oslo, P.O. Box 1033, Blindern, OSLO 3, Norway.

Stavros FARANTOS, School of Molecular Sciences, University of Sussex, Falmer, BRIGHTON BN1 9QJ, United Kingdom.

Werner GANS, Laboratory of Physical Chemistry, Swiss Federal Institute of Technology, Universitätstrasse 16, CH 8092 ZÜRICH, Switzerland.

Jean-Pierre GAZEAU, Laboratoire de chimi physique, 11 rue Pierre et Marie Curie, 75005 PARIS, France.

Genevieve GRENET, Institut de Physique Nucléaire, Université Claude Bernard Lyon 1, 43 bd du 11 Novembre 1918, 69621 VILLEURBANNE, France.

Karl GUSTAFSON, Department of Mathematics, University of Colorado, BOULDER, Colorado 80309, U.S.A.

Peter HABITZ, Institut für Physikalische Chemie und Elektrochemie, (Theoretische Chemie), 75 KARLSRUHE, Kaiserstrasse 12, West Germany.

Klaus D. HÄNSEL, Projektgruppe für Laserforschung der Max-Planck Gesellschaft, D-8046 GARCHING, West Germany.
Richard HEENAN, Department of Chemistry, University of Reading, Whiteknights, READING RG6 2AD, United Kingdom.
Richard M. HEDGES, Jr, Joint Institute for Laboratory Astro-physics, University of Colorado, BOULDER, Colorado, 80309, U.S.A.
Thomas HOFFMAN-OSTENHOF, Institut für Theoretische Chemie und Strahlenchemie der Universität Wien, Währingerstrasse 17, A-1090 VIENNA, Austria.
Brian J. HOWARD, Physical Chemistry Laboratory, University of Oxford, South Parks Road, OXFORD OX1 3QZ, United Kingdom.
Ione IGA, Departamento de Quimica, Universidade Federal de São Carlos, Caixa Postal 676, 13560 SÃO CARLOS-SP, Brazil.
Claude JAUFFRET, U.E.R. Physique Fondamentale, Université des Sciences et Techniques, 59655 VILLENEUVE D'ASCQ CEDEX, France.
Juha M. JAVANAINEN, Research Institute for Theoretical Physics, University of Helsinki, Siltavuorenpenger 20C, 00170 HELSINKI 17, Finland.
Maurice KIBLER, Institut de Physique Nucléaire, 43 bd du 11 Novembre 1918, 69621 VILLEURBANNE, France.
Patrick R.R. LANGRIDGE-SMITH, Physical Chemical Laboratory, University of Oxford, South Parks Road, OXFORD OX1 3QZ, United Kingdom.
Mu-Tao LEE, Departamento de Quimica, Universidade Federal de São Carlos, Caixa Postal 676, 13560 SÃO CARLOS-SP, Brazil.
Elliott H. LIEB, Departments of Mathematics and Physics, Princeton University, P.O. Box 708, PRINCETON, New Jersey 08544, U.S.A.
David E. LOGAN, University Chemical Laboratory, University of Cambridge, Lensfield Road, CAMBRIDGE CB2 1EW, United Kingdom.
Giancarlo MARCONI, Laboratorio Frae CNR, Via Castagnoli 1, BOLOGNA 40126, Italy.
Noémio MACIA MARQUES, Laboratorio de Fisica da Universidade de Lisboa, Faculdade de Ciencias, Rua da Escola Politecnica, 58, 1294 LISBOA CODEX, Portugal.
Elaine Ann MOORE, Faculty of Science, The Open University, Walton Hall, MILTON KEYNES MK7 6AA, United Kingdom.
John MORGAN, Department of Physics, Princeton University, PRINCETON, New Jersey 08544, U.S.A.
André NAUTS, Université Catholique de Louvain, Laboratoire de Chimie Quantique, Bâtiment Lavoisier - Place Louis Pasteur 1, 1348 LOUVAIN-LA-NEUVE, Belgium.
T. Tung NGUYEN DANG, Department of Chemistry, McMaster University, HAMILTON, Ontario L8S 4M1, Canada.
Jose C. NOGUEIRA, Departamento de Quimica, Universidade Federal de São Carlos, Caixa Postal 676, 13560 SÃO CARLOS-SP, Brazil.
Donald W. NOID, Chemistry Division, Oak Ridge National Laboratory, P.O. Box X, OAK RIDGE, Tennessee 37830, U.S.A.
Juan J. NOVOA, Department of Physical Chemistry, Faculty of Chemistry, University of Barcelona, BARCELONA, Spain.

H. Önder PAMUK, Department of Chemistry, Middle East Technical University, ANKARA, Turkey.
Rex PENDLEY, Department of Chemistry, Indiana University, BLOOMINGTON, Indiana 47405, U.S.A.
Carlo PETRONGOLO, Laboratorio di Chimica Quantistica, Via Risorgimento 35, 56100 PISA, Italy.
Peter PFEIFER, Laboratory of Physical Chemistry, Swiss Federal Institute of Technology, Universitätstrasse 16, CH-8092 ZÜRICH, Switzerland.
Roger PRAT, Laboratoire de Photophysique Moléculaire, Batiment 213, Université de Paris Sud, 91405 ORSAY CEDEX, France.
Guido RAGGIO, Laboratory of Physical Chemistry, Swiss Federal Institute of Technology, Universitätstrasse 16, CH-8092 ZÜRICH, Switzerland.
Patricia L. RADLOFF, The James Franck Institute, University of Chicago, 5640 Ellis Avenue, CHICAGO, Illinois 60637, U.S.A.
Paul B. REHMUS, Department of Chemistry, University of Chicago, 5735 South Ellis Avenue, CHICAGO, Illinois 60637, U.S.A.
Joerg REUSS, Faculteit Wiskunde & Natuurwetenshappen, University of Nymegen, Tournooiveld, NYMEGEN, The Netherlands.
William C. ROWLAND, Department of Chemistry, Louisiana State University, BATON ROUGE, Louisiana 70803, U.S.A.
Francesc SAGUES MESTRE, Department of Physical Chemistry, Faculty of Chemistry, University of Barcelona, BARCELONA, Spain.
Eduardo SANHUEZA, Quantum Chemistry Group, University of Uppsala, Box 518, S-751 20 UPPSALA 1, Sweden.
Lutgard SCHEIRE, Seminarie voor Theoretische Vaste Stof - en Lage Energie Kernfysica, Rijksunitersiteit Gent, Krigslaan 271/59, B-9000 GENT, Belgium.
Stephen J. SMITH, Department of Chemistry, University of York, Heslington, YORK YO1 5DD, United Kingdom.
Peter R. TAYLOR, Mathematical Institute, University of Oxford, 24-29 St. Giles, OXFORD OX1 3LB, United Kingdom.
Ad VAN DER AVOIRD, Institute of Theoretical Chemistry, University of Nijmegen, Toernooiveld, NIJMEGEN, The Netherlands.
Ewine F. VAN DISHOECK, Gorlaeus-Laboratories, Department of Physical Chemistry, State University Leiden, P.O. Box 75, LEIDEN, The Netherlands.
Marc C. VAN HEMERT, Gorlaeus-Laboratories, Department of Physical Chemistry, State University Leiden, P.O. Box 75, LEIDEN, The Netherlands.
Antonio J.C. VARANDAS, Departamento de Quimica, Universidade de Coimbra, 3000 COIMBRA, Portugal.
Paul E.S. WORMER, Institute of Theoretical Chemistry, University of Nijmegen, Toernooiveld, NIJMEGEN, The Netherlands.

INDEX

Absolute wavelength measurement, 420-421
Absorption coefficient, 390-391
Acceleration cooling, 405
Action-angle variables, 263-271, 274-276, 302, 314-316, 340, 342, 344
Adiabatic Hamiltonian, 469, 475
Adiabatic nuclei approximation, 513, 522, see fixed nuclei approximation
Ammonia inversion, 369
Anti-Henon-Heiles system, 337
Arnold diffusion, 303
Arnold web, 303
Atom-atom scattering, 234-237

Banach space, 69, 70, 76, 90, 96
Barbanis Hamiltonian, 291, 306
Bennet-hole, 407
β-decay, 360-361
Birkhoff normal form, 299-301, 316
Bloch equations, 243, 247
Bohr-Sommerfeld-Wilson quantization, 305, 311, 312, 316, 319
Boolean algebra, 42-50, 52-56, 65, 68, 71, 73, 74, 83, 85-88, 100
Boolean atlas, 85-86
Boolean frame of reference, 42-44, 50-55, 65, 70, 74, 83-85
Boolean lattice, 48, 73-76, 78, 84, 89
Boolean logic, 44, 68, 84, 85, 87
Boomerang model, 518, 521-522
Borel algebra, 51-55, 66-67
Borel sets, 54-55, 58-60, 65, 67-69, 72, 77, 82
Born approximation, 527-531, 542
Born-Oppenheimer method, 106, 145-148, 154, 155, 157, 159, 164, 180, 197, 219, 222, 240, 250, 369, 435, 458-482, 516, 520, 523
Brumer-Toda-Duff criterion, 320, 331-338, 352

C*-algebra, 71, 76, 90, 95, 96
Canonical coordinates, 436
Channel functions, 486-487
Chirality in atoms, 364, 366
Chiral molecules, 368
Chirikov criterion, 338-348, 352
Circular dichroism, 364-366, 372, 375
Clamped nuclei approximation, 460
Classified limit, 475
Clusters, 184-190
Cluster decomposition, 436, 439, 447, 453, 455-456
Compact operator, 445, 447
Continuum multiple scattering method, 539-541
Coriolis couplina, 12, 156-157
Coriolis force, 9
Correlation, 182
Correlation functions, 485, 487-488, 498
Coupled nonlinear oscillators, 259-303

Detection sensitivity, 388-389
Detector noise, 388
Dilation analytic potentials, 460
Dirichlet-Neumann bracketing, 478-481
Dissociative attachment, 484, 513-526
Doppler-free techniques, 400
Doppler limit, 387
Dunham series, 233

Eckart conditions, 11-13
Eckart frame, 1, 3, 11, 30, 33, 34
Eckart Hamiltonian, 1-9, 31-35, 162
Einstein-Podolsky-Rosen Correlations, 41, 43, 101-102
Einstein quantization condition, 311-312, 321
Electron-molecule collisions, 483-547
Electron-molecular ion collision, 503
Electron scattering by polar molecules, 527-539
Electronic curve, 452, 455-456, 458, 467, see Potential energy curve
Electronic Hamiltonian, 450
Energy gap model, 131, 138
Euler angles, 5, 9-14, 16-19, 34, 203, 210, 223-225

F-G matrix technique, 13
Fixed nuclei approximation, 483-504, 507, 513, 516, 530
Fixed point, 274-280, 283, 285
Frame transformation theory, 483, 506, 514-517, 524, 530

Galitei group, 102-103
Gaussian overlap approximation, 208, 231-232
Generator coordinate method, 197-237
Geometric cooling, 121, 403
Gibbs State, 240, 243, 251
Glauber approximation, 530-531
Greene Orbital Instability Criterion, 350-352

Hamilton-Jacobi representation, 262, 302
Helleman-Bountis periodic orbital construction, 350-352
Hellmann-Frynman theorem, 462, 467, 468
Henon-Heiles Hamiltonian, 280-286, 293, 296, 299-301, 303-305, 309, 316, 319-324, 327, 350, 351
  eigenfunctions of, 309-311
Hilbert space, 30, 69, 72, 77, 80, 82, 90-95, 106, 199, 200, 239, 242, 244, 305, 324-326, 439-442, 451, 466
Hill-Wheeler equation, 200-210, 212, 214-215, 217
HVZ theorem, 448, 450, 452, 454, 455
Hybrid theory, 515-517

Integrals of the motion, 259, 262
  isolating, 260, 262, 272, 282, 288, 302, 303
  non-isolating, 260, 262, 302
Intermodulated fluorescence technique, 410, 421
  and I_2 spectrum, 411-412
Intra cavity absorption, 396-398, 412
Intramolecular dynamics, 257-356
Intramolecular energy transfer, 257-356
Inversion, 357-358, see Parity operation
Ionization
  collision-induced, 399
  field-, 399
  Penning, 399
  Spectroscopy, 398-399

K-matrix, 493, 503, 517, 526
K-system, 63-66, 83
KAM theory, 61, 62, 263-266, 275-279, 282-286, 302-306, 312-321, 328-331, 350
Kato-Rellich theorem, 442, 444, 446
Kernel
  Hamiltonian, 200, 201, 205, 208, 211-217, 227, 230
  overlap, 200, 201, 205, 208, 211-217, 227, 230
Kohn-variational principle, 485, 511
Kolmogorov-Sinai entropy, 64

L^2-methods, 504-513, 522
-Λ-doubling, 232
Lamb-dip, 408, 409, 412, 416, 417
Large amplitude motions, 151, 152
Laser, 122, 128, 155, 329, 377-433
  colour=centre, 379-381, 383
  dye-, 381-385
  ring-, 386
  semi-conductor, 379-381
  spectroscopy, see Spectroscopy
  spin-flip Raman, 379
Lifetime measurements, 378, 426
Linstedt-Poincaré method, 288-289
Liouville equation, 349
Lippmann-Schwinger equation, 510

M-matrix, 503
Mach number, 117-126
Mapping
  area preserving, 273-275, 280, 283
  twist, 275-279
Master equation, 245
Measuring apparatus, 328
Method of averaging, 289
Method of multiple time scales, 289
Molecular dimers, 153
Molecular energies
  asymptotic expansion, 475-482, see Dunham series
Molecular force field, 258
Molecular spectroscopy, see Spectroscopy
Molecular structure, 1, 105, 143, 198, 240, 250, 459
Molecular symmetry group, 148, 153, 158, 162, 169-171, 214

N-body Rinematics, 435-439
N-body Schrödinger operator theory, 435-475
Non-adiabatic effects, 197, 198, 210, 213-215, 221, 232, 428
Non-linear dynamics, 239-255
Non-linear optical mixing techniques, 386
Non-linear Schrödinger equation, 248-252
Nonrigid molecules, 143-190
Norm resolvent continuity, 464, 465, 476
Nuclear magnetic resonance, 150-151

Optical activity, 364, 365, 369-374
Optical-optical double resonance, 422-426
Optical parametric oscillator, 379
Optical pumping, 390

Parity
  conservation of, 357, 359
  operation, 357, 358, 360, 363, 365, 366, 369
  violating neutral currents, 361-364, 369, 372, 375
  violating potential, 361-364, 374, 375
  violations, 357-376
Pendulum motions, 340-343, 353
Permutation group, 145, 157, 159, 170, 172, 174-176

Perturbation theory, 288-301, 324
  regular, 442
  singular, 475
Phase-sensitive detection, 391
Photoacoustic spectroscopy, see Spectroscopy
Photochemistry, 115, 126, 127
Photoionization, 398, 503, see Ionization
Poincaré-Birkhoff theorem, 276, 279, 284
Poincaré surface of section, 268, 270, 272, 273, 276, 283, 299, 301, 322-329, 340, 351
Potential
  correlation, 488, 492, 498, 502
  direct, 487, 489, 491, 498, 516, 540
  energy curve, 525, 537
  exchange, 487, 490, 492, 497, 498, 502, 516, 531
  polarization, 490, 494-497, 502, 517, 531, 540
Pseudorotation, 144, 150
Pseudostates, 490, 491

Quantum defect theory, 503
Quantum dynamical semigroups, 240-248
Quantum efficiency, 393
Quantum logic, 40, 84, 85, 90, 98
Quantum optics, 243
Quasi-periodic motions, 258, 283, 284, 290, 291, 298, 302-304, 309, 312, 319-321, 330, 331, 348, 350, 352

R-matrix method, 484, 491, 502-509, 514, 517, 519, 522-526, 538
Racemization, 371
Raman-effect, 379
Ramsauer-Townsend minimum 497
Resolvent set, 446, 461, 464
Resonance structure, 345-349
Resonance theory, 517-522, 526
Resonance zone, 338-340
Rigidity, 143, 157, 161
Rotational excitation cross-section, 501, 514
Runge-Lenz vector, 183

Semiclassical quantization, 310-320, 330, see Bohr-Sommerfeld-Wilson quantization, see Einstein quantization condition
Sobolev inequalities, 443
Sobolev norm, 442
Sobolev space, 442, 469, 472
Spectral measure, 73
Spectral properties of Schrödinger Hamiltonians, 446-458
Spectraphone, 394-396
Spectroscopy
  excitation, 392-394
  fluorescence, 390
  ionization, 398-399
  laser
    and molecular beams, 400-402
    in fast ion beams, 405
  molecular, 124
    high resolution, 155, 166, 377-433
  photoacoustic, 392, 394-396
  polarization, 378, 413-422
  saturation, 406-413
  sub-Doppler laser, 400-426
Spectrum
  clamped nuclei Hamiltonian, 453
  continuous, 446
  discrete, 446, 450
  essential, 446-450
  irregular, 312, 323, 324, 329
  regular, 312, 323, 324, 329
  Weyl, 446
Stochastic motion, 61, 63, 64, 258, 280, 283, 285, 290, 291, 298, 302-309, 312, 316, 319-321, 325, 330-335, 348-352
Stochastic process, 68

Sub-Doppler spectral resolution, 378, 388, 406, 409, 424
Supersonic expansion, 116, 120, 124
Supersonic jet, 116, 122, 127, 132
Supersonic molecular beams, 115, 122, 124, see Spectroscopy
Symmetric double-well potential, 369, 370, 373
Symmetry breaking, 252

T-matrix method, 484, 506, 510-513
Time-reversal invariance, 350, 368
Tin Woodman, 357, 361
Toda Hamiltonian, 286, 287
Trace class linear operators, 239, 241
Translation operator, 449, 456, 457, 465, 468
Tunable coherent light sources, 378-387

Van der Waals bonds, 115, 131, 133, 139
Van der Waals forces, 115
Van der Waals molecule, 115, 125-132, 137, 139, 153, 154
Vibrational excitation, 484, 513-526, 535-536
Vibronic coupling, 215
Virial theorem, 464
Virtual state, 537-539

W*-algebra, 71, 73-78, 81-84, 90-97
W*-system, 40, 77, 78, 81-84, 90-97
Wave function, 144, 145, 182, 198, 199, 210, 212, 234, 251
  adiabatic, 459, 468-475
  asymptotic Generator coordinate, 236
  atomic, 456-457

Wave function (continued)
  correlated, 182
  electronic, 463
    asymptotic behaviour, 456-457
  of Henon-Heiles Hamiltonian, 309-311
  intrinsic, 210, 225
  molecular, 214, 223, 224, 228
  nodal patterns, 306-309
Wavemeter, 389
Weak coupling limit theorem, 245-248
Weinberg-Salam theory, 361, 362, 368
Wigner function, 319-323, 330
Wizard of Oz, 357